Proceedings in Adaptation, Learning and Optimization

Volume 3

About this Series

The role of adaptation, learning and optimization are becoming increasingly essential and intertwined. The capability of a system to adapt either through modification of its physiological structure or via some revalidation process of internal mechanisms that directly dictate the response or behavior is crucial in many real world applications. Optimization lies at the heart of most machine learning approaches while learning and optimization are two primary means to effect adaptation in various forms. They usually involve computational processes incorporated within the system that trigger parametric updating and knowledge or model enhancement, giving rise to progressive improvement. This book series serves as a channel to consolidate work related to topics linked to adaptation, learning and optimization in systems and structures. Topics covered under this series include:

- complex adaptive systems including evolutionary computation, memetic computing, swarm intelligence, neural networks, fuzzy systems, tabu search, simulated annealing, etc.
- machine learning, data mining & mathematical programming
- hybridization of techniques that span across artificial intelligence and computational intelligence for synergistic alliance of strategies for problem-solving
- aspects of adaptation in robotics
- agent-based computing
- autonomic/pervasive computing
- dynamic optimization/learning in noisy and uncertain environment
- systemic alliance of stochastic and conventional search techniques
- all aspects of adaptations in man-machine systems.

This book series bridges the dichotomy of modern and conventional mathematical and heuristic/meta-heuristics approaches to bring about effective adaptation, learning and optimization. It propels the maxim that the old and the new can come together and be combined synergistically to scale new heights in problem-solving. To reach such a level, numerous research issues will emerge and researchers will find the book series a convenient medium to track the progresses made.

More information about this series at http://www.springer.com/series/13543

Jiuwen Cao · Kezhi Mao
Erik Cambria · Zhihong Man
Kar-Ann Toh
Editors

Proceedings of ELM-2014 Volume 1

Algorithms and Theories

Editors
Jiuwen Cao
Institute of Information and Control
Hangzhou Dianzi University
Zhejiang
China

Kezhi Mao
School of Electrical and Electronic Engineering
Nanyang Technological University
Singapore
Singapore

Erik Cambria
School of Computer Engineering
Nanyang Technological University
Singapore
Singapore

Zhihong Man
Faculty of Engineering and Industrial Sciences
Swinburne University of Technology
Hawthorn Victoria
Australia

Kar-Ann Toh
School of Electrical and Electronic Engineering
Yonsei University
Seoul
Korea, Republic of (South Korea)

ISSN 2363-6084 ISSN 2363-6092 (electronic)
Proceedings in Adaptation, Learning and Optimization
ISBN 978-3-319-36684-5 ISBN 978-3-319-14063-6 (eBook)
DOI 10.1007/978-3-319-14063-6

Springer Cham Heidelberg New York Dordrecht London

Softcover reprint of the hardcover 1st edition 2015

Printed on acid-free paper

Springer International Publishing AG Switzerland is part of Springer Science+Business Media (www.springer.com)

Contents

Algorithms and Theories on ELM

Sparse Bayesian ELM Handling with Missing Data for Multi-class Classification 1
Jiannan Zhang, Shiji Song, Xunan Zhang

A Fast Incremental Method Based on Regularized Extreme Learning Machine 15
Zhixin Xu, Min Yao

Parallel Ensemble of Online Sequential Extreme Learning Machine Based on MapReduce 31
Shan Huang, Botao Wang, Junhao Qiu, Jitao Yao, Guoren Wang, Ge Yu

Explicit Computation of Input Weights in Extreme Learning Machines 41
Jonathan Tapson, Philip de Chazal, André van Schaik

Subspace Detection on Concept Drifting Data Stream 51
Lin Feng, Shenglan Liu, Yao Xiao, Jing Wang

Inductive Bias for Semi-supervised Extreme Learning Machine 61
Federica Bisio, Sergio Decherchi, Paolo Gastaldo, Rodolfo Zunino

ELM Based Efficient Probabilistic Threshold Query on Uncertain Data 71
Jiajia Li, Botao Wang, Guoren Wang

Sample-Based Extreme Learning Machine Regression with Absent Data ... 81
Hang Gao, Xinwang Liu, Yuxing Peng

Two Stages Query Processing Optimization Based on ELM in the Cloud ... 91
Linlin Ding, Yu Liu, Baoyan Song, Junchang Xin

Domain Adaptation Transfer Extreme Learning Machines 103
Lei Zhang, David Zhang

Quasi-Linear Extreme Learning Machine Model Based Nonlinear System Identification 121
Dazi Li, Qianwen Xie, Qibing Jin

A Novel Bio-inspired Image Recognition Network with Extreme Learning Machine 131
Lin Zhang, Yu Zhang, Ping Li

A Deep and Stable Extreme Learning Approach for Classification and Regression 141
Le-le Cao, Wen-bing Huang, Fu-chun Sun

Extreme Learning Machine Ensemble Classifier for Large-Scale Data 151
Haocheng Wang, Qing He, Tianfeng Shang, Fuzhen Zhuang, Zhongzhi Shi

Pruned Annular Extreme Learning Machine Optimization Based on RANSAC Multi Model Response Regularization 163
Lavneet Singh, Girija Chetty

Learning ELM Network Weights Using Linear Discriminant Analysis 183
Philip de Chazal, Jonathan Tapson, André van Schaik

An Algorithm for Classification over Uncertain Data Based on Extreme Learning Machine 193
Ke-yan Cao, Guoren Wang, Donghong Han

Training Generalized Feedforword Kernelized Neural Networks on Very Large Datasets for Regression Using Minimal-Enclosing-Ball Approximation 203
Jun Wang, Zhaohong Deng, Shitong Wang, Qun Gao

An Online Multiple-Model Approach to Univariate Time-Series Prediction 215
Koshy George, Sachin Prabhu, Prabhanjan Mutalik

A Self-Organizing Mixture Extreme Leaning Machine for Time Series Forecasting 225
Hou Muzhou, Chen Ming, Zhang Yangchun

Ensemble Extreme Learning Machine Based on a New Self-adaptive AdaBoost.RT 237
Pengbo Zhang, Zhixin Yang

Machine Learning Reveals Different Brain Activities in Visual Pathway during TOVA Test 245
Haoqi Sun, Olga Sourina, Yan Yang, Guang-Bin Huang, Cornelia Denk, Felix Klanner

Online Sequential Extreme Learning Machine with New Weight-Setting Strategy for Nonstationary Time Series Prediction 263
Jinwan Wang, Wentao Mao, Liyun Wang, Mei Tian

RMSE-ELM: Recursive Model Based Selective Ensemble of Extreme Learning Machines for Robustness Improvement 273
Bo Han, Bo He, Mengmeng Ma, Tingting Sun, Tianhong Yan, Amaury Lendasse

Extreme Learning Machine for Regression and Classification Using L_1-Norm and L_2-Norm 293
Xiong Luo, Xiaohui Chang, Xiaojuan Ban

A Semi-supervised Online Sequential Extreme Learning Machine Method 301
Xibin Jia, Runyuan Wang, Junfa Liu, David M.W. Powers

ELM Feature Mappings Learning: Single-Hidden-Layer Feedforward Network without Output Weight 311
Yimin Yang, Q.M. Jonathan Wu, Yaonan Wang, Dibyendu Mukherjee, Yanjie Chen

ROS-ELM: A Robust Online Sequential Extreme Learning Machine for Big Data Analytics 325
Yang Liu, Bo He, Diya Dong, Yue Shen, Tianhong Yan, Rui Nian, Amaury Lendasse

Deep Extreme Learning Machines for Classification 345
Migel D. Tissera, Mark D. McDonnell

C-ELM: A Curious Extreme Learning Machine for Classification Problems 355
Qiong Wu, Chunyan Miao

Review of Advances in Neural Networks: Neural Design Technology Stack 367
Stanisław Woźniak, Adela-Diana Almási, Valentin Cristea, Yusuf Leblebici, Ton Engbersen

Applying Regularization Least Squares Canonical Correlation Analysis in Extreme Learning Machine for Multi-label Classification Problems 377
Yanika Kongsorot, Punyaphol Horata, Khamron Sunat

Least Squares Policy Iteration Based on Random Vector Basis 397
Lei Zuo, Xin Xu

Identifying Indistinguishable Classes in Multi-class Classification Data Sets Using ELM 407
Felis Dwiyasa, Meng-Hiot Lim

Effects of Training Datasets on Both the Extreme Learning Machine and Support Vector Machine for Target Audience Identification on Twitter 417
Siaw Ling Lo, David Cornforth, Raymond Chiong

Extreme Learning Machine for Clustering 435
Chamara Kasun Liyanaarachchi Lekamalage, Tianchi Liu, Yan Yang, Zhiping Lin, Guang-Bin Huang

Author Index 445

Sparse Bayesian ELM Handling with Missing Data for Multi-class Classification

Jiannan Zhang, Shiji Song*, and Xunan Zhang

Department of Automation, Tsinghua University
Beijing 100084, China
shijis@mail.tsinghua.edu.cn

Abstract. Extreme learning machine (ELM) is a successful machine learning approach for its extremely fast training speed and good generalization performance. The sparse Bayesian ELM (SBELM) approach, which is a variant of ELM, can result in a more accurate and compact model. However, SBELM can not deal with the missing data problem in its standard form. To solve this problem, we design two novel methods, additive models for missing data (AMMD) and self-adjusting neuron state for missing data (SANSMD), by adjusting the calculation of outputs of the hidden layers in SBELM. Experimental results on several data sets from the UCI repository indicate that the proposed modified SBELM methods have significant advantages: high accuracy and good generalization performance compared with several other existing methods. Moreover, the proposed methods enrich ELM with new tools to solve missing data problem for multi-class classification even with up to 50% of the features missing in the input data.

Keywords: ELM, SBELM, Missing data, Multi-class classification.

1 Introduction

Missing data problem [1,2,3] widely exists in many different application fields of science such as in the areas of social, behavioral and medical sciences. Many explanations can be made for why a feature value is unavailable: the data values are simply not measured, human or machine failures during the data acquisition process, or error in transmitting or saving data values into their respective records. Types for missing data are commonly classified as: missing completely at random (MCAR), missing at random (MAR), and missing not at random (MNAR) [3,4]. The missing data for a random variable X are MCAR means that the probability of having missing values doesn't rely on the values of X or any other variables in the data set. We assume the data are MCAR here, and under this assumption, the reason for missing data can be ignored.

The methods for solving missing data problem can be group into four types according to the differences of solutions both on handling missing values and

---

* Corresponding author.

J. Cao et al. (eds.), *Proceedings of ELM-2014 Volume 1*,
Proceedings in Adaptation, Learning and Optimization 3, DOI: 10.1007/978-3-319-14063-6_1

learning procedures [5]. One is deleting the cases with missing values and utilizing the left complete data portion to learn. The second one is imputing missing data with existing information and then learning the models by using the complete data set, e.g., replacing the missing values of one feature with its mean value or values calculated by multiple imputation methods. The third approach is using model-based procedure, such as the expectation-maximization (EM) algorithm [6]. The last one is learning procedures that are designed for handling with missing data without a previous estimation of missing values. Cases deletion and simple imputation may cause bias. Multiple imputation may lead to additional bias in multivariate analysis and model-based procedures are very time consuming when they come up against large data sets.

Nowadays, data sets produced in many fields become very large and it is time consuming for some classic machine leaning methods such as neural networks, support vector machine, etc. [7,8,9,10]. Recently, Huang et al. [11,12,13] proposed an efficient algorithm called extreme learning machine (ELM) which is an elegant learning algorithm for single hidden layer feedforward neural networks (SLFN). This method chooses the input weights and biases of the hidden neurons randomly and calculates the output weights in closed form. Hence, the overall computational time can be reduced by several orders of magnitude, compared to the above-mentioned classical methods.

However, it has been mentioned in [14] and [15] that the ELM algorithm has three drawbacks: 1) it can have some issues when encountering irrelevant or correlated data, 2) the method of obtaining the output weights, a kind of least squares minimization learning, can easily suffer from overfitting and 3) the accuracy of ELM is influenced by the number of hidden neurons. Moreover, ELM or its existing modifications can not handle with missing data for multi-class classification until now.

In this paper, we propose to extend the sparse Bayesian ELM (SBELM) [15] approach, which is a sparse Bayesian learning approach for ELM, to solve the missing data problem for multi-class classification. SBELM retains the advantages of ELM like fast computational speed and universal approximation capability. Moreover, the method can overcome the shortcomings of ELM. SBELM has higher generalization performance with lower computational cost and produces a more compact model than OP-ELM [16] and TROP-ELM [17] which are only for regression problems. In order to deal with missing data for multi-class classification with SBELM, we propose two different methods, which are additive model for missing data (AMMD) and self-adjusting neuron state for missing data (SANSMD) to calculate the output of the hidden neurons . These two methods have respective advantages so that we can choose either one to deal with certain problems.

The rest part of this paper is organized as follows. The following section introduces the details of ELM and SBELM. Into Section 3, we propose two methods to modify SBELM for handling with missing data. Section 4 reports the experimental results obtained on a number of benchmark data sets, and followed by a short conclusion in Section 5.

2 Sparse Bayesian ELM: SBELM

Recently, Luo et al. [15] proposed the sparse Bayesian ELM (SBELM) approach, which uses a sparse Bayesian learning (SBL) method for ELM to find sparse representatives for the output weights assumed with priori distribution. The approach is introduced briefly in this section.

2.1 Extreme Learning Machine (ELM)

Extreme learning machine (ELM) [11,12,13] was originally proposed as a solution for the single hidden layer feedforward neural network (SLFN). The essence of ELM is that all the hidden node parameters are randomly generated without tuning. Thus it is possible to calculate the hidden-layer output matrix and then the output weights.

Consider a given data set of N distinct samples $\mathcal{S}_N = \{\mathbf{x}, t_i\}_{i=1}^{N}$ with $\mathbf{x}_i \in \mathbb{R}^D$ and $t_i \in \mathbb{R}$. An SLFN with m hidden nodes can be modeled as follows

$$\sum_{i=1}^{m} \beta_i g\left(\boldsymbol{\omega}_i \mathbf{x}_j + b_i\right) , \tag{1}$$

for $1 \leq j \leq N$ where $\boldsymbol{\omega}_i$ are the input weights, b_i the hidden layer biases, β_i the output weights and $g(\cdot)$ is an activation function.

If there was an SLFN with activation function g can approximate the N samples perfectly, the model can be written as follows

$$\sum_{i=1}^{m} \beta_i g\left(\boldsymbol{\omega}_i \mathbf{x}_j + b_i\right) = t_j , \tag{2}$$

The above N equations can be written compactly as

$$\mathbf{H}\boldsymbol{\beta} = \mathbf{t} , \tag{3}$$

where

$$\mathbf{H} = \begin{pmatrix} g\left(\boldsymbol{\omega}_1 \mathbf{x}_1 + b_1\right) & \cdots & g\left(\boldsymbol{\omega}_m \mathbf{x}_1 + b_m\right) \\ \vdots & \ddots & \vdots \\ g\left(\boldsymbol{\omega}_1 \mathbf{x}_N + b_1\right) & \cdots & g\left(\boldsymbol{\omega}_m \mathbf{x}_N + b_m\right) \end{pmatrix} , \tag{4}$$

$\boldsymbol{\beta} = (\beta_1, \ldots, \beta_m)^{\mathrm{T}}$ and $\mathbf{t} = (t_1, \ldots, t_N)^{\mathrm{T}}$. In the ELM approach, random values have been assigned to the input weights $\boldsymbol{\omega}_i$ and the hidden layer biases b_i in the beginning of learning. Hence the hidden layer output matrix $\mathbf{H}$ remains unchanged during the learning procedure. For fixed $\boldsymbol{\omega}_i$ and b_i, to train an SLFN is simply equivalent to finding a least-squares solution $\hat{\boldsymbol{\beta}}$ to minimize the cost function

$$E = \sum_{j=1}^{N} \left\| \sum_{i=1}^{m} \beta_i g\left(\boldsymbol{\omega}_i \mathbf{x}_j + b_i\right) - t_j \right\| . \tag{5}$$

By using a Moore-Penrose generalized inverse of the matrix $\mathbf{H}$, denoted as $\mathbf{H}^{\dagger}$[18], the output weights $\boldsymbol{\beta}$ can be calculated from the knowledge of $\boldsymbol{\omega}_i$ and b_i. Therefore the solution mentioned above to the equation $\mathbf{H}\boldsymbol{\beta} = \mathbf{t}$ is denoted as $\hat{\boldsymbol{\beta}} = \mathbf{H}^{\dagger}\mathbf{t}$.

2.2 Sparse Bayesian ELM (SBELM) for Multi-class Classification

The SBELM approach, proposed by Luo et al. [15], uses a sparse Bayesian learning (SBL) method for ELM. The new model, which is benefit from the SBL method, can overcome the shortcomings of ELM by maximizing the marginal likelihood of network outputs and automatically pruning most of the redundant hidden neurons during learning phase. Moreover, as having been verified in [15], SBELM shows remarkable performance on accuracy, model size and generalization ability, compared to other approaches such as relevance vector machine (RVM) [19], support vector machine (SVM) [20], TROP-ELM, and Bayesian ELM (BELM) [21]. When dealing with binary classification problem, SBELM treats it as an independent Bernoulli event with the probability defined as follows

$$p\left(\mathbf{t}|\boldsymbol{\beta},\mathbf{H}\right)=\prod_{i=1}^{N}\sigma(\mathbf{h}_i\boldsymbol{\beta})^{t_i}\left[1-\sigma\left(\mathbf{h}_i\boldsymbol{\beta}\right)\right]^{1-t_i}, \tag{6}$$

where $\sigma(\cdot)$ is sigmoid function and $\mathbf{h}_j=\left(g\left(\boldsymbol{\omega}_1\mathbf{x}_j+b_1\right),\ldots,g\left(\boldsymbol{\omega}_m\mathbf{x}_j+b_m\right)\right)$ for $j=1,\ldots,N$. Then, the SBELM approach makes the assumption that β_i are modeled probabilistically as independent zero-mean Gaussian distribution, with variance α_i^{-1}, so $p\left(\boldsymbol{\beta}|\boldsymbol{\alpha}\right)=\prod_{i=1}^{m}\mathrm{N}\left(\beta_{\mathrm{i}}|0,\alpha_{\mathrm{i}}^{-1}\right)$. Note that $\boldsymbol{\alpha}=\left(\alpha_1,\ldots,\alpha_m\right)^{\mathrm{T}}$ are m independent ARD hyperparameters [22,23]. The values of $\boldsymbol{\alpha}$ can be determined by maximizing $\ln\{p\left(\mathbf{t}|\boldsymbol{\beta},\mathbf{H}\right)p\left(\boldsymbol{\beta}|\boldsymbol{\alpha}\right)\}$ for the reason that the marginal likelihood $p\left(\mathbf{t}|\boldsymbol{\alpha},\mathbf{H}\right)=\int p\left(\mathbf{t}|\boldsymbol{\beta},\mathbf{H}\right)p\left(\boldsymbol{\beta}|\boldsymbol{\alpha}\right)d\boldsymbol{\beta}$, which cannot be integrated out analytically, is proportional to $p\left(\mathbf{t}|\boldsymbol{\beta},\mathbf{H}\right)p\left(\boldsymbol{\beta}|\boldsymbol{\alpha}\right)$. Using the Laplace approximation approach which is simply a quadratic approximation to the log-posterior around its mode, ARD approximates a Gaussian distribution for the marginal likelihood. Thus $\ln\{p\left(\mathbf{t}|\boldsymbol{\beta},\mathbf{H}\right)p\left(\boldsymbol{\beta}|\boldsymbol{\alpha}\right)\}$ is wrote as follows

$$\begin{aligned}&\ln\left\{p\left(\mathbf{t}|\boldsymbol{\beta},\mathbf{H}\right)p\left(\boldsymbol{\beta}|\boldsymbol{\alpha}\right)\right\}\\&=\sum_{i=1}^{N}\left\{t_i\ln y_i+\left(1-t_i\right)\ln\left(1-y_i\right)\right\}-\frac{1}{2}\boldsymbol{\beta}^{\mathrm{T}}\mathbf{A}\boldsymbol{\beta}+\mathrm{const}\,,\end{aligned} \tag{7}$$

where $\mathbf{A}=\mathrm{diag}\left(\boldsymbol{\alpha}\right)$ and $y_i=\sigma\left(\mathbf{h}_i\boldsymbol{\beta}\right)$.

According to the Laplace approximation approach, we can find the Laplace's mode $\hat{\boldsymbol{\beta}}$ by using Newton-Raphson method iterative reweighted least squares (IRLS). $\hat{\boldsymbol{\beta}}$ is obtained by

$$\boldsymbol{\beta}_{\mathrm{new}}=\left(\mathbf{H}^{\mathrm{T}}\mathbf{B}\mathbf{H}+\mathbf{A}\right)^{-1}\mathbf{H}^{\mathrm{T}}\mathbf{B}\hat{\mathbf{t}}\,, \tag{8}$$

where $\hat{\mathbf{t}}=\mathbf{H}\boldsymbol{\beta}_{\mathrm{old}}+\mathbf{B}^{-1}\left(\mathbf{t}-\mathbf{y}\right)$, $\mathbf{B}=\mathrm{diag}\left(\left\{y_1\left(1-y_1\right),...,y_N\left(1-y_N\right)\right\}\right)_{N\times N}$ and $\mathbf{y}=\left(y_1,\ldots,y_N\right)^{\mathrm{T}}$. And we obtain $p\left(\mathbf{t}|\boldsymbol{\beta},\mathbf{H}\right)p\left(\boldsymbol{\beta}|\boldsymbol{\alpha}\right)\sim\mathrm{N}\left(\hat{\boldsymbol{\beta}},\boldsymbol{\Sigma}\right)$, where $\hat{\boldsymbol{\beta}}=\boldsymbol{\Sigma}\mathbf{H}^{\mathrm{T}}\mathbf{B}\hat{\mathbf{t}}$ and $\boldsymbol{\Sigma}=\left(\mathbf{H}^{\mathrm{T}}\mathbf{B}\mathbf{H}+\mathbf{A}\right)^{-1}$.

The log marginal likelihood is as follows:

$$\begin{aligned} L(\boldsymbol{\alpha}) &= \ln p(\mathbf{t}|\boldsymbol{\alpha},\mathbf{H}) \\ &= -\frac{1}{2}\left[N\ln(2\pi) + \ln\left|\mathbf{B}+\mathbf{HAH}^{\mathrm{T}}\right| + (\hat{\mathbf{t}})^{\mathrm{T}}\left(\mathbf{B}+\mathbf{HAH}^{\mathrm{T}}\right)\hat{\mathbf{t}}\right] . \end{aligned} \tag{9}$$

The derivative of the marginal likelihood with respect to α_i is set to zero and α_i^{new} is obtained by

$$\alpha_i^{\mathrm{new}} = \frac{1-\alpha_i^{\mathrm{old}}\boldsymbol{\Sigma}_{i,i}}{\hat{\beta}_i^2} . \tag{10}$$

Through the iterations of (8) and (10) after setting initial values to $\boldsymbol{\beta}$ and $\boldsymbol{\alpha}$ in the beginning, the operation continues to maximize the marginal likelihood function until reaching the convergence criteria. As analyzed in [15], with the iteration's going on, partial α_i's tend to grow to infinity and the corresponding β_i's approximate to zero concomitantly, which implies that their associated hidden neurons are pruned and maintains the sparsity of the model.

Furthermore, on the basis of SBELM for binary classification problem, the model is extended to multi-class classification by using the state-of-the-art method pairwise coupling [24]. The method was also adopted and elaborated in the well-known toolbox LIBSVM [10]. So we do not repeat it here.

3 Improved SBELMs for Dealing with Missing Data

Since missing data happen in almost every field of research and the standard SBELM cannot handle with missing data, we introduce two methods, additive model and self-adjusting neuron state model, to solve the problem for the SBELM in this section.

3.1 Additive Models for Missing Data (AMMD)

Pelckmans et al. proposed a SVM algorithm [25] for handling with missing data in which there was no attempt to recover the missing values in data sets. The key idea of the algorithm is to modify the cost function by using additive models proposed by Hastie and Tibshirani in [26]. It gives us a great illumination for how to handle with the data set with missing values.

Additive models are described as follows:

An input vector $x \in \mathbb{R}^D$ can be written as a combination of Q components of dimension D_q for $q = 1, \ldots, Q$, which means $\mathbf{x} = (\mathbf{x}^{(1)}, \ldots, \mathbf{x}^{(Q)})$ with $\mathbf{x}^{(q)} \in \mathbb{R}^{D_q}$ for $q = 1, \ldots, Q$. The additive models is defined as

$$\begin{aligned} F^D = \{f : \mathbb{R}^D \to \mathbb{R} | f(\mathbf{x}) = \sum_{q=1}^{Q} f^{(q)}(\mathbf{x}^{(q)}) + b, \\ f^{(q)} : \mathbb{R}^{D_q} \to \mathbb{R}, b \in \mathbb{R}, \forall \mathbf{x} = (\mathbf{x}^{(1)}, ..., \mathbf{x}^{(Q)}) \in \mathbb{R}^D\} . \end{aligned} \tag{11}$$

Furthermore, $\mathbf{X}^{(q)}$ is denoted as the random variable (vector) corresponding to the qth components for $q = 1, \ldots, Q$.

We define the sets $\mathcal{A}_q$ and $\mathcal{B}_i$ as follows

$$\mathcal{A}_q = \left\{ i \in \{1, \ldots, N\} | \mathbf{x}_i^{(q)} \text{ is complete} \right\} , \ \forall q = 1, \ldots, Q , \tag{12}$$

$$\mathcal{B}_i = \left\{ q \in \{1, \ldots, Q\} | \mathbf{x}_i^{(q)} \text{ is complete} \right\} , \ \forall i = 1, \ldots, N , \tag{13}$$

and define $\bar{\mathcal{A}}_q = \{1, \ldots, N\} \backslash \mathcal{A}_q$, $\bar{\mathcal{B}}_i = \{1, \ldots, Q\} \backslash \mathcal{B}_i$.

For the purpose of coping with the notational inconvenience caused by the different dependent summands, the index sets $\mathcal{U}_i \in \mathbb{N}^Q$ are defined as follows

$$\mathcal{U}_i = \{(j_1, \ldots, j_Q) | j_q = i \text{ if } \mathrm{q} \in \mathcal{B}_\mathrm{i} \text{ or } \mathrm{j_q} = \mathrm{l}, \ \forall \, \mathrm{l} \in \mathcal{A}_\mathrm{q} \text{ if } \mathrm{q} \in \bar{\mathcal{B}}_\mathrm{i}\} . \tag{14}$$

According to the assumption of the probabilistic model for missing values in [25], the U-statistics of t_i can be obtained as follows

$$\begin{aligned} t_i^* &= \frac{1}{|\mathcal{U}_i|} \sum_{(j_1, \ldots, j_Q) \in \mathcal{U}_i} \left[\sum_{q=1}^{Q} \beta_q g \left(\boldsymbol{\omega}_q \mathbf{x}_{j^q}^{(q)} + b_q \right) \right] \\ &= \sum_{q=1}^{Q} \left\{ \sum_{q \in \mathcal{B}_i} \beta_q g \left(\boldsymbol{\omega}_q \mathbf{x}_i^{(q)} + b_q \right) + \sum_{q \in \bar{\mathcal{B}}_i} \beta_q \frac{1}{|\mathcal{A}_q|} \sum_{j \in \mathcal{A}_q} g \left(\boldsymbol{\omega}_q \mathbf{x}_j^{(q)} + b_q \right) \right\} \\ &= \sum_{q=1}^{Q} \beta_q g_i^{q,*} , \end{aligned} \tag{15}$$

where $g_i^{q,*}$ is equal to $g\left(\boldsymbol{\omega}_q \mathbf{x}_i^{(q)} + b_q\right)$ for $q \in \mathcal{B}_i$ and $\frac{1}{|\mathcal{A}_q|} \sum_{j \in \mathcal{A}_q} g\left(\boldsymbol{\omega}_q \cdot \mathbf{x}_j^{(q)} + b_q\right)$ for $q \in \bar{\mathcal{B}}_i$.

Thus, the marginal likelihood in the standard SBELM can be rewritten as follows

$$\begin{aligned} &\ln \{ p(\mathbf{t}|\boldsymbol{\beta}, \mathbf{H}^*) \, p(\boldsymbol{\beta}|\boldsymbol{\alpha}) \} \\ &= \sum_{i=1}^{N} \{ t_i \ln y_i^* + (1 - t_i) \ln (1 - y_i^*) \} - \frac{1}{2} \boldsymbol{\beta}^{\mathrm{T}} \mathbf{A} \boldsymbol{\beta} + \text{const} , \end{aligned} \tag{16}$$

where $y_i^* = \sigma(t_i^*)$, $\mathbf{H}^* = \left((\mathbf{h}_1^*)^{\mathrm{T}}, \ldots, (\mathbf{h}_\mathrm{N}^*)^{\mathrm{T}} \right)^{\mathrm{T}}$ and $\mathbf{h}_i^* = \left(g_i^{1,*}, \ldots, g_i^{Q,*} \right)$.

Finally, we can get the estimation of $\boldsymbol{\beta}$ with $\mathbf{H}^*$, $\mathbf{t}$ and $\mathbf{y}^*$ by utilizing SBELM. In addition, it is easy to be proved that the modified SBELM reduces to the standard SBELM when no observation is missing.

3.2 Self-adjusting Neuron State for Missing Data (SANSMD)

Recently, Viharos et al. proposed a method for learning artificial neural networks (ANNs) models with missing data [27,28]. So we extend the standard SBELM for missing data by using this method.

The method is based on the main idea of setting protected state to the neurons corresponding to the missing part of certain data vectors and unprotected state to the others. One of the neurons, if having protected state, is not involved in any calculation of the network during the learning procedure. Namely, the neuron is excluded from the network with all of its links to others. However, being protected does not mean the neuron is pruned. If the part of another data vector corresponding to the neuron is complete, the neuron's state will be reset to unprotected state. During the learning process, a flag called validity (binary) is used for indicating whether a value in the data vector is valid or not, that's to say, a validity vector is attached to the data vector to mark the data values valid states in the data. The neurons of the input layer being protected or not depends on the validity vector of the input data vector, namely if a data value is valid in the input vector, the corresponding neuron is set to unprotected, otherwise the corresponding neuron will be protected. Considering an sample $(\mathbf{x}, t)$ and its validity vector $\mathbf{v}$, we denote that $v_j = 1$ if the jth value is complete, otherwise $v_j = 0$. Than we redefine the output of hidden layers as

$$g(\cdot) = g\left(\sum_{j=1}^{D} v_j\left(\omega_j x_j + b_j\right) + b_0\right), \tag{17}$$

where $b_j = c_j/\sqrt{\text{sum}(\mathbf{v}) + 1}$ and c_j are randomly generated from the same Gaussian distribution for $j = 0, \ldots, D$. The basic reason for the new definition is that no matter which part of the data vector is missing, the bias b_j has the same probability distribution. However, they have different values corresponding to the different missing situations, which means the change of the network structure as illustrated in Fig. 1.

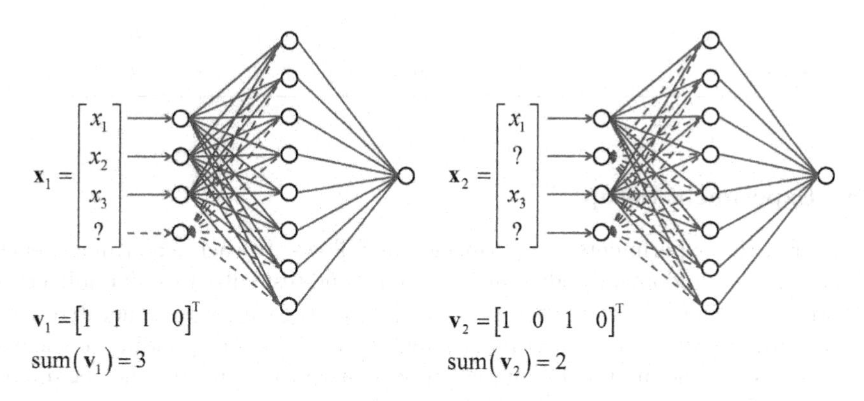

Fig. 1. The network for SANSMD method

Hence, by using the method handing with missing data, the output of hidden layers $\tilde{\mathbf{H}}$ can be computed. Then we can compute the estimation of $\boldsymbol{\beta}$ with $\tilde{\mathbf{H}}$ and $\mathbf{t}$ by using SBELM. The proposed method is also equivalent to the standard SBELM when no missing data occurs.

4 Performance Evaluation

In this section, five data sets are chosen to evaluate the performance of the two methods proposed in this paper, for multi-class classification problem with missing data. We start by introducing the characteristics of the data sets and the method of generating missing data. Then we describe the experimental setup. Finally, the experimental result will be given and analyzed in detail.

4.1 Data Sets

The experiments are performed using five benchmark data sets chosen from the UCI machine learning repository [29]. The characteristic set of data sets contains: name, the number of data samples, attributes and classes. Table 1 shows the description of the selected data sets. Because all the benchmark data sets are complete, it is need to generate missing data artificially. Missing data are generated by using the MCAR mechanism in accordance of the assumption on the type for missing data in Section 1. The percentage of the missing data in the training set ranges from 0% to 50% with a step size of 2%.

Table 1. Properties of training data set

Data sets	Samples	Attributes	Classes
Blood Transfusion Service	747	5	2
Pima Indians Diabetes	767	8	2
Iris	149	4	3
Balance Scale	624	4	3
Seeds	209	7	3

4.2 Experiment Setup

In general, the experiments are performed as follows. In each experiment, each data set is first randomly divided into a training and test subset for five-fold cross validation. During each validation process, the missing values are generated from the training subset with the certain missing percentage corresponding to the experiment. Next, the modified training subset is used to train with the classifiers: the SBELMs combined with AMMD, SANSMD, mean imputation (Mean), 1-nearest neighbor (1-NN) [30,31] and case deletion (CD) [3]. Specifically, in the mean imputation method, the mean of corresponding feature is computed based on the existed feature values to replace the missing values; in the 1-NN imputation method, the missing values are replaced by the corresponding feature of its first nearest neighbor whose value is not missing; and in the case deletion, the samples that have missing values for any feature are omitted from the data set.

Finally, the classification accuracy and root-mean-square error (RMSE) of the classifiers is evaluated by applying the corresponding classification model on the test subset.

Some notes for the experiments are as follows.

1) All the features of each data set are linearly scaled to [-1, 1].

2) The active function of hidden nodes for the SBELMs combined with AMMD, Mean, 1-NN and CD is sigmoid $g(\mathbf{x}) = 1/(1 + \exp(\sum_{j=1}^{D} \omega_j x_j + b_0))$, where ω_j are the input weights randomly generated from uniform distribution within [0, 1] and b_0 from Gaussian distribution $\mathrm{N}(0, 0.5)$ respectively. The active function for the SBELM with SANSMD is defined in equation (17), where ω_j are randomly generated from uniform distribution within [0, 1] and b_j from Gaussian distribution $\mathrm{N}(0, 0.5)$ for $j = 0, \ldots, D$.

3) The samples of the data set are randomly re-ordered and split into five folds in advance and kept fixed in each experiment in which the five classifiers are trained and tested with an identical missing percentage. Because the seed of generating the uniform distribution and Gaussian distribution in 2), so we randomly choose 10 seeds for each data set before training. On each missing percentage, we calculate the generalized accuracy using different combination of the 10 seeds and 6 initial number of hidden neurons chosen from the set [50,70,90,...,150]. Therefore, for each missing percentage, all the five classifiers are trained and tested for 300 ($10 \times 6 \times 5$) times in consideration of the repeating times in the five-fold cross validation process.

4.3 Experimental Result

The mean accuracies of the SBELMs combined with AMMD, SANSMD, Mean, 1-NN and CD are compared under different percentage of the missing data. For each given percentage, each data set is trained and tested on the 10 seeds with 6 initial numbers of hidden neurons. On each seed with 6 initial numbers, the best accuracy is chosen as the final result of the seed. The mean and RMSE of the classification accuracies based on the seeds are chosen for analysis as shown in Fig. 2 and Fig. 3. Observed from Fig. 2, the SBELMs combined with AMMD and SANSMD have higher accuracy and better generalization performance on

Table 2. Comparison of classification accuracy on Iris data

Missing percentage (%)		0	10	20	30	40	50
Accuracy (%)	AMMD	**96.8**	**96.5**	**96.7**	**96.5**	**96.2**	**96.0**
	SANSMD	**96.8**	**96.3**	**96.2**	**95.6**	**95.3**	**95.0**
	Mean	96.8	95.8	94.9	92.8	90.0	86.3
	1-NN	96.8	96.5	96.3	95.7	94.7	93
	CD	96.8	96.3	96.0	94.8	×	×

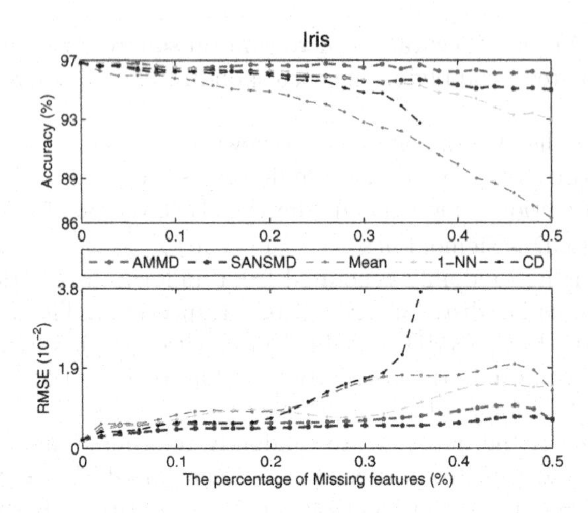

Fig. 2. The classification accuracy and RMSE for the data set

Iris data than those combined with the other methods. All the methods show nearly the same result of accuracy and RMSE at 0% missing percentage which confirms that all the methods are equivalent to the standard SBELM when no missing data occurs from one aspect. From the result in Fig. 2, it shows that when missing percentage is small, all the methods show similar classification results with no significant difference. However, as missing percentage increasing high, the distinction of result becomes significant. When missing percentage is high, case deletion (CD) method is not reliable. One reason is that all the samples of one class may be omitted from the data set. For example, if the samples of one data set have 10 attributes, all the samples nearly have missing values when the missing percentage exceeds 10% which means there will be few samples left by using case deletion method. As the same reason for case deletion, a large proportion of one attribute with high missing percentage will be filled with the same value by using mean imputation. When missing percentage achieves 50%, the classification accuracies of AMMD and SANSMD have a 0.8% decrease and a 1.8% decrease as shown in Table 2, which are superior to the left three methods, compared to the accuracies when no missing occurs. Furthermore, the RMSEs have only a slight rise, which shows the good generalization performance of the proposed two methods. The performance of AMMD is slightly higher than SANSMD, however, which is not significant.

Fig. 3 shows the results for the other data sets. The results are similar, and also, there are some differences. In general, the SBELMs combined with AMMD and SANSMD have higher accuracy except that 1-NN method is better than AMMD on Seeds data when the missing percentage is low. However, when the missing percentage is higher than 36%, the SBELM model is empty in the training process. Therefore, AMMD is more reliable than 1-NN method.

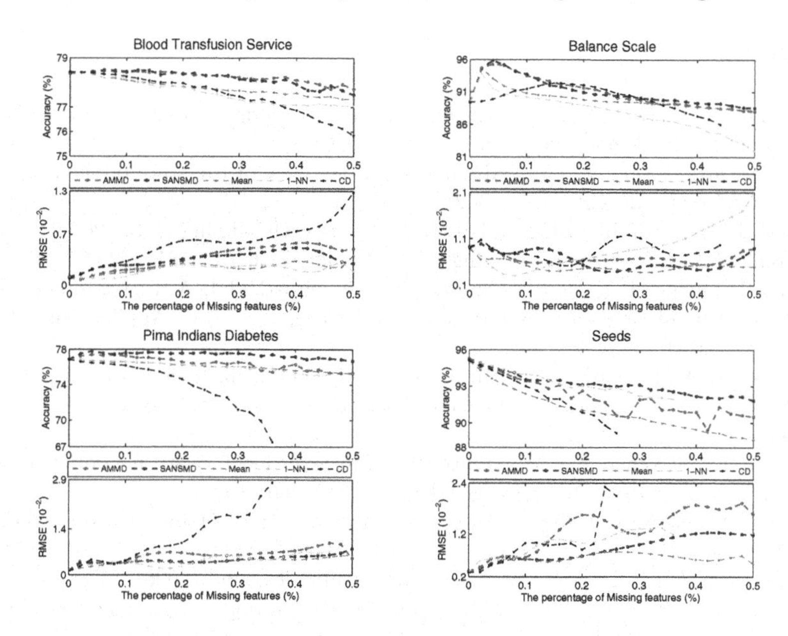

Fig. 3. The classification accuracy and RMSE for the data sets

5 Conclusions

In this paper, we proposed two methods to modify SBELM, a state-of-the-art variant of ELM, to handle missing data for multi-class classification. The standard SBELM can find sparse representatives for the output weights of ELM by utilizing the sparse Bayesian learning approach. The improved SBELMs utilize two different ways to calculate the output of hidden layers on incomplete data with little additional computation. In details, the first method utilizes additive models for missing data (AMMD), and the second method utilizes self-adjusting neuron state model for missing data (SANSMD) to change the input neuron state when encountering missing data. Hence the two methods retain the advantages of SBELM such as fast computational speed, high generalization performance and sparsity. Moreover, the two methods help SBELM to gain the ability of solving the missing data problems. From the experimental results over five benchmark data sets, the proposed methods have better generalization performance than other methods. It is also worth noting that our methods have reliability and good performance even half of the data values are missing. Hence, our methods completes ELM with two new methods to solve missing data problem for multi-class classification. Future work may focus on applying them for multi-output regression.

References

1. Baraldi, A.N., Enders, C.K.: An introduction to modern missing data analyses. Journal of School Psychology 48(1), 5–37 (2010)
2. Schafer, J.L., Graham, J.W.: Missing data: our view of the state of the art. Psychological methods 7(2), 147 (2002)
3. Little, R.J.A., Rubin, D.B.: Statistical Analysis with Missing Data (2002)
4. Rubin, D.B.: Inference and missing data. Biometrika 63(3), 581–592 (1976)
5. Garca-Laencina, P.J., Sancho-Gmez, J.L., Figueiras-Vidal, A.R.: Pattern classification with missing data: a review. Neural Computing and Applications 19(2), 263–282 (2010)
6. Dempster, A.P., Laird, N.M., Rubin, D.B.: Maximum likelihood from incomplete data via the EM algorithm. Journal of the Royal statistical Society 39(1), 1–38 (1977)
7. Hornik, K., Stinchcombe, M., White, H.: Multilayer feedforward networks are universal approximators. Neural networks 2(5), 359–366 (1989)
8. Bishop, C.M.: Neural networks for pattern recognition. Oxford University Press (1995)
9. Rasmussen, C.E.: Gaussian processes for machine learning (2006)
10. Chang, C.C., Lin, C.J.: LIBSVM: a library for support vector machines. ACM Transactions on Intelligent Systems and Technology (TIST) 2(3), 27 (2011)
11. Huang, G.B., Zhu, Q.Y., Siew, C.K.: Extreme learning machine: a new learning scheme of feedforward neural networks. In: 2004 IEEE International Joint Conference on Neural Networks Proceedings, vol. 2, pp. 985–990. IEEE (2004)
12. Huang, G.B., Zhu, Q.Y., Siew, C.K.: Extreme learning machine: theory and applications. Neurocomputing 70(1), 489–501 (2006)
13. Huang, G.B., Chen, L., Siew, C.K.: Universal approximation using incremental constructive feedforward networks with random hidden nodes. IEEE Transactions on Neural Networks 17(4), 879–892 (2006)
14. Miche, Y., Bas, P., Jutten, C., et al.: A Methodology for Building Regression Models using Extreme Learning Machine: OP-ELM[C]. ESANN, pp. 247–252 (2008)
15. Luo, J., Vong, C.M., Wong, P.K.: Sparse Bayesian Extreme Learning Machine for Multi-classification. IEEE Transactions on Neural Networks and Learning Systems 25(4), 836–843 (2014)
16. Miche, Y., Sorjamaa, A., Bas, P., et al.: OP-ELM: optimally pruned extreme learning machine. IEEE Transactions on Neural Networks 21(1), 158–162 (2010)
17. Miche, Y., Van Heeswijk, M., Bas, P., et al.: TROP-ELM: A double-regularized ELM using LARS and Tikhonov regularization. Neurocomputing 74(16), 2413–2421 (2011)
18. Rao, C.R., Mitra, S.K.: Generalized inverse of matrices and its applications. Wiley, New York (1971)
19. Tipping, M.E.: Sparse Bayesian learning and the relevance vector machine. The Journal of Machine Learning Research 1, 211–244 (2001)
20. Cortes, C., Vapnik, V.: Support-vector networks. Machine Learning 20(3), 273–297 (1995)
21. Soria-Olivas, E., Gomez-Sanchis, J., Jarman, I.H., et al.: BELM: Bayesian extreme learning machine. IEEE Transactions on Neural Networks 22(3), 505–509 (2011)
22. Bishop, C.M.: Pattern recognition and machine learning. springer, New York (2006)
23. MacKay, D.J.C.: Bayesian methods for backpropagation networks. Models of neural networks III, pp. 211–254. Springer, New York (1996)

24. Wu, T.F., Lin, C.J., Weng, R.C.: Probability estimates for multi-class classification by pairwise coupling. Journal of Machine Learning Research 5(975-1005), 4 (2004)
25. Pelckmans, K., De Brabanter, J., Suykens, J.A.K., et al.: Handling missing values in support vector machine classifiers. Neural Networks 18(5), 684–692 (2005)
26. Hastie, T.J., Tibshirani, R.J.: Generalized additive models. CRC Press (1990)
27. Viharos, Z.J., Kis, K.B.: Diagnostics of Wind Turbines Based on Incomplete Sensor Data. XX IMEKO World Congress C Metrology for green growth, TC10 on Technical Diagnostics 644 (2012)
28. Viharos, Z.J., Monostori, L., Vincze, T.: Training and application of artificial neural networks with incomplete data. In: Hendtlass, T., Ali, M. (eds.) IEA/AIE 2002. LNCS (LNAI), vol. 2358, p. 649. Springer, Heidelberg (2002)
29. Asuncion, A., Newman, D.: UCI machine learning repository (2007)
30. Chen, J., Shao, J.: Nearest neighbor imputation for survey data. Journal of Official Statistics 16(2), 113–132 (2000)
31. Van Hulse, J., Khoshgoftaar, T.M.: Incomplete-case nearest neighbor imputation in software measurement data. In: IEEE International Conference on Information Reuse and Integration, IRI, pp. 630–637. IEEE (2007)

A Fast Incremental Method Based on Regularized Extreme Learning Machine

Zhixin Xu and Min Yao

School of Computer Science and Technology, Zhejiang University
Zhejiang, Hangzhou, China, 310007

Abstract. Extreme Learning Machine(ELM) proposed by Huang et al is a new and simple algorithm for single hidden layer feedforward neural network(SLFN) with extreme fast learning speed and good generalization performance.When new hidden nodes are added to existing network retraining the network would be time consuming, and EM-ELM is proposed to calculate the output weight incrementally.However there are still two issues in EM-ELM:1.the initial hidden layer output matrix may be nearly singular thus the computation will loss accuracy;2.the algorithms can't always get good generalization performance due to overfitting.So we propose the improved version of EM-ELM based on regularization method called Incremental Regularized Extreme Learning Machine(IR-ELM).When new hidden node is added one by one,IR-ELM can update output weight recursively in a fast way.Empirical studies on benchmark data sets for regression and classification problems have shown that IR-ELM always get better generalization performance than EM-ELM with the similar training time.

Keywords: Extreme Learning Machine, Regularization, Incremental Learning, Neural Networks.

1 Introduction

Recently,a novel and efficient learning method for single hidden layer feedforward neural networks(SLFNs) called Extreme Learning Machine(ELM) has been proposed by Huang et al in [1] [2] [3].Compared with conventional learning algorithms for neural networks(e.g. Back propagation (BP) method [4]) which need adjust learning parameters iteratively,ELM generates parameters of the hidden nodes randomly and then determines the output weights analytically,which makes ELM extremely fast.A number of real world applications [3] [5] [6] [7] [8] have shown the efficiency of ELM algorithm with respect to its good generalization performance and fast learning speed.

Benefiting from its simplicity of structure,ELM can also be applied to incremental learning problem efficiently.Huang et al [9] proposed Incremental Extreme Learning Machine(I-ELM) which add hidden node one by one,and Feng

J. Cao et al. (eds.), *Proceedings of ELM-2014 Volume 1*,
Proceedings in Adaptation, Learning and Optimization 3, DOI: 10.1007/978-3-319-14063-6_2

et al [10] proposed another algorithm called Error Minimized Extreme Learning Machine(EM-ELM) that can add node one by one or group by group(with varying group size) which is more flexible and efficient in some cases.However,there are still some issues in EM-ELM.Firstly,in some applications,the initial hidden layer output matrix is nearly singular,thus the computation of its inverse matrix may loss accuracy.Secondly,the incremental algorithm is based on the original ELM,therefor EM-ELM inherits the problem of overfitting and sometimes can't achieve the expected testing accuracy.

In this paper, to avoid the issues mentioned above, we propose a modified version of EM-ELM called Incremental Regularized Extreme Learning Machine(IR-ELM) based on regularized extreme learning machine(R-ELM) [11]. Under the consideration of structural risk minimization(SRM) principle [12] [13],Deng et al [11] apply regularization method to ELM and get some enhancement of generalization performance. However, when the hidden nodes increase, R-ELM has the same problem as original ELM encountered, which needs much time to retrain the new networks and to recalculate the output weights.In IR-ELM,we firstly initial a SLFN with random hidden nodes of proper size,then we calculate the output weight using R-ELM algorithm,when new hidden node is added,we update the output weights recursively with IR-ELM algorithm which needn't retrain the networks thus can reduce the time complexity efficiently. The experiments are conducted for regression and classification problems on benchmark data sets obtained from UCI Machine Learning Repository [14]. The results have shown that the advantages of EM-ELM and R-ELM can both to a great extent be made use of.

The rest of this paper is organized as follows.Section 2 briefly introduces the original ELM and R-ELM.The details of proposed algorithm are described in Section 3.The experimental results of our proposed algorithm are shown in Section 4.Finally we present the discussions and conclusions in Section 5.

2 Preliminaries

In this section,we briefly introduce the original ELM by Huang et al [1] [2] and R-ELM by Deng et al [11], respectively.In the following,the construction of SLFNs and the theory of ELM are shortly reviewed.

2.1 Extreme Learning Machine

The ELM algorithm was proposed by Huang et al in [1] [2] [3].As a tuning-free learning algorithm,ELM makes full use of the SLFN architecture,and the essence of ELM lies in the random initialization of hidden nodes' weights and biases,then the output weights can be determined analytically.In the following,we first introduce the basic concepts of SLFN.

For N arbitrary input samples $(\mathbf{x}_i, \mathbf{t}_i)$, where $\mathbf{x}_i = [x_{i1}, x_{i2}, \ldots, x_{in}]^{\mathrm{T}} \in \mathbf{R}^n$ and $\mathbf{t}_i = [t_{i1}, t_{i2}, \ldots, t_{im}]^{\mathrm{T}} \in \mathbf{R}^m$,and given activation function $g(x)$,the standard mathematical model of SLFN with $\tilde{N}$ hidden nodes is as follows:

$$\sum_{i=1}^{\tilde{N}} \boldsymbol{\beta}_i g(\mathbf{w}_i \cdot \mathbf{x}_j + b_i) = \mathbf{o}_j, j = 1, \ldots, N \tag{1}$$

where $\mathbf{w}_i = [w_{i1}, w_{i2}, \ldots, w_{in}]^{\mathrm{T}}$ is the input weight vector connecting input nodes and the ith hidden node. $\boldsymbol{\beta}_i = [\beta i1, \beta i2, \ldots, \beta im]^{\mathrm{T}}$ is the output weight vector connecting the ith hidden node and the output nodes,b_i is the basis of ith hidden node.$\mathbf{w}_i \cdot \mathbf{x}_j$ is the inner product of $\mathbf{w}_i$ and $\mathbf{x}_j$.

SLFNs with $\tilde{N}$ hidden nodes can approximate N samples with zero error,which means that $\sum_{j=1}^{N} \|\mathbf{o}_j - \mathbf{t}_j\| = 0$,so there exists $\boldsymbol{\beta}_i, \mathbf{w}_i, b_i$ that:

$$\sum_{i=1}^{\tilde{N}} \boldsymbol{\beta}_i g(\mathbf{w}_i \cdot \mathbf{x}_j + b_i) = \mathbf{t}_j, j = 1, \ldots, N \tag{2}$$

Above N equations can be rewritten compactly as:

$$\mathrm{H}\beta = \mathrm{T} \tag{3}$$

where:

$$\mathrm{H} = \begin{bmatrix} g(\mathbf{w}_1 \cdot \mathbf{x}_1 + b_1) & \cdots & g(\mathbf{w}_{\tilde{N}} \cdot \mathbf{x}_1 + b_{\tilde{N}}) \\ \vdots & \cdots & \vdots \\ g(\mathbf{w}_1 \cdot \mathbf{x}_N + b_1) & \cdots & g(\mathbf{w}_{\tilde{N}} \cdot \mathbf{x}_N + b_{\tilde{N}}) \end{bmatrix}_{N \times \tilde{N}} \tag{4}$$

$$\boldsymbol{\beta} = \begin{bmatrix} \boldsymbol{\beta}_1^{\mathrm{T}} \\ \vdots \\ \boldsymbol{\beta}_{\tilde{N}}^{\mathrm{T}} \end{bmatrix}_{\tilde{N} \times m}, \mathrm{T} = \begin{bmatrix} \mathbf{t}_1^{\mathrm{T}} \\ \vdots \\ \mathbf{t}_N^{\mathrm{T}} \end{bmatrix}_{N \times m} \tag{5}$$

According to Huang [1],H is called hidden layer output matrix of SLFN.

To train SLFN,we need find the special $\hat{\mathbf{w}}_i, \hat{b}_i, \hat{\boldsymbol{\beta}}_i (i = 1, 2, \ldots, \tilde{N})$ that satisfy:

$$\|\mathbf{H}(\hat{\mathbf{w}}_1, \ldots, \hat{\mathbf{w}}_{\tilde{N}}, \hat{b}_1, \ldots, \hat{b}_{\tilde{N}})\hat{\boldsymbol{\beta}} - \mathbf{T}\| = \min_{\mathbf{w}_i, b_i, \boldsymbol{\beta}} \|\mathbf{H}(\mathbf{w}_1, \ldots, \mathbf{w}_{\tilde{N}}, b_1, \ldots, b_{\tilde{N}})\boldsymbol{\beta} - \mathbf{T}\| \tag{6}$$

which is equivalent to minimizing the cost function:

$$E = \sum_{j=1}^{N} \left\| \sum_{i=1}^{\tilde{N}} \boldsymbol{\beta}_i g(\mathbf{w}_i \cdot \mathbf{x}_j + b_i) - \mathbf{t}_j \right\|^2 \tag{7}$$

Traditionally,SLFN is trained by gradient based method,which need determine three parameters iteratively.There are some drawbacks in traditional method such as slow convergence speed,local minimum,and overfitting,so Huang et al [1] [2] proposed a novel learning algorithm for SLFNs referred to as Extreme Learning Machine(ELM).

Unlike conventional method,ELM generates weight vectors $\mathbf{w}_i$ and basis b_i of hidden nodes randomly,then determine the output weight $\boldsymbol{\beta}$ analytically,which is equivalent to find the least square solution $\hat{\boldsymbol{\beta}}$ of linear equation $\mathbf{H}\boldsymbol{\beta} = \mathbf{T}$ according to equation 7:

$$\|\mathbf{H}(\mathbf{w}_1,\ldots,\mathbf{w}_{\tilde{N}},b_1,\ldots,b_{\tilde{N}})\hat{\boldsymbol{\beta}} - \mathbf{T}\| = \min_{\boldsymbol{\beta}} \|\mathbf{H}(\mathbf{w}_1,\ldots,\mathbf{w}_{\tilde{N}},b_1,\ldots,b_{\tilde{N}})\boldsymbol{\beta} - \mathbf{T}\| \tag{8}$$

If the number of hidden nodes $\tilde{N}$ is equal to the number of input samples N,then $\mathbf{H}$ is square matrix and is invertible with probability one,therefor SLFN can approximate the input samples with zero error.However,in most real world applications,the number of hidden nodes $\tilde{N}$ is much less than the number of input samples N,so $\mathbf{H}$ is not square matrix and there are no parameters $\mathbf{w}_i, b_i, \boldsymbol{\beta}_i(i = 1,\ldots,\tilde{N})$ that satisfy the equation $\mathbf{H}\boldsymbol{\beta} = \mathbf{T}$.Huang et al [1] gives the following solution:

$$\hat{\boldsymbol{\beta}} = \mathbf{H}^{\dagger}\mathbf{T} \tag{9}$$

where $\mathbf{H}^{\dagger}$ is Moore-Penrose generalized inverse of matrix $\mathbf{H}$.

ELM algorithm can be summarized as follows.The inputs of algorithm are training data,testing data,number of hidden nodes and activation function;the outputs of algorithm are training time,testing time,training accuracy and testing accuracy.

Algorithm 1. ELM.

For a given training data set $\aleph = \{(\mathbf{x}_i,\mathbf{t}_i)|\mathbf{x}_i \in \mathbf{R}^n,\mathbf{t}_i \in \mathbf{R}^m, i = 1,\ldots,N\}$,the number of hidden nodes $\tilde{\mathbf{N}}$,and activation function $g(x)$:

1. Assign the input weight vectors $\mathbf{w}_i$ and basis $b_i, i = 1,\ldots,\tilde{N}$ randomly;
2. Calculate the hidden layer output matrix $\mathbf{H}$;
3. Calculate the output weight $\boldsymbol{\beta}$:$\boldsymbol{\beta} = \mathbf{H}^{\dagger}\mathbf{T}$,where $\mathbf{T} = [\mathbf{t}_1,\ldots,\mathbf{t}_N]^{\mathrm{T}}$.

2.2 Regularized Extreme Learning Machine

In this subsection,we briefly introduce the regularized extreme learning machine (R-ELM) [11].In 1963,Tikhonov proposed a new method for solving ill-posed problems called regularization,and since then regularization theory has been at the core of many neural network and machine learning algorithms [16].

Deng et al [11] applied regularization method to ELM,and it is pointed out that ELM algorithm is based on ERM principle and tends to overfit.According to statistical learning theory [12] [13],a learning machine with good generalization performance should consider Structural Risk Minimization(SRM) instead of Empirical Risk Minimization(ERM),which need a term that controls the complexity of learning machine.It's proven in [17] that networks tend to have better generalization performance with small norm of output weight $\boldsymbol{\beta}$.Without loss of generality,we assume $m = 1$,thus the objective function is considered to minimize as follows:

$$\begin{aligned} f(\boldsymbol{\beta}) &= \|\mathbf{H}\boldsymbol{\beta} - \mathbf{T}\|^2 + C\|\boldsymbol{\beta}\|^2 \\ &= (\mathbf{H}\boldsymbol{\beta} - \mathbf{T})^T(\mathbf{H}\boldsymbol{\beta} - \mathbf{T}) + C\boldsymbol{\beta}^T\boldsymbol{\beta} \end{aligned} \tag{10}$$

Differentiating with respect to $\boldsymbol{\beta}$ and we obtain:

$$\frac{\partial f(\boldsymbol{\beta})}{\partial \boldsymbol{\beta}} = -2\mathbf{H}^T(\mathbf{T} - \mathbf{H}\boldsymbol{\beta}) + 2C\boldsymbol{\beta} \tag{11}$$

Let the derivative to be zero and we get:

$$2\mathbf{H}^T\mathbf{H}\boldsymbol{\beta} - 2\mathbf{H}^T\mathbf{T} + 2C\boldsymbol{\beta} = \mathbf{0} \tag{12}$$

It is easy to get the following solution:

$$(C\mathbf{I} + \mathbf{H}^T\mathbf{H})\boldsymbol{\beta} = \mathbf{H}^T\mathbf{T} \tag{13}$$

The term $C\mathbf{I}$ makes the $(C\mathbf{I} + \mathbf{H}^T\mathbf{H})$ term nonsingular,so we can get:

$$\boldsymbol{\beta} = (C\mathbf{I} + \mathbf{H}^T\mathbf{H})^{-1}\mathbf{H}^T\mathbf{T} \tag{14}$$

It is obvious that ELM is just the special case of R-ELM when $C \to \infty$.In [5],Huang et al give more thorough and profound demonstration about regularization method combined with ELM.Other theory and applications [15] [18] [19] show that regularization method is efficient.Our proposed algorithm referred to as IR-ELM which updates the output weight β recursively is based on R-ELM.

3 Incremental Regularized Extreme Learning Machine

Incremental learning is necessary in the following two situations:1.for a given error ϵ,to determine the least number of hidden nodes that satisfies testing error $\leq \epsilon$;2.for given data set,to find the best testing performance that the neural network can achieve,which need also determine the number of hidden nodes.In both these two situations,we need add new hidden nodes to the original networks one by one or group by group sequentially as demonstrated in figure 1,thus we may need retrain the networks and recalculate the output weight which is time-consuming.

In [10],Feng et al proposed a fast incremental learning algorithm referred to as error minimized extreme learning machine(EM-ELM).During the growth of networks,the output weight can be updated incrementally.The experimental results have shown that this new approach is faster than other sequential/incremental algorithms with good generalization performance.However there are some drawbacks in EM-ELM:1.the initial hidden layer output matrix may be nearly singular thus the computation will loss accuracy;2.the algorithm can't always get good generalization performance due to overfitting.So we propose the improved version of EM-ELM based on R-ELM.

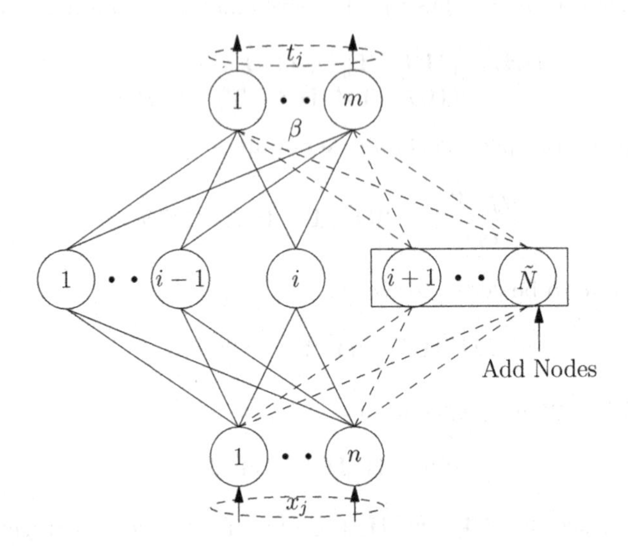

Fig. 1. We add new hidden nodes generated randomly to the original networks one by one or group by group sequentially

The essence of our proposed IR-ELM lies in calculating the output weight in a fast way without losing generalization ability.In the first step of IR-ELM,an initial network with $\tilde{N}$ random hidden nodes is assigned,then we calculate the output weight using R-ELM algorithms.After new hidden node is add to the existing network one by one,we update the output weight recursively until testing accuracy satisfies our requirement.

Assuming there are already $s-1$ hidden nodes in the network,so we can get the output weight $\boldsymbol{\beta}_{s-1} = (C\mathbf{I} + \mathbf{H}_{s-1}^T\mathbf{H}_{s-1})^{-1}\mathbf{H}_{s-1}^T\mathbf{T}$,where $\mathbf{H}_{s-1}$ is the hidden layer output matrix and the subscript $s-1$ indicates the number of hidden nodes.When new node is added,the output weight becomes $\boldsymbol{\beta}_s = (C\mathbf{I} + \mathbf{H}_s^T\mathbf{H}_s)^{-1}\mathbf{H}_s^T\mathbf{T}$,where $\mathbf{H}_s = [\mathbf{H}_{s-1}, \mathbf{v}_s]$ is the hidden layer output matrix of new network and $\mathbf{v}_s$ is the hidden layer output vector associated to the ith hidden node.Let $\mathbf{D}_{s-1} = (C\mathbf{I} + \mathbf{H}_{s-1}^T\mathbf{H}_{s-1})^{-1}\mathbf{H}_{s-1}^T$ and $\mathbf{D}_s = (C\mathbf{I} + \mathbf{H}_s^T\mathbf{H}_s)^{-1}\mathbf{H}_s^T$.The left part of $\mathbf{D}_s$ can be rewritten as:

$$\begin{aligned}(C\mathbf{I} + \mathbf{H}_s^T\mathbf{H}_s)^{-1} &= \left(C\mathbf{I} + \begin{bmatrix}\mathbf{H}_{s-1}^T \\ \mathbf{v}_s\end{bmatrix}[\mathbf{H}_{s-1}, \mathbf{v}_s]\right)^{-1} \\ &= \begin{bmatrix}\mathbf{H}_{s-1}^T\mathbf{H}_{s-1} + C\mathbf{I} & \mathbf{H}_{s-1}^T\mathbf{v}_s \\ \mathbf{v}_s^T\mathbf{H}_{s-1} & \mathbf{v}_s^T\mathbf{v}_s + C\end{bmatrix}^{-1}\end{aligned} \tag{15}$$

because $C\mathbf{I} + \mathbf{H}_s^T\mathbf{H}_s$ is symmetric,the inverse matrix is symmetric too,thus we denote:

$$(C\mathbf{I} + \mathbf{H}_s^T\mathbf{H}_s)^{-1} = \begin{bmatrix}\mathbf{A} & \mathbf{B} \\ \mathbf{B}^T & \mathbf{E}\end{bmatrix}^{-1} = \begin{bmatrix}\mathbf{A}' & \mathbf{B}' \\ \mathbf{B}'^T & \mathbf{E}'\end{bmatrix} \tag{16}$$

where:

$$\begin{cases} \mathbf{A} = \mathbf{H}_{s-1}^T\mathbf{H}_{s-1} + C\mathbf{I} \\ \mathbf{B} = \mathbf{H}_{s-1}^T\mathbf{v}_s \\ \mathbf{E} = \mathbf{v}_s^T\mathbf{v}_s + C \end{cases} \tag{17}$$

So,

$$\begin{aligned} \mathbf{I} &= \begin{bmatrix} \mathbf{A} & \mathbf{B} \\ \mathbf{B}^T & \mathbf{E} \end{bmatrix}^{-1} \begin{bmatrix} \mathbf{A}' & \mathbf{B}' \\ \mathbf{B}'^T & \mathbf{E}' \end{bmatrix} \\ &= \begin{bmatrix} \mathbf{AA}' + \mathbf{BB}'^T & \mathbf{AB}' + \mathbf{BE}' \\ \mathbf{B}^T\mathbf{A}' + \mathbf{EB}'^T & \mathbf{B}^T\mathbf{B}' + \mathbf{EE}' \end{bmatrix} \end{aligned} \tag{18}$$

It is not difficult to get the solution for $\mathbf{A}', \mathbf{B}'$ and $\mathbf{E}'$:

$$\begin{cases} \mathbf{A}' = \dfrac{\mathbf{A}^{-1}\mathbf{B}(\mathbf{A}^{-1}\mathbf{B})^T}{\mathbf{E} - \mathbf{B}^T(\mathbf{A}^{-1}\mathbf{B})} + \mathbf{A}^{-1} \\ \mathbf{B}' = \dfrac{\mathbf{A}^{-1}\mathbf{B}}{\mathbf{B}^T\mathbf{A}^{-1}\mathbf{B} - \mathbf{E}} \\ \mathbf{E}' = \dfrac{1}{\mathbf{E} - \mathbf{B}^T\mathbf{A}^{-1}\mathbf{B}} \end{cases} \tag{19}$$

The following result is obvious:

$$\mathbf{D}_s = \begin{bmatrix} \mathbf{A}' & \mathbf{B}' \\ \mathbf{B}'^T & \mathbf{E}' \end{bmatrix} \cdot \begin{bmatrix} \mathbf{H}_{s-1}^T \\ \mathbf{v}_s^T \end{bmatrix} = \begin{bmatrix} \mathbf{A}'\mathbf{H}_{s-1}^T + \mathbf{B}'\mathbf{v}_s^T \\ \mathbf{B}'^T\mathbf{H}_{s-1}^T + \mathbf{E}'\mathbf{v}_s^T \end{bmatrix} \tag{20}$$

Denote:

$$\mathbf{D}_s = \begin{bmatrix} \mathbf{L} \\ \mathbf{M} \end{bmatrix} \tag{21}$$

Substituting equation 19 and 17 to equation 20 and we can get:

$$\begin{aligned} \mathbf{M} &= \frac{\mathbf{B}^T\mathbf{A}^{-1}\mathbf{H}_{s-1}^T}{\mathbf{B}^T\mathbf{A}^{-1}\mathbf{B} - E} + \frac{\mathbf{v}_s^T}{\mathbf{E} - \mathbf{B}^T\mathbf{A}^{-1}\mathbf{B}} \\ &= \frac{\mathbf{v}_s^T(\mathbf{I} - \mathbf{H}_{s-1}\mathbf{D}_{s-1})}{\mathbf{v}_s^T(\mathbf{I} - \mathbf{H}_{s-1}\mathbf{D}_{s-1})\mathbf{v}_s + C} \end{aligned} \tag{22}$$

and

$$\begin{aligned} \mathbf{L} &= \frac{\mathbf{A}^{-1}\mathbf{B}(\mathbf{A}^{-1}\mathbf{B})^T\mathbf{H}_{s-1}^T}{\mathbf{E} - \mathbf{B}^T(\mathbf{A}^{-1}\mathbf{B})} + \frac{\mathbf{A}^{-1}\mathbf{B}\mathbf{v}_s^T}{\mathbf{B}^T\mathbf{A}^{-1}\mathbf{B} - E} + \mathbf{A}^{-1}\mathbf{H}_{s-1}^T \\ &= \frac{\mathbf{D}_{s-1}\mathbf{v}_s\mathbf{v}_s^T(\mathbf{H}_{s-1}\mathbf{D}_{s-1} - \mathbf{I})}{\mathbf{v}_s^T(\mathbf{I} - \mathbf{H}_{s-1}\mathbf{D}_{s-1})\mathbf{v}_s + C} + \mathbf{D}_{s-1} \\ &= \mathbf{D}_{s-1}(\mathbf{I} - \mathbf{v}_s\mathbf{M}) \end{aligned} \tag{23}$$

Given a set of training data,the number of initial hidden nodes,the maximum number of hidden nodes and the expected learning accuracy,our proposed IR-ELM algorithm can be summarized as follows.

Algorithm 2. IR-ELM.

For a given training data set $\aleph = \{(\mathbf{x}_i, \mathbf{t}_i)|\mathbf{x}_i \in \mathbf{R}^n, \mathbf{t}_i \in \mathbf{R}^m, i = 1, \ldots, N\}$,the initial number of hidden nodes N_0,the maximum number of hidden nodes N_{max} and the expected learning accuracy ϵ:

I Initial the neural network:

1. Assign the input weight vectors $\mathbf{w}_i$ and basis $b_i, i = 1, \ldots, N_0$ randomly;
2. Calculate the hidden layer output matrix $\mathbf{H}_0$;
3. Calculate the output weight $\boldsymbol{\beta}_0$:$\boldsymbol{\beta}_0 = \mathbf{D}_0\mathbf{T} = (\mathbf{H}_0^T\mathbf{H}_0 + C\mathbf{I})^{-1}\mathbf{H}_0^T\mathbf{T}$,where $\mathbf{T} = [\mathbf{t}_1, \ldots, \mathbf{t}_N]^{\mathrm{T}}$;
4. Let $s = 0$,calculate the learning accuracy $\epsilon_s = \epsilon_0$.

II Update the networks recursively.While $N_s < N_{max}$ and $\epsilon_s > \epsilon$:

1. Let $s = s + 1, N_s = N_{s-1} + 1$;
2. Add a new hidden node generated randomly to the existing network and calculate the corresponding output matrix $\mathbf{H}_s = [\mathbf{H}_{s-1}, \mathbf{v}_s]$;
3. The output weight is updated as follows:

$$\mathbf{M}_s = \frac{\mathbf{v}_s^T(\mathbf{I} - \mathbf{H}_{s-1}\mathbf{D}_{s-1})}{\mathbf{v}_s^T(\mathbf{I} - \mathbf{H}_{s-1}\mathbf{D}_{s-1})\mathbf{v}_s + C} \tag{24}$$

$$\mathbf{L}_s = \mathbf{D}_{s-1}(\mathbf{I} - \mathbf{v}_s\mathbf{M}_s) \tag{25}$$

$$\boldsymbol{\beta}_s = \mathbf{D}_s\mathbf{T} = \begin{bmatrix} \mathbf{L}_s \\ \mathbf{M}_s \end{bmatrix} \mathbf{T} \tag{26}$$

4. Calculate the new learning accuracy ϵ_s.

*Remark 1:*Compared with the original ELM whose time complexity is $O(N \times \tilde{N}^2)$,our algorithm need only $O(N \times \tilde{N})$ time complexity to update the output weight recursively,where N is the number of training samples and $\tilde{N}$ is the number of hidden nodes.

*Remark 2:*Similar to the phenomenon we have pointed out in section 2.2 that ELM is just the special case of R-ELM when $C \to 0$,while the hidden node is added one by one EM-ELM [10] is the special case of IR-ELM when $C \to 0$.

*Remark 3:*In [10],the author is able to add new hidden nodes to existing network group by group,which is not available for IR-ELM in a efficient way due to the presence of the regularization term.

4 Performance Verification

In this section,experiments of our proposed IR-ELM are conducted on benchmark data sets for regression and classification problems.In order to investigate the improvement of training time and learning accuracy of the IR-ELM

Table 1. Specifications of benchmark problems

Datasets	Attributes	Class	Training Data	Testing Data	Types
SinC	1	-	5000	5000	Regression
Abalone	8	-	2177	2000	Regression
Boston Housing	13	-	256	250	Regression
Yacht Hydro	6	-	200	108	Regression
Servo	4	-	80	87	Regression
Concrete	8	-	530	500	Regression
Vehicle	18	4	446	400	Classification
Glass	9	7	130	84	Classification
Iris	4	3	100	50	Classification
Blood	4	2	500	248	Classification
Wine	13	3	100	78	Classification
Image Segmentation	19	7	1500	810	Classification

method,original ELM algorithm [1] and EM-ELM algorithm [10] are also evaluated.The experimental environment is Matlab 7.14 running on a desktop PC with Intel 3.2 GHz CPU and 4GB RAM.The activation function of SLFNs is sigmoid function:$g(x) = 1/(1 + exp(-\lambda x))$,where λ is 0.5.All the network inputs have been normalized into the range $[-1, 1]$ and the outputs have been normalized into the range $[0, 1]$.

For regression problems,the performance of these three algorithms is evaluated on artificially synthetic data set (SinC function) and benchmark data sets.The benchmark data sets include Abalone,Boston Housing,Yacht Hydrodynamics,Servo and Concrete Compressive Strength,as shown in Table 1.For classification problems,the benchmark data sets are Vehicle, Glass,Iris,Blood,Wine and Image Segmentation,as shown in Table 1.The benchmark data sets are obtained from UCI Machine Learning Repository [14].

As shown in equation 14 and 22,the parameter C has a great influence on the learning accuracy,the choice of C can enhance the accuracy or reduce the accuracy.To see how C affect the network performance,first we evaluate the testing accuracy of R-ELM on Boston Housing data set with varying C value.The neural network has 100 hidden nodes and C takes value in the set $\{2^{-20}, 2^{-19}, \cdots, 2^{0}, \cdots, 2^{19}, 2^{20}\}$ which has 41 elements.The final results are averaged by 20 trials.Figure 2 shows the training and testing Root Mean Square Error(RMSE) in comparison with original ELM.The red solid line stands for training accuracy of R-ELM,and red dotted line stands for training accuracy of ELM.The blue solid line is testing accuracy of R-ELM,the blue dotted line is ELM.We can see that larger the C is,poorer training performance the network gets.As for testing accuracy,the R-ELM dominates when C takes value around 2^{-10}.In the following experiments,we choose a proper C of IR-ELM for different data sets.

In the following,we compare the time complexity of ELM,EM-ELM and IR-ELM on SinC and Boston Housing data sets.The initial number of hidden nodes for SinC and Boston Housing are 5 and 30,respectively.The hidden node is added one by one,and we retrain the networks for ELM algorithm and update the

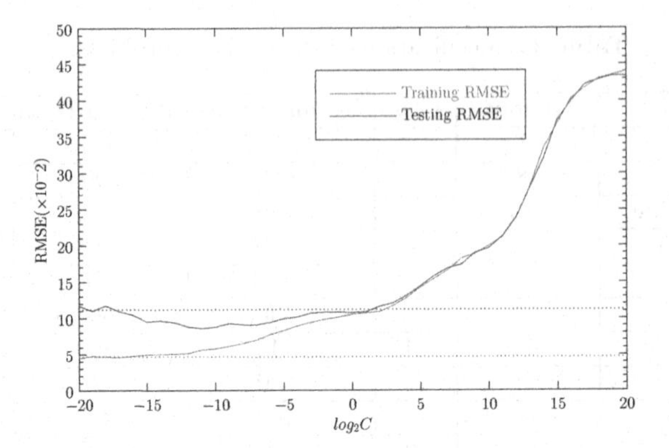

Fig. 2. The result comparison of training and testing accuracy of R-ELM and ELM with 100 hidden nodes on Boston Housing data set,parameter C takes values in $\{2^{-20}, 2^{-19}, \cdots, 2^{0}, \cdots, 2^{19}, 2^{20}\}$

network output weight recursively for EM-ELM and IR-ELM each time.The average results are obtained by 50 trials.From Figure 3 we can see that EM-ELM and IR-ELM have the similar time updating the output weight in accordance with the fact that EM-ELM is a special case of IR-ELM while $C \to 0$ and IR-ELM only has one more operation of scalar addition.Compared with EM-ELM and IR-ELM,the original ELM spends more time retraining the network and recalculating the output weight.And this gap is more obvious when the network has more hidden nodes.

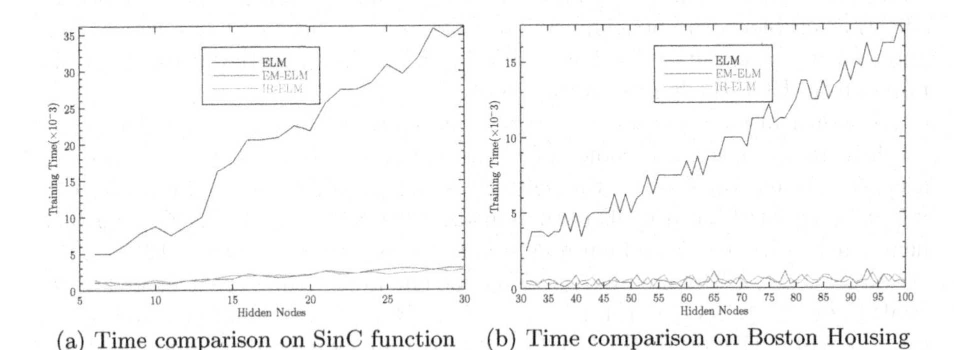

(a) Time comparison on SinC function (b) Time comparison on Boston Housing

Fig. 3. Time comparison of ELM,EM-ELM and IR-ELM on regression data sets while hidden node is added one by one

Next we explore the improvement of learning accuracy of IR-ELM in comparison with EM-ELM.To see this we conduct experiments from two aspects:

1. When hidden node is added one by one,how the learning accuracy of EM-ELM and IR-ELM changes;

Table 2. The settings of IR-ELM and EM-ELM for regression problems

Datasets	Algorithm	Initial Nodes	End Nodes	Stop RMSE	C
Abalone	IR-ELM	15	60	0.078	2^{-12}
	EM-ELM	15	60	0.078	-
Boston Hou	IR-ELM	30	100	0.09	2^{-10}
	EM-ELM	30	100	0.09	-
Yacht Hydro	IR-ELM	40	70	0.035	2^{-30}
	EM-ELM	40	70	0.035	-
Servo	IR-ELM	5	60	0.1	2^{-17}
	EM-ELM	5	60	0.1	-
Concrete	IR-ELM	80	130	0.095	2^{-20}
	EM-ELM	80	130	0.095	-

2. For a given learning accuracy ϵ and the maximum number of hidden nodes, how many hidden nodes are necessary for EM-ELM and IR-ELM to satisfy the expected accuracy.

For the first question,we illustrate the results by averaging 50 trials.For the second question,sometimes the random generated networks can't achieve the expected accuracy no matter how many hidden nodes it has,and we call this "FAIL".We will record the frequency of "SUCCESS",say not "FAIL",out of 50 trials and obtain the average number of hidden nodes for "SUCCESS" case.To eliminate the oscillation of learning accuracy caused by randomness as much as possible,we use the same input weights and bias for both EM-ELM and IR-ELM each time.

As for regression problems,settings of IR-ELM and EM-ELM are listed in Table 2.The total "SUCCESS" number out of 50 trials and the average values(number of hidden nodes,training time,and testing RMSE) when the testing RMSE is exactly satisfied are shown in Table 3.From the results we can see that for a given testing RMSE,IR-ELM gets more "SUCCESS" than EM-ELM in all cases.What's more,for "Abalone" and "Servo" data sets the number of hidden nodes are more than EM-ELM while for the other three cases IR-ELM dominates.In general,IR-ELM can always get better generalization performance than EM-ELM for regression problems in incremental learning scenario.

When the number of hidden nodes ranges from "Initial Nodes" to "End Nodes",the average testing RMSE of IR-ELM and EM-ELM are illustrated in Figure 4,Figure 5 and Figure 6 for "Abalone" data set,"Boston Housing" and "Yacht Hydrodynamics" data sets,"Servo" and "Concrete" data sets,respectively. From the figures we can conclude that in general IR-ELM can always get lower testing RMSE than EM-ELM.When the number of hidden nodes increase,the generalization performance of IR-ELM enhances,however the performance of EM-ELM might get worse,say overfitting which is specially severe for "Servo" data set.In "Abalone","Boston However" and "Servo" cases,the RMSE curve of IR-ELM are smooth,but in "Yacht" and "Conclude" cases oscillation occurs

Table 3. The average number of hidden nodes for EM-ELM and IR-ELM to satisfy the learning accuracy ϵ

Datasets	Algorithm	Hidden Nodes	Time(s)	Testing RMSE	SUCCESS
Abalone	IR-ELM	20.0	0.0222	0.0773	47
	EM-ELM	17.9	0.0177	0.0772	45
Boston Housing	IR-ELM	54.5	0.0406	0.0886	35
	EM-ELM	57.8	0.0390	0.0885	28
Yacht Hydro	IR-ELM	53.8	0.0217	0.0336	46
	EM-ELM	54.1	0.0176	0.0336	46
Servo	IR-ELM	28.1	0.0203	0.0981	23
	EM-ELM	27.3	0.0193	0.0960	17
Concrete	IR-ELM	94.6	0.0449	0.0938	41
	EM-ELM	95.1	0.0569	0.0937	34

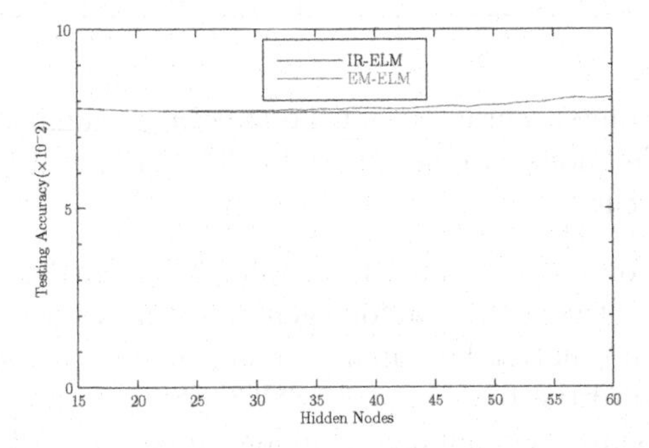

Fig. 4. The average testing RMSE of IR-ELM and EM-ELM on Abalone data set

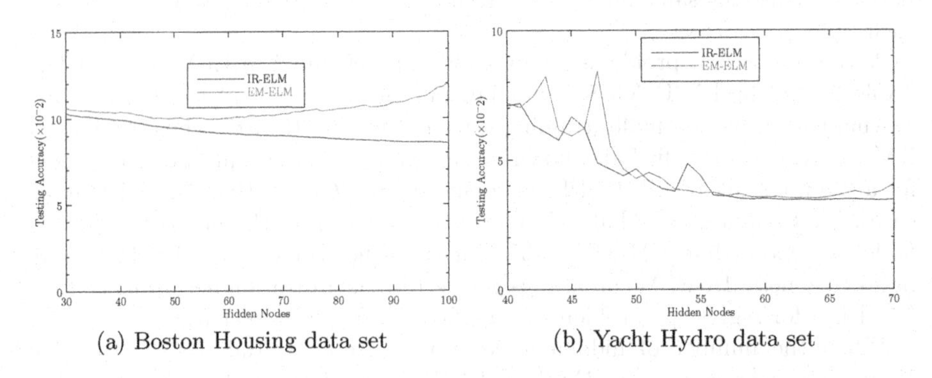

(a) Boston Housing data set (b) Yacht Hydro data set

Fig. 5. The average testing RMSE of IR-ELM and EM-ELM on Boston Housing and Yacht Hydro data sets

when the network is small.And the same phenomenon(even more obvious) happens on EM-ELM in "Yacht","Servo" and "Concrete" cases.

For classification problems,the settings of IR-ELM and EM-ELM are listed in Table 4.The number of "SUCCESS" out of 50 trials and some average val-

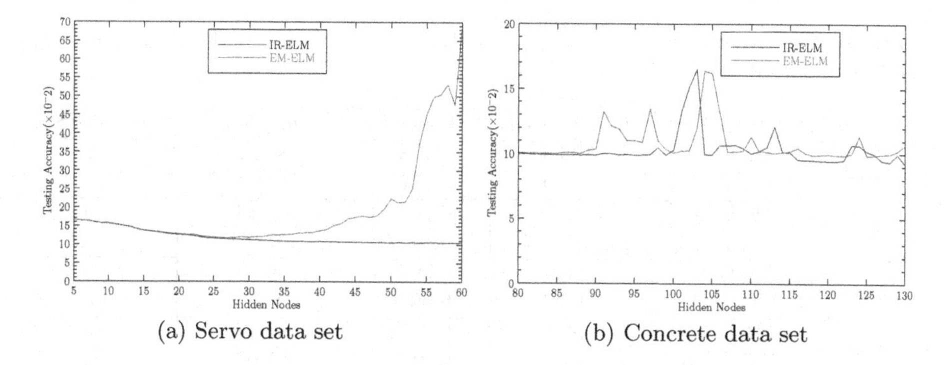

(a) Servo data set (b) Concrete data set

Fig. 6. The average testing RMSE of IR-ELM and EM-ELM on Servo and Concrete data sets

Table 4. The settings of IR-ELM and EM-ELM for classification problems

Datasets	Algorithm	Initial Nodes	End Nodes	Stop Accuracy(%)	C
Vehicle	IR-ELM	70	180	82	2^{-12}
	EM-ELM	100	180	82	-
Glass	IR-ELM	20	70	65	2^{-14}
	EM-ELM	20	70	65	-
Iris	IR-ELM	5	40	95	2^{-19}
	EM-ELM	5	40	95	-
Blood	IR-ELM	10	50	80	2^{-25}
	EM-ELM	10	50	80	-
Wine	IR-ELM	10	50	98.2	2^{-7}
	EM-ELM	10	50	98.2	-
Image Segmentation	IR-ELM	150	200	95	2^{-15}
	EM-ELM	150	200	95	-

ues when the testing accuracy is exactly satisfied are shown in Table 5.We can see that for "Blood" data set,EM-ELM get more "SUCCESS"(25) than IR-ELM(23),but the number of hidden nodes of IR-ELM(10.8) is smaller than that of EM-ELM(13.5).For "Glass" and "Iris" data sets,IR-ELM and EM-ELM get the same frequency of "SUCCESS"(43 and 48 respectively),however the numbers of hidden nodes of IR-ELM are smaller than those of EM-ELM(25.3 versus 27.0 for "Glass" and 11.5 versus 11.6 for "Iris").For "Vehicle","Wine" and "Image Segmentation" data sets,IR-ELM get more "SUCCESS" than EM-ELM with the similar numbers of hidden nodes.Totally IR-ELM performs better than EM-ELM for classification problems for incremental learning.

The average testing accuracy of IR-ELM and EM-ELM when the networks have different number of hidden nodes are shown in Figure 7(for "Vehicle" and "Glass"),Figure 8(for "Iris" and "Blood") and Figure 9(for "Wine" and "Image Segmentation"), respectively.From the figures we can see that IR-ELM

Table 5. The average number of hidden nodes for EM-ELM and IR-ELM to satisty the learning accuracy ϵ

Datasets	Algorithm	Hidden Nodes	Time(s)	Testing Acc(%)	SUCCESS
Vehicle	IR-ELM	105.4	0.1759	82.21	29
	EM-ELM	105.4	0.1776	82.21	26
Glass	IR-ELM	25.3	0.0069	66.83	43
	EM-ELM	27.0	0.0087	67.05	43
Iris	IR-ELM	11.5	0.0033	96.58	48
	EM-ELM	11.6	0.0036	96.58	48
Blood	IR-ELM	10.8	0.0061	81.14	23
	EM-ELM	13.5	0.0137	81.06	25
Wine	IR-ELM	16.6	0.0041	98.89	38
	EM-ELM	16.3	0.0069	99.02	34
Image Seg	IR-ELM	159.5	0.2860	95.28	39
	EM-ELM	157.2	0.2211	95.29	35

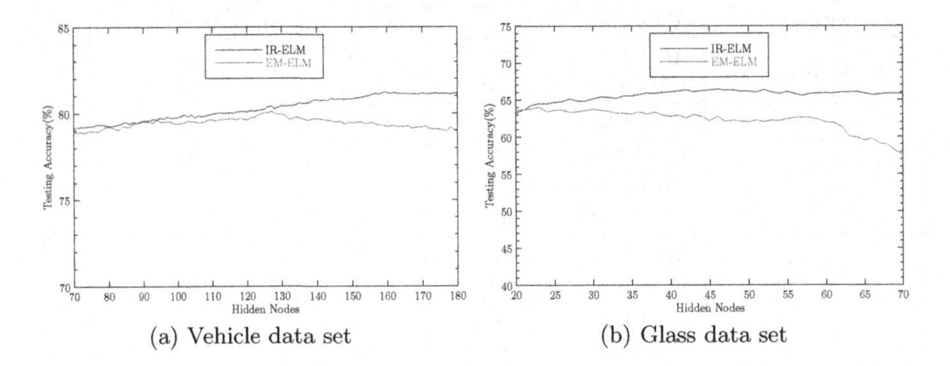

(a) Vehicle data set (b) Glass data set

Fig. 7. The average testing accuracy of IR-ELM and EM-ELM on Vehicle and Glass data sets

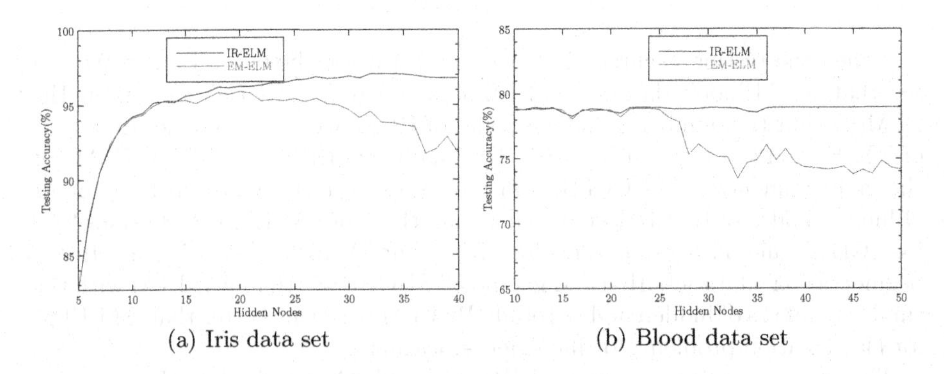

(a) Iris data set (b) Blood data set

Fig. 8. The average testing accuracy of IR-ELM and EM-ELM on Iris and Blood data sets

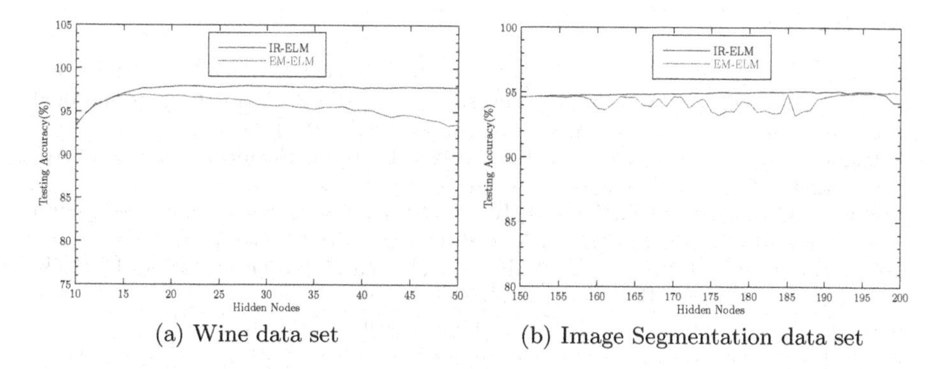

(a) Wine data set (b) Image Segmentation data set

Fig. 9. The average testing accuracy of IR-ELM and EM-ELM on Wine and Image Segmentation data sets

gets higher testing accuracy than EM-ELM for different hidden nodes in all cases.With the increasing of hidden nodes,testing accuracy of IR-ELM enhances or remains unchanged,but that of EM-ELM always decreases,this phenomenon is obvious for "Glass","Iris" and "Blood" data sets.What's more,the accuracy curves of IR-ELM keep smooth,however curves of EM-ELM have some fluctuation in "Iris","Blood" and "Image Segmentation" cases.

5 Conclusions

In this paper,we propose a novel algorithm for incremental learning called Incremental Regularized Extreme Learning Machine(IR-ELM).IR-ELM combines EM-ELM [10] with Regularized Extreme Learning Machine(R-ELM) [11],when new hidden node is added one by one,IR-ELM update the output weight recursively in a very fast way.It has overcome two main issues existing in EM-ELM:1.the regularization term makes the hidden layer output matrix nonsingular thus to obtain more accurate computation results;2.regularization method can control the norm of output weight and prevent the network from overfitting.

The performance of IR-ELM is evaluated for regression and classification problems on benchmark data sets obtained from UCI Machine Learning Repository [14].We have conducted the experiments from two aspects:1.for a given testing accuracy ϵ,to see how many hidden nodes are necessary to meet ϵ;2.when the number of hidden nodes increases,to see how the testing accuracy changes. Empirical studies have shown that IR-ELM performs better than EM-ELM for both questions with a similar training time as EM-ELM. What's more,when the number of hidden nodes changes, the testing accuracy curves of IR-ELM is more smooth than those of EM-ELM which indicates that IR-ELM gets more stable results than EM-ELM.

References

1. Huang, G.B., Zhu, Q.Y., Siew, C.K.: Extreme Learning Machine: a new Learning Scheme of Feedforward Neural Networks. In: Proceedings. 2004 IEEE International Joint Conference on Neural Networks, vol. 2. IEEE (2004)
2. Huang, G.B., Zhu, Q.Y., Siew, C.K.: Extreme Learning Machine: Theory and Applications. Neurocomputing 70, 489–501 (2006)
3. Huang, G.B., Zhu, Q.Y., Siew, C.K.: Real-time Learning Capability of Neural Networks. IEEE Transactions on Neural Networks 17(4), 863–878 (2006)
4. Rumelhart, D.E., Hinton, G.E., Williams, R.J.: Learning Representations by Backpropagating Errors. MIT Press, Cambridge (1988)
5. Huang, G.B., Zhou, H., Ding, X.: Extreme Learning Machine for Regression and Multiclass Classification. IEEE Transactions on Systems, Man, and Cybernetics, Part B: Cybernetics 42(2), 513–529 (2012)
6. Sun, Z.L., Choi, T.M., Au, K.F.: Sales Forecasting Using Extreme Learning Machine with Applications in Fashion Retailing. Decision Support Systems 46(1), 411–419 (2008)
7. Nizar, A.H., Dong, Z.Y., Wang, Y.: Power Utility Nontechnical Loss Analysis with Extreme Learning Machine Method. IEEE Transactions on Power Systems 23(3), 946–955 (2008)
8. Zhang, R., Huang, G.B., Sundararajan, N.: Multicategory Classification Using an Extreme Learning Machine for Microarray Gene Expression Cancer Diagnosis. IEEE/ACM Transactions on Computational Biology and Bioinformatics 4(3), 485–495 (2007)
9. Huang, G.B., Chen, L., Siew, C.K.: Universal Approximation Using Incremental Constructive Feedforward Networks with Random Hidden Nodes. IEEE Transactions on Neural Networks 17(4), 879–892 (2006)
10. Feng, G., Huang, G.B., Lin, Q.: Error Minimized Extreme Learning Machine with Growth of Hidden Nodes and Incremental Learning. IEEE Transactions on Neural Networks 20(8), 1352–1357 (2009)
11. Deng, W., Zheng, Q., Chen, L.: Regularized Extreme Learning Machine. Computational Intelligence and Data Mining. In: IEEE Symposium on. IEEE CIDM 2009, pp. 389–395 (2009)
12. Vapnik, V.: The Nature of Statistical Learning Theory. Springer, 2000
13. Vapnik, V.N.: An Overview of Statistical Learning Theory. IEEE Transactions on Neural Networks 10(5), 988–999 (1999)
14. Frank, A., Asuncion, A.: UCI Machine Learning Repository. Univ. California, Sch. Inform. Comput. Sci., Irvine, CA (2011), `http://archive.ics.uci.edu/ml`
15. Martnez-Martnez, J.M., Escandell-Montero, P., Soria-Olivas, E.: Regularized Extreme Learning Machine for Regression Problems. Neurocomputing 74(17), 3716–3721 (2011)
16. Haykin, S.S.: Neural Networks and Learning Machines. Pearson Education, Upper Saddle River (2009)
17. Bartlett, P.L.: The Sample Complexity of Pattern Classification with Neural Networks: the Size of the Weights is More Important than the Size of the Network. IEEE Transactions on Information Theory 44(2), 525–536 (1998)
18. Girosi, F., Jones, M., Poggio, T.: Regularization Theory and Neural Networks Architectures. Neural Computation 7(2), 219–269 (1995)
19. Jin, Y., Okabe, T., Sendhoff, B.: Neural Network Regularization and Ensembling Using Multi-objective Evolutionary Algorithms. In: Congress on Evolutionary Computation, CEC 2004, vol. 1, pp. 1–8. IEEE (2004)

Parallel Ensemble of Online Sequential Extreme Learning Machine Based on MapReduce

Shan Huang, Botao Wang, Junhao Qiu, Jitao Yao, Guoren Wang, and Ge Yu

College of Information Science and Engineering,
Northeastern University, Liaoning, Shenyang, China 110004
huangshan.neu@gmail.com,{wangbotao,wanggr,yuge}@ise.neu.edu.cn,
lqqiujunhao@163.com,tao00800@126.com,

Abstract. In this era of big data, analyzing large scale data efficiently and accurately has become a challenge problem. Online sequential extreme learning machine is one of ELM variants, which provides a method to analyze data. Ensemble method provides a way to learn data more accurately. MapReduce provides a simple, scalable and fault-tolerant framework, which can be utilized for large scale learning. In this paper, we propose an ensemble OS-ELM framework which supports ensemble methods including Bagging, subspace partitioning and cross validating. Further we design a parallel ensemble of online sequential extreme learning machine (PEOS-ELM) algorithm based on MapReduce for large scale learning. PEOS-ELM algorithm is evaluated with real and synthetic data with the maximum number of training data 5120K and the maximum number of attributes 512. The speedup of this algorithm can reach as high as 40 on a cluster with maximum 80 cores. The accuracy of PEOS-ELM algorithm is at the same level as that of ensemble OS-ELM running on a single machine, which is higher than that of the original OS-ELM.

Keywords: Parallel learning, Ensemble, Extreme Learning Machine, MapReduce, Sequential Learning.

1 Introduction

In this era of big data, analyzing large scale data efficiently and accurately has become a challenge problem. There are often hidden noises behind large scale data. Ensemble methods are proposed to eliminate the influence of the noises. Generally, ensemble methods can reach higher accuracy dealing with the same data set [9]. Ensemble methods usually train several ensemble members and combine the output of these ensemble members to generate the final result. However, this approach would lead in more calculations, and it is hard to analyze large scale data efficiently. Extreme learning machine (ELM) was proposed based on single-hidden layer feed-forward neural networks (SLFNs) [5], and has been verified to have high learning speed as well as high accuracy [3]. It has also been proved that ELM has have universal approximation capability and classification capability [4]. Online sequential extreme learning machine (OS-ELM) [7] is one

J. Cao et al. (eds.), *Proceedings of ELM-2014 Volume 1*,
Proceedings in Adaptation, Learning and Optimization 3, DOI: 10.1007/978-3-319-14063-6_3

of ELM variants that supports online sequential learning. OS-ELM can learn data chunk by chunk with fixed or varying sizes instead of batch learning. There are works researching on combining ensemble methods and ELM [6], [8], [12]. However, these algorithms mainly focus on the accuracy, but they are inefficient to learn large scale data.

MapReduce framework is a well-known framework for large scale data processing and analyzing on a large cluster of commodity machines. There are works research on parallelizing ELM and OS-ELM to improve learning speed [2], [11], [10]. However, ensemble methods are not taken into consideration in these works, so these works are not suitable for large scale data learning due to the accuracy limitation.

In this paper, we present a parallel ensemble of online sequential extreme learning machine (PEOS-ELM) algorithm based on MapReduce for large scale data processing and analyzing. This algorithm supports the most common ensemble methods such as Bagging, subspace partitioning and cross validating. This algorithm splits data according to user customization and calculates hidden layer output matrix of OS-ELM in Map phase. In Reduce phase, the ensemble members finish the remaining training work in parallel. We also test PEOS-ELM algorithm real and synthetic data, and the results show that PEOS-ELM has good scalability and the accuracy of this algorithm is at the same level with that of ensemble OS-ELM running on a single machine.

The remainder of this paper is organized as follows. Section 2 introduces an ensemble OS-ELM framework. Section 3 proposes parallel ensemble of online sequential learning machine algorithm. Section 4 evaluates the PEOS-ELM algorithm with real data and synthetic data, and section 5 concludes the paper.

2 Ensemble OS-ELM Framework

Figure 1 shows the ensemble OS-ELM framework. This framework considers ensemble methods (Bagging, subspace partitioning and cross validation) as well as training phase and testing phase of OS-ELM. $(X_{m,k}, T_{m,k})$ represents data chunks for ensemble member m, where $X_{m,k}$ represents the attributes set and $T_{m,k}$ represents the set of tags in which class the instances belong to corresponding to $X_{m,k}$. There are two ways of using the training data, one is used for training OS-ELM and the other is used for validating OS-ELM. The superscript of $(X_{m,k}, T_{m,k})$ in the figure, marks these two ways of use.

In the framework, data are processed in the following steps.

1. The $(X_{m,k}, T_{m,k})$ used for training and validating are generated by taking with replacement from training data. This procedure is needed by Bagging.
2. The subspace sets of $(X_{m,k}, T_{m,k})$ are generated. This procedure is needed by subspace partitioning.
3. All OS-ELMs are sequentially trained using the subspace sets. This phase follows the way of OS-ELM.
4. Data generated for validating are used to valid the trained OS-ELMs. This procedure is needed by cross validating.

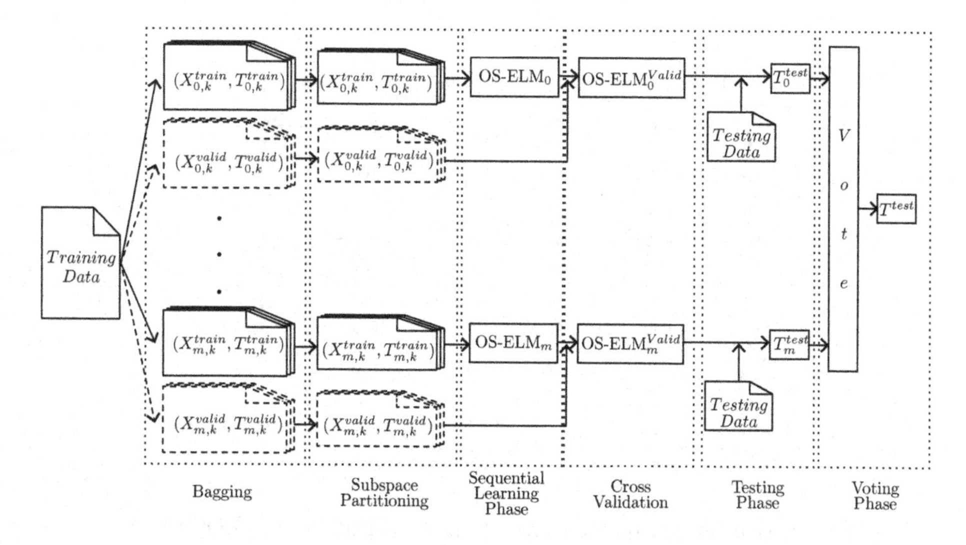

Fig. 1. Ensemble OS-ELM Framework

5. T_m^{test} are generated by OS-ELMs with testing data. This phase follows the way of testing phase of OS-ELM.
6. The T_m^{test} are processed by vote procedure to generated the final T^{test}. This procedure is needed by Bagging, subspace partitioning and cross validating.

The ensemble OS-ELM algorithm can be divided into three mainly phases, initialization phase, sequential learning phase and testing phase. In initialization phase, the initial parameters of OS-ELM and subspace for each ensemble member are generated. In sequential learning phase, several ensemble members are trained in the same way with OS-ELM [7]. In testing phase, the final result is created according to all the results from the ensemble members.

3 Parallel Ensemble of OS-ELM

3.1 Basic Idea

The goal of the parallel ensemble of online sequential extreme learning machine (PEOS-ELM) is to improve the performance of sequential learning phase in EOS-ELM using parallel techniques for large scale learning.

Figure 2 shows the matrix calculation dependency relationships among the matrices in sequential learning phase of EOS-ELM. One dependency is denoted as "$\rightarrow$", which means the matrix at arrow side depends on the matrix at the other side. That is to say, the calculation of matrix at the arrow side cannot start until the calculation of matrix at the other side finished. The matrices which do not depend on any other matrices can be calculated in parallel. Based on the above observations, our basic ideas can be summarized as follows:

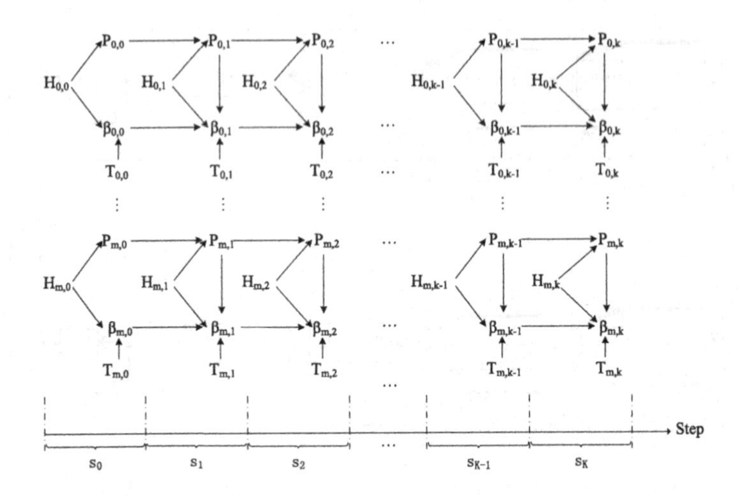

Fig. 2. Dependency relationships of matrix calculations in EOS-ELM

1. The calculation of $\mathbf{H}_{m,k}$ $((0 \leq m \leq M), (0 \leq k \leq K))$ depends on none of $\mathbf{T}_{m,k+1}$, $\mathbf{P}_{m,k+1}$ and $\beta_{m,k}$, besides this matrix can be calculated immediately when its related training data are available. This means that $\mathbf{H}_{m,k}$ can be calculated in parallel when K train data chunks are available for M ensemble members. This can be done in Map phase of MapReduce framework.
2. The calculation of $\beta_{m,k+1}$ is dependent on the calculation of $\mathbf{H}_{m,k+1}$, $\mathbf{T}_{m,k+1}$, $\mathbf{P}_{m,k+1}$ and $\beta_{m,k}$, but not dependent on matrix calculations for another ensemble member $j(0 \leq j \neq m \leq M)$. So the calculation of $\beta_{m,k+1}$ for different ensemble members can also be executed in parallel. This can be done by Reduce phase of MapReduce framework.

3.2 Parallel EOS-ELM

It is preferred to execute sequential learning phase of EOS-ELM in parallel on MapReduce framework as it is the most time consuming phase. The parallel ensemble of online sequential extreme learning machine algorithm uses one MapReduce job to train ensemble members.

Procedure 1 shows the *map*() procedure of PEOS-ELM. For each ensemble member, the input sample is possibly used for normal calculation (line 2-11) and validating (line 12-13).

When the sample is chosen for normal calculation, a buffer is used to store samples and a counter is used to count the number of samples in buffer (line 3-5). There are several data processing steps (line 6-11) when buffer is full to extract subspace, calculate $\mathbf{H}_{m,k}$, calculate $\mathbf{T}_{m,k}$, generate key-value pair, clear the counter and increase k. In the key-value pair, the *key* is composed with ensemble member ID m, blockID k and Tag while the value is made up of $\mathbf{H}_{m,k}$ and $\mathbf{T}_{m,k}$.

Procedure 1. PEOS-ELM map()

```
Input: (Key, Value): Key is the offset in bytes, Value is a sample pair
       (x_i, t_i) ∈ (X_k^train, T_k^train) where 0 ≤ i ≤| (X_k^train, T_k^train) |;
Result: m: Ensemble member ID;
        k: blockID;
        tag: marks whether output is used for normal calculation or validating;
        H_{m,k}: Output weight;
        T_{m,k}: Observation value vector;
1  for m=0 to M do
2      if chooseForThisMember() then
3          add to block_m;
4          count_m ++;
5          if count_m ≥ BLOCK then
6              block_m=GetSpace(block_m);
7              H_{m,k}=calcH(block_m);
8              T_{m,k}=calcT(block_m);
9              output((m,k_m,NormalTag), (H_{m,k}, T_{m,k}));
10             count_m = 0;
11             k_m++;
12     if chooseForValid() then
13         calucateForValid();
```

When the sample is chosen for normal calculation, similar operations are applied with those for normal calculation except that the output *key* is marked with *ValidTag* instead of *NormalTag* to facilitate distinguishing them later in Reduce phase. We briefly express it as *calucateForValid*().

Procedure 2 shows the *reduce*() procedure of PEOS-ELM. The output results of Map which belong to the same ensemble member are partitioned to the same Reducer and then sorted by *tag* and k. When the set of key-value pairs reaches to *reduce*() procedure, the parameters composed in *key* are firstly resolved (line 1-2) Then the key-value pair is processed differently according to the *tag*. If the *tag* is *NormalTag*, the parameters for ensemble member m are initialized if it has not been initialized (line 4-6) and then $\mathbf{H}_{m,k+1}$ and $\mathbf{T}_{m,k+1}$ composed in *value* are resolved (line 7-8). After that, the $\mathbf{P}_{m,k+1}$ and $\beta_{m,k+1}$ are updated according to the equations (line 9-10). If the *tag* is *ValidTag*, it means that this key-value pair is used for validating. The $\mathbf{H}^{valid}_{m,k}$ and $\mathbf{T}^{valid}_{m,k}$ are firstly resolved from *value* (line13-14) and then used for cross validating (line 15).

3.3 Cost Model

The cost of PEOS-ELM algorithm mainly has four parts, (1) cost of starting a MapReduce job, (2) cost of Map procedure, (3) cost of Reduce procedure and (4) cost of data transmitting between Map and Reduce.

As the number of cores in cluster increases while the other parameters keep the same, the cost of Map procedure and cost of Reduce procedure would be

Procedure 2. PEOS-ELM reduce()

Input:
Set of (*key*, *value*): *key* is a combination of m, k and *tag* . *value* is a vector pair (H_{kb}, T_{kb});
Result: β_m: output weight vector (corresponding to $\beta_{m,k}$).

```
1  m = getm(key);
2  tag = gettag(key);
3  if tag=NormalTag then
4  |   if firstRun=true then
5  |   |   initMember(m);
6  |   |   firstRun=false;
7  |   H_{m,k+1} = getH(value);
8  |   T_{m,k+1} = getT(value);
9  |   P_{m,k+1} = P_{m,k} - P_{m,k} H^T_{m,k+1} (I + H_{m,k+1} P_{m,k} H^T_{m,k+1})^{-1} H_{m,k+1} P_{m,k};
10 |   β_{m,k+1} = β_{m,k} + P_{m,k+1} H^T_{m,k+1} (T_{m,k+1} - H_{m,k+1} β_{m,k});
11 if tag=ValidTag then
12 |   for k = 0 to K do
13 |   |   H^{valid}_{m,k} = getH(value);
14 |   |   T^{valid}_{m,k} = getT(value);
15 |   |   CrossValid(H^{valid}_{m,k}, T^{valid}_{m,k});
```

significantly reduced, the cost of starting a MapReduce job and cost of data transmitting between Map and Reduce are all fixed for MapReduce applications. The reason is that all the calculations are equally distributed to the cores and be executed in parallel. So the PEOS-ELM has good scalability.

4 Experimental Evaluation

4.1 Experimental Setup

In this section POS-ELM indicates parallel online sequential learning machine algorithm in our previous work [10] that train each ensemble member one by one. PEOS-ELM-B, PEOS-ELM-S and PEOS-ELM-C represent PEOS-ELM algorithm for Bagging, subspace partitioning and cross validating, respectively.

PEOS-ELM algorithm is evaluated with real data and synthetic data. The real data sets (gisette[1], mnist[1]) are mainly used to test training accuracy and testing accuracy. The specification of real data is shown in Table 1.

The synthetic data are only used for scalability test, which are generated by extending based on Flower[2]. The volume and attributes of training data are extended by duplicating the original data in a round-robin way. The parameters used in scalability test are summarized in Table 2. In the experiments, all the parameters use default values unless otherwise specified.

[1] Downloaded from http://www.csie.ntu.edu.tw/~cjlin/libsvmtools/datasets/
[2] Downloaded from http://www.datatang.com/data/13152

Table 1. Specifications of real data

Data Set	#attributes	#class	#training data	#testing data	Size of test data (KB)
gisette	5000	2	6000	1000	128137.240
mnist	780	10	60000	10000	176001.138

Table 2. Specifications of synthetic data and running parameters for scalability test

Parameter	Value range	Default value
#training data	640k, 1280k, 2560k, 5120k	640k
#attributes	64, 128, 256, 512	64
#cores	10, 20, 40, 80	80
#ensemble member	10, 20, 40, 80	80

In scalability test, all of the data are pushed to each of the ensemble member. That is to say the largest data set in Tabel 2 is as large as 640 instances * (80*512) attributes or (80*5120)instances *64 attributes for normal use. For PEOS-ELM-C, we randomly choose 80% of data for training and another 20% are used for validating.

PEOS-ELM algorithm is implemented in Java 1.6. The universal java matrix package (UJMP) [1] versioned 0.2.5 is used for matrix storage and processing, and Hadoop versioned 0.20.2 is chosen as our MapReduce platform. A Hadoop cluster deployed on 5 servers is used in our experiment. Each server has two Xeon E5-2620 CPUs (6 cores *2 threads), 32G memory, 4*2T hard disk. The servers are connected with Gigabit network. The servers are all running Centos6.4 64 bits Linux operating system. The number of hidden layer node is 25 and the activation function is $g(x) = \frac{1}{1+e^{-x}}$.

4.2 Evaluation Results

Accuracy Test

Table 3 shows the results of accuracy and performance tests with real data. It can be found that the training accuracy and testing accuracy of EOS-ELM are higher than those of OS-ELM. This verifies that ensemble method is useful to increase the learning accuracy. Compared with EOS-ELM, training time of PEOS-ELM reduces while keeps the accuracy at the same level. This result demonstrates that PEOS-ELM can learn large scale data accurately and efficiently.

Scalability Test

Figure 3 shows the scalability (speedup) of PEOS-ELM and POS-ELM. The speedup of PEOS-ELM can reach to as high as 40 whereas the speedup of POS-ELM can only reach to 1.3. The reason for this is that the there are several reduce

Table 3. Evaluation results with real data

Data Set	Algorithm	Training time (s)	Training accuracy	Testing accuracy
gisette	OS-ELM	175.796	0.682	0.643
	EOS-ELM	311.6	0.87	0.869
	PEOS-ELM-B	66.745	**0.882**	0.87
	PEOS-ELM-S	**36.793**	0.876	**0.881**
	PEOS-ELM-C	64	0.767	0.773
mnist	OS-ELM	188.765	0.621	0.634
	EOS-ELM	178.3	0.764	0.78
	PEOS-ELM-B	56.988	**0.789**	**0.798**
	PEOS-ELM-S	**34.633**	0.761	0.778
	PEOS-ELM-C	70.753	0.784	0.796

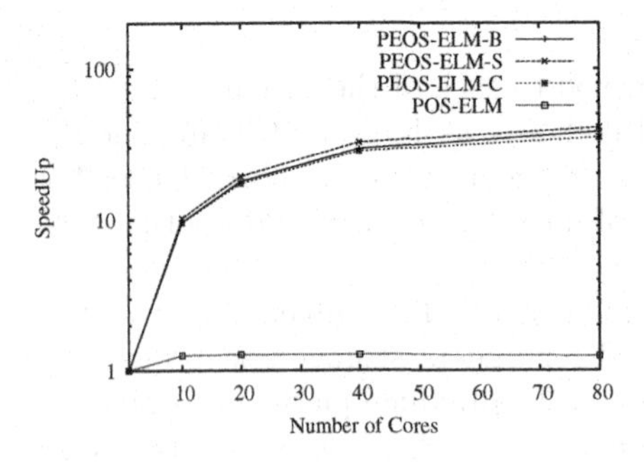

Fig. 3. Speedup of PEOS-ELM with regard to different number of cores

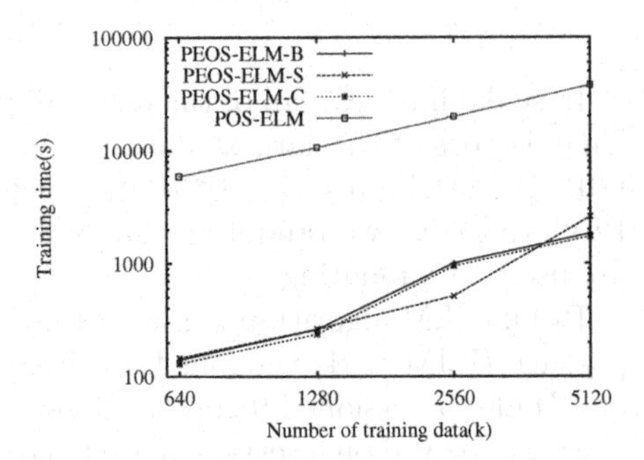

Fig. 4. Scalability of PEOS-ELM with regard to the number of training data

tasks running in parallel to sequentially calculate $\beta_{m,k}$ for different ensemble members whereas there is only one reduce task to calculate β in POS-ELM algorithm. The high speedup is consistent with the cost model.

The speedup decreases as the number of cores increases. This is due to the task scheduling cost and the bottleneck of memory and I/Os.

Figure 4 shows the scalability of PEOS-ELM and POS-ELM algorithm with regard to the number of training data. It can be found that PEOS-ELM has good scalability with regard to the number of training data. This is because the calculations are equally distributed to different map tasks and reduce tasks.

It also can be found from Figure 4 that the performance of PEOS-ELM for different ensemble methods outperforms that of POS-ELM. There are several reasons for this. First, for PEOS-ELM, the calculation of $\beta_{m,k}$ are running in parallel in Reduce phase, while in POS-ELM this calculation is running on one reduce task. Second, the sequential learning data sets are read once and push to each ensemble member in memory, while the POS-ELM read data many times. Third, as there is trade off to run a MapReduce job, the cost of running several MapReduce jobs for POS-ELM is higher than that of running one MapReduce job for PEOS-ELM.

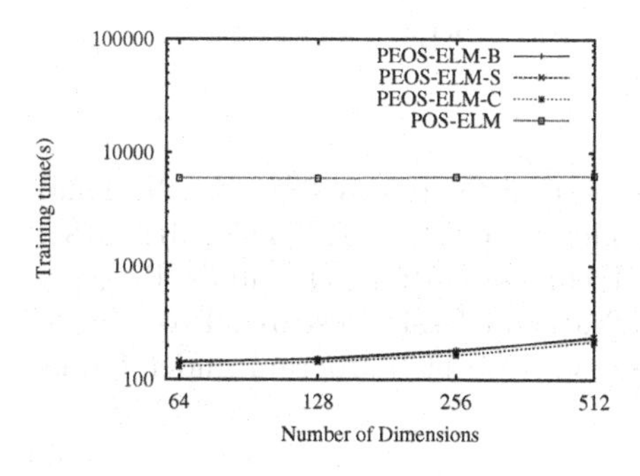

Fig. 5. Scalability of PEOS-ELM with regard to the number of attributes

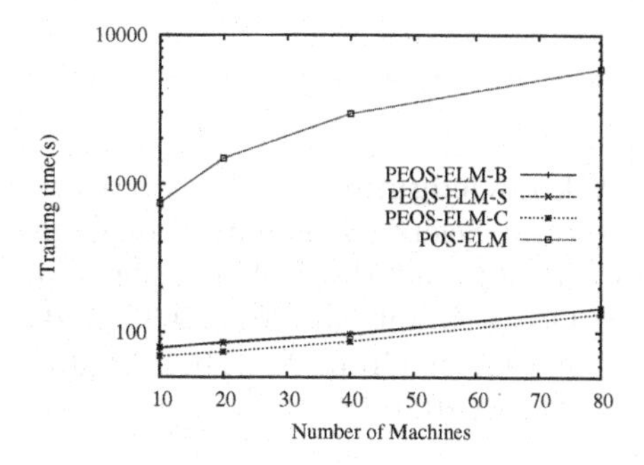

Fig. 6. Scalability of PEOS-ELM with regard to the number of ensemble members

Figure 5 shows the scalability of PEOS-ELM with regard to the number of attributes. It can be found that the training time of PEOS-ELM increased slowly as the number of attributes increases. On reason for this is the equally distributed calculations among map tasks and reduce tasks, while another reason is that the cost for transmitting data between Map phase and Reduce phased does not increases with the number of attributes.

It can also be found from Figure 5 that the training time of PEOS-ELM algorithms for different ensemble methods are nearly the same. This shows that the PEOS-ELM algorithm is suitable for different ensemble methods to analyze large scale data.

Figure 6 shows the scalability of PEOS-ELM with regard to the number of ensemble members. It can be found that the training time scales linearly as the number of ensemble members increases. The reason for this is that all of the matrix calculations are evenly distributed to tasks that running in parallel. This also shows that PEOS-ELM is efficient to train many ensemble members.

5 Conclusions

In this paper, a parallel ensemble of online sequential extreme learning machine (PEOS-ELM) algorithm has been proposed for large scale learning. The basic idea of this algorithm is to parallelize the calculation of $\mathbf{H}_{m,k}$ in Map phase and $\beta_{m,k}$ in reduce phase for different ensemble members.

The algorithm is implemented on MapReduce framework. PEOS-ELM algorithm for Bagging, subspace partitioning and cross validating are evaluated with real and synthetic data with the maximum number of training data 5120k and the maximum number of attributes 512 for each ensemble member. The experimental results show that the accuracy of PEOS-ELM for different ensemble methods are at the same level as that of EOS-ELM, and it has a good scalability with the number of training data and number of attributes. The speedup of PEOS-ELM can reaches as high as 40 on a cluster with maximum 80 cores.

Compared with EOS-ELM and POS-ELM, PEOS-ELM can be used to learn large scale data efficiently and accurately.

Acknowledgments. This research was partially supported by the National Natural Science Foundation of China under Grant No. 61173030, 61272181, 61272182; and the Public Science and Technology Research Funds Projects of Ocean Grant No. 201105033; and the National Basic Research Program of China under Grant No. 2011CB302200-G; and the 863 Program under Grant No.2012AA011004.

References

1. Arndt, H., Bundschus, M., Naegele, A.: Towards a next-generation matrix library for java. In: 33rd Annual IEEE International Computer Software and Applications Conference, COMPSAC 2009, vol. 1, pp. 460–467. IEEE (2009)
2. He, Q., Shang, T., Zhuang, F., Shi, Z.: Parallel extreme learning machine for regression based on mapreduce. Neurocomput. 102, 52–58 (2013)
3. Huang, G.-B., Chen, L.: Convex incremental extreme learning machine. Neurocomputing 70, 3056–3062 (2007)
4. Huang, G.-B., Zhou, H., Ding, X., Zhang, R.: Extreme learning machine for regression and multiclass classification. IEEE Transactions on Systems, Man, and Cybernetics, Part B: Cybernetics 42(2), 513–529 (2012)
5. Huang, G.B., Zhu, Q.Y., Siew, C.K.: Extreme learning machine. In: Technical Report ICIS/03/2004. School of Electrical and Electronic Engineering, Nanyang Technological University, Singapore (January 2004)
6. Lan, Y., Soh, Y.C., Huang, G.-B.: Ensemble of online sequential extreme learning machine. Neurocomputing 72(13), 3391–3395 (2009)
7. Liang, P.N.-Y., Huang, G.-B., Saratchandran, Sundararajan, N.: A fast and accurate online sequential learning algorithm for feedforward networks. IEEE Transactions on Neural Networks 17(6), 1411–1423 (2006)
8. Liu, N., Wang, H.: Ensemble based extreme learning machine. IEEE Signal Processing Letters 17(8), 754–757 (2010)
9. Rokach, L.: Ensemble-based classifiers. Artificial Intelligence Review 33(1-2), 1–39 (2010)
10. Wang, B., Huang, S., Qiu, J., Liu, Y., Wang, G.: Parallel online sequential extreme learning machine based on mapreduce. In: The International Conference on Extreme Learning Machines (ELM 2013), Beijing, China, October 15-17 (2013)
11. Xin, J., Wang, Z., Chen, C., Ding, L., Wang, G., Zhao, Y.: Elm*: distributed extreme learning machine with mapreduce. In: World Wide Web, pp. 1–16 (2013)
12. Zhai, J.-h., Xu, H.-y., Wang, X.-z.: Dynamic ensemble extreme learning machine based on sample entropy. Soft Comput. 16(9), 1493–1502 (2012)

Explicit Computation of Input Weights in Extreme Learning Machines

Jonathan Tapson, Philip de Chazal, and André van Schaik

The MARCS Institute, University of Western Sydney, Penrith NSW, Australia 2751
{j.tapson,p.dechazal,a.vanschaik}@uws.edu.au

Abstract. We present a closed form expression for initializing the input weights in a multilayer perceptron, which can be used as the first step in synthesis of an Extreme Learning Machine. The expression is based on the standard function for a separating hyperplane as computed in multilayer perceptrons and linear Support Vector Machines; that is, as a linear combination of input data samples. In the absence of supervised training for the input weights, random linear combinations of training data samples are used to project the input data to a higher dimensional hidden layer. The hidden layer weights are solved in the standard ELM fashion by computing the pseudoinverse of the hidden layer outputs and multiplying by the desired output values. All weights for this method can be computed in a single pass, and the resulting networks are more accurate and more consistent on some standard problems than regular ELM networks of the same size.

Keywords: Extreme learning machine, machine learning, computed input weights.

1 Introduction

The Extreme Learning Machine has proven its usefulness as a fast and accurate method for building classification and function approximation networks [1]. Its usefulness stems in large part from the fact that it requires no incremental or iterative learning, and has no free parameters which need to be tuned to get the optimal results. Naïve application of a standard ELM to most benchmark problems produces results which are significantly better than most results from highly-tuned iterative methods. In all but a few applications, the ELM can only be outperformed by time-consuming and expertly-applied techniques such as multiple layer deep-learning networks, and recent extension of the ELM to multiple layers suggests that they have the potential to outperform deep learning networks too [2].

One area which has been identified as offering some potential for improvement in ELM is the specification of the input weights, which connect the input neurons to the hidden layer neurons. In a standard ELM these are generally initialized to random values from a uniform distribution on some appropriate range, and thereafter they remain fixed throughout the use of the ELM.

The random input layer projection implemented in ELM is a contrast to almost all other machine learning techniques, which use supervised learning to arrive at explicit

J. Cao et al. (eds.), *Proceedings of ELM-2014 Volume 1*,
Proceedings in Adaptation, Learning and Optimization 3, DOI: 10.1007/978-3-319-14063-6_4

values for the projections at all layers (or in the case of support vector machines (SVM), to choose the support vectors [3]). In this report we proceed from the knowledge that for both multi-layer perceptrons (MLP) trained by standard backpropagation techniques (backprop), and for SVM, the weights produce a projection which is a linear combination of training data samples [3, 4]. We suggest a technique in which randomized linear combinations of input data samples can be systematically produced to provide the input layer weights for ELM, and demonstrate that in some traditionally difficult classification problems, this method results in a superior performance to standard ELM.

2 Weights and Training Samples

The standard ELM is often described as an MLP with an input layer, a hidden layer of neurons with nonlinear activation functions, and an output layer with linear activation functions. The outputs can be described by:

$$y_{n,t} = \sum_{j=1}^{d} w_{nj}{}^{(2)} g\left(\sum_{i=1}^{k} w_{ji}{}^{(1)} x_{i,t}\right) \tag{1}$$

where y_t is the output layer vector corresponding to input data x_t, and t is the sample index (t would be the time step index for time series data). The output y_t is a linear sum of hidden layer outputs weighted by w_{nj}. There are k input neurons and d hidden layer neurons; i is the input layer index, j the hidden layer index and n the output layer index. The hidden layer neurons have a nonlinear activation function $g(\cdot)$ which acts on the sums of linearly weighted inputs $w_{ji}x_i$. The superscripts indicate the layer. If the outputs $y_{n,t}$ were also to be acted on by a nonlinear activation function like $g(\cdot)$, this would be a conventional three-layer MLP. Note that in the development here we are assuming that any bias term would be present in the training data as an input data element.

In a standard ELM the input layer weights are initialized to random values, uniformly distributed in some sensible range – say (*-1, 1*). In this report we describe a closed-form solution for determining these weights.

Consider the conventional backpropagation of error algorithm in MLPs [4]. The training algorithm contains two phases. In the forward phase, the error for each training sample is calculated. In the backward phase, the weights for each layer are updated by multiplying the error for that layer by the activations of that layer (which produces an error gradient) and subtracting a fraction of the gradient from the weights. So, in the network given above, the update for the input layer weights would be of the form

$$\Delta w_{j,t}{}^{(1)} = \alpha \cdot E_t{}^{(1)} \cdot x_t \tag{2}$$

where α is the learning rate (which defines the fraction of the gradient used in updates), and E is the error gradient. It can be seen that each update consists of adding or subtracting some fraction of an input training sample to the weights. If the weights are initialized to zero (which is not unreasonable) then the final value of the weights will be some linear combination of the input samples; for h training samples, after one pass through all training data,

$$w_j{}^{(1)} = \alpha \sum_{t=1}^{h} E_t{}^{(1)} \cdot x_t \tag{3}$$

We find a similar result for the support vector machine [3]. In the linear case, an SVM classifier is designed to find the maximum-margin separating hyperplane between two classes of training data. This optimization problem is generally expressed as follows: given that the separating hyperplane can be expressed as an inner product in the input space such that

$$w \cdot x - b = 0 \tag{4}$$

where b is a bias term, so that the hyperplane does not have to pass through the origin (note that here w is a generic weight vector, not related to w in (1) - (3); we persist with the notation in order to emphasize the commonality in these expressions). The classification margins to this separating hyperplane can be expressed as two parallel hyperplanes at

$$w \cdot x - b = \pm 1 \tag{5}$$

We need to maximize the distance $2/\|w\|$ between these two planes, subject to the constraint that

$$y_t(w \cdot x_t - b) \geq 1 \tag{6}$$

for all t. This constraint requires that all samples remain outside the margins, as is implicit in their meaning. According to the Karush-Kuhn-Tucker condition [5], the solutions can be expressed as a linear combination of input samples

$$w = \sum_{t=1}^{h} \alpha_t \cdot y_t \cdot x_t \tag{7}$$

Note that α here is not the same as the learning rate in the MLP expression above, but is a multiplication factor (a Lagrange multiplier in the optimization process). We have used the same symbol α as this symbol is conventional in both usages in MLP and SVM and serves to emphasize the point of this development, which is that in both MLP and linear SVM cases the weight vectors are linear weighted sums of the input training samples (note the similarities between (3) and (7) above). This should come as little surprise given the acknowledged equivalences between perceptrons and linear SVMs, and MLPs and SVMs (see for example Collobert and Bengio [6]). Similarly, it is consistent with the intuition that the dot product between vectors (such as a training sample and a testing sample) is a measure of their similarity.

Given that the optimal input layer weights in MLPs and the support vectors in SVMs are linear functions of the training samples, it might be the case that an ELM in which the input layer weights were biased towards linear combinations of the input training samples would perform better than one in which the input layer weights were uniformly distributed random weights. There are a number of questions to be addressed here:

- Does an ELM in which the input layer weights are biased towards the input training samples perform better than one with uniformly distributed random values?
- How would we achieve the biasing of the input layer without any of the tedious incremental or stochastic learning of weights which ELM so successfully avoids?
- Is there a rigorous theoretical basis for the belief that an ELM would perform better with appropriately biased weights?

In the work reported here we propose some answers to the first two points, by suggesting a fast closed-form expression for generation of weights biased towards the input training samples, and showing that in some important cases it consistently outperforms a conventional ELM.

3 Methodology

In the method described here, which we refer to as Computed Input Weights ELM (CIW-ELM), we want to generate input weights which have the form of (3) or (7) above; that is to say, they must be (random) weighted sums of the training data samples. The input data are generally normalized to have zero mean and unity standard deviation, and weights are normalized to unity magnitude. ELM hidden layers are often specified in terms of number of neurons, or as a multiplier of the size of the input layer. We divide the number of hidden layer neurons d by the number of classes C, and generate weights for each group of $p = d/C$ hidden layer neurons from input vectors of a single class (we can reasonably define d so that d/C gives an integer number of neurons). We then produce random weights biased to the training samples for the class, by summing the inner products of random binary-valued vectors with the training samples, according to the formula

$$w^{(1)} = R \cdot x \tag{8}$$

where $x_{h \times k}$ is the full set of training data for the class, and $R_{p \times h}$ is a matrix of random binary values (random sign values), i.e. with elements having values in {-1, 1}. Should repeatable weight generation be required, the elements of R could be generated from well-known pseudorandom sequences such as Gold codes, thereby enabling a deterministic weight generation schema. Use of orthogonal pseudorandom codes would also guarantee orthogonality of the input weights, which has been identified to improve the generalization of ELM networks [2, 7].

The outcome of this method is that the weight $w_{ji}^{(1)}$ connecting input i with hidden layer neuron j is the sum of the input sample elements x_i (where i is the vector index, i.e. x_i is the ith element of any training sample x), for all the training samples of one class, where each individual element has had its sign flipped or preserved with random probability.

The columns of weight matrix $w^{(1)}$ are then normalized to unit vectors:

$$w_a^{(1)} = \frac{R_a}{|R_a|} \tag{9}$$

where a is the column index.

In practical terms, the weights are computed according to the following algorithm:

1. Normalize all training data, so as to avoid use of scaling factors.
2. Divide the d hidden layer neurons into C blocks, one for each of C output classes; for data sets where the number of training data samples for each class are equal, the block size is $B = d/C$. We denote the number of training samples per class as K. Where the training data sets for each class are not of equal size, the block size can be adjusted to be proportional to data set size.

3. For each block, generate a random matrix $R_{B\times K}$ of signs.
4. Multiply the input training data set for that class, $x_{K\times k}$, by R to produce $B\times k$ summed inner products, which are the weights for that block of neurons.
5. Normalize the weights to unity magnitude as shown in (9).
6. Concatenate these C blocks of weights for each class, each $B\times k$ in size, into one weight matrix $w_{d\times k}$.
7. Solve for the hidden layer weights of the ELM using standard ELM methods such as singular value decomposition on the computed hidden layer outputs.

4 Results

We have used the CIW-ELM method on several benchmark problems in machine learning. Its performance is perhaps best illustrated in terms of the well-known MNIST handwritten digit recognition (OCR) problem [8]. MNIST is a particularly challenging data set for classification as the dimensionality of the data is high (784 pixels for input, 10 classes for output, and 60000 training samples). We have previously reported good results with conventional ELM on MNIST [9]. Here we show that the use of CIW-ELM gives significantly greater accuracy for similar-sized networks than conventional ELM (see Fig. 1).

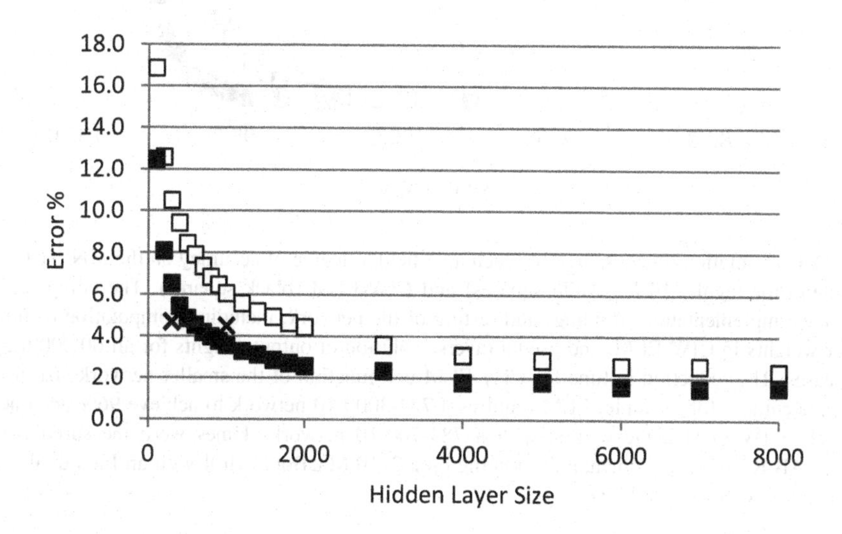

Fig. 1. Accuracy of standard ELM (white squares) and CIW-ELM (black squares) on the MNIST data set, for various hidden layer sizes. It can be seen that CIW-ELM consistently produces more accurate results for a given hidden layer size. Input data were not deskewed or centered, or in any way preprocessed except for normalization. The pseudoinverse solution was not regularized. The points indicated by the two crosses are MLP results from LeCun et al. [8] for 784-300-10 and 784-1000-10 networks, which represent the best results likely for backpropagation on equivalent networks (backpropagation is of course much more time-consuming than either ELM technique here).

One of the most significant advantages of the ELM method is in speed of implementation, because there is no incremental learning or optimization required. The most time-consuming computation in ELM is the calculation of the Moore-Penrose pseudoinverse solution to output weights, and this computation generally scales as d^2 where d is the hidden layer size. For very large datasets, such as MNIST and in modern "big data" problems, the size of the pseudoinverse computation can become problematic. As such, the potential to use CIW-ELM to obtain high accuracy with smaller hidden layers has real benefits. This is illustrated in Figure 2 which compares execution times for the ELM and CIW-ELM; note that these were carried out on a very modest computational platform, illustrating the high speed of ELM implementation.

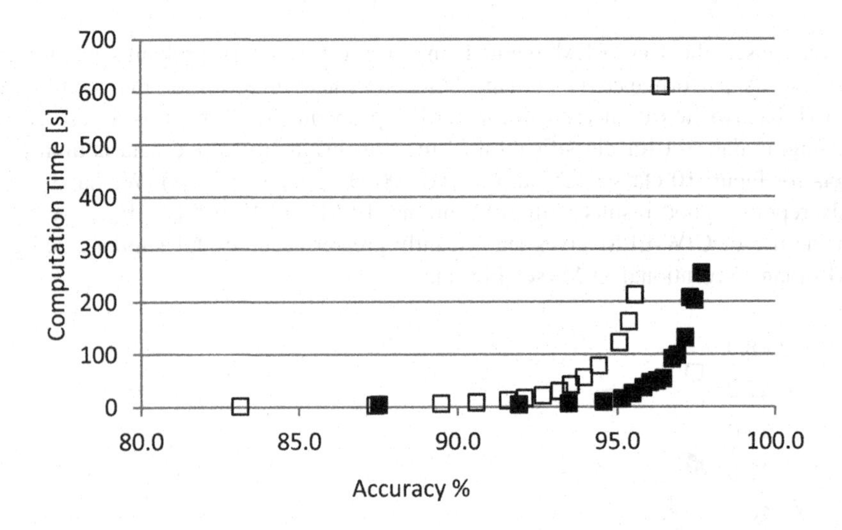

Fig. 2. Computation time required to reach a particular degree of accuracy on the MNIST OCR problem, for regular ELM (white squares) and CIW-ELM (black squares). The computation involves implementation (training) and testing of the network, including computation of input layer weights in CIW-ELM, and pseudoinverse solution of output weights for all 60 000 training cases. The reduction in time for CIW-ELM is a function of the smaller network size for a given accuracy; for example, ELM requires a 784-3000-10 network to achieve 96% accuracy, whereas CIW-ELM achieves this with a 784-700-10 network. Times were measured for a MATLAB R2012B implementation running on a 2010 MacBook Air 4 with an Intel dual core i5 processor clocked at 1.6GHz.

We have also used CIW-ELM on other standard classification problems, such as the well-known *abalone*, *iris*, and *wine* data sets [10]. The performance of CIW-ELM was superior to ELM for small network sizes in *wine* and *iris*, with convergence between the two methods at larger network sizes. There was no clear difference between the two methods for the *abalone* classification problem, but both methods produced results similar to the state of the art. Results for these data sets are illustrated in Figures 3-5.

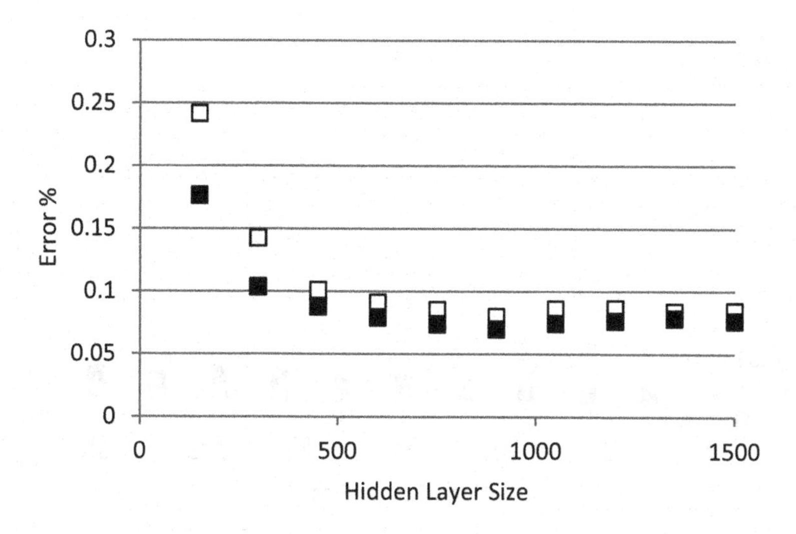

Fig. 3. Comparison of ELM (white squares) and CIW-ELM (black squares) on the *iris* classification problem. It can be seen that CIW-ELM shows more accuracy at all network sizes. There is some suggestion of overfitting for network sizes above 5-900-3, which is not surprising given the low dimensionality of the data (150 × 5 training data elements).

5 Discussion

The results illustrated in Figs. 1-5 suggest that for certain problems, the method may offer considerable improvement in performance over standard ELM. In cases where the data dimensionality is small, the performance of standard ELM converges rapidly onto the CIW-ELM result (however, it is for large data sets, where computation time becomes an issue, that the method shows the most promise). It is not yet clear whether the CIW-ELM method will improve results in function approximation tasks, as the algorithm as currently specified is intrinsically based on class information; we could in principle treat all input data as a single case for the purpose of a regression implementation. It is significant that the only problem so far on which CIW-ELM has not convincingly outperformed standard ELM is the *abalone* problem, which is really a regression problem (estimate the age of the abalone) recast as a classification problem (bin the abalone into one of three age bins). We note that Huang has demonstrated that SVM intrinsically requires the bias term b in (4) – (6) and that this forces optimization within a more constrained space than standard ELM, which was designed from the beginning for regression or function optimization [11]. Our use of expressions (3) and (7) as models for the input weights will similarly have constrained the solution space, and this may also contribute to a lack of improvement in regression problems, although this may be difficult to evaluate.

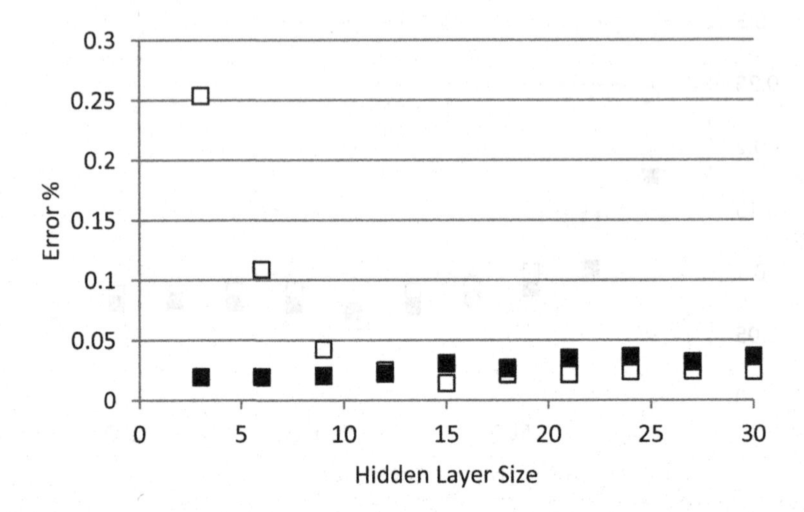

Fig. 4. Comparison of ELM (white squares) and CIW-ELM (black squares) performance on the *wine* classification benchmark. It can be seen that CIW-ELM performs extremely well for small networks, with both methods showing overfitting for larger network sizes.

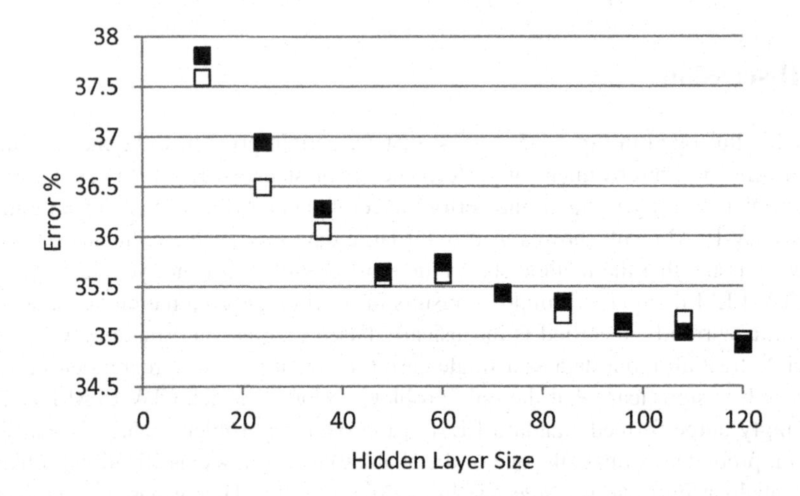

Fig. 5. Comparison of ELM (white squares) and CIW-ELM (black squares) performance on the *abalone* classification benchmark. It can be seen that there is little difference in performance (note the fineness of the vertical scale). This is well–known to be an underdetermined problem with the best results for all standard classification methods being no lower than 34%.

6 Conclusions

We have presented a schema for generation of input layer weights in ELM which offers reduced computation time and increased accuracy on some standard classification problems. The method is quick to implement and can be used in a deterministic and repeatable fashion. It still remains to evaluate this method on a wide range of benchmark problems and establish the circumstances under which it is a good choice for machine learning.

References

1. Huang, G.-B., Zhu, Q.-Y., Siew, C.-K.: Extreme Learning Machine: Theory and Applications. Neurocomputing 70, 489–501 (2006)
2. Kasun, L.L.C., Zhou, H., Huang, G.-B., Vong, C.M.: Representational Learning with Extreme Learning Machine for Big Data. IEEE Intelligent Systems 28(6), 31–34 (2013)
3. Cortes, C., Vapnik, V.: Support-Vector Networks. Machine Learning 20, 273–297 (1995)
4. Rumelhart, D.E., Hinton, G.E., Williams, R.J.: Learning representations by back-propagating errors. Nature 323, 533–536 (1986)
5. Kuhn, H.W., Tucker, A.W.: Nonlinear programming. In: Proceedings of 2nd Berkeley Symposium, pp. 481–492. University of California Press, Berkeley (1951)
6. Collobert, R., Bengio, S.: Links between perceptrons, MLPs and SVMs. In: Proc. ICML 2004: Proceedings of 21st International Conference on Machine Learning, p. 23 (2004)
7. Widrow, B., Greenblatt, A., Kim, Y., Park, D.: The No-Prop algorithm: a new learning algorithm for multilayer neural networks. Neural Networks 37, 182–188 (2013)
8. LeCun, Y., et al.: Gradient-Based Learning Applied to Document Recognition. Proc. IEEE 86(11), 2278–2324 (1998)
9. Tapson, J., van Schaik, A.: Learning the pseudoinverse solution to network weights. Neural Networks 45, 94–100 (2013)
10. Data sets from the UCI Machine Learning Repository, `http://archive.ics.uci.edu/ml/`
11. Huang, G.-B.: An Insight into Extreme Learning Machines: Random Neurons, Random Features and Kernels. Cognitive Computation (in press 2014), doi:10.1007/s12559-014-9255-2

Subspace Detection on Concept Drifting Data Stream

Lin Feng[1,2,*], Shenglan Liu[1,2], Yao Xiao[1], and Jing Wang[1]

[1] School of Computer Science and Technology, Faculty of Electronic Information and Electrical Engineering, Dalian University of Technology, Dalian, China, 116024
[2] School of Innovation Experiment, Dalian University of Technology, Dalian, China, 116024
fenglin@dlut.edu.cn

Abstract. In recent years, data stream mining has become a hot spot in machine learning. Network data and sensor data are both data stream. However, there is concept drift problem in data stream so that traditional machine learning methods no longer work. Meanwhile, real-time learning is required in data stream and most of concept detection methods can't support real-time demand. For solving this problem, this paper proposes a data stream learning framework which improves the classical Linear Discriminant Analysis (LDA) method based on a robust subspace learning method. It can not only detect concept drift in data stream quickly, but also classify data stream in real-time. The experimental results of sensor data and UCI repository validate the effectiveness of our method.

Keywords: Concept Driftingsubspace learning, data stream.

1 Introduction

In recent years, with the rapid development of internet and multimedia technologies, data stream mining has becoming a hot topic in the domain of data mining. The domains of the applications of data stream mining include sensor networks, social networks, Web logs and Web pages click streams, network monitoring and traffic engineering, telecommunication calling records, etc. Comparing with traditional static data sets, data stream has a few new characteristics. For example, the data in data stream is dynamic in real-time and in overwhelming volume; the labels and distribution of data may change in trends over time known as concept drift. Researchers have proposed many algorithms to construct different kinds of learning models to detect concept drift and classify data. These methods can be roughly divided into two categories: (1) Adjust the learning strategies to adapt the new data at the position which is detected when concept drift happens. (2) Retrain the learning machine in real-time without considering the concept drift to deal with the new data. Referece[1-11] proposed some efficient methods for learning concept drifting data streams. With the motivation

* Corresponding author.

J. Cao et al. (eds.), *Proceedings of ELM-2014 Volume 1*,
Proceedings in Adaptation, Learning and Optimization 3, DOI: 10.1007/978-3-319-14063-6_5

of learning from concept drifting data streams with unlabeled data, Reference [12] proposes a semi-supervised method based on decision tree algorithm—SUN. For detecting a concept drift, SUN produces concept clusters at leaves based on a concept set. It searches the decision tree from bottom to up to evaluate the deviation and distance between history concept and current concept by setting a variable. Reference [13] analyzes the statistical meaning of Bayes formula in data stream applications from statistical aspect. Through analyzing the relationship between the input distribution $P(X)$ and the posterior probabilities of the classes $P(\omega_i|X)$, they find the concept drift can be detected by detecting the change of $P(X)$, thus we can detect the concept drift in the case of unlabeled data by calculating the posterior probability and the probability density functions.

When processing the problem of concept drift, using ensemble learning methods is usually a reasonable option. Thence, many researchers design data stream learning methods based on learning machine ensembles. Reference [14] discusses the advantages of building classifier ensembles for non-stationary environments where the classification task changes during the operation of the ensembles. The methods with multi classifier ensembles exceed that with one classifier. Reference [15] and [16] deal with concept drift by adjusting the weight of sub-classifier and process new concepts by adding and deleting sub-classifiers dynamically. But classifier ensembles are based on the prior knowledge in most cases, while not based on the basic characteristics of data. So it is not robust for new classes detecting. Moreover, traditional incremental and online learning methods learn with a single model. To solve this problem, reference [17] proposes improved learning machine ensembles. They integrate the clusters and the classifiers. And then they utilize the cluster methods to reduce the amount of calculation in processing labels. Through giving a weight to classifiers, they can update the learning machine ensembles dynamically and process the concept drift.

Detecting concept drifts in high dimensional data streams is also an important subject in data stream learning; the curse of dimensionality brought by high dimensional data makes the traditional machine learning method to be not appropriate. One classical detecting concept drifts method is proposed by Dasu et al. in [18]. They take an information-theoretic approach to the high dimensional data concept changes detection problem. The method measures the differences between two given distributions with Kullback-Leibler (KL) distance. However, this method needs a procedure of discretization to get the probability density. Moreover, it can only detect two concepts and after a discretization the time complex is relative high in the bootstrap step.

All methods mentioned above are time consuming as detecting concept drift. Concept drift detection is often an independent step without considering the effects on classification of data stream. Subspace learning is a hot topic in machine learning, such as classical manifold learning methods [19-23]. For handling the problem of classification some researchers propose supervised dimension reduction methods. For example, LDA[24] considers minimizing within-class scatter and maximizing between-class scattersimultaneously.However, small size sample

(SSS) problem is hard to solve in LDA. For solving SSS problem, Maximum Margin Criterion (MMC)[25] constructs a new optimization function to avoid SSS problem based on LDA. This paper designs an Angle Robust LDA(ARLDA) learning method which is motivated by classical to detect the concept drift. ARLDA is a geometry detection method; it detects the concept drift based on the changes of projection variances and projection angles in current window which is needless knowing the distribution of current data and has a better time complex. Moreover, when the concept drift is detected, ARLDA will recalculate the projection subspace. Subspace data can be more effective learning (classification). Meanwhile, this paper proposes a robust supervised dimension reduction method which can calculate the subspace in ARLDA to reduce the dimension of current window data for adapting the learning in current window and later windows.

2 A Brief of LDA and MMC

The dataset is denoted as $X = [x_1, \cdots, x_N] \in R^{D\times N}$ which satisfies $N = \sum_{i=1}^{c} N_i$, where N_i is the sample size of each class, c is the number of class. Linear projection is denoted as $W = [w_1, w_2, \cdots, w_d] \in \mathrm{R}^{D\times d}$, where d is the dimension of target subspace. The optimal process can be realized with the optimization function below:

$$W_{opt}^{LDA} = \arg\max_{W} \frac{tr(WS_bW^T)}{tr(WS_wW^T)} \tag{1}$$

Where $S_b = X_bX_b^T$ is the within-class scatter matrix, $X_b = \frac{1}{\sqrt{N}}[\sqrt{N_1}\hat{\mu}_1, \cdot\cdot\cdot, \sqrt{N_c}\hat{\mu}_c]\hat{\mu}_i = \mu_i - \mu$, μ is the global mean vector; $S_w = \frac{1}{N}\sum_{i=1}^{c}\sum_{j=1}^{N_i}(x_{ij} - \mu_i)(x_{ij} - \mu_i)^T = X_wX_w^T$,$X_i = [x_{i1}, \cdots, x_{iN_i}]$ is the sample matrix of the i-th class,$X_w = \frac{1}{\sqrt{N}}[X_1 - \mu_1e_1, \cdots, X_c - \mu_ce_c]e_i = [1, ..., 1] \in \mathrm{R}^{1\times N_i}W_{opt}^{LDA}$ can be got from eigenvectors corresponding to the d largest eigenvalues through the eigendecomposition of $S_w^{-1}S_b$.

In MMC, define the distance of two different classes $d(c_i, c_j)$ as:

$$d(c_i, c_j) = d(\mu_i, \mu_j) - (S(c_i) + S(c_j)) \tag{2}$$

Where $d(\mu_i, \mu_j)$ is the distance between μ_i and μ_j, $S(c_i)$ and $S(c_j)$ are scatter matrix trace of each class. According to formula (2), the low-dimension subspace W_{opt}^{MMC} of MMC can be get through the optimization function as follows:

$$W_{opt}^{MMC} = \arg\max_{W} tr\left(W\left(S_b - S_w\right)W^T\right) \tag{3}$$

W_{opt}^{MMC} can be got from eigenvectors corresponding to the d largest eigenvalues through the eigendecomposition of $S_b - S_w$. Details can be seen in reference [25].

3 A New Dimensionality Reduction Method-ARLDA

In LDA, for any point x, the minimum scatter is equal to $W_{opt} = \arg\min_{W} \left\|WW^T(x-\mu)\right\|_2^2$, where $WW^T(\cdot)$ is orthogonal projection of $(\cdot)$, $(\cdot)_e$ is the unitization form of vector $(\cdot)$, we have:

$$\begin{aligned}\cos^2\alpha &= \frac{\left[(x-\mu)^T \cdot WW^T(x-\mu)\right]^2}{\|x-\mu\|_2^2 \left\|WW^T(x-\mu)\right\|_2^2} \\ &= \frac{(x-\mu)^T}{\|x-\mu\|_2} WW^T \frac{(x-\mu)}{\|x-\mu\|_2} \\ &= (x-\mu)_e^T \cdot WW^T(x-\mu)_e \end{aligned} \tag{4}$$

From formula (4) we can know that the minimum within-class scatter is equal to $\min_{W}\cos^2\alpha$ in geometry. For the same reason maximum between-class scatter $\max_{W} d_b$ (fig. 5(c)) is equal to $\max_{W}\cos^2\beta$ in geometry.

If utilizing the thought of LDA, we can get W_{opt} from optimization function $\max_{W}\frac{\cos^2\beta}{\cos^2\alpha}$ with constraint condition $W^TW = I$ to construct the optimal function of ARLDA. Make Co-Angle $\alpha_{ij} = \alpha_{ij}(W) = \left\langle x_{ij}-\mu_i, WW^T(x_{ij}-\mu_i)\right\rangle \beta_i = \beta_i(W) = \left\langle \mu_i-\mu, WW^T(\mu_i-\mu)\right\rangle$where $i = 1,2,\cdots,cj = 1,2,\cdots,N_i$. The within-class scatter can be written as follows:

$$\begin{aligned}\sum_{i=1}^{c}\left[\frac{N_i}{N}\frac{1}{N_i}\sum_{j=1}^{N_i}\cos^2\alpha_{ij}\right] &= \frac{1}{N}\left[\sum_{i=1}^{c}\sum_{j=1}^{N_i}\frac{(x_{ij}-\mu_i)^T WW^T(x_{ij}-\mu_i)}{\|x_{ij}-\mu_i\|_2^2}\right] \\ &= \frac{1}{N}\left[\sum_{i=1}^{c}\sum_{j=1}^{N_i}(x_{ij}-\mu_i)_e^T WW^T(x_{ij}-\mu_i)\right] \\ &= tr\left[\frac{1}{N}W^T\left[\sum_{i=1}^{c}\sum_{j=1}^{N_i}(x_{ij}-\mu_i)_e(x_{ij}-\mu_i)_e^T\right]W\right] \\ &= tr\left[W^T S'_w W\right]\end{aligned} \tag{5}$$

Where $S'_w = \frac{1}{N}\sum_{i=1}^{c}\sum_{j=1}^{N_i}(x_{ij}-\mu_i)_e(x_{ij}-\mu_i)_e^T = X_{we}X_{we}^T X_{we}$
$= \frac{1}{\sqrt{N}}\left[(x_{11}-\mu_1)_e, \cdots, (x_{1N_1}-\mu_1)_e, \cdots, (x_{c1}-\mu_c)_e, \cdots, (x_{cN_c}-\mu_c)_e\right]$

The between-class scatter can be written as follows:

$$\sum_{i=1}^{c}\left(\frac{N_i}{N}\cos^2\beta_i\right)=\frac{N_i}{N}\sum_{i=1}^{c}\frac{(\mu_i-\mu)^T WW^T(\mu_i-\mu)}{\|\mu_i-\mu\|_2^2}$$

$$=\frac{N_i}{N}\sum_{i=1}^{c}(\mu_i-\mu)_e^T WW^T(\mu_i-\mu)$$

$$=tr\left[\frac{N_i}{N}W^T\sum_{i=1}^{c}(\mu_i-\mu)_e(\mu_i-\mu)_e^T W\right]$$

$$=tr\left[W^T S_b' W\right] \tag{6}$$

Where $S_b'=\frac{1}{N}\sum_{i=1}^{c}N_i(\mu_i-\mu)_e(\mu_i-\mu)_e^T=X_{be}X_{be}^T, X_{be}=\frac{1}{\sqrt{N}}[\sqrt{N_1}(\mu_1-\mu)_e,\cdots,$ $\sqrt{N_c}(\mu_c-\mu)_e]$

The optimization function of ARLDA can be written as: $J(W)=tr\left(\frac{W^TS_b'W}{W^TS_w'W}\right)$.

So, ARLDA can be transformed into an optimization problem with constraints:

$$\max_W tr\frac{W^TS_b'W}{W^TS_w'W}$$
$$s.t.\quad W^TW=I$$

The optimization problem above can be transformed into a generalized eigenvalue problem.

$$S_b'W=\lambda S_w'W \tag{7}$$

The optimal low-dimension subspace of ARLDA can be got from eigenvectors $w_1,\cdots,w_d$ corresponding to the d largest eigenvalues $\lambda_1\geq\lambda_2\cdots\geq\lambda_d$ through the eigendecomposition of $S_b^{'}S_{w^{-1}}^{'}$.

4 A New Concept Drift of Data Stream Learning Method

In this section, we propose a novel concept drift detection method for data stream learning. When there is concept drift in data stream, generally researchers will design an independent detection method, and then according to detection results they adjust learning strategy. Concept drift detection is time consuming and has no positive effects on subsequent classification of data stream. It only retrains the classifier (such as SVM) or makes some adjustment.

For example, utilizing KL distance to detect concept drift is an effective method, but there are some shortcomings: 1) it needs discretization to calculate probability density. 2) It can only handle concept drift between two classes; multiclass can only be decided based on results of two classes. 3) The process of bootstrap and discretization is time consuming. The time complex of MMD in concept drift detection is close to cubic class. However, data stream needs real-time processing, so MMD is not appropriate to calculate stream data. Generally, we can use statistics of samples to judge concept drift quickly. For example, if the variance of dataset changes, we can say there is a concept drift. However, there is a significant drawback: it only describes the dispersion degree of data with size but without direction. So variance cannot detect the rotational concept drift of dataset. For solving this problem, we consider to combine variance, angle and mean of projected data in subspace to detect concept drift. Obviously, index above can detect changes in geometric distribution simply. The framework of concept drift detection in data stream can be seen in Fig. 1 as follows:

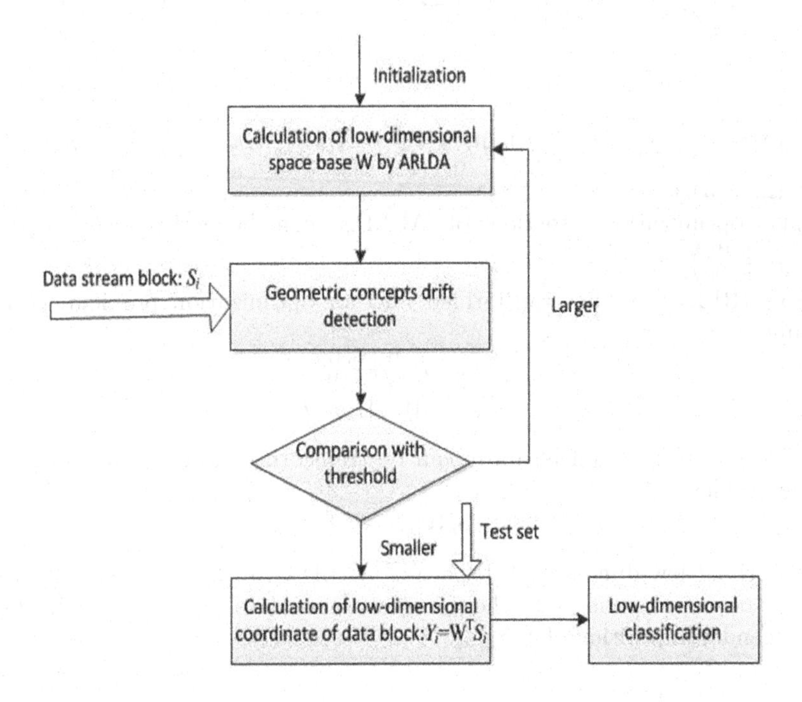

Fig. 1. The learning framework of subspace data stream

We can conclude that learning method using subspace not only detects concept drift, but also reduces the dimension of data stream. Thus we will have a better classification performance.

5 Experiment

In this section, we will perform the experiment on the synthetic and real-word datasets. To evaluate the performance of ARLDA, we compare it with a No-Detection method. SVM[26] and ELM[27] classification methods are used in our experiments.

In this section, we conduct experiments with the Sensor (The sensor data is available in reference [28]), Gas, Forest and Mnist datasets which are commonly used in data streams. The description of datasets and experiment setting is as follows (Table 1).

Table 1. The description of datasets and the experiment parameters setting

	Sample size	Dimension	Class number	Window size	d
Sensor	1017676	4	4	2000	2
Gas	10310	128	6	600	2
Forest	369168	54	7	1000	5
Mnist	56102	784	9	1000	6

We set the window size according to the classification results and the sample size. The d is set based on the dimension of datasets and the dimension reduction performance.

The classification accuracy on three datasets are as follows(table 2,3).

Table 2. The average classification accuracy (%) on different datasets and classification methods

	No-Detection		**ARLDA**	
	SVM	**ELM**	**SVM**	**ELM**
Sensor	42.97±14.78	41.98±17.15	**67.56±18.15**	**57.74±16.84**
Gas	21.26±5.59	45.80±24.86	**39.54±21.12**	**70.18±17.21**
Forest	61.72±19.81	69.40±17.60	**77.05±9.02**	**81.20±6.46**
Mnist	79.13±6.38	67.75±11.56	**84.54±2.86**	**84.13±2.96**

Table 3. Specification of tested binary classification problems

	No-detection		**ARLDA**	
	SVM	**ELM**	**SVM**	**ELM**
Sensor	0.73102	**0.061174**	0.098332	0.080314
Gas	0.19343	0.031013	0.051049	**0.023345**
Forest	0.10213	**0.012175**	0.029078	0.022987
Mnist	2.6698	0.18812	0.033689	**0.16412**

From the results above we can see that, ARLDA outperforms No-Detection method in 3 datasets. ARLDA detects the concept drift with angle-cosine and projection variance, when the drift is detected, it will train the model again and classify with the new model to get a good result.

Comparing with low-dimensional data, it costs more time in calculation of high-dimensional data. In this paper, ARLDA conducts dimensionality reduction first and then detects concept drifts; therefore, it costs less time than methods of no dimensionality reduction. Besides, ELM is an efficient classification method and costs less time than SVM.

In Forest dataset, we divide the data into two parts and construct obvious concept drift. In the first part, we remove the samples of the first class, and in the second part, remove samples of the third class. Both ARLDA and KLLDA detect drifts more than real drifts. Comparing with KLLDA, ARLDA detects drift 42 times which is far less than 99 times of KLLDA.

6 Conclusion

This paper utilizes subspace learning to realize concept drift detection. Experimental results in different datasets demonstrate the effectiveness of our method. Subspace learning can not only detect concept drift, but also reduce the dimension of data stream. The time complex of our method is mainly on searching low-dimension subspace. In future research, searching fast and stable low-dimension subspace method will be the focus.

References

[1] Ouyang, Z.-Z., Gao, Y.-H., Zhao, Z.-P., Wang, T.: Study on the Classification of Data Streams withConcept Drift. In: 2011 Eighth International Conference on Fuzzy Systems and Knowledge Discovery (FSKD), vol. 3, pp. 1673–1677 (2011)

[2] Gama, J., Castillo, G.: Learning with Local Drift Detection. Advanced Data Mining and Applications 4093, 42–55 (2006)

[3] Ikonomovska, E., Gama, J., Sebasti, R., Gjorgjevik, D.: Regression trees from data streams with drift detection. Discovery Science 5808, 121–135 (2009)

[4] Stanley, K.-O.: Learning Concept Drift with a Committee of Decision Trees, Informe tcnico: UT-AI-TR-03-302, Department of Computer Sciences. University of Texas at Austin, USA (2003)

[5] Klinkenberg, R.: Using labeled and unlabeled data to learn drifting concepts. In: Workshop notes of the IJCAI-01 Workshop on Learning from Temporal and Spatial Data, pp. 16–24 (2001)

[6] Lindstrom, P., Delany, S.-J., Namee, B.-M.: Handling concept drift in a text data stream constrained by high labelling cost. In: FLAIRS Conference (2010)

[7] Dries, A., Ruckert, U.: Adaptive Concept Drift Detection. Analysis and Data Mining 2(5-6), 311–327 (2009)

[8] Ross, G.-J., Adams, N.-M., Tasoulisand, D.-K., Hand, D.-J.: Exponentially weighted moving average charts for detecting concept drift. 33(2), 91–198 (2012)

[9] Gama, J., Medas, P., Castillo, G., Rodrigues, P.: Learning with drift detection. Advances in Artificial Intelligence(SBIA) 3171, 286–295 (2004)

[10] Baena-Garcira, M., del Campo-Avila, J., Fidalgo, R., Bifet, A., Gavalda, R., Morales-Bueno, R.: Early Drift Detection Method. Knowledge Discovery from Data Streams (2006)

[11] Sobhani, P., Beigy, H.: New Drift Detection Method for Data Streams. Adaptive and Intelligent Systems 6943, 88–97 (2011)

[12] Wu, X.-D., Li, P.-P., Hu, X.-G.: Learning from concept drifting data streams with unlabeled data. Neurocomputing 92, 145–155 (2012)
[13] Zliobaite, I.: Change with delayed labeling when is it detectable. In: Data Mining Workshops (ICDMW), pp. 843–850 (2010)
[14] Kuncheva, L.-I.: Classifier ensembles for changing environments. Multiple Classifier Systems 3077, 1–15 (2004)
[15] Kolter, J.-Z., Maloof, M.-A.: Dynamic Weighted Majority: A New Ensemble Method for Tracking Concept Drift. Data Mining, 123–130 (2003)
[16] Wang, H.-X., Fan, W., Yu, P.-S., Han, J.-W.: Mining Concept Drifting Data Streams Using Ensemble Classifiers. In: Proceedings of the Ninth ACM SIGKDD International Conference on Knowledge Discovery and Data Mining, pp. 226–235 (2003)
[17] Zhang, P., Zhu, X.-Q., Tanand, J.-L., Guo, L.: Classifier and Cluster Ensembles for Mining Concept Drifting Data Streams. Data Mining (ICDM), 1175–1180 (2010)
[18] Dasu, T., Krishnan, S., Venkatasubramanian, S., Yi, K.: An Information-Theoretic Approach to Detecting Changes in Multi-Dimensional Data Streams. In: Proc. Symp. on the Interface of Statistics, Computing Science, and Applications (2006)
[19] Roweis, S.T., Saul, L.K.: Nonlinear dimensionality reduction by locally linear embedding. Science 290, 2323–2326 (2000)
[20] Zhang, Z., Zha, H.: Principal Manifolds and Nonlinear Dimensionality Reduction via Tangent Space Alignment. SIAM Journal of Scientific Computing 26(1), 313–338 (2004)
[21] He, X., Cai, D., Yan, S., Zhang, H.: Neighborhood preserving embedding. In: 2010 The 2nd IEEE International Conference on Information Management and Engineering (ICIME), pp. 1208–1213 (2005)
[22] Pang, Y.-W., Zhang, L., Liu, Z.-K., Yu, N.-H., Li, H.-Q.: Neighborhood preserving projections(NPP): a novel linear dimension reduction method. Advances in Intelligent Computing 3644, 117–125 (2005)
[23] Min, W.-L., Lu, K., He, X.-F.: Locality pursuit embedding. Pattern Recognition 37(4), 781–788 (2004)
[24] Zheng, W.-M., Zhao, L., Zou, C.-R.: An efficient algorithm to solve the small sample size problem for LDA. Pattern Recognition 37(5), 1077–1079 (2004)
[25] Li, H., Jiang, T., Zhang, K.: Efficient and Robust Feature Extraction by Maximum Margin Criterion. IEEE Trans. on Neural Networks 17(1), 157–165 (2006)
[26] Cortes, C., Vapnik, V.: Support vector networks. Machine Learning 20(3), 273–297 (1995)
[27] Guangbin, H., Zhuqin, Y.: CheeKheongSiew. Extreme learning machine-Theory and applications. Neurocomputing 70(1-3), 489–501 (2006)
[28] Zhu, X., Ding, W., Philip, S.Y., Zhang, C.: One-class learning and concept summarization for data streams. Knowledge and Information Systems 28(3), 523–553 (2011)

Inductive Bias for Semi-supervised Extreme Learning Machine

Federica Bisio[1], Sergio Decherchi[2], Paolo Gastaldo[1], and Rodolfo Zunino[1]

[1] Dept. of Electric, Electronic and Telecommunications Engineering and Naval Architecture (DITEN), University of Genoa, Genova, Italy
federica.bisio@edu.unige.it,
{paolo.gastaldo,rodolfo.zunino}@unige.it
[2] Istituto Italiano di Tecnologia, IIT, Morego, Genova, Italy
sergio.decherchi@iit.it

Abstract. This research shows that inductive bias provides a valuable method to effectively tackle semi-supervised classification problems. In the learning theory framework, inductive bias provides a powerful tool, and allows one to shape the generalization properties of a learning machine. The paper formalizes semi-supervised learning as a supervised learning problem biased by an unsupervised reference solution. The resulting semi-supervised classification framework can apply any clustering algorithm to derive the reference function, thus ensuring maximum flexibility. In this context, the paper derives the biased version of Extreme Learning Machine (bELM). The experimental session involves several real world problems and proves the reliability of the semi-supervised classification scheme.

Keywords: Extreme Learning Machine, semi-supervised learning, inductive bias.

1 Introduction

Inductive bias is of fundamental importance in learning theory, as it influences heavily the generalization ability of a learning system [1]. From a mathematical point of view, the inductive bias can be formalized as the set of assumptions that determine the choice of a particular class of functions to support the learning process. Therefore, it represents a powerful tool to embed the prior knowledge on the applicative problem at hand.

In literature, modifications to the original Extreme Learning Machine (ELM) [2] scheme have been proposed [3, 4]. This paper addresses the advantages and the issues of introducing an inductive bias in the ELM when semi-supervised classifications problems are being tackled. In semi-supervised classification, one exploits both unlabeled and labeled data to learn a classification rule/function empirically [5]; the semi supervised approach should improve over the classification rule that is learnt by only using labeled data. The interest in semi-supervised learning has increased recently,

J. Cao et al. (eds.), *Proceedings of ELM-2014 Volume 1*,
Proceedings in Adaptation, Learning and Optimization 3, DOI: 10.1007/978-3-319-14063-6_6

especially because application domains exist (e.g., text mining, natural language processing, image and video retrieval, and bioinformatics) [6, 7], in which large datasets are available but labeling is difficult, expensive, or time consuming.

The research presented here shows that semi-supervised learning can benefit from biased regularization, which provides a viable approach to implement an inductive bias in a kernel machine, as confirmed by the generalized 'Representer Theorem' [8]. Biased regularization of Support Vector Machines (SVMs) has been adopted in [9] for a malware detection system. The research presented here shows that semi-supervised learning can benefit from biased regularization, too. First, a novel, general biased-regularization scheme is introduced that encompasses the biased version of ELM. Then, the paper proposes a semi-supervised learning model, which is based on that biased-regularization scheme and follows a two-step procedure. In the first step, an unsupervised clustering of the whole dataset (including both labeled and unlabeled data) obtains a reference solution; in the second step, the clustering outcomes drive the learning process in a biased ELM (bELM) to acquire the class information provided by labels. The ultimate result is that the overall learned function exploits both labeled and unlabeled data. The integrated framework applies to both linear and non linear data distributions: in the former case, one works under a cluster assumption on data; in the latter case, one works under a manifold hypothesis [10]. As a consequence, for a successful semi-supervised learning, unlabeled data are assumed to carry some intrinsic geometric structure, e.g., in the ideal case, a low-dimensional, non-linear manifold.

The proposed biased semi-supervised approach exhibits several features, such as modularity in the procedure that generates a biasing solution, convexity of the cost function, predictable complexity, and out-of-sample extension.

The experimental verification of the method involved the USPS [12] dataset. Experimental results confirmed the effectiveness of bELM and proved that the proposed semi-supervised learning scheme compares positively with state-of-the-art algorithms, such as LapRLS [10], LapSVM [10], Transductive SVM (TSVM) [13], and SS-ELM [11].

The paper is organized as follows. Section 2 gives a brief theoretical background on regularization based learning. Section 3 formalizes the biased regularization based learning scheme, and introduces biased ELM. Section 4 presents the semi-supervised classification framework based on biased regularization. Section 5 discusses experimental results and proposes a comparison with LapRLS, LapSVM, TSVM, and SS-ELM. Finally, Section 6 gives some concluding remarks.

2 Theoretical Background

2.1 Regularization-Based Learning

Modern classification methods often rely on regularization theory, which was initially introduced in [14] and generalized to the non linear case by kernel methods [15]. In a regularized functional, a positive parameter, λ, rules the tradeoff between the empirical risk, $R_{emp}[f]$, (loss function) of the decision functions f (i.e., regression or classification) and a regularizing term. The cost to be minimized can be expressed as:

$$R_{reg} = R_{emp}[f] + \lambda\Omega[f] \tag{1}$$

where the regularization operator, $\Omega[f]$, quantifies the complexity of the class of functions from which f is drawn. When dealing with maximum-margin algorithms, $\Omega[f]$ is implemented by the term $\| f \|$, which supports a square norm in the feature space. The Representer Theorem [8] proves that, when $\Omega[f] = \| f \|$, the solution of the regularized cost (1) can be expressed as a finite summation over a set of labeled training patterns $\mathcal{T} = \{(\mathbf{x},y)_i;\ i = 1,\ldots,P\}$, with $y \in \{-1,1\}$.

The ELM framework indeed belongs to the class of regularized learning methods. In principle, the ELM model implements a single-hidden layer feedforward neural networks (SLFN) with N_h mapping neurons. The neuron's response to an input stimulus, $\mathbf{x}$, is implemented by any nonlinear piecewise continuous functions $a(\mathbf{x},\boldsymbol{\zeta})$, where $\boldsymbol{\zeta}$ denotes the set of parameters of the mapping function. The overall output function is then expressed as

$$f(\mathbf{x}) = \sum_{j=1}^{Nh} w_j h_j(\mathbf{x}) \tag{2}$$

where w_j denotes the weight that connects the jth neuron with the output, and $h_j(\mathbf{x}) = a(\mathbf{x},\boldsymbol{\zeta}_j)$. In ELM the parameters $\boldsymbol{\zeta}_j$ are set randomly. As the training process reduces to the adjustment of the output layer, training ELMs is equivalent to solving a regularized least squares problem. Hence, the minimization problem can be expressed as

$$\min_f \left\{ \sum_{i=1}^{P} \left(y_i - f(\mathbf{x}_i)\right) + \lambda \|f\|^2 \right\} \tag{3}$$

The vector of weights $\mathbf{w}$ is then obtained as follows:

$$\mathbf{w} = (\mathbf{H}^t\mathbf{H} + \lambda\mathbf{I})^{-1}\mathbf{H}^t\mathbf{y} \tag{4}$$

Here, $\mathbf{H}$ is a $P \times N_h$ matrix with $h_{ij} = h_j(\mathbf{x}_i)$.

3 A Unifying Framework for Biased Learning

The general biased regularization model consists in biasing the solution of a regularization-based learning machine by a reference function (e.g., a hyperplane). The nature of this reference function is a crucial aspect that concerns the learning theory in general. This section discusses two main aspects, that is, the formal definition of a general biased-regularization scheme, and the formalization of the biased ELM (bELM) within this scheme.

3.1 Biased Regularization

In the linear domain one can define a generic convex loss function, $L(\mathcal{X}, \mathcal{Y}, \mathbf{w})$, and a biased regularizing term; the resulting cost function is:

$$L(\mathcal{X},\mathcal{Y},\mathbf{w})+\frac{\lambda_1}{2}\left\|\mathbf{w}-\lambda_2\mathbf{w}_0\right\|^2 \tag{5}$$

where $\mathbf{w}_0$ is a “reference” hyper-plane, λ_1 is the classical regularization parameter that controls smoothness (i.e., λ in (1)), and λ_2 controls the adherence to the reference solution $\mathbf{w}_0$. Expression (3) is a convex functional and thus admits a global solution. From (3) one gets:

$$L(\mathcal{X},\mathcal{Y},\mathbf{w})+\frac{\lambda_1}{2}\left\|\mathbf{w}-\lambda_2\mathbf{w}_0\right\|^2=L(\mathcal{X},\mathcal{Y},\mathbf{w})+\frac{\lambda_1}{2}\left\|\mathbf{w}\right\|^2-\lambda_1\lambda_2\mathbf{w}\mathbf{w}_0 \tag{6}$$

The role played by parameter λ_2 is of fundamental importance both from the theoretical and the practical point of view. This parameter allows the cost function (6) to implement the model discussed in [16], which analyzed the implication of using a strong bias or a weak bias on the hypothesis space. In this regard, Figure 1 explicates the role played by parameter λ_2 by illustrating two different examples; for the sake of clarity, and without losing generalization, all the examples assume that λ_1 is set to a fixed value (i.e., $\lambda_1 = 1$).

Both Fig. 1(a) and Fig. 1(b) sketches an ideal space of hypothesis, where $\mathbf{w}$ is set as the origin, $\mathbf{0}$ and a grey square indicates the 'true' optimal solution $\mathbf{w}^*$. Fig. 1(a) refers to the situation in which the reference $\mathbf{w}_0$ is closer to the true solution $\mathbf{w}^*$ than $\mathbf{w}_{\lambda 2=0}$; the black line connecting these two points actually shows where $\lambda_2\mathbf{w}_0$ would lie, as a function of $\lambda_2 \in [0,1]$. In this figure, a bold dashed line limits the portion of the space, $F_{\lambda 1,1}$, to be explored to include $\mathbf{w}^*$ when $\lambda_2 = 1$; a thin dashed line limits the portion of the space, $F_{\lambda 1,0}$, to be explored to include $\mathbf{w}^*$ when $\lambda_2 = 0$. The example clarifies that under the assumption “$\mathbf{w}_0$ closer to $\mathbf{w}^*$ than $\mathbf{w}_{\lambda 2=0}$” one can take full advantage of biased regularization, as the portion of space to be explored would shrink as long as $\lambda_2 \rightarrow 1$ (i.e., strong bias).

Fig. 1(b) illustrates the opposite case: the reference $\mathbf{w}_0$ is more distant from the true solution $\mathbf{w}^*$ than $\mathbf{w}_{\lambda 2=0}$ (it is worth to note that the relative position of $\mathbf{w}^*$ and $\mathbf{w}_{\lambda 2=0}$ with respect to the origin $\mathbf{w}=0$ remain unchanged when compared with Fig. 1(a)). In this situation, one would obtain the best outcome by setting $\lambda_2 = 0$, thus neutralizing the contribution of the biased regularization. Hence, as $\mathbf{w}_0$ does not represent a helpful reference, one would benefit from a weak bias.

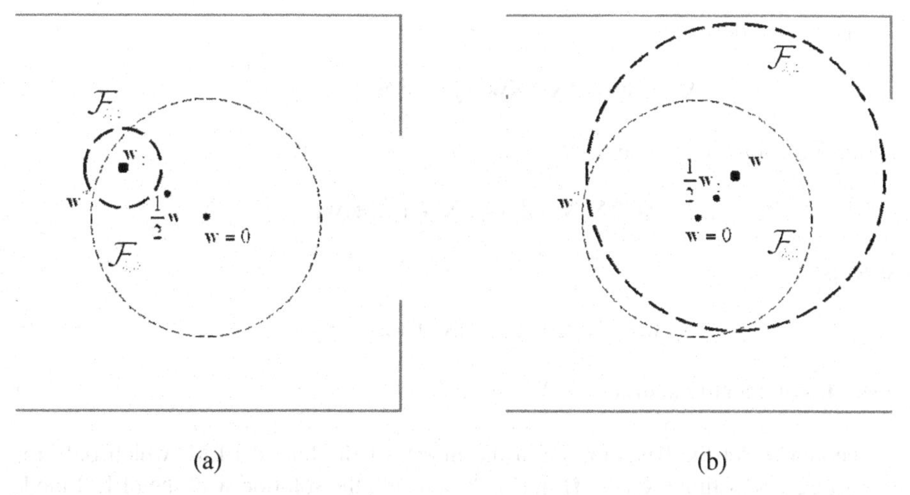

(a) (b)

Fig. 1. The role played by parameter λ_2 in the proposed problem setting. Two different situations are analyzed: (a): the reference $\mathbf{w}_0$ is closer to the true solution $\mathbf{w}^*$ than $\mathbf{w}_{\lambda 2=0}$; (b): the reference $\mathbf{w}_0$ is more distant from the true solution $\mathbf{w}^*$ than $\mathbf{w}_{\lambda 2=0}$.

In practice, the limit cases addressed in Fig. 1 (a) and (b) show that, in general, by adjusting λ_2 one can modulate the contribution provided by $\mathbf{w}_0$ to the cost function (6). Hence, one can take advantage of biased regularization even when the reference solution is not optimal. The crucial aspect is represented by the ability to shrink the space to be explored to get the optimal solution, which in turn means the ability to reduce the complexity of the hypothesis space. The next section extends (6) to the ELM model, thus obtaining the biased ELM (bELM).

3.2 Biased ELM (bELM)

The following theorem formalizes the linear biased version of a regularized least squares problem.

Theorem 1. *Given a reference hyperplane* $\mathbf{w}_0$, *a regularization constant* λ_1, *and a biasing constant* λ_2, *the problem:*

$$\min_{\mathbf{w}}\left\{\|\mathbf{X}\mathbf{w}-\mathbf{y}\|^2+\frac{\lambda_1}{2}\|\mathbf{w}-\lambda_2\mathbf{w}_0\|^2\right\} \tag{7}$$

has solution

$$\mathbf{w}=(\mathbf{X}^t\mathbf{X}+\lambda_1\mathbf{I})^{-1}\left(\mathbf{X}^t\mathbf{y}+\lambda_1\lambda_2\mathbf{w}_0\right) \tag{8}$$

Proof
Equation (7) can be conveniently rewritten as:

$$\min_{\mathbf{w}}\left\{\|\mathbf{X}\mathbf{w}-\mathbf{y}\|^2+\frac{\lambda_1}{2}\|\mathbf{w}\|^2-\lambda_1\lambda_2\mathbf{w}\mathbf{w}_0\right\}$$

The associated derivative is

$$\nabla_{\mathbf{w}} L(\mathbf{w}) = 2\mathbf{X}^t(\mathbf{X}\mathbf{w} - \mathbf{y}) + \lambda_1\mathbf{w} - \lambda_1\lambda_2\mathbf{w}_0$$

Setting $\nabla_{\mathbf{w}} L(\mathbf{w})=0$ one obtains

$$\mathbf{w}(2\mathbf{X}^t\mathbf{X} + \lambda_1) = 2\mathbf{X}^t\mathbf{y} + \lambda_1\lambda_2\mathbf{w}_0$$

As a result

$$\mathbf{w} = (\mathbf{X}^t\mathbf{X} + \lambda_1\mathbf{I})^{-1}\left(\mathbf{X}^t\mathbf{y} + \lambda_1\lambda_2\mathbf{w}_0\right)$$

where $\mathbf{I}$ is an identity matrix $N_h \times N_h$. □

Theorem 1 set the basis to obtain the model of the biased ELM, which only requires one to substitute $\mathbf{X}$ with $\mathbf{H}$ in (7). As a result, the solution $\mathbf{w}$ of the bELM model can be expressed as follows:

$$\mathbf{w} = (\mathbf{H}^t\mathbf{H} + \lambda_1\mathbf{I})^{-1}\left(\mathbf{H}^t\mathbf{y} + \lambda_1\lambda_2\mathbf{w}_0\right) \tag{9}$$

4 Semi-supervised Learning by Using bELM

This Section formalizes the semi-supervised classification scheme, and discusses the appealing advantages provided by adopting the bELM model as a learning machine in that framework.

4.1 A Semi-supervised Learning Scheme Based on Biased Regularization

The proposed formalization of the semi-supervised learning scheme applies the following notation:

- $\mathcal{G}$ is a dataset composed by P patterns; the first l patterns are labeled, the remaining u patterns are unlabeled.
- $\mathcal{G}^L$ is the subset of $\mathcal{G}$ composed by the labeled patterns; thus, $\mathcal{G}^L = \{(\mathbf{x},y)_i;\ i = 1,\ldots,l\}$
- $\mathcal{G}^U$ is the subset of $\mathcal{G}$ composed by the remaining unlabeled patterns; thus, $\mathcal{G}^U = \{\mathbf{x}_i;\ i = l+1,\ldots,u\}$

The semi-supervised learning scheme requires one to apply a four-step procedure:

1. *Clustering*. Complete an unsupervised clustering (bi-partite in the simplest case) of the dataset $\mathcal{G}$ by adopting any algorithm supporting that task.
2. *Calibration*. For each cluster: first, set the cluster label by adopting a majority voting scheme that exploits the labeled samples; second, assign that label to each sample belonging to the cluster. Let $\mathbf{y}^0$ denote this new vector of labels.

3. *Mapping*. Obtain $\mathbf{w}^0$ as the result of the training of a standard ELM on the dataset $\mathcal{G}^0 = \{(\mathbf{x},y^0)_i;\ i = 1,\ldots,P)\}$ (as per (4)).
4. *Biasing*. Obtain $\mathbf{w}$ as the result of the training of a bELM biased by $\mathbf{w}_0$ on the dataset $\mathcal{G}^L$. The solution $\mathbf{w}$ eventually endows information from both the labeled data $\mathcal{G}^L$ and the unlabeled data $\mathcal{G}^U$.

In this procedure, Step 4, 'Biasing,' is fully supported by the biased-regularization scheme introduced in Sec. 3. The procedure has similarities to that adopted in deep learning architectures [17]. In that case the training algorithm performs a preliminary unsupervised stage, then uses labels only to adjust the network for the specific classification task; the eventual representation mostly reflects the outcome of the learning process completed in pre-training phase. Likewise, in the proposed framework, a pre-training phase builds $\mathbf{w}_0$ and a final adjustment derives the final $\mathbf{w}$.

Overall, the proposed semi-supervised learning scheme possesses some interesting features. First, the semi-supervised learning task is tackled by separating the two actions: clustering, and biasing. As a result, one can control and adjust each specific action separately, e.g., by adopting a particular solution or by designing a new algorithm. This may be the case for the clustering task: every clustering method can be used to build the reference solution. Therefore, one may take advantage from methodologies that address effectively complex, non linear domains. Second, a global solution is granted, as the framework preserves the convexity of the original ELM cost function.

5 Experimental Results

The experimental section aims at evaluating the accuracy performances of the proposed method on unseen data, i.e. to assess induction performances. Sec 5.1 proposes a comparison between bELM and state-of-the-art methods for semi-supervised learning: LapRLS [10], LapSVM [10], TSVM [13], and SS-ELM [11].

5.1 Comparison with State-of-the-Art Methods

The experimental session was designed to compare the proposed semi-supervised scheme with LapRLS, LapSVM, TSVM, and SS-ELM. To perform a fair comparison, the experiments involved a dataset already addressed by those approaches: USPS [12].

As publicly available Matlab code is provided for LapRLS, LapSVM and TSVM [19], experiments involving the proposed semi-supervised learning scheme were developed by embedding the both bELM and SS-ELM into those routines. As a result, the experiments also exploited the data preprocessing designed by the authors of those approaches. SS-ELM has been implemented in Matlab according to the algorithm presented in [11]; the corresponding graph Laplacian has been computed by exploiting the dedicated routine already provided in [19]. A publicly available Matlab version of spectral clustering [18] has been used to support the clustering step in bELM. The Matlab code of bELM is freely available at: http://www.sealab.dibe.unige.it/biased_learning.

The experiment addressed the USPS dataset, which is an OCR dataset collecting digits images. The experiment involved all the 45 bi-class problems that can be generated from the dataset. For each problem, the experimental set up followed that adopted in [10]: the first 400 images are inserted in the training set and pre-processed by PCA, which is exploited to obtain a feature space with dimension 100; the remaining images compose the test set. For each class, 2 samples randomly selected were labeled, while the others were left unlabeled. Results have been averaged over 10 random choices of labeled examples.

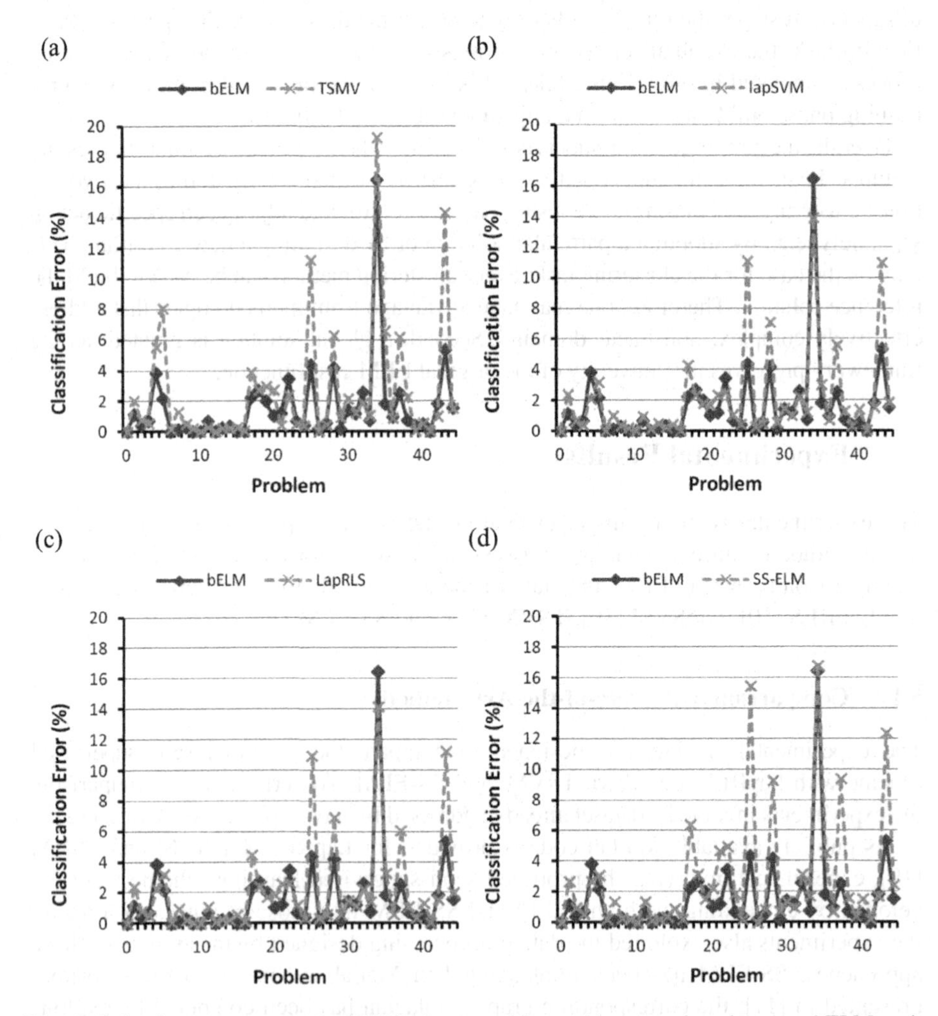

Fig. 2. Inductive accuracy on the USPS binary problems: (a) comparison between bELM and TSVM; (b) comparison between bELM and LapSVM; (c) comparison between bELM and LapRLS; (c) comparison between bELM and SS-ELM

Polynomial kernel was used of degree ‘3’ in LapSVM, LapRLS, and TSVM. The parameters setting in [10] denotes that geometry information is very important to correctly address this problem, as a 9:1 ratio in the regularizers settings results in a predominance of the geometric term induced by the Graph Laplacian. Therefore, parameters λ_1 and λ_2 were set accordingly in all the experiments involving bELM: $\lambda_1 = 0.1$ (i.e., $C = 10$ in SS-ELM) and $\lambda_2 = 1$. These settings result in a configuration that gives high confidence to the reference solution provided by clustering. The number of neurons was set as $N_h = 2000$ both in bELM and SS-ELM.

Figure 2 compares the performances of bELM with those attained by LapRLS, LapSVM, TSVM and SS-ELM. In particular: fig. 2(a) compares bELM with LapRLS; fig. 2(b) compares bELM with LapSVM; fig. 2(c) compares bELM with TSVM; fig. 2(d) compares bELM with SS-ELM. All the results refer to the accuracy values obtainable at the break-even points in the precision-recall curves; such set up follows the one adopted in [10] and allows a fair comparison between the different approaches. Numerical results show that in several cases the proposed semi-supervised scheme improves over the other methods. Indeed, in a few cases the gain in classification error obtained with bSVM and SS-ELM is significant.

6 Conclusions

The present research proves that using biasing techniques in regularization-based learning can lead to an effective, yet simple learning scheme for semi-supervised classification. The eventual framework is characterized by several appealing features. First, the semi-supervised learning task is tackled by separating the two actions: clustering, and biasing. Therefore, one can control and adjust a specific action separately, e.g., by adopting a particular solution or by designing a new algorithm. This may be the case for the clustering task, which can take advantage from methodologies that address effectively complex, non linear domains. Second, if the employed learning machine supports a convex optimization problem, the learning scheme preserves convexity. In this paper, the proposed biased ELM indeed preserves the convexity of the original ELM cost function.

Empirical evidence showed that bELM compares favorably with state-of-the-art semi-supervised learning schemes. In particular, bELM can attain reliable performance even when the number of labeled samples is very small (i.e., five samples or less).

References

1. Mitchell, T.M.: The need for biases in learning generalizations, CBM-TR 5-110. Rutgers University, New Brunswick (1980)
2. Huang, G.-B., Zhou, H., Ding, X., Zhang, R.: Extreme Learning Machine for Regression and Multiclass Classification. IEEE Transactions on Systems, Man, and Cybernetics - Part B: Cybernetics 42(2), 513–529 (2012)

3. Decherchi, S., Gastaldo, P., Zunino, R., Cambria, E., Redi, J.: Circular-ELM for the reduced-reference assessment of perceived image quality. Neurocomputing 102, 78–89 (2013)
4. Gastaldo, P., Zunino, R., Cambria, E., Decherchi, S.: Combining ELM with Random Projections. IEEE Intelligent Systems 28(6), 46–48 (2013)
5. Chapelle, O., Schölkopf, B., Zien, A.: Semi-Supervised Learning. MIT Press (2006)
6. Poria, S., Gelbukh, A., Hussain, A., Bandyopadhyay, S., Howard, N.: Music genre classification: A semi-supervised approach. In: Carrasco-Ochoa, J.A., Martínez-Trinidad, J.F., Rodríguez, J.S., di Baja, G.S. (eds.) MCPR 2012. LNCS, vol. 7914, pp. 254–263. Springer, Heidelberg (2013)
7. Poria, S., Gelbukh, A., Cambria, E., Das, D., Bandyopadhyay, S.: Enriching senticnet polarity scores through semi-supervised fuzzy clustering. In: 2012 IEEE 12th International Conference on Data Mining Workshops (ICDMW), pp. 709–716. IEEE (December 2012)
8. Schölkopf, B., Herbrich, R., Smola, A.J.: A Generalized Representer Theorem. In: Helmbold, D.P., Williamson, B. (eds.) COLT 2001 and EuroCOLT 2001. LNCS (LNAI), vol. 2111, pp. 416–426. Springer, Heidelberg (2001)
9. Bisio, F., Gastaldo, P., Zunino, R., Decherchi, S.: Semi-supervised machine learning approach for unknown malicious software detection. In: Proceedings of the 2014 IEEE International Symposium on Innovations in Intelligent Systems and Applications (INISTA), pp. 52–59. IEEE (June 2014)
10. Belkin, M., Niyogi, P., Sindhwani, V.: Manifold regularization: A geometric framework for learning from labeled and unlabeled examples. Journal of Machine Learning Research 7, 2399–2434 (2006)
11. Huang, G., Song, S.D., Gupta, J.N., Wu, C.: Semi-Supervised and Unsupervised Extreme Learning Machines. IEEE Trans, on Cybernetics, in print (accepted)
12. http://www.csie.ntu.edu.tw/~cjlin/libsvmtools/datasets/multiclass.html
13. Vapnik, V.N.: Statistical Learning Theory. Wiley-Interscience Pub (1998)
14. Tikhonov, A.N., Arsenin, V.Y.: Solution of Ill-posed Problems. Winston & Sons, Washington (1977)
15. Schlkopf, B., Smola, A.J.: Learning With Kernels. MIT Press (2001)
16. Utgoff, P.E.: Shift of bias for inductive concept learning. PhD Dissertation (1984)
17. Erhan, D., Bengio, Y., Courville, A., Manzagol, P.A., Vincent, P., Bengio, S.: Why Does Unsupervised Pre-training Help Deep Learning? Journal of Machine Learning Research 11, 625–660 (2010)
18. Chen, W.Y., Song, Y., Bai, H., Lin, C.J., Chang, E.Y.: Parallel Spectral Clustering in Distributed Systems. IEEE Transactions on Pattern Analysis and Machine Intelligence 33(3), 568–586 (2011)
19. http://manifold.cs.uchicago.edu/manifold_regularization/data.html

ELM Based Efficient Probabilistic Threshold Query on Uncertain Data

Jiajia Li, Botao Wang, and Guoren Wang

College of Information Science & Engineering, Northeastern University, P.R. China
jiajia4487@gmail.com, {wangbotao,wangguoren}@ise.neu.edu.cn

Abstract. The *probabilistic threshold query* (PTQ), which returns all the objects satisfying the query with probabilities higher than a probability threshold, is widely used in uncertain database. Most previous work focused on the efficiency of query process, but paid no attention to the setting of thresholds. However, setting the thresholds too high or too low may lead to empty result or too many results. For a user, it is too difficult to choose a *suitable threshold* for a query. In this paper, we propose a new framework for PTQs based on threshold classification using ELM, where the probability threshold is replaced by the *number range* of results which is more intuitive and easier to choose. We first introduce the features selected for the probabilistic threshold nearest neighbor query (PTNNQ), which is one of the most important PTQ types. Then a threshold classification algorithm (TCA) using ELM is proposed to set a suitable threshold for the PTNNQ. Further, the whole PTNNQ processing integrated with TCA are presented, and a dynamic classification strategy is proposed subsequently. Extensive experiments show that compared with the thresholds those the users input directly, the thresholds chosen by ELM classifiers are more suitable, which further improves the performance of PTNNQ. In addition, ELM outperforms SVM with regards to both the response time and classification accuracy.

Keywords: Probabilistic threshold query, Nearest neighbor query, Classification, Extreme learning machine.

1 Introduction

Uncertain data are inherent in numerous emerging applications due to limitations of measuring equipment, delayed data updates, or privacy protection. As pointed out in [1], queries based on these imprecise data may produce erroneous results. In order to provide more accurate and informative answers for queries on these uncertain data, the idea of *probabilistic threshold query* (PTQ) is proposed, which returns all the objects satisfying the query with probabilities higher than some probability threshold. The *probability threshold nearest neighbor query* (PTNNQ) [1–3], as one of the most important types of PTQs, has attracted a lot of research attention in the past decade. Though many interesting and efficient pruning strategies are designed to answer the PTNNQ in previous work, PTQ has inherent problems in choosing a suitable threshold. Particularly, setting the

J. Cao et al. (eds.), *Proceedings of ELM-2014 Volume 1*,
Proceedings in Adaptation, Learning and Optimization 3, DOI: 10.1007/978-3-319-14063-6_7

threshold too high may lead to empty result, and hence the query needs to be restarted with a lower threshold. And setting the threshold too low may produce too many results and increase query time. As far as we know, there is no previous work paying attention to the threshold setting problem as done in this paper.

We observed that for a user, setting a suitable threshold is difficult, while directly setting the number of results which they want is more intuitive and easier. Therefore, we propose a transformation method, which transforms the *number range* input by a user to a suitable threshold. That is, given a query and a *number range*, the transformation method chooses a suitable probabilistic threshold such that the number of results based on this threshold is located in the *number range*. The relationship between *number range* and threshold is very complex, so the classification method is utilized to realize this transformation. Since most queries require real-time processing, one of the main challenges for the classification method is the speed. Both the learning and classification process should be fast enough. To suit our needs of speed, Extreme Learning Machine (ELM) [4, 5] is adopted for classifying queries into their respective threshold classes, because that ELM has more fast learning speed and better generalization performance than neural networks and Support Vector Machines (SVMs) [6].

The main contributions in this paper are: (1) A new framework based on threshold classification is proposed for PTQs (Sec. 4.1). It only needs a parameter of the required number of results, and the user doesn't have to input the probabilistic threshold which is difficult to choose. (2) Some useful features are selected for the PTNNQ and a threshold classification algorithm (TCA) using ELM is proposed(Sec. 4.2). (3) The whole PTQ processing integrated with the proposed TCA for NN query is presented, and a dynamic classification strategy is proposed (Sec. 4.3). (4) Extensive experiments are devised to study the performance of the classification and query algorithms (Sec. 5).

2 Related Work

In recent years, a number of studies have been focused on the processing of NN query on uncertain data. The *probabilistic queries* [7, 8] return all the objects with non-zero probabilities of being NN query results, as well as their probabilities. The *top-k probable queries* [9] return k objects with the highest probabilities of being query results. Our work concentrate on the *probabilistic threshold queries*, which return objects whose probabilities exceed a given threshold. In this section, the existing work on PTNNQ are briefly reviewed.

Cheng et al. [1] first used the probability threshold as an answering criterion for NN queries on uncertain data. Efficient verification methods, which utilized an object's uncertainty information and its relationship with other objects, were proposed for deriving lower and upper bounds of an object's probability being query result. Later, the probabilistic threshold kNN ($k > 1$) query was studied by Cheng et al. [2]. Since the evaluation of kNN faces the additional problem of examining a large number of k-subsets, new methods to reduce the number of candidate k-subsets were proposed. Yang et al. [3] studied the probabilistic threshold kNN query in symbolic indoor space. The minimal indoor walking distance (MIWD) metric was proposed and used to prune objects that have no chance to be query results.

It can be seen that, although many pruning techniques were proposed for probabilistic threshold NN query, there is no work paying attention to the probabilistic threshold setting problem as done in this paper.

3 Background

3.1 Problem Definition

The definition of the suitable threshold is given below. Subsequently, the uncertain model and the definition of PTNNQ are introduced.

Definition 1 (Suitable Threshold, s-threshold). *Given a set of uncertain objects $U = \{U_1, ..., U_n\}$, a query object q of any type, and a user-specified number range $[nlb, nub]$, the probabilistic threshold λ is called a s-threshold iff the number of objects that satisfy q with probabilities greater than λ is in $[nlb, nub]$.*

There are two common ways in describing an uncertain object, either *continuously* which uses a probability density function (*pdf*) [1, 9] or *discretely* which uses a discrete set of alternative values associated with assigned probabilities [10–12]. In this paper, we adopt the discrete one.

Definition 2 (Probabilistic Threshold Nearest Neighbor Query, PTNNQ). *Given a set of n uncertain data objects $U = \{U_1, ..., U_n\}$, an uncertain query object $q = \{q_1, ..., q_m\}$ and a probabilistic threshold $\lambda \in (0, 1]$, a PTNNQ retrieves objects $U_i \in U$, such that they are expected to be the nearest neighbor of query q with a probability higher than λ , that is,*

$$PTNNQ_q = \{U_i | \sum_{\omega \in \Omega, q_j \in q} P(\omega) \cdot \delta(NN^{\omega}_{q_j}(U_i)) \geq \lambda\} \tag{1}$$

where $\delta(NN^{\omega}_{q_j}(U_i))$ is 1 if U_i is the NN of q_j in possible world ω and 0 otherwise.

3.2 Review of Extreme Learning Machine (ELM)

ELM [4, 5] has originally been developed based on Single-hidden Layer Feedforward Neural Networks (SLFNs) and then extended to the "generalized" SLFNs, where the hidden layer need not be neuron alike [13, 14]. ELM is less sensitive to user specified parameters, and can be deployed faster and more conveniently than conventional learning algorithms for classification [15, 16].

For N arbitrary distinct samples $(\mathbf{x}_j, \mathbf{t}_j)$, where $\mathbf{x}_j = [x_{j1}, x_{j2}, ..., x_{jn}]^T \in \mathbf{R}^n$ and $\mathbf{t}_j = [t_{j1}, t_{j2}, ..., t_{jm}]^T \in \mathbf{R}^m$, standard SLFNs with hidden nodes L and activation function $g(\mathbf{x})$ are mathematically modeled as: $\sum_{i=1}^{L} \beta_i g_i(\mathbf{x}_j) = \sum_{i=1}^{L} \beta_i g(\mathbf{w}_i \cdot \mathbf{x}_j + b_i) = \mathbf{o}_j$, where L is the number of hidden layer nodes, $\mathbf{w}_i$ is the input weight vector, β_i is the output weight vector, and b_i is the threshold of the ith hidden node, and $\mathbf{o}_j$ is the jth output vector of the SLFNs [5].

The standard SLFNs can approximate these N samples with zero error. The error of ELM is $\sum_{j=1}^{L} ||\mathbf{o}_j - \mathbf{t}_j|| = 0$ and there exist β_i, $\mathbf{w}_i$ and b_i such that $\sum_{i=1}^{L} \beta_i g(\mathbf{w}_i \cdot \mathbf{x}_j + b_i) = \mathbf{t}_j$, where $j = 1, 2, \ldots, N$.

The equation above can be expressed compactly by: $\mathbf{H}\beta = \mathbf{T}$, where
$\mathbf{H}(\mathbf{w}_1, \mathbf{w}_2, \ldots, \mathbf{w}_L, b_1, b_2, \ldots, b_L, \mathbf{x}_1, \mathbf{x}_2, \ldots, \mathbf{x}_L)$

$$= \begin{bmatrix} h(x_1) \\ h(x_2) \\ \vdots \\ h(x_N) \end{bmatrix} = \begin{bmatrix} g(\mathbf{w}_1 \cdot \mathbf{x}_1 + b_1) & g(\mathbf{w}_2 \cdot \mathbf{x}_1 + b_2) & \ldots & g(\mathbf{w}_L \cdot \mathbf{x}_1 + b_L) \\ g(\mathbf{w}_1 \cdot \mathbf{x}_2 + b_1) & g(\mathbf{w}_2 \cdot \mathbf{x}_2 + b_2) & \ldots & g(\mathbf{w}_L \cdot \mathbf{x}_2 + b_L) \\ \vdots & \vdots & \vdots & \vdots \\ g(\mathbf{w}_1 \cdot \mathbf{x}_N + b_1) & g(\mathbf{w}_2 \cdot \mathbf{x}_N + b_2) & \ldots & g(\mathbf{w}_L \cdot \mathbf{x}_N + b_L) \end{bmatrix}_{N \times L} \tag{2}$$

with $\beta = [\beta_1^T, \ldots, \beta_L^T]_{m \times L}^T$ and $\mathbf{T} = [\mathbf{t}_1^T, \ldots, \mathbf{t}_L^T]_{m \times N}^T$. The smallest norm least-squares solution of the above multiple regression system is: $\hat{\beta} = \mathbf{H}^\dagger \mathbf{T}$, where $\mathbf{H}^\dagger$ is the Moore-Penrose generalized inverse of matrix $\mathbf{H}$. Then the output function of ELM can be modeled as: $f(\mathbf{x}) = \mathbf{h}(\mathbf{x})\beta = \mathbf{h}(\mathbf{x})\mathbf{H}^\dagger\mathbf{T}$

4 Probabilistic Threshold Query (PTQ)

In this section, the threshold classification based framework for PTQ is described firstly. Then the features used to classify the threshold for PTNNQ are introduced, and the ELM based threshold classification algorithm (TCA) is proposed. Finally, the whole PTNNQ processing integrated with TCA are presented.

4.1 Framework and Basic Idea

The general framework for a PTQ on uncertain data usually consists of three phases: *filtering*, *verification* and *refinement*, which is widely adopted in processing probabilistic NN query [1, 2]. In this paper, a new framework is proposed, which contains another phase named *setting* before the *verification* phase.

Figure 1 shows the details of the new framework for PTQs based on threshold classification. The gray parts are new proposed in this paper, and the rest parts are compatible with traditional algorithms. The main four phases are: (1) The *filtering* phase is mainly to remove all objects who have no chance to be a query result. The objects can't be removed are inserted into *cndSet*. (2) The *setting* phase is to set a *s-threshold* λ for query q based on the threshold classification, which is the main work of this paper. Firstly, the useful feature values of q are selected and computed. Then the corresponding threshold class of q is predicted by ELM classifiers, and the predicted threshold of q is transferred to the next phase. (3) The objective of the *verification* phase is to decide which objects satisfy or fail the PTQs. Some probabilistic pruning algorithms may be proposed and the objects that can't be accepted or rejected are inserted into $rfnSet$. (4) The last phase is *refinement*, in which the exact probability of each object in

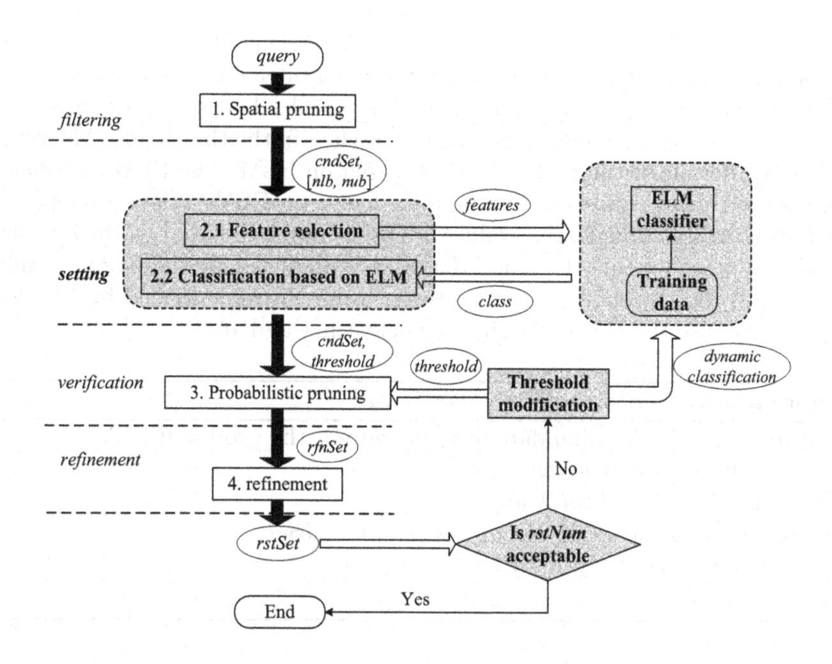

Fig. 1. Framework of PTQ based on threshold classification

$rfnSet$ is computed to decide the final results set $rstSet$. If the number of $rstSet$ is acceptable, the query is finished; or else, the threshold should be modified and recall the third phase. In addition, if the dynamic environment is detected based on a counting method, ELM classifiers should be retrained.

4.2 Feature Selection and Threshold Classification Algorithm

Feature selection plays an important role in the process of classification and affects the accuracy to some exten. In this subsection, we discuss the problem that which features should be selected for PTNNQ and the ELM based threshold classification algorithm (TCA).

According to Definition 2, the probabilities of objects being PTNNQ results may be affected by the attributes of the query q, the attributes of objects in *cndSet* and the spatial relationship between q and these objects. The attributes of q contain the *coordinate* of the center of uncertain region, the *number of instances* contained in q, the user's required *number range* of the results and the *difference* between them. The attributes of the objects in *cndSet* contain the *number of objects*, the *sum number of the instances* and the *average radius* of the uncertain regions of the objects. The spatial relationship between q and these objects contains the *number of objects around* q, and the *sum number of instances around* q. In order to collect the information about the objects around q, we use the regular grid decomposition method [17] to divide the region into many cells. And then count the number of objects which located in the 9 cells around q. And the *length* of each grid are also considered as the features.

Since ELM requires less training time and provides better performance than traditional learning machines, it is adopted to classify the thresholds. Algorithm 1 shows the details of the proposed TCA based on ELM. The ELM classifier is pre-computed using different training data based on ELM. When a query q is issued, its corresponding feature values are first computed based on the information of *cndSet* and q (line 1). Then the class is predicted by the ELM classifier (line 2). And the threshold of q is set to the value of $class$ divided by ten (line 3), that's because the $class$ is an integer between 1 and 10.

Algorithm 1. TCA(q, *cndSet*, $classifier$)

Input: query q, the candidate objects *cndSet*, the produced ELM $classifier$ in advance;

Output: the *s-threshold* λ for q

1 compute the feature values x of q based on *cndSet*;

2 predict the *class* of x using ELM $classifier$;

3 $\lambda = class \div 10$;

4.3 Threshold Classification Based Query Processing

Since the data objects are not static in most applications for NN query, the ELM classifier based on the old training data may bring incorrect classification. Therefore, a dynamic classification strategy of ELM is proposed. And the whole PTNNQ processing integrated with the proposed TCA is presented.

Dynamic classification means collecting training data dynamically and retraining ELM classifiers dynamically. Each time the modification threshold method is called, it's means that a classification error occurs, a new sample training data based on this query can be obtained. It is known that the retraining ELM processing is also a kind of overhead here. And there's no need to retrain the new data once a classification error occurs. To balance the trade-off between the cost of retraining ELM and recomputing the query results, a counting method is required to detect this dynamical environment. If the number of times that error occurs is up to some specified number, ELM classifiers are retained immediately using these new sample training data. And the need of online retraining is another motivation for us to adopt ELM as our classify method, due to that ELM has more fast training time than other classifiers.

The pseudo code of the overall algorithm (namely FSVR) for PTNNQ is shown in Algorithm 2, which mainly consists of four phases: ***filtering***, ***setting***, ***verification***, ***refinement***. As introduced in Sec. 4.1, the *setting* phase is the focus of our work. Therefore, the details of other three phases are omitted, and it is worth noting that any spatial pruning algorithm, probabilistic pruning algorithm and refinement algorithm can be applied to FSVR. Firstly, some variants are initialized which will be used in the query processing (lines 1-3). Next, FSVR first executes the corresponding algorithms of the four phases one by one (lines 4-7).

Algorithm 2. FSVR(q, nlb, nub)

Input: query q, the given number range $[nlb, nub]$;
Output: the results set $rstSet$ being NN query of q

1 pre-compute the $classifier$ using ELM;
2 set the amount of error tolerance $etNum$;
3 bool $errorFlag = 0$;
4 execute the spatial pruning algorithms on q and get $cndSet$;
5 $\lambda =$ TCA(q, $cndSet$, $classifier$);
6 execute the probabilistic pruning algorithms on $cndSet$ using λ and get $rfnSet$ and a subset of $rstSet$;
7 compute the exact probability of each object in $rfnSet$, and put the results into $rstSet$;
8 $rstNum =$ Length($rstSet$);
9 **if** $(rstNum > nub) \mid\mid (rstNum < nlb)$ **then**
10   **if** $rstNum > nub$ **then**
11         $\lambda = \lambda + 0.1$;
12   **else**
13         $\lambda = \lambda$ - 0.1;
14     $errorFlag = 1$;
15   goto Line 3;
16 **if** $errorFlay$==*1* **then**
17   insert the corresponding feature values and *s-threshold*λ of q into the $errorList$;
18   **if** *the error number of the* $errorList > etNum$ **then**
19   retrain the ELM classifier with the new samples in $errorList$;

Then, if the number of results is unacceptable, the threshold will be modified and the probabilistic pruning and refinement algorithms will be re-executed (lines 8-15). At last, if the classification of the current query generates a *non-s-threshold*, FSVR will check whether the ELM classifiers should be retrained (lines 16-19).

5 Experimental Evaluation

5.1 Experiment Setup

In this section, we compare our ELM classification based algorithm with other probabilistic threshold query algorithms in a simulated environment[12]. The examined algorithms are: (1) TPT, which is the traditional PTNNQ algorithm that the thresholds are input by users. (2) TCSVM, which follows our framework but using different classier SVM. (3) TCELM, which is our proposed algorithm as illustrated in Algorithm 2. (4) sTCELM, which is the static version of TCELM. That is, ELM classifier are never retrained even if a number of errors occur.

It's worth mentioning that the three phases, *filtering, verification* and *refinement*, are required by all the above algorithms. And the algorithms proposed in [2] are applied. All the algorithms are implemented in C++ and complied with GNN GCC. The hardware platform is a IBM X3500 server with 2 Quad Core 1.3GHz CPUs and 16GB memory under linux (Red Hat 4.1.2-42). The parameters used in our experiments are: (1) The number of uncertain objects. The values are 20k, 40k, 60k, 80k, 100k, and 60k is the default one. (2) The maximum number of instances per object. The values are 50, 100, 200, 400, and 100 is the default one. (3) The number of updates (deleted, added or moved). The values are 200, 400, 600, 800, and 600 is the default one. For each setting, 2000 PTNNQs are issued, and the response time are averaged. In addition, data objects are updated randomly and the length of the grid index is set to 200. The size of training data set is 2000 and 50 hidden layer nodes are used for ELM.

5.2 Evaluation Results

Figure 2 illustrates the total numbers of modifications, that made by algorithms for 2000 NN queries under different numbers of uncertain objects. It is interesting to see that the number of objects has little effect on the number of modifications. This's because that for the TPT algorithms, the threshold and the *number range* are both input randomly, so whether the threshold is a *s-threshold* or not is independent on the number of objects. And for the threshold classification based TCSVM and TCELM, their classification accuracy have high stability with the growth of the number of objects, which makes the number of the required modifications keep invariable. It can also be seen that the numbers of modifications of TCSVM and TCELM are far smaller than TPT, which shows that the threshold classification based algorithms have more greater advantage in the threshold setting problem. TCELM needs more smaller modifications than TCSVM, which shows that the classification accuracy of ELM is higher than SVM.

Figure 3 illustrates the response time of the algorithms with different number of uncertain objects. The response time increases as the number of objects gets larger, this's because more objects mean more required calculation operations on distances and probabilities, which brings more expensive cost. The response time of TCSVM and TCELM is far less than TPTs, this's because the required modifications of TCSVM and TCELM are far smaller than TPT, such that the numbers of recalling *verification* and *refinement* are smaller. So, a large amount of recalling time is saved in the query process.

Figure 4 illustrates the response time with regard to the maximum number of instances per object. It can be seen that the response time increases as the number of instances gets larger. Though the number of instances has little effect on the threshold classification algorithm, the larger number of instances may bring more calculations of distances and probabilities, which leads to the increase of the whole response time. The response time of TCSVM and TCELM increase more smoothly than TPT, due to that the smaller number of modifications save a large amount of recalling time.

Figure 5 investigates the effect of the number of updates of objects within 2000 queries on sTCELM and TCELM. It can be seen that the modifications of sTCELM increase as the number of updates gets larger, while TCELM is

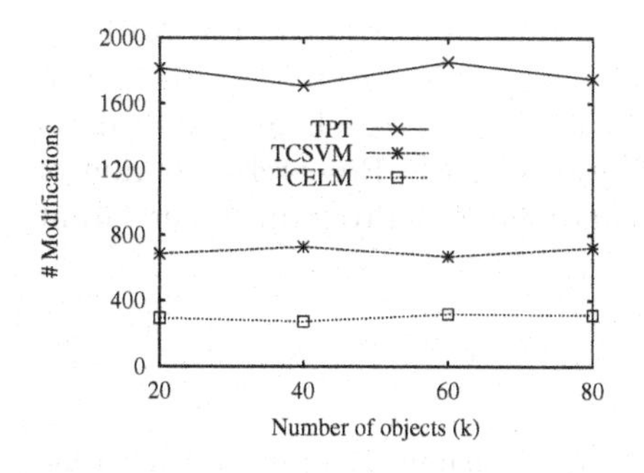

Fig. 2. Number of modifications vs. Number of objects

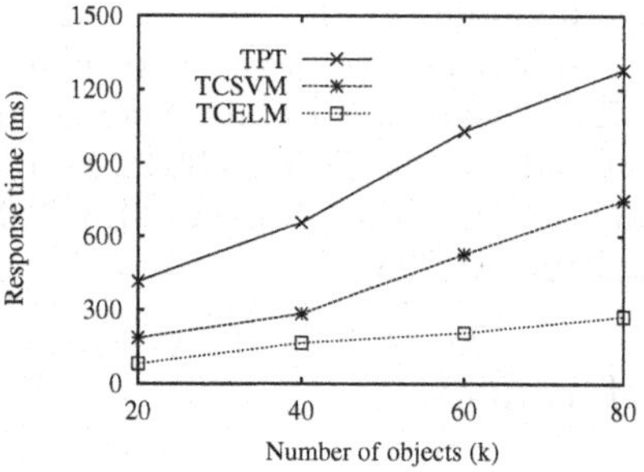

Fig. 3. Response time vs. Number of objects

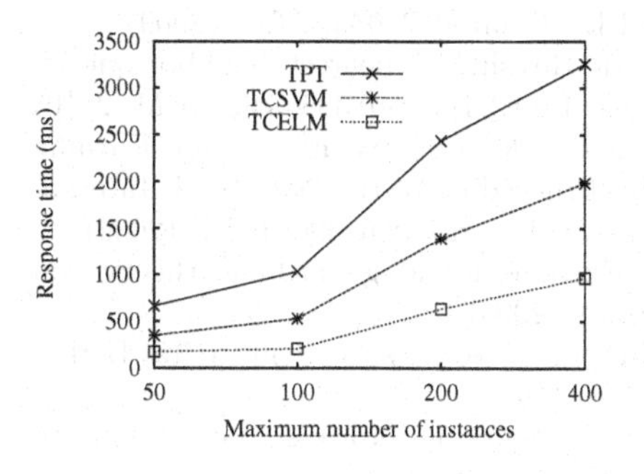

Fig. 4. Response time vs. Maximum number of instances per object

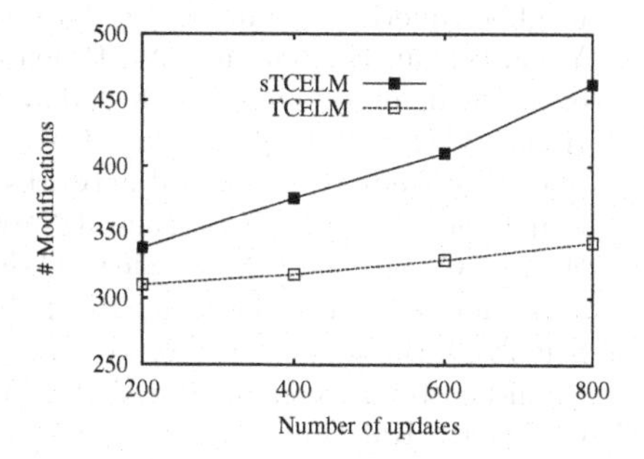

Fig. 5. Number of modifications vs. Number of updates of objects

not sensitive to the updates. This's because when a larger number of updates occur, the ELM classifier may not be applicable to the new environment any longer. sTCELM never retrains the ELM classifier and the classification accuracy decreases as updates increase, while TCELM retrains the ELM classifier based on a counting method which increases the classification accuracy.

6 Conclusion

In this paper, we address the threshold setting problem for the widely used *probabilistic threshold queries* (PTQs). A new framework, which is based on threshold classification method and can be used to answer any type of PTQs, is proposed. In particular, a threshold classification algorithm using ELM is proposed to set a *suitable threshold* for queries. Additionally, in order to adapt to the changes of environment, a dynamic classification strategy is proposed to retrain ELM classifier dynamically. The results show that the thresholds set by ELM classifier needs smaller modifications than those the user input directly.

For future work, we will study the evaluation of other PTQs utilizing the proposed framework, e.g., reverse nearest neighbor query.

Acknowledgement. This research was partially supported by the National Natural Science Foundation of China under Grant No. 61173030, 61272181, 61272182; and the Public Science and Technology Research Funds Projects of Ocean Grant No.201105033; and the National Basic Research Program of China under Grant No.2011CB302200-G; and the 863 Program under Grant No.2012AA011004.

References

1. Cheng, R., Chen, J., Mokbel, M., Chow, C.Y.: Probabilistic verifiers: Evaluating constrained nearest-neighbor queries over uncertain data. In: ICDE, pp. 973–982. IEEE (2008)
2. Cheng, R., Chen, L., Chen, J., Xie, X.: Evaluating probability threshold k-nearest-neighbor queries over uncertain data. In: EDBT, pp. 672–683. ACM (2009)
3. Yang, B., Lu, H., Jensen, C.S.: Probabilistic threshold k nearest neighbor queries over moving objects in symbolic indoor space. In: EDBT, pp. 335–346. ACM (2010)
4. Huang, G.-B., Zhu, Q.-Y., Siew, C.-K.: Extreme learning machine: a new learning scheme of feedforward neural networks. In: Proceedingsof the 2004 IEEE International Joint Conference on Neural Networks, vol. 2, pp. 985–990. IEEE (2004)
5. Huang, G.-B., Zhu, Q.-Y., Siew, C.-K.: Extreme learning machine: theory and applications. Neurocomputing 70(1), 489–501 (2006)
6. Schölkopf, B.: Support Vector LearningPhD thesis, Technischen Universität Berlin. Published by: R. Oldenbourg Verlag, Munich (1997)
7. Cheng, R., Kalashnikov, D.V., Prabhakar, S.: Querying imprecise data in moving object environments. TKDE 16(9), 1112–1127 (2004)
8. Cheng, R., Xie, X., Yiu, M.L., Chen, J., Sun, L.: Uv-diagram: A voronoi diagram for uncertain data. In: ICDE, pp. 796–807. IEEE (2010)
9. Beskales, G., Soliman, M.A., Ilyas, I.F.: Efficient search for the top-k probable nearest neighbors in uncertain databases. VLDB 1(1), 326–339 (2008)
10. Cheema, M.A., Lin, X., Wang, W., Zhang, W., Pei, J.: Probabilistic reverse nearest neighbor queries on uncertain data. TKDE 22(4), 550–564 (2010)
11. Bernecker, T., Emrich, T., Kriegel, H.P., Renz, M., Zankl, S., Züfle, A.: Efficient probabilistic reverse nearest neighbor query processing on uncertain data. VLDB 4(10), 669–680 (2011)
12. Li, J., Wang, B., Wang, G.: Efficient probabilistic reverse k-nearest neighbors query processing on uncertain data. In: Meng, W., Feng, L., Bressan, S., Winiwarter, W., Song, W. (eds.) DASFAA 2013, Part I. LNCS, vol. 7825, pp. 456–471. Springer, Heidelberg (2013)
13. Huang, G.-B., Chen, L.: Letters: Convex incremental extreme learning machine. Neurocomputing 70(16-18), 3056–3062 (2007)
14. Huang, G.-B., Chen, L.: Enhanced random search based incremental extreme learning machine. Neurocomputing 71(16-18), 3460–3468 (2008)
15. Huang, G.-B., Ding, X., Zhou, H.: Optimization method based extreme learning machine for classification. Neurocomputing 74(1-3), 155–163 (2010)
16. Huang, G.-B., Zhou, H., Ding, X., Zhang, R.: Extreme learning machine for regression and multiclass classification. IEEE Transactions on Systems, Man, and Cybernetics, Part B: Cybernetics 42(2), 513–529 (2012)
17. Nievergelt, J., Hinterberger, H., Sevcik, K.C.: The grid file: An adaptable, symmetric multikey file structure. TODS 9(1), 38–71 (1984)

Sample-Based Extreme Learning Machine Regression with Absent Data

Hang Gao, Xinwang Liu, and Yuxing Peng

Science and Technology on Parallel and Distributed Processing Laboratory
National University of Defense Technology, Changsha, China, 410073
{hanggao1821,1022xinwang.liu}@gmail.com,
pengyuxing@aliyun.com

Abstract. Regression is of great importance in machine learning community. As one of state-of-the-art algorithms on this regard, extreme learning machine (ELM) has been received intensive attention and successfully applied into regression tasks. However, existing ELM regression algorithms cannot effectively handle the problem of absent data, which is relatively common in practical applications. In this paper, we propose a sample-based ELM regression algorithm to tackle this issue. The corresponding optimization problem is reformulated as a convex one, which can be readily implemented via off-the-shelf optimization packages. We conduct comprehensive experiments on synthetic and UCI benchmark datasets to compare the proposed algorithm with the widely used imputation approaches, including zero-filling and mean-filling. As shown, our algorithm demonstrates superior performance over the compared ones, especially when the absent ratio is relatively intensive.

Keywords: Extreme learning machine, Absent data regression, ϵ-insensitive regression.

1 Introduction

As a promising unified learning framework, extreme learning machine (ELM) has been an active research topic in the last few years[1]. It was originally developed for single hidden-layer feedforward neural networks (SLFNs), and achieves better generalization than conventional neural networks[2]. With the advantage of high efficiency, easy-implementation and universality, ELM has been successfully applied in many real world applications[3–5]. As for regression, a fundamental problem in machine learning, ELM provides general model for standard setting, it achieves better predictive accuracy than traditional SLFNs[6]. In addition, many variants and extensions were proposed. Following is a brief review. New Fuzzy ELM (NF-ELM) takes the difference of data importance into account, makes ELM less sensitive to outliers[7]. Considering irrelevant and correlate data, optimally pruned ELM (OP-ELM) using pruned neurons leads to a more robust algorithm[8]. Systematic two-stage algorithm (TS-ELM) was introduced for regression to determine the network structure of basic ELM[9]. MapReduce-based Parallel ELM was designed for large scale regression[10]. Based on the

J. Cao et al. (eds.), *Proceedings of ELM-2014 Volume 1*,
Proceedings in Adaptation, Learning and Optimization 3, DOI: 10.1007/978-3-319-14063-6_8

basic ELM, multiple regression machine system(MRMS) algorithm is aim to establish the soft sensor for overcoming shortage of soft sensors in practical production process[11]. Although there are many progress and extensions for ELM regression, ELM cannot be applied directly in absent data regression. Nowadays, due to equipment corruption or limit, missing values and unobserved data are ubiquitous in regression analysis[12]. Widely used treatments are imputation and omitting. Both of them preprocess the absent data and eliminate incompleteness, then use standard regression algorithms in complete dataset. This can be harmful when some useful information be deleted or inaccurate value be introduced by imputation[13]. In this paper, we propose Sample-based ELM (S-ELM) algorithm. As an extension of ELM, it can be directly applied in absent data regression. We compare S-ELM and common imputation approaches in synthetic and real world datasets. As experiments show, S-ELM achieves better predictive accuracy, especially when absent ratio is relatively intensive. This paper is organized as follows. Section 2 discuss the absent data problem in regression and gives a brief review of ELM regression. Section 3 proposed sample-based ELM regression algorithm to handle absent data in regression. Performance evaluations are presented in Section 4. Conclusions are given in Section 5.

2 Related Works

In this section, we give a brief explanation of absent data problem and illustrate its consequences in regression. Then, ELM for ϵ-insensitive regression is introduced for the preliminary of latter proposed algorithm.

2.1 The Problem of Absent Data in Regression

Nowadays, with ever increasing data velocity and volume, absent data are commonly encountered in practice[14]. Absence means that there are some unobserved value in datasets. In general, there are two types of absence, namely absent feature and absent target. In this paper, we focus on the former one. Besides, both training and testing phase are in our considerations. Absence can occur when data collecting equipment malfunction. For example, sensors in a remote sensor network may be damaged and cease to transmit data. Certain regions of a gene microarray may fail to yield measurements of the underlying gene expressions due to scratches, finger prints or dust[14]. Regression is widely used for estimating the relationships among variables. Standard regression assumes estimation are based on complete dataset (i.e. all the samples' feature are complete). Unfortunately, this kind of completeness is not always guaranteed in real application. Standard regression algorithms cannot be directly applied on absent datasets[15]. For dealing with this problem, there are three types of treatments[13]. First is simply omitting the samples with absent features, then carrying out standard regressions. Although the absent sample are not complete, its remaining features contain some useful information for regression. So omitting leads to error ignorance[16]. The second type of treatmeant is imputation, which is also conducted before standard regression. Imputation can be done in several ways[12]. A simple way is filling the absent features with some default value such as zero or mean value of other samples. Another complex way is using some probabilistic generative models to find most reasonable completion. Obviously, imputation methods introduce extra processing steps before regression and worse still the estimation deviation may result in a certain degree of prediction error and have side effect on regression accuracy. The third type of treatment is extending standard regression algorithm[13]. Different from first two, this elegant approach neither fill absent features with hypothetical value nor delete incomplete samples. It makes sure no useful information lost and no extra error introduced before regression. Our way follows last approach.

2.2 ELM for ϵ-Insensitive Regression

ELM, proposed by Huang, was originally invented for single hidden layer feed forward neural networks (SLFNs). ELM tends to reach the minimum training error as well as the minimum generalization error[2]. It has been successfully applied in many domains including regression and classification[3, 17]. Its' general optimization formula is as

$$\min_{\boldsymbol{\beta}} ||\boldsymbol{\beta}||_p^{\sigma_1} + C \times ||h(\boldsymbol{x})\boldsymbol{\beta} - \boldsymbol{y}||_q^{\sigma_2}, \tag{1}$$

where $\boldsymbol{\beta}$ is output weight vector, $\sigma_1 > 0$, $\sigma_2 > 0$, p,q=0,1/2,1,2, $h(.)$ is mapping function, C is the regularization parameter which trades off the training error and the norm of output weight. With different types of constraints (i.e. equality and inequality constraints) and different value of parameters in Eq.(1), ELM has variety of formulations. Inspired by Vapnik's epsilon insensitive loss function, [18] proposed ϵ-insensitive ELM. Its' optimization formula is as follows,

$$\min_{\boldsymbol{\beta}} \left(\frac{1}{2} \times ||\boldsymbol{\beta}||^2 + \frac{C}{2} * ||h(\boldsymbol{x})\boldsymbol{\beta} - \boldsymbol{y}||_\varepsilon^2 \right), \tag{2}$$

where ϵ is insensitive factor and the error loss function is calculated by Eq.(3)

$$\begin{aligned} ||h(\boldsymbol{x})\boldsymbol{\beta} - \ y||_\varepsilon{}^2 &= \sum_{i=1}^{n} |f(x_i) - y_i|_\varepsilon{}^2, \\ with \ \ |f(x_i) - y_i|_\varepsilon &= \max\{0, |f(x_i) - y_i| - \varepsilon\} \end{aligned} \tag{3}$$

Compared with the basic ELM regression, ELM with ϵ-insensitive loss function is less sensitive to different levels of noise[18]. It uses margin ϵ to measure the empirical risk. ELM not only minimizes the training error but also controls the sparsity of the solution. Meanwhile, with ϵ-insensitive loss function, regression are not so sensitive to noisy data and outliers[18]. However, ELM with ϵ-insensitive loss function cannot handle the issue of absent data. In specific, as can be seen in Eq.(2), sample vectors are assumed complete and required to have corresponding components with output weight vector. Next, we present our extension of ELM to tackle this issue.

3 Proposed Sample-Based ELM Regression

In this section, we start by explaining the sample-based subspaces. Based on this, we define the Sample-based ELM for ϵ-insensitive regression and explain its rationality in specific. After that, we reformulate its optimization formulation as a convex one and solved it.

3.1 S-ELM Formulation

In basic ELM regression, each sample can be seen as a vector. Each component of vector is correspondent to a sample feature[2]. In the standard setting, all samples' vectors have complete components and lie in the same space. While in the case of absent data, sample vectors with absent components lie in their own relevant subspaces. Considering the absent case in standard ELM optimization, absence only affects the optimization through inner product with β. Intuitively, the intuitive way is merely skipping the absent components. However, the difference of sample-based output weight vector is not considered. Referring to Eq.(4), this method is equivalent to consider minimizing output weight in full space while constraints are in subspaces.

In this paper, we propose that the output weight vectors in optimization are sample-based and should be normalized in their own relevant subspaces. The core of our proposed S-ELM is that the generalization performance is guaranteed by the minimization of $\boldsymbol{\beta}_i$ rather than $\boldsymbol{\beta}$. Then, the S-ELM's optimization formula can be derived as Eq.(4). In this paper, we use 1-norm constraint ϵ-insensitive. Actually, 2-norm formulation can be induced similarly.

$$\begin{aligned} \min_{\boldsymbol{\beta}^{(i)},\xi_i,\xi_i^*} \quad & \frac{1}{2}\|\boldsymbol{\beta}^{(i)}\|^2 + C\sum_{i=1}^{n}(\xi_i + \xi_i^*) \\ s.t. \quad & \mathbf{y}_i - {\boldsymbol{\beta}^{(i)}}^T \mathbf{x}_i \leq \epsilon + \xi_i \\ & {\boldsymbol{\beta}^{(i)}}^T \mathbf{x}_i - \mathbf{y}_i \leq \epsilon + \xi_i^* \end{aligned} \tag{4}$$

In Eq.(4), different from basic ELM, it substitutes the basic output weight vector $\boldsymbol{\beta}$ with sample-based output weight vector $\boldsymbol{\beta}_i$. The components of $\boldsymbol{\beta}_i$ are taken from $\boldsymbol{\beta}$ and its norm is calculated by Eq.(5).

$$||\boldsymbol{\beta}^{(i)}|| = \sum_{p=1}^{m_i} \beta_p^2, \tag{5}$$

where β_p is the pst component of $\boldsymbol{\beta}^{(i)}$, m_i is the number of features sample i contains. In line with basic ELM, samples without absent feature reside in full space and their corresponding $\boldsymbol{\beta}^{(i)}$ are actually $\boldsymbol{\beta}$.

3.2 S-ELM Optimization

Since the variance of different $\boldsymbol{\beta}^{(i)}$s, it is not feasible to solve Eq.(4) by using standard method. With the observation of relation between $\boldsymbol{\beta}$ and $\boldsymbol{\beta}^{(i)}$, we manage to remove this variance. First, with the observation that minimizing all $\boldsymbol{\beta}^{(i)}$ is equal to minimizing the maximum of all $\boldsymbol{\beta}^{(i)}$, we change the formulation of Eq.(4) as follows,

$$\begin{aligned} \min_{\beta^{(i)},\xi_i,\xi_i^*} \quad & (\max_i \frac{1}{2}||\boldsymbol{\beta}^{(i)}||^2 + C\sum_{i=1}^{n}(\xi_i + \xi_i^*)) \\ s.t. \quad & \mathbf{y}_i - {\boldsymbol{\beta}^{(i)}}^T \mathbf{x}_i \leq \epsilon + \xi_i \\ & {\boldsymbol{\beta}^{(i)}}^T \mathbf{x}_i - \mathbf{y}_i \leq \epsilon + \xi_i^* \end{aligned} \tag{6}$$

Next, we introduce an absent matrix $\mathbf{S}$. Each of its element $\mathbf{S}(i,p)$ represents whether the $\mathbf{x}_i$'s pst component is absent or not. 0 represents absent while 1 represents not absent. Then, we use an auxiliary variable θ as follows,

$$\theta = \max_i \frac{1}{2}||\boldsymbol{\beta}^{(i)}||^2 = \max_i \frac{1}{2}\sum_{p=1}^{m}\mathbf{S}(i,p)\beta_p. \tag{7}$$

where m is the number of complete features. Then, Eq.(6) is rewritten as Eq.(8).

$$\begin{aligned} \min_{\beta^{(i)},\xi_i,\xi_i^*} \quad & \theta + C\sum_{i=1}^{n}(\xi_i + \xi_i^*) \\ s.t. \quad & \mathbf{y}_i - {\boldsymbol{\beta}^{(i)}}^T \mathbf{x}_i \leq \epsilon + \xi_i \\ & {\boldsymbol{\beta}^{(i)}}^T \mathbf{x}_i - \mathbf{y}_i \leq \epsilon + \xi_i^* \\ & \theta \geq \max_i \frac{1}{2}\sum_{p=1}^{m}\mathbf{S}(i,p)\beta_p \end{aligned} \tag{8}$$

Finally, S-ELM formulation is a convex optimization. It can be conveniently solved by off the shelf convex optimization tools. In this paper, we use cvx[19].

3.3 Discussion

To further explain S-ELM, we discuss the differences among the proposed S-ELM and a widely used imputation method zero-filling. The main difference is the way of minimizing output weight vector. Zero-filling impute the absent value with zero and optimizes $\boldsymbol{\beta}$ for all samples. Obviously, if the unobserved value of absent feature is seriously derived from zero, zero-filling introduces considerable error. While S-ELM take into account the variance of output weights, it minimize the maximum of sample-based $\boldsymbol{\beta}^{(i)}$. This way will not introduce extra error and make training and prediction more accurate. For the same reason, S-ELM is different from other imputation methods. Next, we demonstrate it in our experiments.

4 Experiments

All our simulations are implemented on MATLAB 2013b environment running in Core(TM) 3.0 GHz CPU and 8 GB RAM. Before each algorithm run, there are some preprocessing. The UCI benchmark datasets are normalized to [0, 1]. For synthetic datasets, all independent variables' domain are in [0, 1], and their intervals are randomly different. Both the synthetic datasets and UCI datasets are permutated randomly and were spitted into training data and testing data ($\frac{\#train}{\#test}$ equals to $\frac{3}{2}$). For each algorithm and each dataset, ε is set to be 0.01 empirically. The regularization factor C is chosen by five folds cross validations for each dataset. Root mean square error (RMSE) is used as the criterion to evaluate to prediction accuracy of algorithms. It is calculated as Eq.(9).

$$RMSE = \sqrt[2]{\frac{\sum_{t=1}^{n}(\widehat{y_t} - y_t)}{n}} \tag{9}$$

In order to eliminate randomness, final RMSEs are the average of fifty times repetitions. We also calculate the standard deviation to show the stability of algorithms.

In the end of this subsection, we explain how to generate the absent matrix S in Eq.(4). At first, S is a matrix with all its elements are set to be one. The dimension of S is the same with X on the training data. Then, according to absent ratio, we randomly choose the absent elements and set them to zero. Note that there is no single row or column to be set all zeroes. Besides, absent matrix on testing data are generated similarly.

4.1 Synthetic Dataset

To illustrate performance of S-ELM, we considered four synthetic datasets produced by four basic math functions list as Table. 1. As Fig.1 shows, It can be observed that S-ELM achieves better prediction accuracy in different absent ratio. Specifically, S-ELM performs better than ZF-ELM, which proofs that S-ELM reduces generalization error by using sample-based output weight vector norm. Compared with MF-ELM, S-ELM wins in most cases. A small case is MF-ELM predicts more accurate than S-ELM. The reason is mean value sometimes close to absent value. However, in case of higher absent ratio, S-ELM beats other two methods.

4.2 Realworld Benchmark Dataset

To further illustrate the advantage of S-ELM, we use UCI benchmark datasets for comparison. Table 2 specifies the UCI benchmark datasets used in our

Table 1. Synthetic datasets and corresponding functions

Dataset	Function
Synthetic dataset1	$f_1(x) = x_1 + x_2 + x_3 + x_4 + x_5 + x_6$
Synthetic dataset2	$f_2(x) = x_1 \times x_2 \times x_3 + x_2 \times x_3 \times x_4 + x_3 \times x_4 \times x_5 + x_4 \times x_5 \times x_6$
Synthetic dataset3	$f_3(x) = 2^{x_1} + 2^{x_2} + 2^{x_1} + 2^{x_4} + 2^{x_1} + 2^{x_6}$
Synthetic dataset4	$f_4(x) = \frac{x_1+x_2+x_3}{x_4+x_5+x_6}$

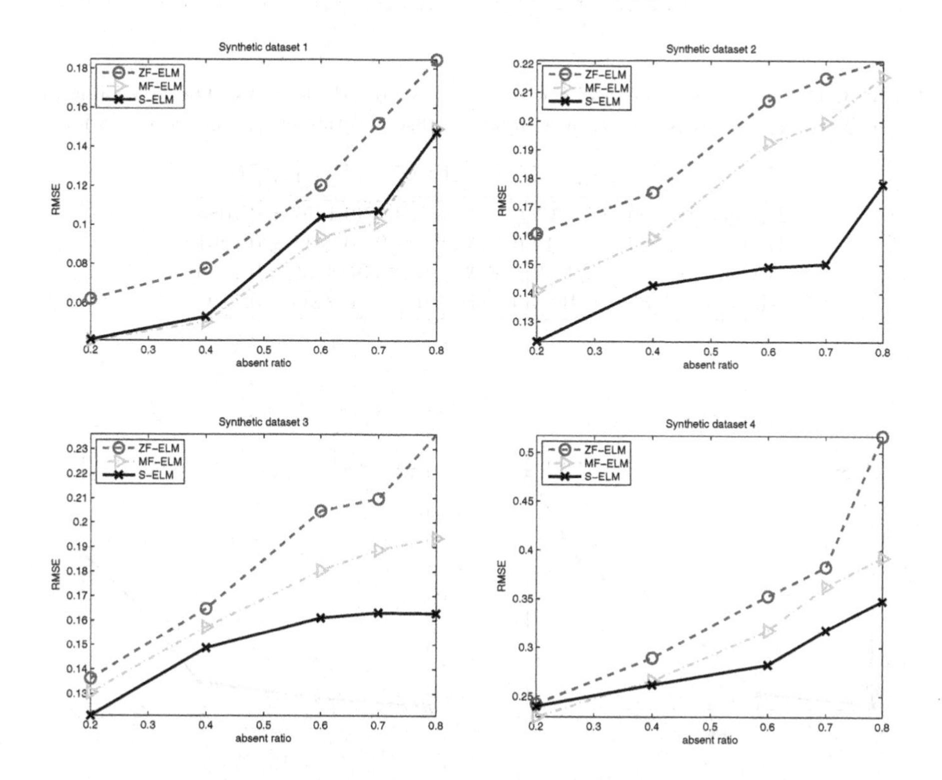

Fig. 1. RMSE comparisons on the synthetic datasets

experiments. Those four datasets are varied in the number of samples and features. Mean RMSE and standard dev are reported in Table 3. We can see S-ELM performs better than zero-filling and mean-filling imputation methods in mean RMSE and dev over different absent ratio. As Fig.2 shows, with the absent ratio increasing, S-ELM's prediction accuracy decreases mildly. Especially in Pyrim dataset, when the number of samples is relatively small, imputation methods are more inaccurate.

Table 2. The UCI benchmark datasets used in our experiments

Dataset	# train	# test	# features
Bodyfat	152	100	14
Housing	304	202	13
Pyrim	45	29	27
Abalone	2507	1670	8

Table 3. Performance comparisons with UCI benchmark datasets. The two value of each cell represents mean RMSE over different absent ratios and standard deviation.

	S-ELM	ZF-ELM	MF-ELM
Bodyfat	0.1299 ± 0.0905	0.1781 ± 0.1007	0.1457 ± 0.1088
Housing	0.2229 ± 0.1019	0.3234 ± 0.1697	0.2511 ± 0.1261
Pyrim	0.3816 ± 0.0842	0.5603 ± 0.2253	0.4537 ± 0.1396
Abalone	0.2336 ± 0.1083	0.3004 ± 0.1537	0.2643 ± 0.1193

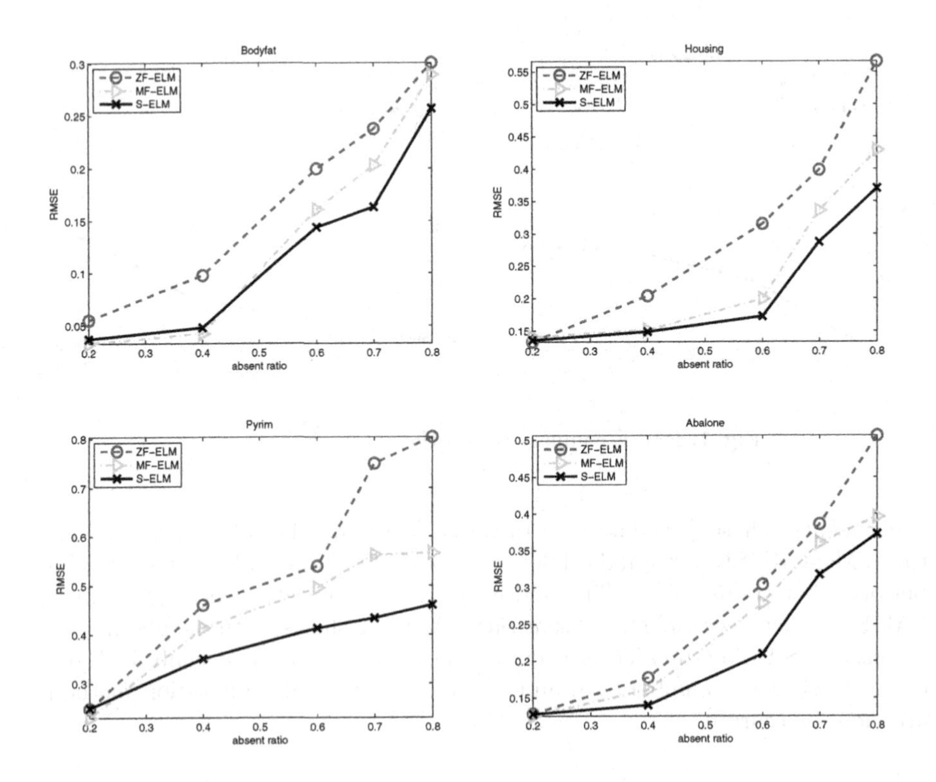

Fig. 2. RMSE comparisons on the UCI benchmark datasets

5 Conclusion

In this paper, we propose S-ELM algorithm in order to address the problem of absent data in regression. Specifically, we extend the ϵ-insensitive ELM regression by sample-based way and transform its formulation to a convex optimization. As the experiments show, S-ELM achieves better performance than widely used imputation methods, especially when the absent ratio is relatively intensive. Many following work are worth exploring. Next, we are going to extend S-ELM with kernel method or some mapping space.

Acknowledgement. This work is supported by the Major State Basic Research Development Program of China (973 Program) under the Grant No.2014CB340303, and the National High Technology Research and Development Program of China (863 Program) under Grant No.2013AA01A213.

References

1. Huang, G.-B., Wang, D.H., Lan, Y.: Extreme learning machines: a survey. International Journal of Machine Learning and Cybernetics 2(2), 107–122 (2011)
2. Huang, G.-B., Zhu, Q.-Y., Siew, C.-K.: Extreme learning machine: a new learning scheme of feedforward neural networks. In: Proceedings. 2004 IEEE International Joint Conference on Neural Networks, vol. 2, pp. 985–990. IEEE (2004)
3. Huang, G.-B., Zhu, Q.-Y., Siew, C.-K.: Extreme learning machine: theory and applications. Neurocomputing 70(1), 489–501 (2006)
4. Liu, Q., Yin, J., Leung, V.C., Zhai, J.-H., Cai, Z., Lin, J.: Applying a new localized generalization error model to design neural networks trained with extreme learning machine. Neural Computing and Applications, 1–8 (2014)
5. Xie, P., Liu, X., Yin, J., Wang, Y.: Absent extreme learning machine algorithm with application to packed executable identification. Neural Computing and Applications, 1–8 (2013)
6. Huang, G.-B., Zhou, H., Ding, X., Zhang Extreme, R.: learning machine for regression and multiclass classification. IEEE Transactions on Systems, Man, and Cybernetics, Part B: Cybernetics 42(2), 513–529 (2012)
7. Zheng, E., Liu, J., Lu, H., Wang, L., Chen, L.: A new fuzzy extreme learning machine for regression problems with outliers or noises. In: Motoda, H., Wu, Z., Cao, L., Zaiane, O., Yao, M., Wang, W. (eds.) ADMA 2013, Part II. LNCS, vol. 8347, pp. 524–534. Springer, Heidelberg (2013)
8. Miche, Y., Sorjamaa, A., Bas, P., Simula, O., Jutten, C., Lendasse, A.: Op-elm: optimally pruned extreme learning machine. IEEE Transactions on Neural Networks 21(1), 158–162 (2010)
9. Lan, Y., Soh, Y.C., Huang, G.-B.: Two-stage extreme learning machine for regression. Neurocomputing 73(16), 3028–3038 (2010)
10. He, Q., Shang, T., Zhuang, F., Shi, Z.: Parallel extreme learning machine for regression based on mapreduce. Neurocomputing 102, 52–58 (2013)
11. Chang, Y., Wang, S., Tian, H., Zhao, Z.: Multiple regression machine system based on ensemble extreme learning machine for soft sensor. Sensor Letters 11(4), 710–714 (2013)

12. Royston, P.: Multiple imputation of missing values. Stata Journal 4, 227–241 (2004)
13. Chechik, G., Heitz, G., Elidan, G., Abbeel, P., Koller, D.: Max-margin classification of data with absent features. The Journal of Machine Learning Research 9, 1–21 (2008)
14. Marlin, B.M.: Missing data problems in machine learning. Thesis (2008)
15. Horton, N.J., Lipsitz, S.R.: Multiple imputation in practice: comparison of software packages for regression models with missing variables. The American Statistician 55(3), 244–254 (2001)
16. Yuan, Y.C.: Multiple imputation for missing data: Concepts and new development (version 9.0). SAS Institute Inc., Rockville (2010)
17. Wang, Y., Cao, F., Yuan, Y.: A study on effectiveness of extreme learning machine. Neurocomputing 74(16), 2483–2490 (2011)
18. Balasundaram, S.: On extreme learning machine for -insensitiive regression in the primal by newton method. Neural Computing and Applications 22(3-4), 559–567 (2013)
19. Inc. CVX Research. CVX: Matlab software for disciplined convex programming, version 2.0 (August. 2012), http://cvxr.com/cvx

Two Stages Query Processing Optimization Based on ELM in the Cloud

Linlin Ding[1], Yu Liu[1], Baoyan Song[1], and Junchang Xin[2]

[1] School of Information, Liaoning University, Shenyang Liaoning, China, 110036
{dinglinlin,bysong}@lnu.edu.cn
[2] College of Information Science & Engineering, Northeastern University, Shenyang Liaoning, China, 110819

Abstract. As one variant of MapReduce framework, ComMapReduce adds the lightweight communication mechanisms to improve the performance of query processing programs. Although the existing research work has already solved the problem of how to identify the communication strategy of ComMapReduce, there are still some drawbacks, such as relative simple model and too much user participation. Therefore, in this paper, we propose a two stages query processing optimization model based on ELM, named ELM to ELM (*E2E*) model. Then, we develop efficient sample training strategy, predicting and execution algorithm to construct the *E2E* model. Finally, extensive experiments are conducted to verify the effectiveness and efficiency of the *E2E* model.

Keywords: ELM, MapReduce, ComMapReduce, Prediction Model.

1 Introduction

Nowadays, MapReduce [1] has emerged as a famous programming framework for big data analysis. MapReduce and its variants are used for big data applications, such as Web indexing, data mining, machine learning, financial analysis [2–5]. As one of a successful improvements of MapReduce, ComMapRedcue [2, 3] adds simple lightweight communication mechanisms to generate the certain *shared information* and executes the query processing applications with large scale datasets in the Cloud. In ComMapReduce framework, three basic and two optimization communication strategies are proposed to illustrate how to communicate and obtain the *shared information* of different applications. During further analyzing ComMapReduce execution course and the abundant experiments, we find out that different communication strategies of ComMapReduce can substantially affect the performance of query processing applications.

The existing work [6] proposes a query processing model named *ELM_CMR* based on ELM [7] which has the classification performance at an excellent performance. *ELM_CMR* can identify the communication strategy of ComMapReduce according to the characteristics of query processing programs. However, *ELM_CMR* still has the following drawbacks. First, to the MapReduce or

J. Cao et al. (eds.), *Proceedings of ELM-2014 Volume 1*,
Proceedings in Adaptation, Learning and Optimization 3, DOI: 10.1007/978-3-319-14063-6_9

ComMapReduce, a program only consists of black_box Map and Reduce functions, without knowing the distributed details about the framework. But the configuration parameters of the framework can fully influence the performance of query processing. Finding the most suitable configuration parameters setting itself is difficult. Second, *ELM_CMR* only adopts simple training method to gain the classification model.

Therefore, in this paper, for solving the above drawbacks, we propose the two stages query processing optimization model based on ELM, named ELM to ELM (*E2E*) model. The first stage gains the *feature parameters* according to the program of users by using ELM algorithm. After that, according to the results of the first stage, the second stage can identify the final classification result by ELM too. Furthermore, an efficient training strategy, predicting and execution algorithm are presented. The contributions of this paper can be summarized as follows.

- We propose an efficient two stages query processing optimization model based on ELM, *E2E* model, which can realize the most optimal executions of query processing programs in MapReduce or ComMapReduce framework.
- We develop a sample training strategy, predicting and execution algorithm to construct the *E2E* model.
- The experimental studies using synthetic data show the effectiveness and efficiency of the *E2E* model.

The remainder of this paper is organized as follows. Section 2 briefly introduces the background, containing the ELM and *ELM_CMR*. Our *E2E* model is proposed in Section 3. The experimental results to show the performance of *E2E* model are reported in Section 5. Finally, we conclude this paper in Section 6.

2 Background

2.1 Review of ELM

Nowadays, Extreme Learning Machine (ELM) [7] and its variants [8–18] have the characteristics of excellent generalization performance, rapid training speed and little human intervene, which have attracted increasing attention from more and more researchers. ELM is originally designed for single hidden-layer feedforward neural networks (SLFNs [19]) and is then extended to the "generalized" SLFNs. ELM algorithm first randomly allocates the input weights and hidden layer biases and then analytically computes the output weights of SLFNs. Contrary to the other conventional learning algorithms, ELM reaches the optimal generalization performance with a very fast learning speed. ELM is less sensitive to the user defined parameters, so it can be deployed fast and convenient.

For N arbitrary distinct samples $(\mathbf{x}_j, \mathbf{t}_j)$, where $\mathbf{x}_j = [x_{j1}, x_{j2}, \ldots, x_{jn}]^T \in \mathbb{R}^n$ and $\mathbf{t}_j = [t_{j1}, t_{j2}, \ldots, t_{jm}]^T \in \mathbb{R}^m$, standard SLFNs with hidden nodes L and activation function $g(x)$ are mathematically modeled as

$$\sum_{i=1}^{L} \beta_i g_i(\mathbf{x}_j) = \sum_{i=1}^{L} \beta_i g(\mathbf{w}_i \cdot \mathbf{x}_j + b_i) = \mathbf{o}_j \qquad (j = 1, 2, \ldots, N) \tag{1}$$

where L is the number of hidden layer nodes, $\mathbf{w}_i = [w_{i1}, w_{i2}, \ldots, w_{in}]^T$ is the weight vector between the ith hidden node and the input nodes, $\beta_i = [\beta_{i1}, \beta_{i2}, \ldots, \beta_{im}]^T$ is the weight vector connecting the ith hidden node and the output nodes, b_i is the threshold of the ith hidden node, and $\mathbf{o}_j = [o_{j1}, o_{j2}, \ldots, o_{jm}]^T$ is the jth output vector of the SLFNs.

The standard SLFNs can approximate these N samples with zero error. The error of ELM is $\sum_{j=1}^{L} ||\mathbf{o}_j - \mathbf{t}_j|| = 0$ and there exist β_i, $\mathbf{w}_i$ and b_i such that

$$\sum_{i=1}^{L} \beta_i g(\mathbf{w}_i \cdot \mathbf{x}_j + b_i) = \mathbf{t}_j \qquad (j = 1, 2, \ldots, N) \tag{2}$$

Equation (2) can be expressed compactly as follows:

$$\mathbf{H}\beta = \mathbf{T} \tag{3}$$

where $\mathbf{H}(\mathbf{w}_1, \mathbf{w}_2, \ldots, \mathbf{w}_L, b_1, b_2, \ldots, b_L, \mathbf{x}_1, \mathbf{x}_2, \ldots, \mathbf{x}_L)$

$$= \begin{bmatrix} h(x_1) \\ h(x_2) \\ \vdots \\ h(x_N) \end{bmatrix} = \begin{bmatrix} g(\mathbf{w}_1 \cdot \mathbf{x}_1 + b_1) & g(\mathbf{w}_2 \cdot \mathbf{x}_1 + b_2) & \ldots & g(\mathbf{w}_L \cdot \mathbf{x}_1 + b_L) \\ g(\mathbf{w}_1 \cdot \mathbf{x}_2 + b_1) & g(\mathbf{w}_2 \cdot \mathbf{x}_2 + b_2) & \ldots & g(\mathbf{w}_L \cdot \mathbf{x}_2 + b_L) \\ \vdots & \vdots & \vdots & \vdots \\ g(\mathbf{w}_1 \cdot \mathbf{x}_N + b_1) & g(\mathbf{w}_2 \cdot \mathbf{x}_N + b_2) & \ldots & g(\mathbf{w}_L \cdot \mathbf{x}_N + b_L) \end{bmatrix}_{N \times L} \tag{4}$$

$$\beta = \begin{bmatrix} \beta_1^T \\ \beta_2^T \\ \vdots \\ \beta_L^T \end{bmatrix}_{L \times m} \quad \text{and} \quad \mathbf{T} = \begin{bmatrix} t_1^T \\ t_2^T \\ \vdots \\ t_N^T \end{bmatrix}_{N \times m} \tag{5}$$

$\mathbf{H}$ is set as the hidden layer output matrix of the neural network. The ith column of $\mathbf{H}$ is called the ith hidden node output with respect to inputs $\mathbf{x}_1, \mathbf{x}_2, \ldots, \mathbf{x}_N$. The smallest norm least-squares solution of the above multiple regression system is shown as follows:

$$\hat{\beta} = \mathbf{H}^{\dagger}\mathbf{T} \tag{6}$$

where $\mathbf{H}^{\dagger}$ is the Moore-Penrose generalized inverse of matrix $\mathbf{H}$. Then the output function of ELM can be modeled as follows.

$$f(\mathbf{x}) = \mathbf{h}(\mathbf{x})\beta = \mathbf{h}(\mathbf{x})\mathbf{H}^{\dagger}\mathbf{T} \tag{7}$$

2.2 *ELM_CMR* Model

ComMapReduce [2, 3] is an improved MapReduce framework with lightweight communication mechanisms. A new node, named the Coordinator node, is added to store and generate the certain *shared information* of different applications. In ComMapReduce, three basic communication strategies, LCS, ECS and HCS, and two optimization communication strategies, PreOS and PostOS are proposed to identify how to receive and generate the *shared information*. In short, without affecting the existing characteristics of the original MapReduce framework, ComMapReduce is a successful parallel programming framework with global *shared information* to filter the unpromising data of query processing programs.

Figure 1 shows the architecture of *ELM_CMR* model. The four components of *ELM_CMR* are respectively the *Feature Selector*, the *ELM Classifier*, the *Query Optimizer* and the *Execution Fabric*.

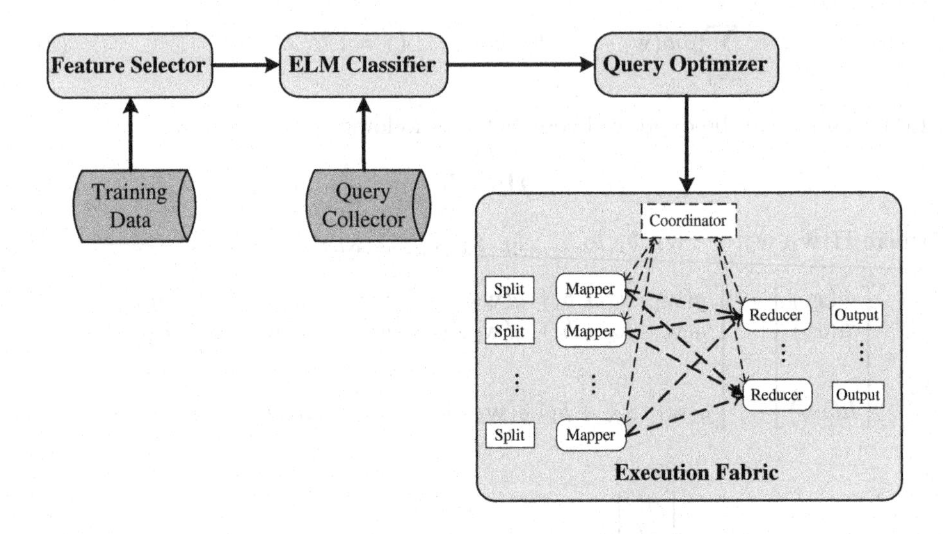

Fig. 1. Architecture of ELM_CMR Model

The *Feature Selector* mainly examines the training query processing programs and selects the configuration parameters that can wholly affect the query performance by the job profiles. Naturally, the parameters of program p can be divided into three types, parameters that predominantly affect Map task execution; parameters that predominantly affect Reduce task execution and the cluster parameters. Then, in each cluster, we adopt the minimum-redundancy-maximum-relevance (mRMR) [20] feature selection to find the optimal parameters sharply affecting the performance. And then, we generate the globally optimal configuration parameter settings by combining the results of each subspace. Therefore, the near-optimal configuration parameter setting can be generated.

After selecting the features of training data, the *Feature Selector* sends the extracted training data to the *ELM Classifier*. It uses the training data to construct the ELM model by the traditional ELM algorithm. After that, when there are one or multiple queries to be processed, the *ELM Classifier* can rapidly obtain the classification results of the queries, and then sends them to the *Query Optimizer*.

The *Query Optimizer* applies the classification results of the *ELM Classifier* and combines the implementation patterns to choose an optimized *execution order*. After gaining the *execution order*, the query is sent to the *Execution Fabric*.

The *Execution Fabric* implements the program in ComMapReduce framework. When there is one query to be processed, the *Execution Fabric* implements the query according to the classification result of the *Query Optimizer* in *ELM_CMR*. When there are multiple queries to be processed, the multiple queries can be classified by *ELM Classifier* and gain the best communication strategy of each program. Then, a Task Scheduler Simulator is used to simulate the execution time of queries. According to the execution time and the classification results of the queries, the *Query Optimizer* designs an *execution order* following the common principle of Shortest Job First (*SJF*) to implement multiple queries.

3 *E2E* Model

3.1 Overview of *E2E* Model

Our *E2E* model can identify the optimal communication strategies of query processing programs in MapReduce or ComMapReduce, which contains three main phases, respectively the *training phase*, the *prediction phase* and the *execution phase*. Figure 2 shows the whole workflow of query processing in the *E2E* model. The main workflow is as follows.

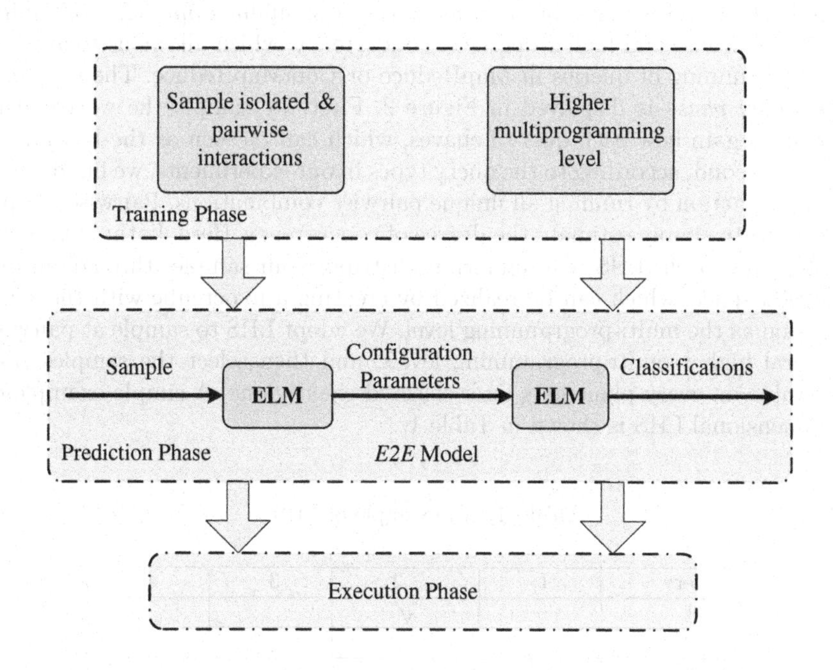

Fig. 2. Workflow of *E2E* Model

First, the *training phase* is responsible for extracting the training query processing programs that have a large affect on the *E2E* model. In the *training phase*, we use sample-based training strategies to run our workload, in isolation, pairwise and at several higher multi-programming levels. After gaining the training samples, they can be used to generate the *E2E* model in the *prediction phase*.

Second, in the *prediction phase*, by using the training samples from the *training phase*, the two stages *E2E* model can be generated based on the traditional ELM algorithm. According to the user's programs, the first stage can obtain the most optimal *feature parameters* of the programs by ELM. And then, using the *feature parameters* of the first stage, the second stage can identify the classification results of the query processing programs by ELM.

Third, after gaining the *E2E* model, for the query processing program submitted to MapReduce or ComMapReduce, we can predict the optimal communication strategy in the *execution phase.*

3.2 Training *E2E* Model

To obtain the prediction model more accurately, we need to train the *E2E* model. Different from the simple training course of *ELM_CMR* model, the *training phase* of our *E2E* model consists of running the queries in isolation, pairwise as well as at several higher multi-programming levels.

The *E2E* model realizes the *training phase* by sampling approach, containing isolation, pairwise and higher degree of concurrency, which allows us to approximate the running of queries in MapReduce or ComMapReduce. The course of the *training phase* is displayed in Figure 2. First, we sample the workload in isolation to gain how each query behaves, which can be seen as the baseline of training. Second, according to the query types in our experiments, we build a matrix of interaction by running all unique pairwise combinations. Pairwise sample can help us to simply estimate the degree of concurrency. Here, Latin hypercube sampling approach (LHS) can uniformly distribute our samples throughout our prediction space, which can be realized by creating a hypercube with the same dimension as the multi-programming level. We adopt LHS to sample at pairwise or several higher multi-programming levels, and then select the samples that every value on every plane gets inter selected exactly one. A simple example of two dimensional LHS is shown in Table 1.

Table 1. An example of LHS

Query	1	2	3	4
1		√		
2			√	
3	√			
4				√

3.3 Predicting *E2E* Model

Then, we introduce the details of the *prediction phase.* A MapReduce job j can be expressed by a MapReduce program p running on input data d and cluster r, which can be expressed as j=$< p, d, r, c >$ in short. We call the d, r, c the *feature parameters* of p. The users only submit their jobs to MapReduce or ComMapReduce without knowing the internal configuration details of the system.

However, a number of choices can be made in order to fully specify how the job should be executed. These choices, represented by c in $< q, d, r, c >$, stand for a high dimensional space of configuration parameter settings, such as the number of Map and Reduce task, the block size, the amount of memory, and so on. The performance changes a lot in different configuration parameters. For any parameter, its value is not specified explicitly during job submission, either the default values shipped with the system or specified by the system administrator. However, the normal users don't understand the running details of MapReduce. That is to say, finding good configuration settings for MapReduce job is time consuming and requires extensive knowledge of system internals. Because the users have little information of the parallelization details of MapReduce, it is necessary to gain the suitable configuration parameters. The first stage of our *E2E* model is to generate the model for identifying the suitable configuration parameters of query processing programs by ELM. The main goal of the first stage is to construct a black_box *feature parameters* of queries, containing $< d, r, c >$. The configuration parameters wholly affecting the performance adopted by *E2E* are the same as our *ELM_CMR* approach. The corresponding information of the job can be obtained from sampling a few tasks of the job. After the first stage, the *feature parameters* setting can be obtained by ELM algorithm. After gaining the configuration parameters, the second stage is to use them to predict the communication strategy of ComMapReduce or MapReduce by ELM algorithm too, and then generates the classification results.

4 Execution of *E2E* Model

When there is one new query being submitted, the *E2E* model generates its classification result as soon as possible. According to the characteristics of the coming query, the first stage of *E2E* model can generate the *feature parameters* of this query based on ELM. After that, the user can make a decision that whether adopting ComMapReduce or adopting which communication strategy by the second stage of *E2E* model, and then implement the query processing application.

The course of execution is shown in Algorithm 1. First, the most optimal *feature parameters* of query processing job j are extracted in the first stage of *E2E* model (Line 1). Second, after obtaining the *feature parameters* of job j, the *E2E* generates the classification result of j (Line 2). Third, according to the classification of j, the *E2E* ensures how to implement the query and

Algorithm 1. Execution of *E2E* Model

1: Generate the *feature parameters* of j by the first stage of *E2E* model;
2: Generate the execution plan of j by the second stage of *E2E* model;
3: Execute j with its communication strategy;

sends it to MapReduce or ComMapReduce framework. The framework uses the optimization result to execute the query program (Line 3).

For example, for a submitted skyline query, the first stage of *E2E* can obtain the most optimal *feature parameters* of this skyline query. After abstracting its *feature parameters*, the *E2E* can generate its classification and then identifies the communication strategy of this skyline query, such as PostOS. After that, the skyline query will be implemented in ComMapReduce with PostOS.

5 Performance Evaluation

5.1 Experimental Setup

The experimental setup is the same as *ELM_CMR* model as follows. The experimental setup is a Hadoop cluster running on 9 nodes in a high speed Gigabit network, with one node as the Master node and the Coordinator node, the others as the Slave nodes. Each node has an Intel Quad Core 2.66GHZ CPU, 4GB memory and CentOS Linux 5.6. We use Hadoop 0.20.2 and compile the source codes under JDK 1.6. The ELM algorithm is implemented in MATLABR2009a.

The data in our experiments are synthetic data. The classification results of *E2E* model contains 7 types, respectively ECS, HCS-0.5, HCS-1, HCS-2, PreOS, PostOS and MapReduce (MR). HCS-0.5 means the preassigned time interval of HCS is 0.5s. We evaluate the performance of *E2E* model by comparing with *ELM_CMR*. Four typical query processing applications are adopted to evaluate the peformance, respectively top-k, kNN, skyline and join.

5.2 Experimental Results

Figure 3 shows the performance of top-k queries (k=1000), with the data size is 2G, 4G, 6G and 8G in uniform distribution. We can see that the performance of top-k queries in the classification results of *E2E* model is better than *ELM_CMR*. The reason is that *E2E* model can effectively evaluate the optimal configuration parameters of the queries than *ELM_CMR* model. When k is much smaller than the original data, the global *shared information* of ComMapReduce can reach the most optimal one quickly, so the Mappers can retrieve the *shared information* in the initial phase to filter the unpromising data.

Figure 4 shows the performance of kNN queries with different data size of uniform distribution. The performance of *E2E* is also optimal to *ELM_CMR* model, where the reason is that *E2E* model can gain the suitable classification results.

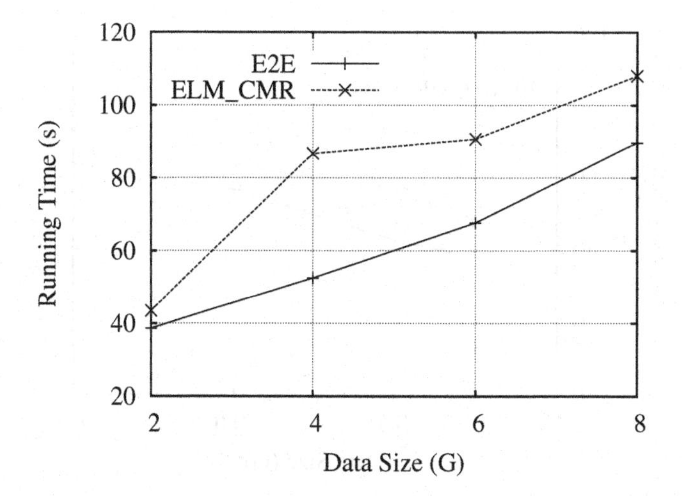

Fig. 3. Performance of top-k Query

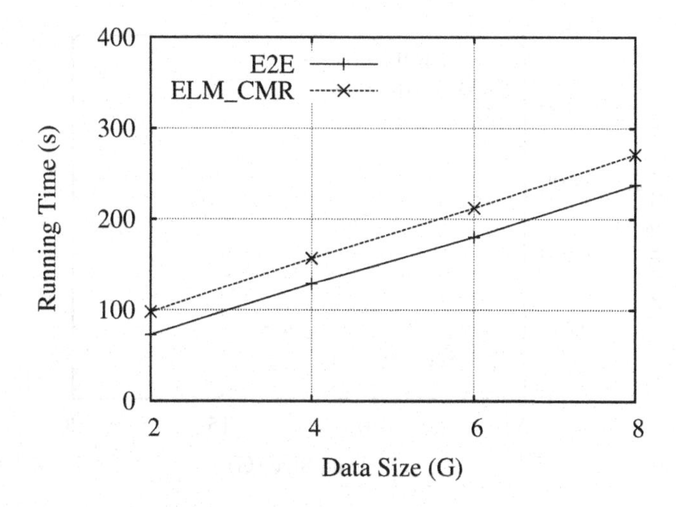

Fig. 4. Performance of kNN Query

Figure 5 shows the performance of skyline queries in anti-correlated distribution with different data size. We can see that the performance of different execution plans is not obviously different, but PostOS is a little better. The reason is that the original data are skewed to the final results in anti-correlated distribution. The percentage of filtering is low, so the performance difference is not obvious. In this situation, although *E2E* and *ELM_CMR* can obtain the classification, it can also choose the other communication strategies.

Figure 6 shows the performance of join queries in different data size of small-big tables, with the same data size of the small table 2G, and the different data

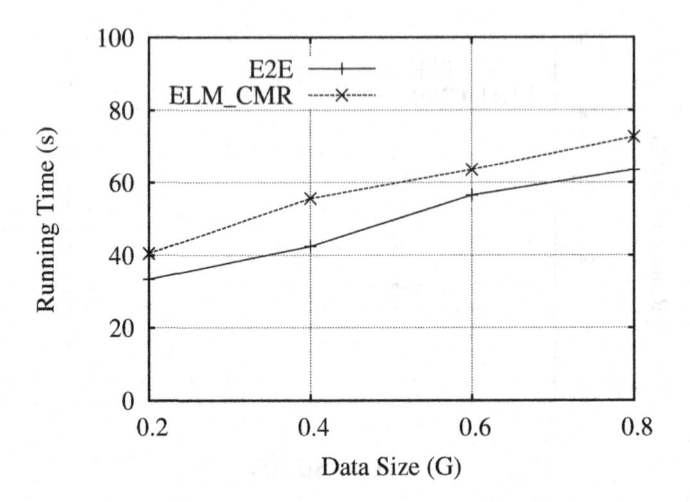

Fig. 5. Performance of Skyline Query

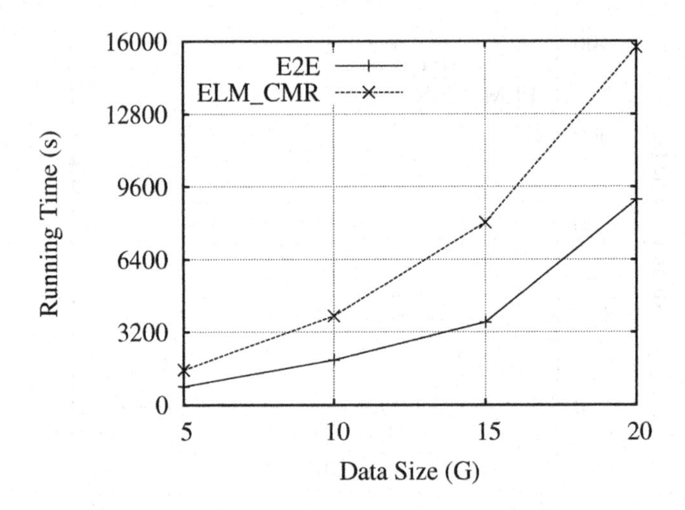

Fig. 6. Performance of join Query

sizes of big table are shown in Figure 6. The performance of *E2E* is much better than *ELM_CMR*. In ComMapReduce, the join attributes of the small table can be set as the *shared information* to filter the unpromising intermediate results.

6 Conclusions

In this paper, we propose an efficient query processing optimization model based on ELM, *E2E* model. Our *E2E* can effectively analyze the query processing applications, and then generates the most optimized executions of query processing applications. After analyzing the problems of the former *ELM_CMR* model, we use two stages model to classify the query processing applications in ComMapReduce framework. Then, we propose an efficient training approach to train our model. We also give the predicting and query executions. The experiments demonstrate that the *E2E* model is efficient and the query processing applications can reach an optimal performance.

Acknowledgement. This research is supported by National Natural Science Foundation of China (NO. 60873068, 61003003), Talent Projects of the Educational Department of Liaoning Province (NO. LR201017), the National Natural Science Foundation of China under Grant Nos. 61100022.

References

1. Dean, J., Ghemawat, S.: Mapreduce: simplified data processing on large clusters. Communications of the ACM 51(1), 107–113 (2008)
2. Ding, L., Xin, J., Wang, G., Huang, S.: ComMapReduce: An improvement of mapReduce with lightweight communication mechanisms. In: Lee, S.-g., Peng, Z., Zhou, X., Moon, Y.-S., Unland, R., Yoo, J. (eds.) DASFAA 2012, Part II. LNCS, vol. 7239, pp. 150–168. Springer, Heidelberg (2012)
3. Ding, L., Wang, G., Xin, J., Wang, X., Huang, S., Zhang, R.: Commapreduce: an improvement of mapreduce with lightweight communication mechanisms. Data Knowledge Engineering 88, 224–247 (2013)
4. Deng, D., Li, G., Hao, S., Wang, J., Feng, J.: Massjoin: A mapreduce-based method for scalable string similarity joins. In: ICDE, pp. 340–351. IEEE (2014)
5. Qin, L., Yu, J.X., Chang, L., Cheng, H., Zhang, C., Lin, X.: Scalable big graph processing in mapreduce. In: SIGMOD Conference, pp. 827–838. ACM (2014)
6. Ding, L., Xin, J., Wang, G.: An efficient query processing optimization based on elm in the cloud. Neural Computing and Applications (2014)
7. Huang, G.-B., Zhu, Q.-Y., Siew, C.-K.: Extreme learning machine: a new learning scheme of feedforward neural networks. In: Proceedings of the 2004 IEEE International Joint Conference on Neural Networks, vol. 2, pp. 985–990. IEEE (2004)
8. Lendasse, A., He, Q., Miche, Y., Huang, G.-B.: Advances in extreme learning machines (elm2012). Neurocomputing 128, 1–3 (2014)
9. Zong, W., Huang, G.-B.: Learning to rank with extreme learning machine. Neural Processing Letters 39(2), 155–166 (2014)
10. Basu, A., Shuo, S., Zhou, H., Lim, M.-H., Huang, G.-B.: Silicon spiking neurons for hardware implementation of extreme learning machines. Neurocomputing 102, 125–134 (2013)
11. Xi-Zhao, W., Qing-Yan, S., Qing, M., Jun-Hai, Z.: Architecture selection for networks trained with extreme learning machine using localized generalization error model. Neurocomputing 102, 3–9 (2013)
12. Zong, W., Huang, G.-B., Chen, Y.: Weighted extreme learning machine for imbalance learning. Neurocomputing 101, 229–242 (2013)
13. He, Q., Shang, T., Zhuang, F., Shi, Z.: Parallel extreme learning machine for regression based on mapreduce. Neurocomputing 102, 52–58 (2013)

14. Zhang, R., Lan, Y., Huang, G.-B., Xu, Z.-B., Soh, Y.C.: Dynamic extreme learning machine and its approximation capability. IEEE T. Cybernetics 43(6), 2054–2065 (2013)
15. Zhai, J.-H., Xu, H.-Y., Wang, X.-Z.: Dynamic ensemble extreme learning machine based on sample entropy. Soft Computing 16(9), 1493–1502 (2012)
16. Sun, Y., Yuan, Y., Wang, G.: An os-elm based distributed ensemble classification framework in p2p networks. Neurocomputing 74(16), 2438–2443 (2011)
17. Zhao, X.-g., Wang, G., Bi, X., Gong, P., Zhao, Y.: Xml document classification based on elm. Neurocomputing 74(16), 2444–2451 (2011)
18. Jun, W., Shitong, W., Chung, F.-l.: Positive and negative fuzzy rule system, extreme learning machine and image classification. International Journal of Machine Learning and Cybernetics 2(4), 261–271 (2011)
19. Huang, G.-B., Zhu, Q.-Y., Siew, C.-K.: Extreme learning machine: theory and applications. Neurocomputing 70(1), 489–501 (2006)
20. Peng, H., Long, F., Ding, C.: Feature selection based on mutual information criteria of max-dependency, max-relevance, and min-redundancy. IEEE Transactions on Pattern Analysis and Machine Intelligence 27(8), 1226–1238 (2005)

Domain Adaptation Transfer Extreme Learning Machines

Lei Zhang[1,2] and David Zhang[2]

[1] College of Computer Science, Chongqing University, No.174 Shazheng Street, ShapingBa District, Chongqing, China
leizhang@cqu.edu.cn
[2] Department of Computing, The Hong Kong Polytechnic Unviersity, Hung Hom, Hong Kong
csdzhang@comp.polyu.edu.hk

Abstract. Extreme learning machines (ELMs) have been confirmed to be efficient and effective learning techniques for pattern recognition and regression. However, ELMs primarily focus on the supervised, semi-supervised and unsupervised learning problems in single domain and the generalization ability in multiple domains based learning issues is hardly studied. This paper aims to propose a unified framework of ELMs with domain adaptation and improve their transfer learning capability in cross domains without loss of the computational efficiency of traditional ELMs. We integrate domain adaptation into ELMs and two algorithms including source domain adaptation transfer ELM (TELM-SDA) and target domain adaptation transfer ELM (TELM-TDA) are proposed. For insight of the difference among ELM, TELM-SDA and TELM-TDA, two *remarks* are provided. Experiments on the popular sensor drift big data with multiple batches in machine olfaction, the results clearly demonstrate the characteristics of the proposed domain adaptation transfer ELMs that they can not only copy with sensor drift efficiently without cumbersome measures comparable to state-of-the-art methods but also bring new perspectives for ELM.

Keywords: Extreme learning machine, domain adaptation, transfer learning, semi-supervised learning.

1 Introduction

Extreme learning machine (ELM), proposed for solving a single layer feed-forward network (SLFN) by Huang et al [1, 2], has been proven to be effective and efficient algorithms for pattern classification and regression in different fields. ELM can analytically determine the output weights between the hidden layer and output layer using Moore-Penrose generalized inverse by adopting the square loss of prediction error, which then turns into solving a regularized least square problem efficiently in closed form. The output of the hidden layer is activated by an infinitely differentiable function with randomly selected input weights and biases of the hidden layer. Huang [3] rigorously prove that the input weights and hidden layer biases can be randomly

J. Cao et al. (eds.), *Proceedings of ELM-2014 Volume 1*,
Proceedings in Adaptation, Learning and Optimization 3, DOI: 10.1007/978-3-319-14063-6_10

assigned if the activation function is infinitely differentiable, who also showed that single SLFN with randomly generated additive or RBF nodes with such activation functions can universally approximate any continuous function on any compact subspace of Euclidean space [4].

In recent years, ELM has witnessed a number of improved versions in models, algorithms and real-world applications. ELM shows a comparable or even higher prediction accuracy than that of SVMs which solves a quadratic programming problem. In [3], their differences have been discussed. Some specific examples of improved ELMs have been listed as follows. As the output weights are computed with prefixed input weights and biases of hidden layer, a set of non-optimal input weights and hidden biases may exist. Additionally, ELM may require more hidden neurons than conventional learning algorithms in some special applications. Therefore, Zhu et al [5] proposed an evolutionary ELM for more compact networks that speed the response of trained networks. In terms of the imbalanced number of classes, a weighted ELM was proposed for binary/multiclass classification tasks with both balanced and imbalanced data distribution [6]. Due to that the solution of ELM is dense which will require longer time for training in large scale applications, Bai et al [7] proposed a sparse ELM for reducing storage space and testing time. Besides, Li et al [8] also proposed a fast sparse approximation of ELM for sparse classifiers training at a rather low complexity without reducing the generalization performance. For all the versions of ELM mentioned above, supervised learning framework was widely explored in application which limits its ability due to the difficulty in obtaining the labeled data. Therefore, Huang et al [9] proposed a semi-supervised ELM, in which a manifold regularization with graph Laplacian was set, and under the formulation of semi-supervised ELM, an unsupervised ELM was also explored.

In the past years, the contributions to ELM theories and applications have been made substantially by researchers from various fields. However, with the rising of big data, the data distribution obtained in different stages with different experimental conditions may change, i.e. from different domains. It is also well know that collection of labeled instances is tedious and labor ineffective, while the classifiers trained by a small number of labeled data are not robust and therefore lead to weak generalization, especially for large-scale application. Though ELM performs better generalization in a number of labeled data, the transferring capability of ELM may be reduced with very little number of labeled training instances from different domains. Domain adaptation methods have been proposed for classifiers learning with a few labeled instances from target domains by leveraging a number of labeled samples from the source domains [10-14]. Domain adaptation methods have also been employed for object recognition and sensor drift compensation [15, 16]. It is worth noting that domain adaptation is different from semi-supervised learning which assumes that the labeled and unlabeled data are from the same domain in classifier training.

In this paper, we extend ELMs to handle domain adaptation problems for improving the transferring capability of ELM between multiple domains with very few labeled guide instances in target domain, and overcome the generalization disadvantages of ELM in multi-domains application. Inspired by domain adaptation theory, two domain adaptation ELMs with similar structures but different knowledge adaptation characteristics are proposed for domain adaptation learning and knowledge transfer. The proposed domain adaptation ELMs are named as source domain adaptation transfer ELM (TELM-SDA) and target domain adaptation transfer ELM (TELM-TDA), respectively. Specifically, TELM-TDA learns a classifier using the very few labeled instances from target domain, while the remaining numerous unlabeled data are also fully exploited by approximating the prediction of the base classifier which can be trained in the source domain by regularized ELM or SVM. TELM-SDA is a more instinct framework which learns a classifier by using the large number of labeled data from the source domain, and very few labeled instances from target domain as regularization. From the learning mechanism of both domain adaptation ELMs, TELM-TDA has larger computation than TELM-SDA, due to the base classifier training and the numerous unlabeled data from target domain considered in learning. It is worth noting that both TELM-TDA and TELM-SDA can be formed into a unified ELM framework which refers to two stages including random feature mapping and output weights training.

The rest of this paper is organized as follows. In Section 2, a brief review of ELM is presented. In Section 3, the proposed TELM-SDA is illustrated in principle and algorithm. In Section 4, the proposed TELM-TDA is presented with its principle and algorithm. In Section 5, we present the experiments and results on the popular sensor drift data with multiple batches collected by electronic nose with 3 years for gas recognition. The conclusion of this paper is given in Section 6.

2 Related Work: A Brief Review of ELM

Given N samples $[\mathbf{x}_1, \mathbf{x}_2, \cdots, \mathbf{x}_N]$ and their corresponding target$[\mathbf{t}_1, \mathbf{t}_2, \cdots, \mathbf{t}_N]$, where $\mathbf{x}_i = [x_{i1}, x_{i1}, \cdots, x_{in}]^{\mathrm{T}} \in \mathbb{R}^n$ and $\mathbf{t}_i = [t_{i1}, t_{i1}, \cdots, t_{im}]^{\mathrm{T}} \in \mathbb{R}^m$. The output of the hidden layer is denoted as $h(\mathbf{x}_i) \in \mathbb{R}^{1\times L}$, where L is the number of hidden nodes and $h(\cdot)$ is the activation function (e.g. RBF function, sigmoid function). The output weights between the hidden layer and the output layer being learned is denoted as $\boldsymbol{\beta} \in \mathbb{R}^{L\times m}$, where O is the number of output nodes.

Regularized ELM aims to solve the output weights by minimizing the squared loss summation of prediction errors and the norm of the output weights for over-fitting control, which results in the following formulation

$$\begin{cases} \min_{\boldsymbol{\beta}} \mathcal{L}_{ELM} = \frac{1}{2}\|\boldsymbol{\beta}\|^2 + C \cdot \frac{1}{2} \cdot \sum_{i=1}^{N} \boldsymbol{\xi}_i^2 \\ s.t.\ \mathrm{h}(\mathbf{x}_i)\boldsymbol{\beta} = \mathbf{t}_i - \boldsymbol{\xi}_i, i = 1, \ldots, N \end{cases} \tag{1}$$

where $\boldsymbol{\xi}_i$ denotes the prediction error w.r.t. the i-th training pattern, and C is a penalty constant on the training errors.

By substituting the constraint term in (1) into the objective function, an equivalent unconstrained optimization problem can be obtained as follows

$$\min_{\boldsymbol{\beta}\in\mathbb{R}^{L\times m}} \mathcal{L}_{ELM} = \frac{1}{2}\|\boldsymbol{\beta}\|^2 + C \cdot \frac{1}{2} \cdot \|\mathbf{T} - \mathbf{H}\boldsymbol{\beta}\|^2 \tag{2}$$

where $\mathbf{H} = [h(\mathbf{x}_1); h(\mathbf{x}_2); \ldots; h(\mathbf{x}_N)] \in \mathbb{R}^{N\times L}$ and $\mathbf{T} = [\mathbf{t}_1, \mathbf{t}_2, \ldots, \mathbf{t}_N]^{\mathrm{T}}$.

The optimization problem (2) is a well known regularized least square problem. The closed form solution of $\boldsymbol{\beta}$ can be easily solved by setting the gradient of the subjective function (2) w.r.t. $\boldsymbol{\beta}$ to zero.

There are two cases when solving $\boldsymbol{\beta}$, i.e. if the number N of training patterns is larger than L, the gradient equation is over-determined, and the closed form solution can be obtained as

$$\boldsymbol{\beta}^* = \left(\mathbf{H}^{\mathrm{T}}\mathbf{H} + \frac{\mathbf{I}_L}{C}\right)^{-1}\mathbf{H}^{\mathrm{T}}\mathbf{T} \tag{3}$$

where $\mathbf{I}_L$ denotes the identity matrix with size of L.

If the number N of training patterns is smaller than L, an under-determined least square problem would be handled. In this case, the solution of (2) can be obtained as

$$\boldsymbol{\beta}^* = \mathbf{H}^{\mathrm{T}}\left(\mathbf{H}\mathbf{H}^{\mathrm{T}} + \frac{\mathbf{I}_N}{C}\right)^{-1}\mathbf{T} \tag{4}$$

where $\mathbf{I}_N$ denotes the identity matrix with size of N.

Therefore, in classifier training of ELM, the output weights can be computed by using (3) or (4) which depends on the number of training instances and the number of hidden nodes.

3 Proposed Domain Adaptation Transfer ELM

3.1 Source Domain Adaptation Transfer ELM (TELM-SDA)

Suppose that the source domain and target domain are represented D_S and D_T. In this paper, we assume that all the samples in the source domain are labeled data. The proposed *TELM-SDA* aims to learn a classifier β_S using a number of labeled instances from the source domain, and set the very few labeled data from the target domain as an appropriate regularizer for adapting to the source domain. The TELM-SDA can be formulated as

$$\min_{\beta_S,\xi_S^i,\xi_T^i} \frac{1}{2}\|\beta_S\|^2 + C_S\frac{1}{2}\sum_{i=1}^{N_S}\left(\xi_S^i\right)^2 + C_T\frac{1}{2}\sum_{j=1}^{N_T}\left(\xi_T^j\right)^2 \tag{5}$$

$$\text{s.t.}\begin{cases} H_S^i\beta_S = t_S^i - \xi_S^i, i = 1,\ldots,N_S \\ H_T^j\beta_S = t_T^j - \xi_T^j, j = 1,\ldots,N_T \end{cases}$$

where $H_S^i \in \mathbb{R}^{1\times L}, \xi_S^i \in \mathbb{R}^{1\times m}, t_S^i \in \mathbb{R}^{1\times m}$ denote the output of hidden layer, the prediction error and the label with respect to the i-th training instance x_S^i from the source domain, $H_T^j \in \mathbb{R}^{1\times L}, \xi_T^j \in \mathbb{R}^{1\times m}, t_T^j \in \mathbb{R}^{1\times m}$ denote the output of hidden layer, the prediction error and the label with respect to the j-th guide samples x_T^j from the target domain, $\beta_S \in \mathbb{R}^{L\times m}$ is the output weights being solved, N_S and N_T denote the number of training instances and guide samples from the source domain and target domain, respectively, C_S and C_T are the penalty coefficients on the prediction errors of the labeled training data from source domain and target domain, respectively. Note that we call the very few labeled samples in target domain as "guide samples" in this paper.

From (5), we can find that the very few labeled guide samples from target domain can assist the learning of β_S and realize the knowledge transfer between source domain and target domain by introducing the third term as regularization with the second constraint, which makes the feature mapping of the guide samples from target domain approximate the labels with the output weights β_S learned by the training data from the source domain. The structure of the proposed TELM-SDA algorithm is illustrated in Fig.1.

To solve the optimization (5), the Largange multiplier equation is formulated as

$$L(\beta_S, \xi_S^i, \xi_T^j, \alpha_S, \alpha_T) = \frac{1}{2}\|\beta_S\|^2 + C_S\frac{1}{2}\sum_{i=1}^{N_S}(\xi_S^i)^2 + C_T\frac{1}{2}\sum_{j=1}^{N_T}(\xi_T^j)^2 \quad (6)$$
$$-\alpha_S(H_S^i\beta_S - t_S^i + \xi_S^i) - \alpha_T(H_T^i\beta_T - t_T^i + \xi_T^i)$$

By setting the partial derivation with respect to $\beta_S, \xi_S^i, \xi_T^j, \alpha_S, \alpha_T$ as zero, we have

$$\begin{cases} \frac{\partial L}{\partial \beta_S} = 0 \to \beta_S = H_S^{\mathrm{T}}\alpha_S + H_T^{\mathrm{T}}\alpha_T \\ \frac{\partial L}{\partial \xi_S} = 0 \quad \to \quad \alpha_S = C_S\xi_S \\ \frac{\partial L}{\partial \xi_T} = 0 \quad \to \quad \alpha_T = C_T\xi_T \\ \frac{\partial L}{\partial \alpha_S} = 0 \to H_S\beta_S - t_S + \xi_S = 0 \\ \frac{\partial L}{\partial \alpha_T} = 0 \to H_T\beta_S - t_T + \xi_T = 0 \end{cases} \quad (7)$$

where H_S and H_T are the output matrix of hidden layer with respect to the labeled data from source domain and target domain, respectively.

To solve β_S, α_S and α_T should be solve first. For the case that the number of training samples N_S is smaller than L ($N_S<L$), we substitute the 1st, 2nd, and 3rd equations into the 4th and 5th equations in (7), there is

$$\begin{cases} H_T H_S^{\mathrm{T}}\alpha_S + \left(H_T H_T^{\mathrm{T}} + \frac{I}{C_T}\right)\alpha_T = t_T \\ H_S H_T^{\mathrm{T}}\alpha_T + \left(H_S H_S^{\mathrm{T}} + \frac{I}{C_S}\right)\alpha_S = t_S \end{cases} \quad (8)$$

Let $H_T H_S^{\mathrm{T}} = A, H_T H_T^{\mathrm{T}} + \frac{I}{C_T} = B, H_S H_T^{\mathrm{T}} = C, H_S H_S^{\mathrm{T}} + \frac{I}{C_S} = D$, then eq.(8) can be written as

$$\begin{cases} A\alpha_S + B\alpha_T = t_T \\ C\alpha_T + D\alpha_S = t_S \end{cases} \to \begin{cases} B^{-1}A\alpha_S + \alpha_T = B^{-1}t_T \\ C\alpha_T + D\alpha_S = t_S \end{cases} \quad (9)$$

Then α_S and α_T can be solved as

$$\begin{cases} \alpha_S = (CB^{-1}A - D)^{-1}(CB^{-1}t_T - t_S) \\ \alpha_T = B^{-1}t_T - B^{-1}A(CB^{-1}A - D)^{-1}(CB^{-1}t_T - t_S) \end{cases} \quad (10)$$

According to the 1st equation in (7), we can obtain the output weights as

$$\begin{aligned}\beta_S &= H_S^T\alpha_S + H_T^T\alpha_T \\ &= H_S^T(CB^{-1}A - D)^{-1}(CB^{-1}t_T - t_S) + \\ &\quad H_T^T[B^{-1}t_T - B^{-1}A(CB^{-1}A - D)^{-1}(CB^{-1}t_T - t_S)]\end{aligned} \quad (11)$$

where $A = H_T H_S^T, B = H_T H_T^T + \frac{I}{C_T}, C = H_S H_T^T, D = H_S H_S^T + \frac{I}{C_S}$, I is the identity matrix with size of N_S.

For the case that the number of training samples N_S is larger than L ($N_S>L$), we can obtain from the 1st equation in (7) that $\alpha_S = (H_S H_S^T)^{-1}(H_S\beta_S - H_S H_T^T\alpha_T)$, which is substituted into the 4th and 5th equations, then we calculate the output weights β_S as follows

$$\begin{cases} H_S\beta_S + \xi_S = t_S \\ H_T\beta_S + \xi_T = t_T \end{cases} \to \begin{cases} H_S\beta_S + \dfrac{I}{C_S}\alpha_S = t_S \\ H_T\beta_S + \dfrac{I}{C_T}\alpha_T = t_T \end{cases} \to \begin{cases} H_S^T H_S\beta_S + \dfrac{I}{C_S}H_S^T\alpha_S = H_S^T t_S \\ H_T\beta_S + \dfrac{I}{C_T}\alpha_T = t_T \end{cases}$$

$$\to \begin{cases} H_S^T H_S\beta_S + \dfrac{I}{C_S}H_S^T(H_S H_S^T)^{-1}(H_S\beta_S - H_S H_T^T\alpha_T) = H_S^T t_S \\ \alpha_T = C_T(t_T - H_T\beta_S) \end{cases}$$

$$\to \begin{cases} \left[H_S^T H_S + \dfrac{I}{C_S}H_S^T(H_S H_S^T)^{-1}H_S\right]\beta_S - \dfrac{I}{C_S}H_S^T(H_S H_S^T)^{-1}H_S H_T^T\alpha_T = H_S^T t_S \\ \alpha_T = C_T(t_T - H_T\beta_S) \end{cases}$$

$$\to \left(H_S^T H_S + \frac{I}{C_S} + \frac{C_T}{C_S}H_T^T H_T\right)\beta_S = H_S^T t_S + \frac{C_T}{C_S}H_T^T t_T$$

$$\to \beta_S = (I + C_S H_S^T H_S + C_T H_T^T H_T)^{-1}(C_S H_S^T t_S + C_T H_T^T t_T) \quad (12)$$

where I is the identity matrix with size of L.

In fact, the optimization (5) can be reformulated an equivalent unconstrained optimization problem in a matrix form by substituting the constraints into the objective function as

$$\min_{\beta_S} L_{TELM-SDA}(\beta_S) = \frac{1}{2}\|\beta_S\|^2 + C_S\frac{1}{2}\|t_S - H_S\beta_S\|^2 + C_T\frac{1}{2}\|t_T - H_T\beta_S\|^2 \quad (13)$$

By setting the gradient of $L_{TELM-SDA}$ with respect to β_S as zero, there is

$$\nabla L_{TELM-SDA} = \beta_S - C_S H_S^T(t_S - H_S\beta_S) - C_T H_T^T(t_T - H_T\beta_S) = 0 \quad (14)$$

Then, we can easily solve the β_S as formulated in (12).

For recognition of the numerous unlabeled data in target domain, we calculate the final output using the following

$$y_{Tu}^{k} = H_{Tu}^{k} \cdot \beta_S, k = 1, \ldots, N_{Tu} \tag{15}$$

where H_{Tu}^{k} denote the hidden layer output with respect to the k-th unlabeled vector in target domain, and N_{Tu} is the number of unlabeled vectors in target domain.

In terms of the above discussion, the *TELM-SDA* algorithm is summarized as **Algorithm 1**.

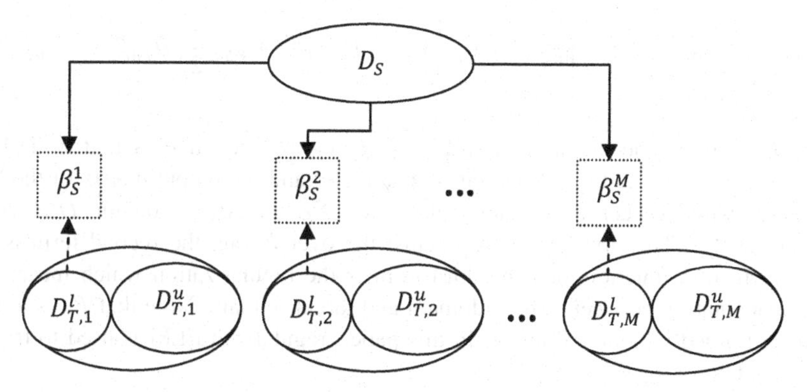

Fig. 1. Structure of TELM-SDA algorithm with M target domains (M tasks). The solid arrow denotes the training data from source domain D_S and the dashed arrow denotes the tiny labeled guide data from target domain D_T^l for classifier learning. The unlabeled data from D_T^u are not used.

Algorithm 1. TELM-SDA algorithm

Input:

Training samples $\{X_S, t_S\} = \{x_S^i, t_S^i\}_{i=1}^{N_S}$ of the source domain S;

Labeled guide samples $\{X_T, t_T\} = \{x_T^j, t_T^j\}_{j=1}^{N_T}$ of the target domain T;

The tradeoff parameter C_S and C_T for source and target domain T.

Output:

The output weights β_S;

The predicted output y_{Tu} of unlabeled data in target domain.

Procedure:

1. Initialize the ELM network of L hidden neurons with random input weights W and hidden bias B.
2. Calculate the output matrix H_S and H_T of hidden layer with source and target domains as $H_S = h(W \cdot X_S + B)$ and $H_T = h(W \cdot X_T + B)$.
3. **If** $N_S < L$, compute the output weights β_S using (11);
   Else, compute the output weights β_S using (12).
4. Calculate the predicted output y_{Tu} using (15).

Return The output weights β_S and predicted output y_{Tu}.

3.2 Target Domain Adaptation Transfer ELM (TELM-TDA)

In the proposed TELM-SDA, the classifier β_S is learned on the source domain with the very few labeled guide samples from the target domain as regularization. Study demonstrates that unlabeled data can also improve the performance of classification [17]. While the proposed TELM-TDA aims to learn a classifier β_T on the very few labeled guide samples from the target domain, and fully explore the numerous unlabeled data in the target domain with a base classifier β_S trained in source domain. As illustrated, the proposed TELM-SDA is formulated as

$$\min_{\beta_T} L_{TELM-TDA}(\beta_T) = \frac{1}{2}\|\beta_T\|^2 + C_T\frac{1}{2}\|t_T - H_T\beta_T\|^2 + C_{Tu}\frac{1}{2}\|H_{Tu}\beta_S - H_{Tu}\beta_T\|^2 \tag{16}$$

where β_T denotes the learned classifier, C_T, H_T, t_T are the same as that in TELM-SDA, C_{Tu}, H_{Tu} denote the regularization parameter and the output matrix of the hidden layer with respect to the unlabeled data X_{Tu} in target domain D_T, where $D_T = X_T \cup X_{Tu}$. The first term is to against the over-fitting, the second term is the least square loss function, and the third term is the regularization which means the domain adaptation between source domain and target domain. Note that β_S is a base classifier trained in source domain. In this paper, regularized ELM is used to train a base classifier β_S by solving

$$\min_{\beta_S} L_{TELM-TDA}(\beta_S) = \frac{1}{2}\|\beta_S\|^2 + C_S\frac{1}{2}\|t_S - H_S\beta_S\|^2 \tag{17}$$

where C_S, t_S, H_S denote the same meaning as that in TELM-SDA.

The structure of the proposed TELM-SDA is described in Fig.2, from which we can see that the unlabeled data in target domain have also been explored. To solve the optimization (16), by setting the gradient of $L_{TdaTELM}$ with respect to β_T as zero, we then have

$$\nabla L_{TELM-TDA} = \beta_T - C_T H_T^{\mathrm{T}}(t_T - H_T\beta_T) - C_{Tu}H_{Tu}^{\mathrm{T}}(H_{Tu}\beta_S - H_{Tu}\beta_T) = 0 \tag{18}$$

If the number of training samples $N_T > L$, then we can have from (18)

$$\beta_T = (I + C_T H_T^{\mathrm{T}} H_T + C_{Tu} H_{Tu}^{\mathrm{T}} H_{Tu})^{-1}(C_T H_T^{\mathrm{T}} t_T + C_{Tu} H_{Tu}^{\mathrm{T}} H_{Tu}\beta_S) \tag{19}$$

where I is the identity matrix with size of L.

If the number of training samples $N_T < L$, we would like to obtain β_S of the proposed TELM-TDA by a unified ELM framework. Let $t_{Tu} = H_{Tu}\beta_S$, the model (16) can be re-written as

$$\min_{\beta_T,\xi_T^i,\xi_{Tu}^i} \frac{1}{2}\|\beta_T\|^2 + C_T\frac{1}{2}\sum_{i=1}^{N_T}\left(\xi_T^i\right)^2 + C_{Tu}\frac{1}{2}\sum_{j=1}^{N_{Tu}}\left(\xi_{Tu}^j\right)^2 \tag{20}$$

$$\text{s.t.} \begin{cases} H_T^i\beta_T = t_T^i - \xi_T^i, i = 1, \ldots, N_T \\ H_{Tu}^j\beta_T = t_{Tu}^j - \xi_{Tu}^j, j = 1, \ldots, N_{Tu} \end{cases}$$

The Lagrange multiplier equation of (20) can be written as

$$L(\beta_T, \xi_T^i, \xi_{Tu}^i, \alpha_T, \alpha_{Tu}) = \frac{1}{2}\|\beta_T\|^2 + C_T\frac{1}{2}\sum_{i=1}^{N_T}(\xi_T^i)^2 + C_{Tu}\frac{1}{2}\sum_{j=1}^{N_{Tu}}(\xi_{Tu}^j)^2$$

$$-\alpha_T(H_T^i\beta_T - t_T^i + \xi_T^i) - \alpha_{Tu}(H_{Tu}^i\beta_T - t_{Tu}^i + \xi_{Tu}^i) \tag{21}$$

By setting the partial derivation with respect to $\beta_T, \xi_T^i, \xi_{Tu}^j, \alpha_T, \alpha_{Tu}$ as zero, we have

$$\begin{cases} \frac{\partial L}{\partial \beta_T} = 0 & \rightarrow \beta_T = H_T^{\mathrm{T}}\alpha_T + H_{Tu}^{\mathrm{T}}\alpha_{Tu} \\ \frac{\partial L}{\partial \xi_T} = 0 & \rightarrow \alpha_T = C_T\xi_T \\ \frac{\partial L}{\partial \xi_{Tu}} = 0 & \rightarrow \alpha_{Tu} = C_{Tu}\xi_{Tu} \\ \frac{\partial L}{\partial \alpha_T} = 0 & \rightarrow H_T\beta_T - t_T + \xi_T = 0 \\ \frac{\partial L}{\partial \alpha_{Tu}} = 0 & \rightarrow H_{Tu}\beta_T - t_{Tu} + \xi_{Tu} = 0 \end{cases} \tag{22}$$

To solve β_T, let $H_{Tu}H_T^{\mathrm{T}} = O, H_{Tu}H_{Tu}^{\mathrm{T}} + \frac{I}{C_{Tu}} = P, H_T H_{Tu}^{\mathrm{T}} = Q, H_T H_T^{\mathrm{T}} + \frac{I}{C_T} = R$, with similar calculation of (8), (9), and (10),we can get

$$\begin{cases} \alpha_T = (QP^{-1}O - R)^{-1}(QP^{-1}t_{Tu} - t_T) \\ \alpha_{Tu} = P^{-1}t_{Tu} - P^{-1}O(QP^{-1}O - R)^{-1}(QP^{-1}t_{Tu} - t_T) \end{cases} \tag{23}$$

Therefore, the output weights if $N_T<L$ can be obtained as

$$\begin{aligned} \beta_T &= H_T^{\mathrm{T}}\alpha_T + H_{Tu}^{\mathrm{T}}\alpha_{Tu} \\ &= H_T^{\mathrm{T}}(QP^{-1}O - R)^{-1}(QP^{-1}t_{Tu} - t_T) \\ &\quad + H_{Tu}^{\mathrm{T}}[P^{-1}t_{Tu} - P^{-1}O(QP^{-1}O - R)^{-1}(QP^{-1}t_{Tu} - t_T)] \end{aligned} \tag{24}$$

where $t_{Tu} = H_{Tu}\beta_S, O = H_{Tu}H_T^{\mathrm{T}}, P = H_{Tu}H_{Tu}^{\mathrm{T}} + \frac{I}{C_{Tu}}, Q = H_T H_{Tu}^{\mathrm{T}}, R = H_T H_T^{\mathrm{T}} + \frac{I}{C_T}$, I is the identity matrix with size of N_T.

For recognition of the numerous unlabeled data in target domain, we calculate the final output using the following

$$y_{Tu}^k = H_{Tu}^k \cdot \beta_T, k = 1, \ldots, N_{Tu} \tag{25}$$

where H_{Tu}^k denote the hidden layer output with respect to the k-th unlabeled vector in target domain, and N_{Tu} is the number of unlabeled vectors in target domain.

In terms of the above discussion, the *TELM-TDA* algorithm is summarized as **Algorithm 2**.

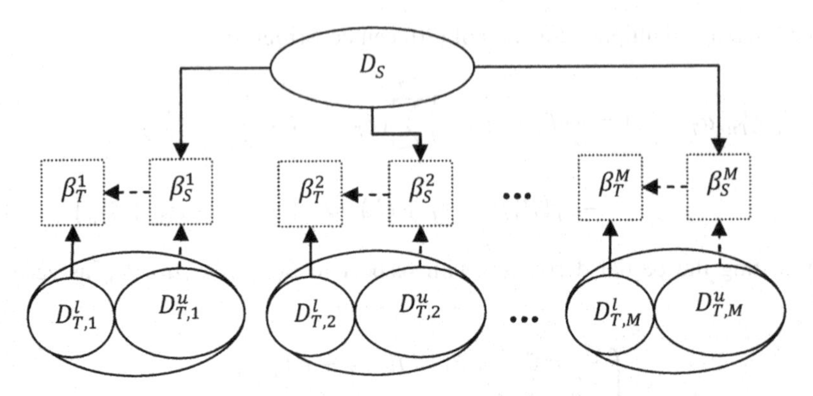

Fig. 2. Structure of TELM-TDA algorithm with M target domains (M tasks). The solid arrow connected with D_S denotes the training for base classifier β_S, the dashed line connected with D_T^u denotes the tentative test of base classifier using the unlabeled data from target domain, the solid arrow connected with D_T^l denotes the terminal classifier learning of β_T, and the dashed arrow connected between β_S and β_T denotes the regularization for learning β_T.

Algorithm 2. TELM-TDA algorithm

Input:

Training samples $\{X_S, t_S\} = \{x_S^i, t_S^i\}_{i=1}^{N_S}$ of the source domain S;

Labeled guide samples $\{X_T, t_T\} = \{x_T^j, t_T^j\}_{j=1}^{N_T}$ of the target domain T;

Unlabeled samples $\{X_{Tu}\} = \{x_{Tu}^k\}_{k=1}^{N_{Tu}}$ of the target domain T;

The tradeoff parameters C_S, C_T and C_{Tu}.

Output:

The output weights β_T;

The predicted output y_{Tu} of unlabeled data in target domain.

Procedure:

1. Initialize the ELM network of L hidden neurons with random input weights W_1 and hidden bias B_1.
2. Calculate the output matrix H_S of hidden layer with source domain as $H_S = h(W_1 \cdot X_S + B_1)$.
3. **If** $N_S < L$, compute the output weights β_S of the base classifier using (4);
   Else, compute the output weights β_S of the base classifier using (3).
4. Initialize the ELM network of L hidden neurons with random input weights W_2 and hidden bias B_2.
5. Calculate the output matrix H_T and H_{Tu} of hidden layer with labeled and unlabeled data in target domains as $H_T = h(W_2 \cdot X_T + B_2)$ and $H_{Tu} = h(W_2 \cdot X_{Tu} + B_2)$.
6. **If** $N_T < L$, compute the output weights β_T using (24);
   Else, compute the output weights β_T using (19).
7. Calculate the predicted output y_{Tu} using (25).

Return The output weights β_T and predicted output y_{Tu}.

Remark 1: From the proposed source domain adaptation transfer ELM (TELM-SDA) and target domain adaptation transfer ELM (TELM-TDA), we can observe that two stages are included namely feature mapping with random selected weights and biases and output weights learning which are the main parts in ELM. For ELM, the only information in source domain is considered. However, for domain adaptation transfer ELM, very few labeled samples from target domain are explored without changing the unified ELM framework. The common differences of the ELMs lie in the calculation of output weights. The unified framework for TELM-SDA and TELM-TDA might draw some new perspectives for developing the ELMs.

Remark 2: We can observe that the TELM-SDA and TELM-TDA have similar structure in model and algorithm, except for the base classifier learning in TELM-TDA. However, the essential difference lies in that the numerous unlabeled data which may be useful for improving generalization performance are also explored in TELM-TDA. Specifically, TELM-SDA trains a classifier using the information of source domain but draw some knowledge with labeled guide samples from the target domain. In this way, the knowledge from target domain can be effectively transferred to source domain through appropriate models. Instead, TELM-TDA aims to train a classifier using the guide data from target domain but introduce a regularizer through exploring the unlabeled data and a base classifier trained from source domain.

4 Experiments

In this section, we will employ the proposed TELM-SDA and TELM-TDA algorithms on olfactory data collected by electronic nose for sensor drift compensation. Electronic nose is an artificial olfaction system, which is developed for gas recognition [18, 19], tea quality assessment [20, 21], medical diagnosis [22], environmental monitor and gas concentration estimation [23, 24], etc. by using pattern recognition and gas sensor array with cross-sensitivity and broad spectrum characteristics. However, gas sensor drift will be caused due to the change of internal component and aging, which would reduce the generalization performance of well trained classifier [25]. Therefore, researchers have to retrain the classifier using a number of new samples in a period regularly. The tedious work for classifier retraining and acquisition of new labeled samples regularly seems to be impossible, due to the complicated experiments of electronic nose. Though researchers have paid more attention to sensor drift and aim to find some effective ways for drift compensation through classifier ensembles and drift prediction [16, 26-29], sensor drift is still a challenging issue in machine olfaction community and sensory field. To our best knowledge, there are no very effective methods for dealing with sensor drift. Therefore, we aim to enhance the adaptive performance of classifiers to drifted data with very low complexity and little work. It would be very meaningful and interesting to train a classifier using very few labeled new samples (target domain) as guide samples without giving up the recognized “useless” old data (source domain), and make the new trained classifier adapt to the new patterns in target domain.

4.1 Experimental Data

For verification of the proposed TELM-SDA and TELM-TDA algorithms, the long-term sensor drift big data of three years which was released in UCI Machine Learning Repository [31] by Vergara *et al.* [26, 30] has been explored in this paper.

This dataset contains 13,910 measurements (observation samples) from an electronic nose system with 16 gas sensors exposed to 6 kinds of pure gaseous substances including acetone, acetaldehyde, ethanol, ethylene, ammonia, and toluene at different concentration levels. The sensor drift big dataset was gathered during the period of January 2008 to February 2011 with 36 months in a gas delivery platform. For each sensor, 8 features were extracted, and results in a 128-dimensional feature vector (8 features × 16 sensors) for each measurement. We refer readers to [26] for specific technical details on how to select the 8 features for single sensor. Totally, 10 batches of data are included in the dataset which was divided according to months. The details of the dataset have been presented in Table 1.

Table 1. Number of samples for each subject in the sensor drifted big data

Batch ID	Month	Acetone	Acetaldehyde	Ethanol	Ethylene	Ammonia	Toluene	Total
Batch 1	1, 2	90	98	83	30	70	74	445
Batch 2	3-10	164	334	100	109	532	5	1244
Batch 3	11-13	365	490	216	240	275	0	1586
Batch 4	14, 15	64	43	12	30	12	0	161
Batch 5	16	28	40	20	46	63	0	197
Batch 6	17-20	514	574	110	29	606	467	2300
Batch 7	21	649	662	360	744	630	568	3613
Batch 8	22, 23	30	30	40	33	143	18	294
Batch 9	24, 30	61	55	100	75	78	101	470
Batch 10	36	600	600	600	600	600	600	3600

4.2 Experimental Setup

We follow the experimental setup in [26] to evaluate the proposed domain adaptation transfer ELM models. The number of hidden neurons L is set as 1000. The features are scaled appropriately to lie between -1 and +1. The RBF function is used as the activation function in the hidden layer (i.e. feature mapping function) in which the kernel width is set as 1. In TELM-SDA model, the penalty coefficients C_S and C_T are set as 0.01 and 10 throughout the experiments, respectively. In TELM-TDA model, the penalty coefficient C_S for base classifier is set as 0.001, C_T and C_{Tu} are set as 0.001 and 100 throughout the experiments, respectively. For effective verification of the proposed methods, two experimental settings according to [16] are given as follows:

Setting 1: Take batch 1 (source domain) as fixed training set and tested on other 9 batches (target domains);

Setting 2: The training set (source domain) is dynamically changed with batch K-1 and tested on batch K (target domain), K=2,…,10.

For studying the relation between the number k of labeled samples in target domain and the recognition accuracy, k is tried in the set of {5, 10, 15, 20, 25, 30, 35, 40, 45, 50}. In addition, for comparisons, we have compared with multi-class SVM with RBF kernel (SVM-rbf), the geodesic flow kernel (SVM-gfk), and the combination kernel (SVM-comgfk). Besides, we also compared with the semi-supervised methods such as manifold regularization with RBF kernel (ML-rbf) and manifold regularization with combination kernel (ML-comgfk), which have been presented in [16, 26] for the same sensor drift data. Additionally, the regularized ELM with RBF function in hidden layer (ELM-rbf) from [29] is also compared in our experiments. In experiments, we run the ELM-rbf, TELM-SDA and TELM-TDA 10 times, and the average value for each item is provided.

4.3 Results

Under the consideration above, we employ the experiments on **Setting 1** and **Setting 2**, respectively. The comparisons under setting 1 with recognition accuracy of 9 batches for different methods are presented in Table 2. We have shown two conditions of TELM-SDA with 20 labeled guide samples and 30 labeled guide samples. For TELM-TDA, 40 and 50 labeled samples from the target domain are used, respectively, considering that TELM-TDA trains a classifier using the labeled samples from the target domain, therefore, more labeled samples would be necessary which is slightly different from TELM-SDA. From Table 2, it can be obviously seen that the proposed TELM-SDA and TELM-TDA are much better than other existing methods including SVM with different kernels, manifold regularization with different kernels. For TELM-SDA and TELM-TDA, the testing accuracies on batch 2-10 with a training classifier using the data in batch 1 can still be feasible without performance reduction. This means that the sensor drift can be compensated very well with domain adaptation knowledge transfer. For visually observing the change of performance with sensor drift, we show the recognition accuracy on batches successively as Fig. 3. Through the results of the regularized ELM, we can see that the generalization performance and knowledge transfer capability have been well improved by the proposed TELM-SDA and TELM-TDA with domain adaptation. Comparison between TELM-SDA and TELM-TDA, the latter needs more labeled samples than the former. From the computational complexity, due to that there is a base classifier in TELM-TDA, TELM-SDA would be more appropriate in real-world applications which considers the data in source domain and very few labeled guide samples from target domain for classifier learning. The specific comparisons between TELM-SDA and TELM-TDA will be employed later.

Table 2. Comparisons of recognition accuracy (%) under the experimental **Setting 1**, i.e. trained on batch 1 and tested on other successive 9 batches

Batch ID	Batch 2	Batch 3	Batch 4	Batch 5	Batch 6	Batch 7	Batch 8	Batch 9	Batch 10
SVM-rbf	74.36	61.03	50.93	18.27	28.26	28.81	20.07	34.26	34.47
SVM-gfk	72.75	70.08	60.75	75.08	73.82	54.53	55.44	69.62	41.78
SVM-comgfk	74.47	70.15	59.78	75.09	73.99	54.59	55.88	70.23	41.85
ML-rbf	42.25	73.69	75.53	66.75	77.51	54.43	33.50	23.57	34.92
ML-comgfk	80.25	74.99	78.79	67.41	77.82	71.68	49.96	50.79	53.79
ELM-rbf	70.63	66.44	66.83	63.45	69.73	51.23	49.76	49.83	33.50
TELM-SDA(20)	87.57	96.53	82.61	81.47	84.97	71.89	78.10	87.02	57.42
TELM-SDA(30)	87.98	95.74	85.16	95.99	94.14	83.51	86.90	100.0	53.62
TELM-TDA(40)	83.52	96.34	88.20	99.49	78.43	80.93	87.42	100.0	56.25
TELM-TDA(50)	97.96	95.34	99.32	99.24	97.03	83.09	95.27	100.0	59.45

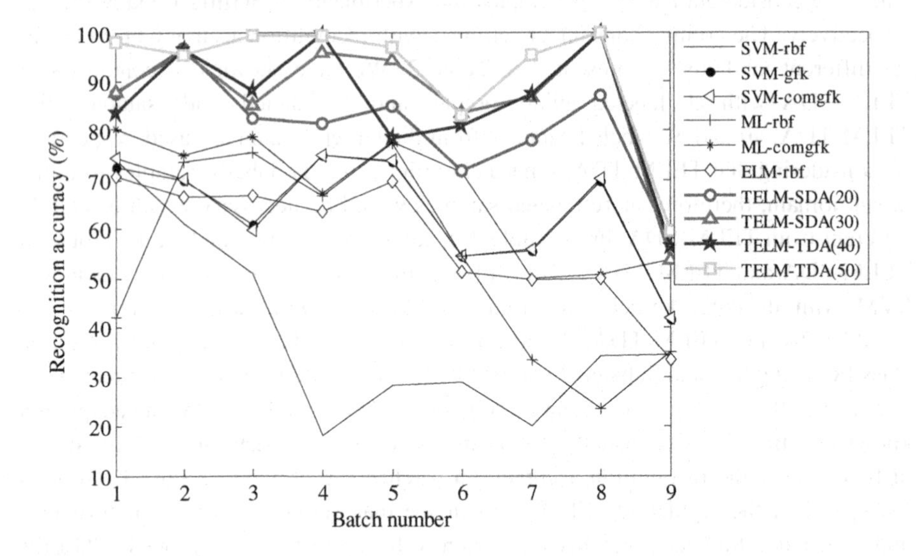

Fig. 3. Comparisons of different methods in **Setting 1**

From the experimental results in **Setting 1**, the proposed methods perform better results, and the sensor drift can be well compensated. We have also employed the experiments by following Setting 2 i.e. trained on batch *K*-1 and tested on batch *K*, for which the results are presented in Table 3. We can find that the proposed domain adaptation transfer ELM performs much better than other baseline methods for sensor drift big data. The visual insight of these methods in setting 2 has been described in Fig.4 which shows the robust performance of the proposed methods in sensor drift compensation and knowledge transfer.

Table 3. Comparisons of recognition accuracy (%) under the experimental **Setting 2**. i.e. trained on batch K-1, and tested on batch K ($2 \leq K \leq 10$).

Batch ID	1→2	2→3	3→4	4→5	5→6	6→7	7→8	8→9	9→10
SVM-rbf	74.36	87.83	90.06	56.35	42.52	83.53	91.84	62.98	22.64
SVM-gfk	72.75	74.02	77.83	63.91	70.31	77.59	78.57	86.23	15.76
SVM-comgfk	74.47	73.75	78.51	64.26	69.97	77.69	82.69	85.53	17.76
ML-rbf	42.25	58.51	75.78	29.10	53.22	69.17	55.10	37.94	12.44
ML-comgfk	80.25	98.55	84.89	89.85	75.53	91.17	61.22	95.53	39.56
ELM-rbf	70.63	40.44	64.16	64.37	72.70	80.75	88.20	67.00	22.00
TELM-SDA(20)	87.57	96.90	85.59	95.89	80.53	91.56	88.71	88.40	45.61
TELM-SDA(30)	87.98	96.58	89.75	99.04	84.43	91.75	89.83	100.0	58.44
TELM-TDA(40)	83.52	96.41	81.36	96.45	85.13	80.49	85.71	100.0	56.81
TELM-TDA(50)	97.96	95.62	99.63	98.17	97.13	83.10	94.90	100.0	59.88

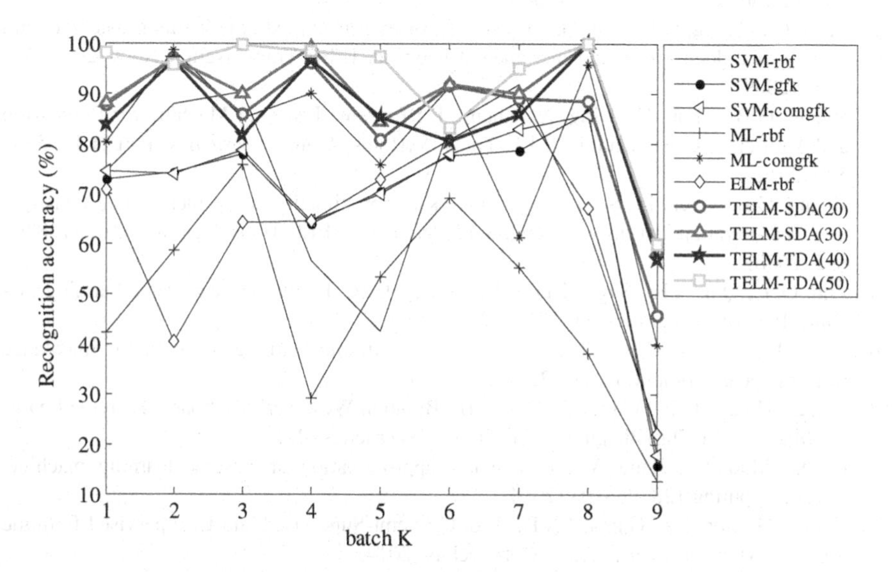

Fig. 4. Comparisons of different methods in **Setting 2**

5 Conclusion

In this paper, two ELM based algorithms, TELM-SDA and TELM-TDA have been proposed to extend the ELMs for learning tasks with multi-domains, respectively. Through the sensor drift big data analysis in machine olfaction, the proposed domain adaptation transfer ELMs consistently outperform the existing methods such as SVMs, semi-supervised manifold regularizations and ELMs for sensor drift compensation. For dealing with large scale sensor drift data collected by an electronic nose, the proposed methods have also the advantages of ELMs including high efficiency of classifier/predictor learning and straightforward implementation of multi-class classification. The adaptation of multi-domain sensor drift big data can be efficiently and effectively implemented by using the proposed domain adaptation transfer ELMs. More importantly, the proposed methods can also provide new perspectives for exploring ELM theory. Experimental results demonstrate that the proposed methods can obvious improve the transfer capability of ELMs in real-world applications.

Acknowledgement. This work was supported by National Natural Science Foundation of China (61401048), Hong Kong Scholar Program (No.XJ2013044) and also funded by China Postdoctoral Science Foundation (No. 2014M550457).

References

1. Huang, G.B., Zhu, Q.Y., Siew, C.K.: Extreme learning machine: Theory and applications. Neurocomputing 70, 489–501 (2006)
2. Feng, G., Huang, G.B., Lin, Q., Gay, R.: Error minimized extreme learning machine with growth of hidden nodes and incremental learning. IEEE Trans. Neural Netw. 20, 1352–1357 (2009)
3. Huang, G.B., Zhou, H., Ding, X., Zhang, R.: Extreme Learning Machine for Regression and Multiclass Classification. IEEE Trans. Systems, Man, Cybernetics: Part B 42, 513–529 (2012)
4. Huang, G.B., Chen, L., Siew, C.K.: Universal approximation using incremental constructive feedforward networks with random hidden nodes. IEEE Trans. Neural. Netw. 17, 879–892 (2006)
5. Zhu, Q.Y., Qin, A.K., Suganthan, P.N., Huang, G.B.: Evolutionary extreme learning machine. Pattern Recognition 38, 1759–1763 (2005)
6. Zong, W., Huang, G.B., Chen, Y.: Weighted extreme learning machine for imbalance learning. Neurocomputing 101, 229–242 (2013)
7. Bai, Z., Huang, G.B., Wang, D., Wang, H., Brandon Westover, M.: Sparse Extreme Learning Machine for Classification. IEEE Trans. Cybernetics (2014)
8. Li, X., Mao, W., Jiang, W.: Fast sparse approximation of extreme learning machine. Neurocomputing 128, 96–103 (2014)
9. Huang, G., Song, S., Gupta, J.N.D., Wu, C.: Semi-Supervised and Unsupervised Extreme Learning Machines. IEEE Trans. Cybernetics (2014)
10. Blitzer, J., McDonald, R., Pereira, F.: Domain adaptation with structural correspondence learning. In: Proc. Conf. Emp. Methods Natural Lang. Process, pp. 120–128 (2006)
11. Yang, J., Yan, R., Hauptmann, A.G.: Cross-domain video concept detection using adaptive SVMs. In: Proc. Int. Conf. Multimedia, pp. 188–197 (2007)
12. Pan, S.J., Tsang, I.W., Kwok, J.T., Yang, Q.: Domain adaptation via transfer component analysis. IEEE Trans. Neural Netw. 22, 199–210 (2011)
13. Duan, L., Tsang, I.W., Xu, D., Chua, T.S.: Domain adaptation from multiple sources via auxiliary classifiers. Proc. Int. Conf. Mach. Learn., 289–296 (2009)
14. Duan, L., Xu, D., Tsang, I.W.: Domain Adaptation from Multiple Sources: Domain-Dependent Regularization Approach. IEEE Trans. Neur. Netw. Learn. Syst. 23, 504–518 (2012)
15. Gopalan, R., Li, R., Chellappa, R.: Domain adaptation for object recognition: An unsupervised approach. In: Proc. ICCV, pp. 999–1006 (2011)

16. Liu, Q., Li, X., Ye, M., Sam Ge, S., Du, X.: Drift Compensation for Electronic Nose by Semi-Supervised Domain Adaptation. IEEE Sensors Journal 14, 657–665 (2014)
17. Belkin, M., Niyogi, P., Sindhwani, V.: Manifold regularization: A geometric framework for learning from labeled and unlabeled examples. J. Mach. Learn. Res. 7, 2399–2434 (2006)
18. Zhang, L., Tian, F.C.: A new kernel discriminant analysis framework for electronic nose recognition. Analytica Chimica Acta 816, 8–17 (2014)
19. Zhang, L., Tian, F., Nie, H., Dang, L., Li, G., Ye, Q., Kadri, C.: Classification of multiple indoor air contaminants by an electronic nose and a hybrid support vector machine. Sens. Actu. B. 174, 114–125 (2012)
20. Brudzewski, K., Osowski, S., Dwulit, A.: Recognition of coffee using differential electronic nose. IEEE Trans. Instru. Meas. 61, 1803–1810 (2012)
21. Tudu, B., Metla, A., Das, B., Bhattacharyya, N., Jana, A., Ghosh, D., Bandyopadhyay, R.: Towards Versatile Electronic Nose Pattern Classifier for Black Tea Quality Evaluation: An Incremental Fuzzy Approach. IEEE Trans. Instru. Meas. 58, 3069–3078 (2009)
22. Gardner, J.W., Shin, H.W., Hines, E.L.: An electronic nose system to diagnose illness. Sens. Actu. B. 70, 19–24 (2000)
23. Zhang, L., Tian, F., Kadri, C., Pei, G., Li, H., Pan, L.: Gases concentration estimation using heuristics and bio-inspired optimization models for experimental chemical electronic nose. Sens. Actu. B. 160, 760–770 (2011)
24. Zhang, L., Tian, F.: Performance Study of Multilayer Perceptrons in a Low-Cost Electronic Nose. IEEE Trans. Instru. Meas. 63 (2014)
25. Di Carlo, S., Falasconi, M.: Drift Correction Methods for Gas Chemical Sensors in Artificial Olfaction Systems: Techniques and Challenges. Advances in Chemical Sensors, pp. 305–326 (2012)
26. Vergara, A., Vembu, S., Ayhan, T., Ryan, M.A., Homer, M.L., Huerta, R.: Chemical gas sensor drift compensation using classifier ensembles. Sens. Actu. B. 167, 320–329 (2012)
27. Romain, A.C., Nicolas, J.: Long term stability of metal oxide-based gas sensors for e-nose environmental applications: An overview. Sens. Actu. B. 146, 502–506 (2010)
28. Zhang, L., Tian, F., Liu, S., Dang, L., Peng, X., Yin, X.: Chaotic time series prediction of E-nose sensor drift in embedded phase space. Sens. Actu. B. 182, 71–79 (2013)
29. Arul Pon Daniel, D., Thangavel, K., Manavalan, R., Chandra, S., Boss, R.: ELM-Based Ensemble Classifier for Gas Sensor Array Drift Dataset. Computational Intelligence, Cyber Security and Computational Models. Advances in Intelligent Systems and Computing 246, 89–96 (2014)
30. Lujan, I.R., Fonollosa, J., Vergara, A., Homer, M., Huerta, R.: On the calibration of sensor arrays for pattern recognition using the minimal number of experiments. Chemometrics and Intelligent Laboratory Systems 130, 123–134 (2014)
31. `http://archive.ics.uci.edu/ml/datasets/Gas+Sensor+Array+Drift+Dataset+at+Different+Concentrations`

Quasi-Linear Extreme Learning Machine Model Based Nonlinear System Identification

Dazi Li, Qianwen Xie, and Qibing Jin

Institute of Automation, Beijing University of Chemical Technology,
Beijing 100029, China

Abstract. A regression algorithm of quasi-linear model with extreme learning machine (QL-ELM) and its applications for nonlinear system identification are presented. The distinctive feature of the proposed method is that the Quasi-linear model is constructed as a linear ARX model with a complicate nonlinear coefficient. It not only has various linearity properties but also shows some good approximation ability. The complicated coefficients are separated into two parts. The linear part is determined by recursive least square, while the nonlinear part is identified through extreme learning machine. The whole methodology is presented in detail. The effectiveness and accuracy of the proposed method is extensively verified in two nonlinear system identification, including a chemical continuously stirred tank reactor (CSTR) process.

Keywords: Quasi-linear model, Extreme learning machine (ELM), Nonlinear process, Identification, Recursive least squares (RLS).

1 Introduction

Nonlinear process is very common in actual industrial processes. However, in many cases it is difficult to obtain a dynamic model via the physical processes [1-2]. Therefore, nonlinear black-box structure is often considered to describe an uncertainly nonlinear system.

In recent years, some block-oriented models have been proposed and applied widely, such as Wiener model [3] and Hammerstein model [4-6]. Both of them are quite simple structures, but they have limitation in systems which cannot be easily separated into a linear dynamic block and a memoryless nonlinear one. Volterra model [7] provides an elaborate mathematical description for a great many of nonlinear systems. But the obvious shortcoming is its high complexity to identify the kernel function. Meanwhile, many methods combining the nonlinear nonparametric models with some conventional statistical models have achieved some great results. McLoone et al. [8] proposes an off-line hybrid training algorithm for feed-forward neural networks. Peng et al. [9-10] proposes hybrid pseudo-linear RBF-AR, RBF-ARX models, and Marjan Golob et al.[11] proposes a decomposed neuro-fuzzy ARX model supported by a neural network-based learning algorithm with reduced number of rules in the rule base.

J. Cao et al. (eds.), *Proceedings of ELM-2014 Volume 1*,
Proceedings in Adaptation, Learning and Optimization 3, DOI: 10.1007/978-3-319-14063-6_11

However, these models show highly nonlinear characteristics, which are difficult to analysis in theory. Linear expression will be more convenient and valuable for control design as well as the control law derivation. Hu et al. [12-13] proposes a quasi-linear model, which is constructed as a linear structure from a macro standpoint with non-linear coefficients. Such model has a great flexibility to deal with the nonlinearity of the system.

The model represents as a linear one with a nonlinear hybrid structure, which often utilizes diversity and flexibility of neural networks to identify. In the past years, some artificial neural networks such as feed-forward [14-15] and feedback [16-17] are used in nonlinear system modeling. Furthermore, it is known that the neural network and support vector regression (SVR) [18] have some criticisms on their slow learning speed and parameters adjustment problem. To deal with the above problems, extreme learning machine (ELM) proposed by Huang et al. [19-20] shows great advantages. With the input layer weights and hidden biases chosen randomly, it can also obtain a smaller training error via a canonical equation. The advantage of ELM is its low computational effort and high generalization. It has been successfully applied to pattern recognition [21], fault diagnosis [22], function approximation [23] and reinforcement learning [24].

In this paper, a novel quasi-linear ELM (QL-ELM) structure is proposed, which is capable of effective identification of nonlinear dynamic systems. The model uses not only the easy-to-express feature of the linear model, but also the ELM incorporated to the nonlinear part which increases the flexibility and the learning speed for a better generalization performance. Furthermore, this identification method has less human interference.

2 Extreme Learning Machine

Unlike other traditional implementations, extreme learning machine is a single hidden layer feed-forward neural networks (SLFNs) which randomly choose the input weights and biases. For the input nodes $x_i = [x_{i1}, x_{i2}, ..., x_{in}]$ and output nodes $y_i = [y_{i1}, y_{i2}, ..., y_{im}]$ $(i \in (1,2,...,N))$ SLFNs with M hidden nodes and activation function $g(x)$ are expressed as

$$\sum_{i=1}^{M} \beta_i g(w_i \cdot x_i + b_i) = y_i \tag{1}$$

where $\beta_i = [\beta_{i1}, \beta_{i2}, ..., \beta_{in}]^T$ are the weights between hidden layer and the output nodes. $w_i = [w_{i1}, w_{i2}, ..., w_{in}]^T$ are the weights between input vectors and hidden layer. In addition, b_i are the biases of the hidden nodes. The output of the hidden layer is written as a matrix H , Eq. (1) can be rewritten as

$$H\beta = Y \tag{2}$$

where $H = \begin{bmatrix} g(w_1 x_1 + b_1) & \dots & g(w_M x_1 + b_M) \\ \dots & \dots & \dots \\ g(w_1 x_N + b_1) & \dots & g(w_M x_N + b_M) \end{bmatrix}_{N \times M}$.

With these theorems proposed in [23-24], the main idea of the ELM is that training problem is simplified to find a least square solution. According to the Moore-Penrose generalized inverse theory, the output can be calculated in one step using the equation

$$\hat{\beta} = H^{\dagger} Y \tag{3}$$

This one step algorithm can produce fast speed. It also avoids producing local optimum results, and poor generalization.

3 The Quasi-Linear ELM Model Treatment

A quasi linear model can be seen as a neural network embedded in the coefficients of a linear model. For a nonlinear SISO described as

$$y(t) = f(X^T(t)) + e(t) \tag{4}$$

where $X(t) = [y(t-1), y(t-2), \dots, y(t-n_a), u(t), u(t-1), \dots, u(t-n_b)]^T$, $X(t)$ is the regression vector. n_a, n_b are the order of the system $n_a + n_b = d, \varphi \in R^d$. $e(t)$ is stochastic noise with zero-mean. Using Taylor equation expands the $f(X^T(t))$ around the zero region

$$y(t) = f(0) + f'(0)X(t) + \frac{1}{2}X^T(t)f''(0)X(t) + \dots + \delta_n \tag{5}$$

Set $y_0 = f(0)$; $\Theta(X(t))^T = (f'(0) + \frac{1}{2}X^T(t)f''(0) + \dots)$.

Eq. (5) can be rewritten as

$$y(t) = y_0 + X^T(t)\Theta(X(t)) + e(t) = X^T(t)(\theta + \Theta(X(t))) + e(t) \tag{6}$$

The quasi-linear has a linear structure which can be processed as a functional-coefficient ARX model. It can be separated into a nonlinear and a linear part described as

$$y(t) = y_L(t) + y_N(t) + e(t) \tag{7}$$

$$y_L(t) = X^T \theta \tag{8}$$

$$y_N(t) = X^T(t)\Theta(X(t)) \tag{9}$$

with $\Theta = [a_1, a_2, \dots, a_n, b_1, b_2, \dots, b_m]^T$.

For near linear cases, nonlinear part is supplement for nonlinear feature, so linear part can achieve a good regression result. For the nonlinear cases, using nonlinear part as interpolated coefficient expends the regression space.

The coefficient in Eq. (9) can be seen as a multi-dimensional input space X to a one-dimensional scalar space $\Theta(X(t))$ .Using ELM to estimate nonlinear part parameters will be more convenient and concise. The model is rewritten as

$$y(t) = X^T(t)\sum_{i=1}^{M} W_i^2 \Gamma(W_i^1 X(t) + B) + X^T\theta + b + e(t) \tag{10}$$

define $T(t) = W_2\Gamma(W_1 X(t) + B)$.

In fact, θ is seen as the biases vector of the output nodes, and W_1, W_2, B as the nonlinear parameters which present the input, output weights and the node biases respectively. Γ is the activation function. The proposed QL-ELM model is shown in Fig. 1, where $X(t)$ is the input of the ELM. The QL-ELM model partially disperses the complexity of the model using its linear properties, so it needs not have many centers to achieve similar prediction accuracy. It means that the QL-ELM model may require a smaller number of hidden nodes than normal ELM model.

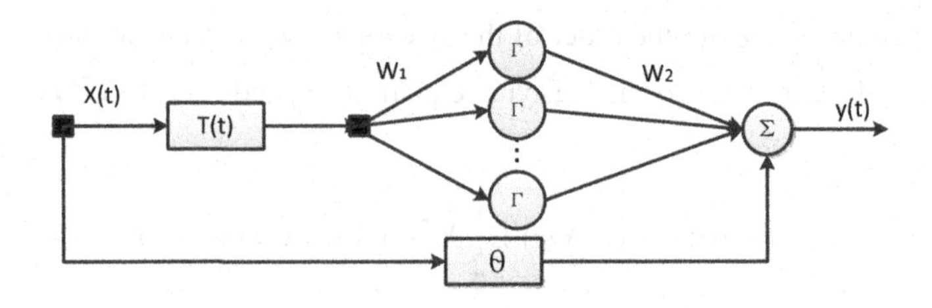

Fig. 1. The structure of quasi linear-ARX model

4 Quasi Linear-ELM Model Learning Algorithm

The whole process is described in Fig. 2. n and m are the orders of the input and output. The parameters are updated during each iterative process until the ultimate goal to make the error between the output of actual model and the QL-ELM model minimum. The nonlinear part is updated by the deviation between $y(t)$ and $y_L(t)$ .The linear part is updated by the deviation between $y(t)$ and $y_N(t)$.

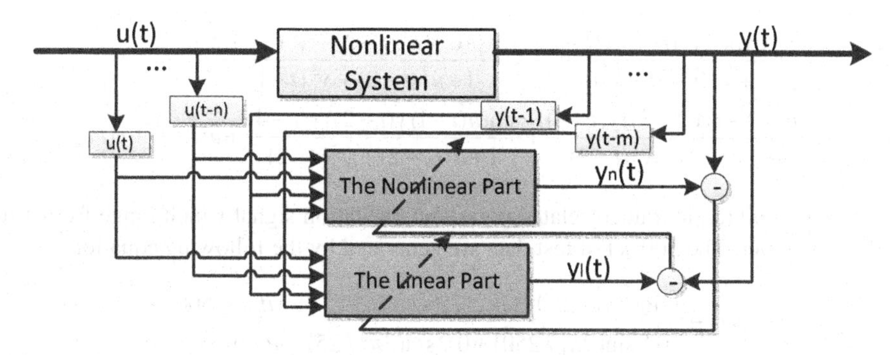

Fig. 2. The identification process

Linear part: at each iteration, RLS is used to estimate linear parameters.

Nonlinear part: fixing the weights of input layer and biases, the training error $e = \|HW_2 - T\|^2$ is minimized by ELM. Where the activation function is chosen as: $\Gamma(x) = \dfrac{1}{1+e^{-x}}$.

$\{(y(i), u(i)); i \in 1,.., N\}$ is the training set. The QL-ELM learning algorithm is implemented by the following steps:

Step1. Initialization: Choose the order of the regression vector m, n . Set θ to zero, the number of nodes in hidden layer M , and nonlinear parameters W_1, W_2, B to some small values randomly. The number of iterations is set to n=1.

Step2. Calculate the linear part and estimate θ_L using Eq. (8).

Step3. Calculate the nonlinear part and estimate $\theta_N(W_2, t)$ using Eq. (9)

Step4. Turn to step 2, set n=n+1 until the conditions satisfied.

5 Simulation Study

In this section, two simulation examples are used to evaluate the nonlinear system identification power of the QL-ELM model. The proposed method is compared with other identification methods, such as the linear and RBF kernel SVR, Quasi linear SVR [13] and back propagation (BP) neural network. Performance is measured by the Root Mean Square Error (RMSE), the indicator is expressed by

$$P = \sqrt{\frac{1}{N}\sum_{k=1}^{N}(y(k) - y(k \mid \theta))} \tag{11}$$

5.1 Case 1. Numerical Experiments

Consider a nonlinear system with 7 input vectors, $e(t)$ is the white noise range from 0 to 0.5.

$$y(t)=\frac{\exp(-y^2(t-2)y(t-1))}{1+u^2(t-3)+y^2(t-2)}+\frac{\exp(0.5(u^2(t-2)+y^2(t-3))}{1+u^2(t-2)+y^2(t-1)} \\ +\frac{\sin(u(t-1)y(t-3))y(t-3)}{1+u^2(t-1)+y^2(t-3)}+\frac{\sin(u(t-1)y(t-2))y(t-4)}{1+u^2(t-2)+y^2(t-2)}+u(t-1)+e(t) \quad (12)$$

Using 1000 random uncorrelated persistent excitation signal which range from 1 to -1 as the input. The 800 set of test data are generated by the following equation

$$u(t)=\begin{cases}\sin(2\pi t/250) & \text{if } t<500 \\ 0.8\sin(2\pi t/250)+0.2\sin(2\pi t/25) & \text{otherwise}\end{cases} \quad (13)$$

Then proposed method is used to identify the model. Obtaining the optimal parameters of SVR via cross-validation methodology, the scale parameter σ (The variance in RBF kernel function) is set to 0.01 and the penalty factor is set to 800. Fig. 3 indicates the testing result of different methods. The red solid line is the output of the proposed method. It can be seen that it has better approximation ability in unsmooth section. In addition, it can be seen from Table. 1 that the proposed QL-ELM obtains a smaller test error and faster convergence speed.

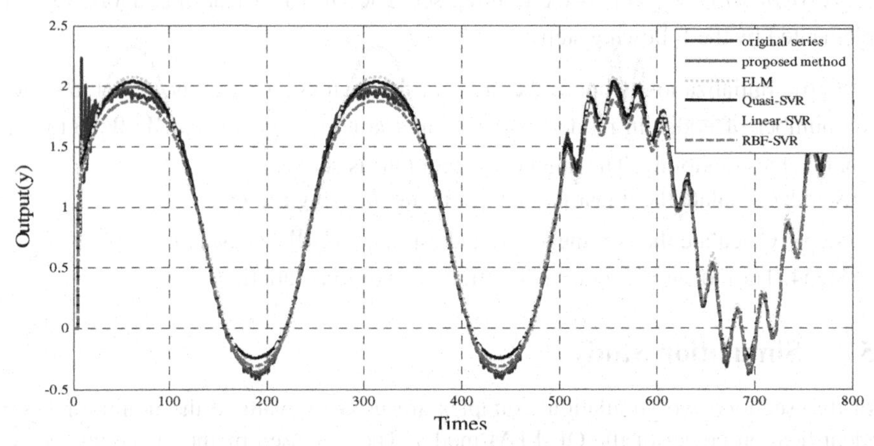

Fig. 3. The identification results with different methods

Table 1. A comparison performance of different methods in case 1

Algorithm	Hidden Nodes	Times(seconds)	RMSE
ELM	60	0.541496	0.0739
QL+ELM	**10**	**0.581019**	**0.0427**
Linear-SVR	None	9.336846	0.1503
RBF-SVR	None	13.652457	0.0876
Quasi-SVR	None	0.582523	0.0615

5.2 Case 2. CSTR Process

A typical nonlinear system in chemical processes is CSTR process. In this study, the dynamic behavior is described by the following equation [25].

$$\begin{aligned} \dot{x}_1 &= -x_1 + D_a(1-x_1)\exp\left(\frac{x_2}{1+x_2/\varphi}\right) + d_1. \\ \dot{x}_2 &= -(1+\delta)x_2 + B\cdot D_a(1-x_2)\exp\left(\frac{x_2}{1+x_2/\varphi}\right) + \delta\cdot u + d_2. \\ y &= x_1 . \end{aligned} \quad (14)$$

where x_1 and x_2 represent the dimension-less reactant concentration and reactor temperature; d_1 and d_2 denote the system's disturbances. The input u is the cooling jacket temperature. The physical parameters in the CSTR model is B=8, δ =0.3,Da=0.072, φ =20.

Under the initial condition $[x_1(0), x_2(0)] = [-0.1, 0.1]$, $u(t) = 0.4 * randn(1)$ and the integral step size $\Delta t = 0.1$, it generates 3000 groups of samples, 501-1800 groups of data as training samples and 1801-2200 as test samples. Firstly, we use the AIC value [26] to decide the order of the regression vectors. From Fig.4 below, it can be seen that in order to reduce the redundant feature vector, we choose the smallest AIC value as the order of the system, so $n = 4, m = 5$.

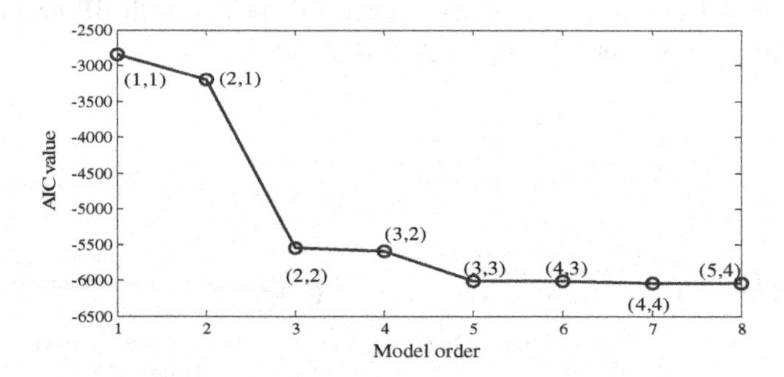

Fig. 4. Values of the order determination based on AIC

The result of identifying the model above with the QL-ELM and its corresponding error are shown in Fig. 6(a). Comparative method ELM and the result are shown in the Fig. 6(b). It shows that there is a smaller error and a more stable test result in the proposed method.

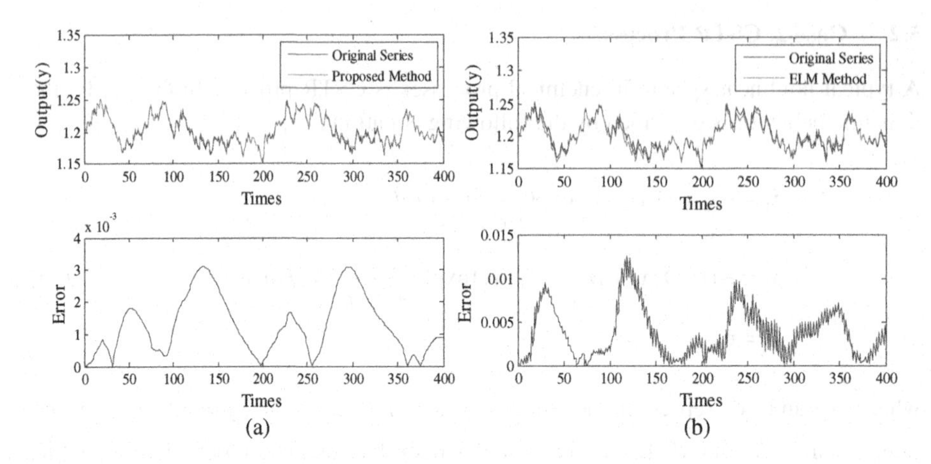

Fig. 5. Identification results and error of CSTR

Fig. 6 shows the convergence of the linear part in different interaction. (a)- (d) represent convergence of the linear parameters when the number of iterations n = 1, 3, 5, 10, respectively. With the increasing number of iterations, the parameters can get a faster convergence to the true value ultimately.

The different identification methods of CSTR system are compared in items of the number of nodes in the hidden layer, training and learning times, identification measurable indicator, which is listed in Table. 2. It is easy to see that the proposed method is the trade-off in terms of time and accuracy compared with BP neural network and SVR. It needs less time and gains higher accuracy.

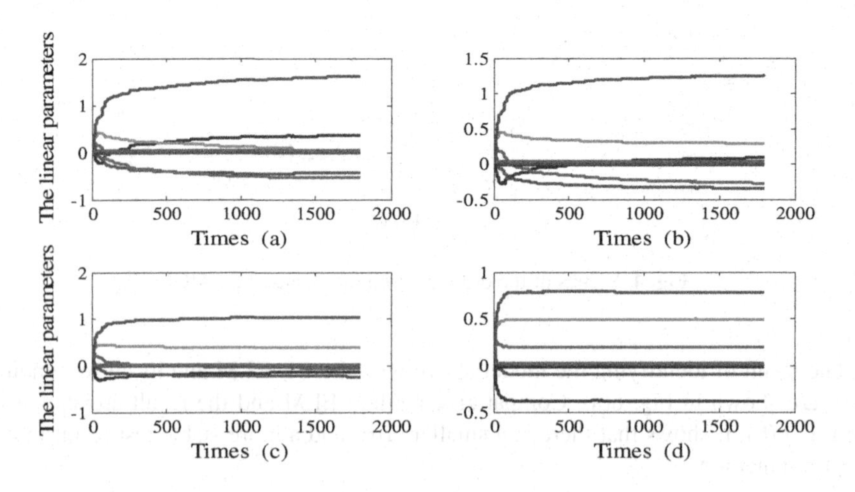

Fig. 6. The convergence of the linear part parameters

Table 2. A comparison performance of different methods for CSTR

Algorithm	Hidden nodes	Times(seconds)	RMSE
ELM	80	0.255673	0.0041
QL+ELM	**10**	**0.342147.**	**0.0015**
Linear-SVM	None	2.192229	0.0019
RBF-SVM	None	9.698744	0.0038
BP	40	19.094721	0.0071
RLS+BP	40	26.127525	0.0043

6 Conclusion

In this paper, a novel QL-ELM algorithm is proposed, which is computationally efficient for identifying some nonlinear systems. In the QL-ELM structure, the nonlinear SLFN expends the space of the regression results. In two examples of nonlinear systems identification, the advantages of QL-ELM are its reduction of computational complexity, improvement of convergence rate and identification accuracy. The iterative learning process speeds up the convergence of parameters. It also has less unknown parameters to adjust. Simulation results indicate that performance of the QL-ELM is superior to the conventional BP neural networks, SVR with linear and RBF kernel.

References

1. Madár, J., Abonyi, J., Szeifert, F.: Genetic programming for the identification of nonlinear input-output models. Industrial & engineering chemistry research 44, 3178–3186 (2005)
2. Sjöberg, J., Zhang, Q., Ljung, L., Benveniste, A., Delyon, B., Glorennec, P.-Y., Hjalmarsson, H., Juditsky, A.: Nonlinear black-box modeling in system identification: a unified overview. Automatica 31, 1691–1724 (1995)
3. Savaresi, S.M., Bittanti, S., Montiglio, M.: Identification of semi-physical and black-box non-linear models: the case of MR-dampers for vehicles control. Automatica 41, 113–127 (2005)
4. Tang, Y.G., Li, Z.H., Guan, X.P.: Identification of nonlinear system using extreme learning machine based Hammerstein model. Communications In Nonlinear Science And Numerical Simulation 199, 3171–3183 (2014)
5. Wills, A., Schön, T.B., Ljung, L., Ninness, B.: Identification of Hammerstein–Wiener models. Automatica 49, 70–81 (2013)
6. Wang, D., Ding, F.: Extended stochastic gradient identification algorithms for Hammerstein–Wiener ARMAX systems. Computers & Mathematics with Applications 56, 3157–3164 (2008)
7. Pearson, R., Ogunnaike, B.A.: Identification and control using Volterra models. Springer (2002)

8. McLoone, S., Brown, M.D., Irwin, G., Lightbody, G.: A hybrid linear/nonlinear training algorithm for feed forward neural networks. IEEE Transactions on Neural Networks 9, 669–684 (1998)
9. Peng, H., Ozaki, T., Haggan-Ozaki, V., Toyoda, Y.: A parameter optimization method for radial basis function type models. IEEE Transactions on Neural Networks 14, 432–438 (2003)
10. Peng, H., Ozaki, T., Toyoda, Y., Shioya, H., Nakano, K., Haggan-Ozaki, V., Mori, M.: RBF-ARX model-based nonlinear system modeling and predictive control with application to a NOx decomposition process. Control Engineering Practice 12, 191–203 (2004)
11. Golob, M., Tovornik, B.: Input–output modelling with decomposed neuro-fuzzy ARX model. Neurocomputing 71, 875–884 (2008)
12. Hu, J., Kumamaru, K., Inoue, K., Hirasawa, K.: A hybrid quasi-ARMAX modeling scheme for identification of nonlinear systems. Transactions-Society of Instrument and Control Engineers 34, 977–985 (1998)
13. Cheng, Y., Hu, J.: Nonlinear system identification based on SVR with quasi-linear kernel. In: The International Joint Conference on Neural Networks (IJCNN) (IEEE 2012), pp. 1–8 (2012)
14. Wu, X.-J., Huang, Q., Zhu, X.-J.: Thermal modeling of a solid oxide fuel cell and micro gas turbine hybrid power system based on modified LS-SVM. International Journal of Hydrogen Energy 36, 885–892 (2011)
15. Cisse, Y., Kinouchi, Y., Nagashino, H., Akutagawa, M.: Identification of homeostatic dynamics for a circadian signal source using BP neural networks. ITBM-RBM 21, 24–32 (2000)
16. Atencia, M., Joya, G., Sandoval, F.: Identification of noisy dynamical systems with parameter estimation based on Hopfield neural networks. Neurocomputing 121, 14–24 (2013)
17. Wei, H.-L., Billings, S.A., Zhao, Y., Guo, L.: An adaptive wavelet neural network for spatio-temporal system identification. Neural Networks 23, 1286–1299 (2010)
18. Lu, Z., Sun, J.: Non-Mercer hybrid kernel for linear programming support vector regression in nonlinear systems identification. Applied Soft Computing 9, 94–99 (2009)
19. Huang, G.-B., Zhu, Q.-Y., Siew, C.-K.: Extreme learning machine: theory and applications. Neurocomputing 70, 489–501 (2006)
20. Huang, G.-B., Chen, L.: Convex incremental extreme learning machine. Neurocomputing 70, 3056–3062 (2007)
21. Song, Y., Zhang, J.: Automatic recognition of epileptic EEG patterns via Extreme Learning Machine and multiresolution feature extraction. Expert Systems with Applications 40, 5477–5489 (2013)
22. Wong, P.K., Yang, Z., Vong, C.M., Zhong, J.: Real-time fault diagnosis for gas turbine generator systems using extreme learning machine. Neurocomputing (2013)
23. Han, F., Huang, D.-S.: Improved extreme learning machine for function approximation by encoding a priori information. Neurocomputing 69, 2369–2373 (2006)
24. Escandell-Montero, P., Martínez-Martínez, J.M., Martín-Guerrero, J.D., Soria-Olivas, E., Gómez-Sanchis, J.: Least-squares temporal difference learning based on an extreme learning machine. Neurocomputing (April 5, 2014)
25. Chen, C.T., Peng, S.T.: Intelligent process control using neural fuzzy techniques. Journal of Process Control 9, 493–503 (1999)
26. Akaike, H.: A new look at the statistical model identification. IEEE Transactionson Automatic Control 19(6), 716–723 (1974)

A Novel Bio-inspired Image Recognition Network with Extreme Learning Machine

Lin Zhang[1], Yu Zhang[1,*], and Ping Li[1,2]

[1] School of Aeronautics and Astronautics, Zhejiang University, Hangzhou, China
[2] Department of Control Science and Engineering, Zhejiang University, Hangzhou, China
{zhanglin0123,zhangyu80}@zju.edu.cn, pli@iipc.zju.edu.cn

Abstract. This paper presents a novel bio-inspired network for image recognition. The HMAX model and the extreme learning machine (ELM) are combined, to construct a five-layer feed-forward network: S1-C1-S2-C2-H. The previous four layers, originating from HMAX, provide robust feature representation of specific object, and the feature classification stage at the H layer is implemented with ELM. The HMAX model simulates the hierarchical processing mechanism in primate visual cortex, to calculate feature representation. As a biological learning algorithm for SLFNs, ELM learns much faster with good generalization, and performs well in classification applications. Our experimental results show effective accuracy performance with fast learning speed.

Keywords: Image recognition, Extreme learning machine, HMAX, Feature representation.

1 Introduction

Object recognition has been a popular area of intense research, and is also a very challenging task in computer vision, while human vision with unique processing mechanism has the ability to recognize objects rapidly, accurately, and effortlessly. The difficulty of object recognition in images is due to different illumination, viewpoints, occlusions, scale and shift transforms. Thus achieving robust object recognition would be beneficial for many fields and applications, such as security surveillance, robot navigation, clinical image understanding and many others.

Many research works have been done for object recognition. Mohan [1] built Haar wavelets parts detector to represent the image, and then use a support vector machine (SVM) for classification. Lowe [2] developed the scale-invariant feature transform (SIFT). Bio-inspired features based on Gabor filters and MAX operations have been developed [3, 4]. Although the performance of object recognition increases, none of the existing algorithms available today can surpass the performance of the human brain. Object recognition in human brain is largely invariant to changes of the object, which may give us inspirations to design algorithms for object recognition. A hierarchical cortical based model, named HMAX [5, 6], focuses on designing simple and complex operations inspired by the visual cortex, and it can provide robust

* Corresponding author.

J. Cao et al. (eds.), *Proceedings of ELM-2014 Volume 1*,
Proceedings in Adaptation, Learning and Optimization 3, DOI: 10.1007/978-3-319-14063-6_12

representations of specific images, outperforming SIFT under various invariance tasks [7]. Recently, a novel learning algorithm for single hidden layer feed-forward networks (SLFNs), namely, extreme learning machine (ELM), proposed by Huang et al.[8], can be applied to classification problems [9]. Also it has been successfully applied in the face recognition [10], which improves the recognition accuracy.

This paper brings together two bio-inspired algorithms, HMAX and ELM, and insights to construct a novel bio-inspired network for image recognition. Since the HMAX features have better scale and translation invariance, the four-layer HMAX model, is employed for providing robust feature representation of specific object image. As it has better performance than traditional methods, such as SVM, and it learns extremely fast, which is akin to the fast learning mechanism of the higher cortical areas, ELM is introduced for feature representation classification. Several experiments will be performed to demonstrate the superiority of our proposed network.

The rest of the paper is organized as follows. In Section 2, preliminary information about HMAX and ELM are presented. Section 3 details the proposed ELM-based image recognition algorithm. In Section 4, several experiments are performed, and followed by results and discussions. The paper is concluded in Section 5.

2 Preliminaries

2.1 Brief of the HMAX

The HMAX model, proposed by Riesenhuber and Poggio [5], summarizes the basic facts about the ventral visual stream, a hierarchy of brain areas thought to mediate object recognition in cortex. Serre et al.[6] extend the original HMAX model by adding multi-scale representations as well as more complex visual features. The general HMAX follows a basic alternating convolution/pooling scheme. Each convolution step yields a set of feature maps and each pooling step provides robustness to variations in these feature maps. And the operations of each layer are detailed in [6]. The lowest levels correspond to orientation-selective cells in primary visual cortex, while the highest levels correspond to object-selective cells in inferotemporal cortex. Many researchers have improved the HMAX in different aspects [4, 11, 12].

2.2 Brief of Extreme Learning Machine

Extreme learning machine (ELM) [8] was originally developed for the SLFNs and then extended to the "generalized" SLFNs which may not be neuron alike [13]. Here, take SLFNs with L hidden nodes as an example. The output function of ELM for generalized SLFNs is

$$f_L(\mathrm{x}) = \sum_{i=1}^{L} \beta_i h_i(\mathrm{x}) = \sum_{i=1}^{L} \beta_i G(\mathbf{a}_i, b_i, \mathrm{x}) = \mathbf{h}(\mathrm{x})\boldsymbol{\beta} \tag{1}$$

where $G(\mathbf{a}_i, b_i, \mathrm{x})$ is the activation function of the ith hidden node, $\mathbf{a}_i$ and b_i are the parameters which are randomly generated and fixed. $\boldsymbol{\beta} = [\beta_1, \beta_2, ..., \beta_L]^{\mathrm{T}}$ represents the vector of the output weights between the hidden layer and the output node. $\mathbf{h}(x) = [h_1(x), h_1(x),\ldots, h_L(x)]^T$ is the output vector of the hidden layer with respect to the input $\mathbf{x}$. $\mathbf{h}(x)$ actually maps data from the M-dimensional input space to the L-dimensional hidden-layer feature space $\mathbf{H}$, and thus, $\mathbf{h}(x)$ is indeed a feature mapping.

ELM does not only aim at reaching the minimum training error but also the smallest norm of the output weights, which yields better generalization performance [9]. So the cost function is expressed as

$$\text{Minimize}: \ ||\mathbf{H}\boldsymbol{\beta} - \mathbf{T}|| \ \text{and} \ ||\boldsymbol{\beta}|| \tag{2}$$

where $\mathbf{T} = [\mathbf{t}_1, \mathbf{t}_2,\ldots, \mathbf{t}_N]^T$ contains the training target value and $\mathbf{H}$ is the hidden-layer output matrix.

$$\mathbf{H} = \begin{bmatrix} \mathbf{h}(x_1) \\ \vdots \\ \mathbf{h}(x_N) \end{bmatrix} = \begin{bmatrix} h_1(x_1) \cdots h_L(x_1) \\ \vdots \quad \cdots \quad \vdots \\ h_1(x_N) \cdots h_L(x_N) \end{bmatrix}_{N \times L} \tag{3}$$

The minimal norm least square method [5] instead of the standard optimization method was used in the original implementation of ELM [4, 18].

$$\boldsymbol{\beta} = \mathbf{H}^{\dagger}\mathbf{T} \tag{4}$$

where $\mathbf{H}^{\dagger}$ is the Moore–Penrose generalized inverse of matrix $\mathbf{H}$.

3 A Novel ELM Based Image Recognition Algorithm Design

With the study of the primate visual cortex in great detail, some facts about the object recognition are known to us. Core object recognition is well described by a largely feed forward cascade of nonlinear filtering operations and we humans can recognize objects in about 100~200 ms [14]. That is to say, a remarkable aspect of primate visual system is that the recognition process can be very fast. In accordance with the fact, we design a feed-forward image recognition network, which utilizes the ELM for classification using high-level feature representation that is generated though the HMAX scheme. The architecture of our proposed network is shown in Fig. 1. The feed-forward network consists of five layers: S1-C1-S2-C2-H. And two main stages of the network are the feature representation with HMAX, simulating the biological feature building mechanism, and the feature classification with ELM, focusing on the biological learning mechanism.

3.1 Feature Representation with HMAX

The HMAX model is employed to calculate the image feature representation, which exhibits a better trade-off between invariance and selectivity than template-based or histogram based approaches. The image layer is the input layer, where the color image is converted to the grayscale image. And then, the feature construction, selection, and extraction are performed along the hierarchy from the S1 layer to C2 layer. The S1 layer extracts the simple features with Gabor filters. The C1 layer selects the local maximum value of S1 simple features, which increases the invariance and reduces the dimension. The S2 layer combines the C1 features into more complex features using a Radial Basis Filter or a Normalized Dot Product, where the prototypes are defined during a training phase which impacts the type of complex features representation. At the C2 layer, the final feature vector is computed by selecting the maximum output of S2 across all positions and scales, to gain global invariance. The high-level C2 feature is a feature vector with M dimensions, which is shown in Fig 1. And the M elements of the C2 feature vector are defined as M input nodes of the ELM for classification.

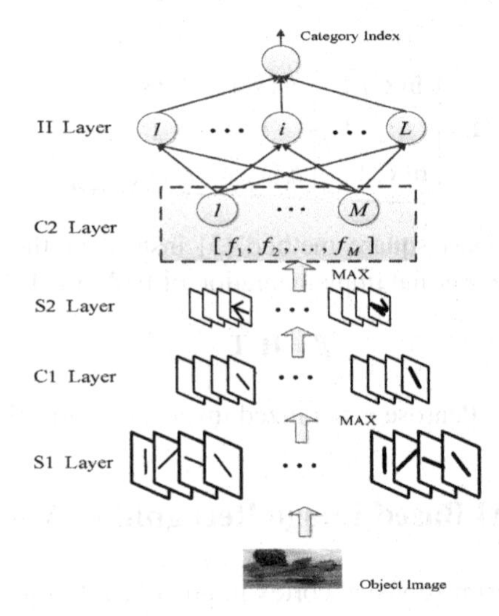

Fig. 1. The architecture of our proposed image recognition network

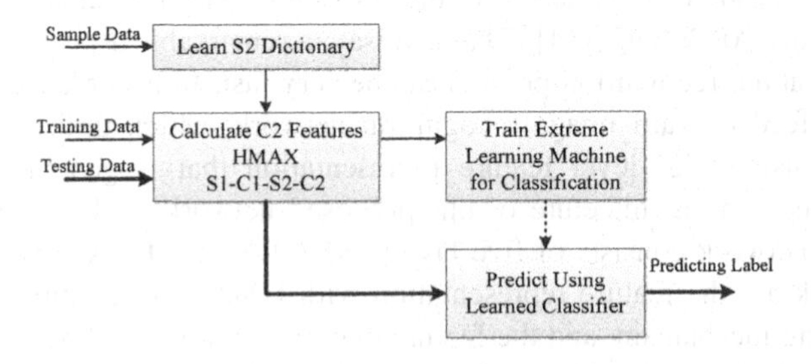

Fig. 2. The schematic diagram of the image recognition system

3.2 Feature Classification with ELM

The biologically inspired classification algorithm ELM is selected as the proper classifier with good generalization, in conjunction with the HMAX feature representation for object recognition. The output of the ELM classifier $\mathbf{h}(\mathrm{x})\boldsymbol{\beta}$ can be as close to the class labels in the corresponding regions as possible, m-class of classifiers have m output nodes. The predicted class label of a given testing sample is the index number the output node which has the highest output value for the given testing sample [9]. The size of the hidden layer matrix **H** is only decided by the number of training examples N and the number of hidden nodes L, which is irrelevant to the number of output nodes (number of classes). ELM has better scalability, faster learning speed, and smaller training error than traditional SVM. Thus, ELM can be utilized easily and effectively without tedious and time-consuming parameter tuning.

The input of the ELM classifier is the HMAX C2 feature vector, and its output is the predicting category index. If the feature mapping is unknown to users, kernels can be applied in ELM [9]. Thus whether the feature mapping is known or unknown, we can always use the ELM for feature classification, to perform the image recognition. And Fig. 2 demonstrates the schematic diagram of our proposed image recognition procedure, including data preparation, S2 dictionary generation, C2 feature representation calculation, ELM classifier training, and category index predicting.

4 Experiments and Results

4.1 Datasets and Settings

In order to evaluate our proposed network, several experiments were performed on two image datasets: (1) Fifteen Scenes [15]: It is composed of 15 natural categories of urban and rural scenes for a total of 4885 images. (2) Still Actions [16]: There are about 1200 images in total for six action queries. Some sample images are shown in Fig. 3. All the following experiments are carried out in MATLAB 2010b environment running in the Core i5-3470 3.20GHz CPU with 4-GB RAM computer.

4.2 Results and Discussions

For validating the performance of our proposed network, the non-kernel based ELM classifier and the kernel based ELM classifier were implemented, to compare with the SVM classifier. For non-kernel based ELM classifier, in which the feature mapping is known to us, the number of hidden nodes L and the stable scalar C are the tunable parameters. The output function in our experiment, is the Sigmoid function, define in Eq.(5). For kernel based ELM classifier, in which the feature mapping is unknown to us, the radial basis function (RBF), define as in Eq.(6), where **u** and **v** are feature vectors, is used as the corresponding kernel in our experiment. Thus, the tunable parameters of RBF-kernel based ELM classifier are the parameter γ of the kernel and the stable scalar C. In the SVM, we also use the RBF as the kernel function, as in Eq.(6). The parameter γ of the kernel and the soft margin parameter C can be tuned via evaluating the generalization performance of each pair (C, γ) on the training data.

$$h_i(\mathrm{x}) = G(\mathbf{a}_i, b_i, \mathrm{x}) = \frac{1}{1+\exp(\mathbf{a}_i \cdot b_i + \mathrm{x})} \tag{5}$$

$$K(u,v) = exp(-\gamma |\mathbf{u}-\mathbf{v}|^2) \tag{6}$$

Fig. 3. Sample images from two datasets (Left: Fifteen Scenes, Right: Still Actions)

Table 1. The tunable parameters of SVM(C, γ), ELM(C, L), and k-ELM(C, γ)

k	1	2	3	4	5	6	7	8	9	10
C(k)	0.01	0.02	0.05	0.1	0.2	0.5	1	2	5	10
γ(k)	0.001	0.002	0.005	0.01	0.02	0.05	0.1	0.2	0.5	1
L(k)	10	30	50	70	100	200	300	400	500	600

11	12	13	14	15	16	17	18	19	10	21
20	50	100	200	500	700	1000	1200	1500	2000	
2	5	10	20	50	100	200	500	700	1000	1500
700	800	900	1000	1500	2000	2500	3000	3500	4000	4500

* C(k) and γ(k) are the same for SVM, k-ELM.

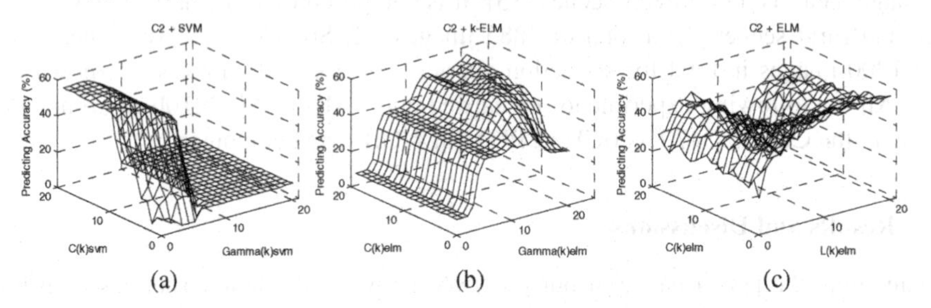

Fig. 4. Parameters selection effect on the predicting accuracy rate of Fifteen Scenes dataset (a) C2+SVM (C, γ), (b) C2+ELM (C, L), (c) C2+k-ELM (C, γ)

1) Fifteen Scenes

On the Fifteen Scenes dataset, we randomly select 10 sample images from each category, and then extract one feature per image, to generate the S2 dictionary. Therefore, the dimension of the feature representation is 150. The average time for S2 dictionary generation is 0.4360s per feature and the average time for C2 feature representation calculation is 1.0757s per image. Then, the feature representation is passed into the ELM for classification, and the number of input nodes is 150. As stated above, non-kernel based ELM (ELM) and kernel based ELM (k-ELM) are implemented for task, while the SVM for comparison.

The relationship of the predicting accuracy rate and parameter selection for the above three approaches is shown in Fig. 4, with 60 training examples and 30 testing examples per category. Totally 420 different pairs of (C, γ) for SVM and k-ELM, and

420 different pairs of (C, L) for ELM, which can be seen from Table 1, were tried to evaluate the sensitivity of parameter. The C2+ELM may less sensitive to the parameters when the number of the hidden nodes L, that is, the dimension of the feature space, is becoming larger. Although the kernel of the SVM and the k-ELM are both the RBF kernel, the tunable parameter C may have different meanings. Thus, the sensitivity of the parameters of SVM and k-ELM are different, and the k-ELM performs better than the SVM.

Next, we carried out three tests. The optimal parameters of each case are listed in Table 2. Also the average recognition results of each case were obtained based on 20 trials, and in each trial, the training data and testing data are randomly generated from the beginning and the end of each category in the dataset, respectively. And three approaches are employed on the same training set and testing set. From Table 2, the average predicting accuracy of three approaches increases when the rate Tr/Ts is becoming larger. Meanwhile, the average predicting accuracy rate of the C2+ELM and the C2+k-ELM is higher than that of the C2+SVM. And the average predicting accuracy rate of C2+k-ELM is a little higher than C2+ELM, because the ELM always randomly generates the coefficients according to the Sigmoid function in each trail while in k-ELM, the RBF kernel coefficients is determined. The key advantage of the C2+ELM and the C2+k-ELM is the small training time, and the average training time of the C2+k-ELM is much smaller than that of the C2+ELM, which indicates the C2+ELM and the C2+k-ELM learns quite fast. Comparing with SVM, the average predicting time of the C2+k-ELM is also small, which is benefical for fast image recognition.. Hence, the C2+ELM and the C2+k-ELM are quite appealing approaches with their less learning time and higher average predicting accuracy rate.

2) Still Actions

For this dataset, our goal to recognize the human actions in the still web images. When generating the S2 dictionary, 10 sample images are randomly selected from each category with 10 features per image. Thus the dimension of the feature vector is 600. We use the C2+ELM and C2+k-ELM to recognize the actions. The results of confusion tables are shown in Fig 5. With the techniques in [16], the mean accuracy is 86.67%. And we acheieve the average accuracy rate about 85% with our proposed network. Although it doesn't match the performance of theirs, our algorithm learns fast, and the dimension of feature vector is 600, which is much smaller than theirs.

Finally, the results of our proposed bio-inspired networks, the C2+ELM and the C2+k-ELM, on the two datasets for image recognition, demonstrate different average predicting accuracy performances. And what they have in common is the small average training time, and less sensitivity to parameters. Meanwhile, the factors that affect the average predicting accuracy performance, includes the Tr/Ts rates, the dimension of the feature vector, the parameters of the previous layers, and so on. For the Fifteen Scenes dataset, the results show that our proposed network is suitable for general scene recognition, and it manages to match the algorithm in [12], while achieving fast learning performance. For the Still Actions dataset, it is seen that our proposed

Table 2. Average performance and optimal parameters of three tests on Fifteen Scenes dataset

Case Tr/Ts	Approach	Predicting Accuracy/%	Standard Deviation	Training Time/s	Predicting Time/s	C	L/γ
15/30	C2+SVM	42.12	0.0219	0.1201	0.0301	1000	0.01
	C2+ELM	**43.10**	0.0221	**0.0292**	0.0302	0.02	2000
	C2+k-ELM[1]	**43.13**	0.0179	**0.0064**	0.0088	50	500
30/30	C2+SVM	47.11	0.0239	0.3502	0.0516	1000	0.01
	C2+ELM	**48.72**	0.0206	**0.0532**	0.0300	0.02	2000
	C2+k-ELM	**49.16**	0.0225	**0.0159**	0.0100	5	1000
60/30	C2+SVM	52.17	0.0235	1.0540	0.0940	1000	0.01
	C2+ELM	**53.34**	0.0263	**0.1494**	0.0312	0.02	2000
	C2+k-ELM	**54.02**	0.0226	**0.0683**	0.0222	2	50

*Tr/Ts = training examples/testing examples per category, L for ELM, γ for SVM and k-ELM

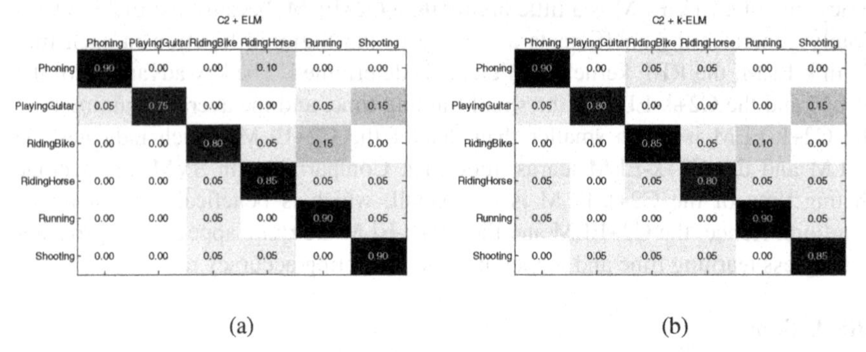

(a) (b)

Fig. 5. Confusion tables for Still Actions dataset (a) C2+ELM (C, L) = (0.02, 2000), (b) C2+k-ELM (C, γ) = (20, 50)

network can be used for recognize the human actions in the images, although the different backgrounds of the human actions may introduce some redundancy information. These results showcase that our proposed bio-inspired image recognition networks can be practical in many applications, with impressive predicting accuracy and fast learning speed.

5 Conclusion

In this paper, we have proposed a novel bio-inspired image recognition network based on the HMAX and the extreme learning machine. The network consists of five layers: S1-C1-S2-C2-H, to complete the object recognition task. It can potentially benefit both military and industrial applications because of its fast learning, fast processing and high precision. Our proposed network, combining two bio-inspired mechanisms together, seems a little step towards human brain alike recognizing and learning.

[1] The k-ELM is short for kernel based ELM, for distinguishing with non-kernel based ELM.

Acknowledgements. This work was supported by the National Natural Science Foundation of China (Grant No. 61005085) and Fundamental Research Funds for the Central Universities (2012QNA4024).

References

1. Mohan, A., Papageorgiou, C., Poggio, T.: Example-based object detection in images by components. IEEE Transactions on Pattern Analysis and Machine Intelligence 23, 349–361 (2001)
2. Lowe, D.G.: Distinctive image features from scale-invariant keypoints. International Journal of Computer Vision 60, 91–110 (2004)
3. Serre, T., Wolf, L., Poggio, T.: Object recognition with features inspired by visual cortex. In: IEEE Computer Society Conference on Computer Vision and Pattern Recognition CVPR 2005, pp. 994–1000. IEEE (2005)
4. Mutch, J., Lowe, D.G.: Object class recognition and localization using sparse features with limited receptive fields. International Journal of Computer Vision 80, 45–57 (2008)
5. Riesenhuber, M., Poggio, T.: Hierarchical models of object recognition in cortex. Nature Neuroscience 2, 1019–1025 (1999)
6. Serre, T., Wolf, L., Bileschi, S., Riesenhuber, M., Poggio, T.: Robust object recognition with cortex-like mechanisms. IEEE Transactions on Pattern Analysis and Machine Intelligence 29, 411–426 (2007)
7. Pinto, N., Barhomi, Y., Cox, D.D., DiCarlo, J.J.: Comparing state-of-the-art visual features on invariant object recognition tasks. In: 2011 IEEE workshop on Applications of computer vision (WACV), pp. 463–470. IEEE (2011)
8. Huang, G.-B., Zhu, Q.-Y., Siew, C.-K.: Extreme learning machine: a new learning scheme of feedforward neural networks. In: Proceedings of the 2004 IEEE International Joint Conference on Neural Networks, pp. 985–990. IEEE (2004)
9. Huang, G.-B., Zhou, H., Ding, X., Zhang, R.: Extreme learning machine for regression and multiclass classification. IEEE Transactions on Systems, Man, and Cybernetics, Part B: Cybernetics 42, 513–529 (2012)
10. Zong, W., Huang, G.-B.: Face recognition based on extreme learning machine. Neurocomputing 74, 2541–2551 (2011)
11. Huang, Y., Huang, K., Tao, D., Tan, T., Li, X.: Enhanced biologically inspired model for object recognition. IEEE Transactions on Systems, Man, and Cybernetics, Part B: Cybernetics 41, 1668–1680 (2011)
12. Theriault, C., Thome, N., Cord, M.: Extended coding and pooling in the HMAX model. IEEE Transactions on Image Processing 22, 764–777 (2013)
13. Huang, G.-B., Chen, L.: Enhanced random search based incremental extreme learning machine. Neurocomputing 71, 3460–3468 (2008)
14. Hung, C.P., Kreiman, G., Poggio, T., DiCarlo, J.J.: Fast readout of object identity from macaque inferior temporal cortex. Science 310, 863–866 (2005)
15. Lazebnik, S., Schmid, C., Ponce, J.: Beyond bags of features: Spatial pyramid matching for recognizing natural scene categories. In: 2006 IEEE Computer Society Conference on Computer Vision and Pattern Recognition, pp. 2169–2178. IEEE (2006)
16. Li, P., Ma, J., Gao, S.: Actions in still web images: Visualization, detection and retrieval. In: Wang, H., Li, S., Oyama, S., Hu, X., Qian, T. (eds.) WAIM 2011. LNCS, vol. 6897, pp. 302–313. Springer, Heidelberg (2011)

A Deep and Stable Extreme Learning Approach for Classification and Regression*

Le-le Cao, Wen-bing Huang, and Fu-chun Sun

Tsinghua National Laboratory for Information Science and Technology (TNList),
Department of Computer Science and Technology, Tsinghua University,
Beijing 100084, P.R. China
{caoll12,huangwb12}@mails.tsinghua.edu.cn,
fcsun@mail.tsinghua.edu.cn

Abstract. The random-hidden-node based extreme learning machine (ELM) is a much more generalized cluster of single-hidden-layer feed-forward neural networks (SLFNs) whose hidden layer do not need to be adjusted, and tends to reach both the smallest training error and the smallest norm of output weights. Deep belief networks (DBNs) are probabilistic generative modals composed of simple, unsupervised networks such as restricted Boltzmann machines (RBMs) or auto-encoders, where each sub-network's hidden layer serves as the visible layer for the next. This paper proposes an approach: DS-ELM (a deep and stable extreme learning machine) that combines a DBN with an ELM. The performance analysis on real-world classification (binary and multi-category) and regression problems shows that DS-ELM tends to achieve a better performance on relatively large datasets (large sample size and high dimension). In most tested cases, DS-ELM's performance is generally more stable than ELM and DBN in solving classification problems. Moreover, the training time consumption of DS-ELM is comparable to ELM.

Keywords: extreme learning machine (ELM), deep belief networks (DBNs), classification, regression, deep-and-stable ELM.

1 Introduction

In the research field of machine learning, the capability of classification and regression is often evaluated from perspectives such as accuracy, time cost, stability, and statistical significance. The research introduced in this paper will focus on the performance of learning machines with respect to accuracy and stability of both classification and regression tasks. In particular, the neural networks approaches are our major focus. On one hand, Extreme Learning Machine (ELM) [1–3] is within our scope because of its simple architecture with proven potential in solving classification and regression problems [1, 4]; on the other hand, we emphasize deep neural networks, specifically on deep belief networks

* This work was supported by grants from China National Natural Science Foundation under Project 613278050 and 61210013.

J. Cao et al. (eds.), *Proceedings of ELM-2014 Volume 1*,
Proceedings in Adaptation, Learning and Optimization 3, DOI: 10.1007/978-3-319-14063-6_13

(DBNs) composed of several layers of restricted Boltzmann machines (RBMs), which "seek to learn concepts instead of recognizing objects" [5].

Both ELM and DBNs have gained widespread popularity these years; and many sound and successful applications built upon ELM and DBNs have been reported. Generally speaking, ELM has high scalability and less computational complexity, while DBNs are known to have good modeling ability for higher-order and highly non-linear statistical structure in the input [6]. It is commonly accepted that the first layers of DBNs are expected to extract relatively low-level features out of the input space while the upper layers are expected to gradually refine previously learnt concepts to generate more abstract ones. Hence, it is natural to think of the possibility of combining ELM and DBNs, so we can have advantages from both methodologies in one.

Because the output of the higher DBN layers can easily be used as the input of a supervised classifier, Ribeiro et al. [5] used an ELM classifier for classify the deep concepts and lower the training cost of DBNs by applying adaptive learning rate technique and Graphics Processing Units (GPU) implementation of DBNs [7]. The recognition rate of their proposed approach (named DBN-ELM) is competitive (and often better) than other successful approaches in well-known benchmarks. The so called DBN-ELM approach is making use of full unsupervised learning power of DBN (as an auto encoder), the output of which is fed into a typical ELM classifier with randomly generated hidden neurons. This method tends to consume noticeable more time than a standalone ELM classifier; and the performance is not stable, meaning the test accuracy of a single trial (given exactly the same training and testing split) might very well likely be different from others trials. Although this kind of performance fluctuation is acceptable and limited within a certain interval, this phenomenon makes it mandatory to carry out multiple trials and perform statistical significance analysis.

The authors of [8] proposed a method called multilayer ELM (ML-ELM) using ELM as an auto encoder for learning feature representations. ML-ELM is composed of ELM auto encoders (ELM-AE) which performs layer-by-layer unsupervised learning like a typical DBN. ELM-AE can be regarded as a special case of ELM, where the input is equal to output. In a nutshell, the proposed method [8] initialize ELM-AE hidden layer weights within a deep structure in a random manner, adjust the weights using layer-wise unsupervised training, and finally fine-tune the entire network with BP algorithm. Although the reported testing accuracy of ML-ELM outperforms DBNs and ELMs, the problem of performance fluctuation still exists as in DBN-ELM [5].

In short, DBN-ELM [5] concatenates a DBN with an ELM (on top layer as a supervised classifier), while ML-ELM [8] stacks multiple ELM-AE together to form a deep network. Because of the random hidden nodes in both approaches, the performance (testing accuracy) is not stable. One of our objectives is to propose a new way of assembling DBN and ELM together, obtaining an ELM learning machine with deep structure, which is expected to have a relatively stable performance. DBN-ELM was only tested towards image reconstruction and classification tasks in [5]. ML-ELM was tested merely on MINIST data set which is commonly used for testing deep network performance. The other objective of our research is testing our approach on classification (binary and multiple class) and regression datasets to obtain a full picture of its performance.

2 Extreme Learning Machine

Extreme learning machine (ELM) [1–3] was developed specifically for single-hidden-layer feed-forward neural networks (SLFNs) at the very beginning. Huang et al. then "generalized" it to a kind of SLFN which may not be neuron alike [9, 10]. ELM was also extended to kernel learning in [4] showing that ELM can make use of various feature mappings such as random hidden nodes and kernels. ELM tends to reach both the smallest training error and the smallest norm of output weights [1, 4, 11]. It has been proved in [4] that ELM can achieve fast learning speed and good generalization performance on both regression and classification tasks. ELM is fast, which may be attributed to its single-hidden-layer structure requiring no iterative process. ELM also requires less human intervention and supervision, which makes it an efficient algorithm especially when facing large datasets where training time and easy parameter tuning are critical.

The decision function of ELM for generalized SLFNs is shown in the following equation. For simplicity, we take the case of one output node as an example.

$$f_L(x) = \sum_{i=1}^{L} \beta_i G(a_i, b_i, x) = \beta \cdot h(x) \tag{1}$$

where L denotes the number of hidden-layer node; β_i represents the weight connecting the i-th hidden node and the output; notation $G(a_i, b_i, x)$ is the activation function of the i-th hidden node; $h(x) = [G(a_1, b_1, x), \cdots, G(a_L, b_L, x)]^T$ is the output vector of the hidden layer with respect to the input x [11]. $h(x)$ maps the feature dimension(s) from N to L. It is worth mentioning that parameters for hidden node (i.e. $\{a_i, b_i\}_{i=1 \cdots L}$) can be randomly generated obeying any continuous probability distribution [4, 11]. As a result, ELM could generate the hidden node parameters before seeing the training data. As long as the output functions of hidden neurons are nonlinear piecewise continuous, neural networks with random hidden neurons attain both universal approximation and classification capabilities, and the changes in finite number of hidden neurons and their related connections do not affect the overall performance of the networks. Equation (1) is equivalent to $H\beta = T$, where

$$H = \begin{bmatrix} G(a_1, b_1, x_1) & \cdots & G(a_L, b_L, x_1) \\ \vdots & \cdots & \vdots \\ G(a_1, b_1, x_N) & \cdots & G(a_L, b_L, x_N) \end{bmatrix}, \beta = \begin{bmatrix} \beta_1^T \\ \vdots \\ \beta_L^T \end{bmatrix}, T = \begin{bmatrix} t_1^T \\ \vdots \\ t_N^T \end{bmatrix} \tag{2}$$

As a result, the hidden layer output matrix H is also called ELM feature space [4] mapped from input layer to hidden layer. The i-th column of H is the output of the i-th hidden node with respect to inputs $x_1, x_2, \cdots, x_N$. Given a training set $\aleph = \{(x_i, t_i) | x_i \in R^n, t_i \in R^m, i = 1, \cdots, N\}$, hidden node output function $G(a, b, x)$, and the number of hidden nodes L, the output weight β can be calculated by equation (3) [4]:

$$\beta = H^{\dagger}T = \begin{cases} H^T(\frac{I}{C} + HH^T)^{-1}T, & \text{when training set is not huge} \\ (\frac{I}{C} + H^TH)^{-1}H^TT, & \text{when training set is huge} \end{cases} \tag{3}$$

where $H^{\dagger}$ is the Moore-Penrose generalized inverse of hidden layer output matrix H [1]. The positive value $\frac{I}{C}$ is added to the diagonal of H^TH or HH^T to make the resulting solution stabler and obtain better generalization performance [4, 12]. Unlike traditional gradient-based learning algorithms facing several issues like local minima, improper learning rate and overfitting, etc, ELM tends to reach the solutions straightforward without such trivial issues [13].

3 Deep Belief Networks

DBNs are probabilistic generative modals, or alternatively a kind of deep neural network, composed of multiple latent variables (hidden units). DBNs were initially introduced in [14], addressing three problems that exist in traditional deeply layered neural networks: (1) large demand for training examples; (2) time consuming to reach convergence; (3) prone to local optima [15]. Recent key findings on neocortex of mammal brain such as [16] motivated the emergence of deep learning machine even further. Many researchers have affiliated the fact that DBNs can reach an equivalent modeling capability as SLFNs using a lot less nodes in each hidden layer.

DBNs can be viewed as a composition of simple, unsupervised networks such as restricted Boltzmann machines (RBMs) [14] or autoencoders [17], where each sub-network's hidden layer serves as the visible layer for the next. As is illustrated in Fig. 1 (a), each layer tries to model the distribution of its input. Every RBM has a layer containing visible nodes v that represent the data and a layer containing hidden nodes h that learn to represent input features capturing higher-order correlations in the data. [5] The topology of DBNs depicts a joint distribution based on observation input v and multiple hidden units $h_1, h_2, \cdots, h_L$:

$$P(v, h_1, h_2, \cdots, h_L) = (\prod_{k=0}^{L-2} P(h_k | h_{k+1}))P(h_{L-1}, h_L) \tag{4}$$

The key idea behind DBNs is that the weights, W, connecting two layers have no connections within a layer. This matrix of symmetrically weighted connections is learned by an RBM which defines both $p(v|h, W)$ and the prior distribution over hidden vectors, $p(h|W)$, so the probability of generating the visible vector, v, can be written as:

$$p(v) = \sum_h p(h|W)p(v|h, W) \tag{5}$$

By starting with the data vector on the visible units and alternating several times between sampling from $p(h|v, W)$ and $p(v|h, W)$, it is easy to get the learning weights W. The learning algorithm for DBNs proposed by Hinton et al. [14, 18] has two training phases: (1) a greedy learning algorithm for transforming representations (unsupervised learning), and (2) Back-Fitting with the up-down algorithm (fine-tune). The term "epochs" is used to represent iterations or sweeps of unsupervised pre-training (per layer) and supervised fine-tune.

4 Proposed Approach

This section presents a new machine learning approach named deep and stable extreme learning machine (DS-ELM) inspired by the ELM and DBN methodology. Our overall intention is to use a quick-and-dirty DBN to generate a relatively stable feature space H that is fed into an ELM to calculate the output weights. The details of DS-ELM approach are explained in two steps below (Fig. 1):

Step 1. Setup a DBN structure fed with input vector x; and perform a quick-and-dirty training based on the pre-defined DBN structure [cf. Fig. 1(a)]. The term "quick-and-dirty" means that both "unsupervised pre-training" and "supervised fine-tune" are accomplished within only a few iterations rather than sufficient iterations. Because the separating hyper-plane of ELM feature space goes through the origin in theory [4, 11]; hence bias b is not needed in training this quick-and-dirty DBN.

Step 2. The nodes in the top hidden layer can be viewed equal to hidden nodes in a typical ELM network; those hidden layer output matrix H is feature space of the input vector. the feature space H initiated via Step 1 with help of a DBN is then fed into a typical ELM solver to calculate the output weights β with equation (3). [cf. Fig. 1(b)]

By integrating a DBN with an ELM in this manner, DS-ELM adds the following potential advantages to a standalone ELM:

- **Auto-abstraction of deep concepts.** Most of the classification and regression problems have input examples which are usually represented by a set of manually extracted features. In many cases, the challenging nature of many problems lie on the difficulty of extracting features such as "behavioral characteristics like mood, fatigue, energy, etc." [5] A typical example is object image classification problem which is especially challenging due to the fact that same object might appear differently because of pose and illumination conditions; the low level visual features are far detached from the semantics of the scene, making it problem-prone when used to infer object presence. [19] Human-crafted features [20] is very hard to embody complex functions hidden in input data, but the unsupervised pre-training of DBNs allows learning those complex functions by mapping the input to the output directly. Specifically speaking, the bottom layers are expected to extract and represent low-level features from the input data while the upper layers are expected to gradually refine previously learnt concepts [5].

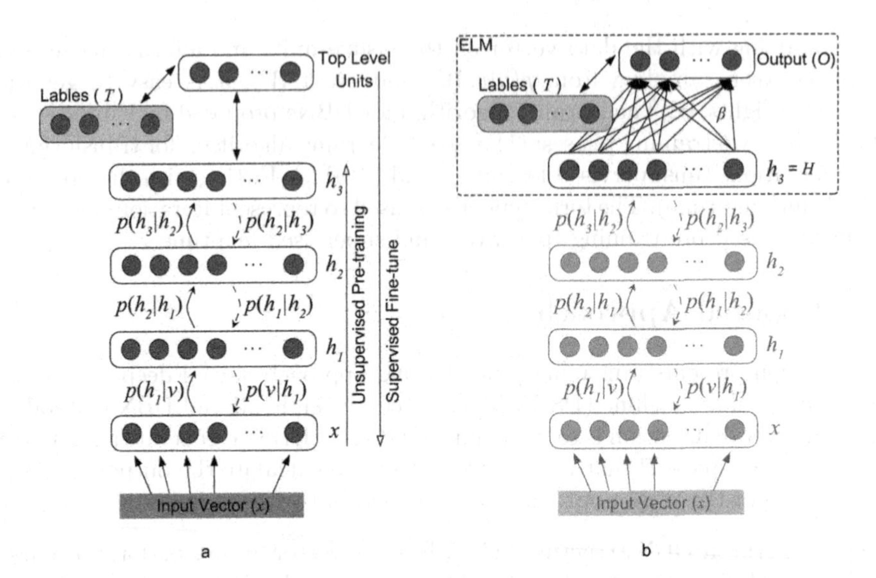

Fig. 1. An example (three hidden layers) of two-step training process of DS-ELM. (a) Step 1: initialize ELM feature space H with a quick-and-dirty DBN. (b) Step 2: calculate output weights β from H with a typical ELM solver.

- **Stable feature space and performance.** The parameters for ELM hidden node (i.e. $\{a_i, b_i\}_{i=1\cdots L}$) are randomly generated obeying any continuous probability distribution [4, 11]. Hence the ELM feature space H (or called feature mapping matrix) defined in equation (2) does not stays the same even for the same input data. According to ELM universal approximation capability (of approximating any target continuous function) theorems [9, 3], we can prove the classification capability of ELM [4]; but the performance of ELM with random hidden nodes is not quite stable. The most straightforward impact is on the test accuracy of a single trial (given exactly the same training and testing split) which might very likely be different from others trials. Although this kind of performance fluctuation is acceptable and controlled within certain limits, this behavior makes it necessary to carry out groups of trials and perform statistical significance analysis. DS-ELM, on the other hand, stabilize the ELM performance (test accuracy) by initialize ELM feature space H with a quick-and-dirty DBN.

5 Experiments and Analysis

In order to extensively verify the performance of DS-ELM, a variety type of real-world data was chosen for each problem category (regression, binary classification, and multi-category classification). Seen from Table 1, the simulations involve 17 datasets ranging from small to large data size; and from low to high dimension. Fixed training/testing division is applied for all datasets.

Most of the simulated experiments of DS-ELM, ELM[1], and DBN[2] are carried out with Matlab 2012b (maci64) run on Intel Core i7, 2.3-GHz CPU with 16-GB, 1333-MHz RAM and 250-GB, SATA 6GB/s SSD. A few datasets (i.e. Shuttle, CTslice [21], and Protein [22]) require larger memory, so we have to execute algorithms for these datasets with a Matlab 2013a installation on a high-performance server with dual Xeon E7-4820 2.266GHz CPU and 4x64G RAM.

Table 1. Dimension and size of selected benchmark data sets: binary classification problems (noted as "bincls"), multiple-category classification problems (noted as "multicls"), and regression problems (noted as "reg")

Dim. \ Size	Small	Large
Low	**bincls**: Diabetes, Liver [21] **multicls**: Iris, Segment [21], Vowel [23] **reg**: Pyrim, Housing [21]	**bincls**: Mushroom, Musk2 [21] **multicls**: Shuttle [21] **reg**: Abalone [21]
High	**bincls**: Leukemia [24] **multicls**: DNA [21] **reg**: Crime [21, 26]	**bincls**: Gisette [25] **multicls**: Protein [22] **reg**: CTslice [21]

The code of ELM classifier is originally obtained from [4]. The source code is then adjusted to fit in the needs of data pre-processing and feature preparation. The toolbox containing DBN implementation is retrieved in accordance with [27]. Our implementation of DS-ELM approach is a combination of ELM [4] and DBN [27] based on the procedure explained in section 4. In all simulations, Sigmoidal hidden layer activation function is used, and 20 trials are executed for each dataset.

The experimental results for classification problems and regression problems are put together in Table 2 and 3 respectively. The best results among the three tested approaches are highlighted in bold. Generally speaking, the three approaches are capable of achieving similar generation performance for most tested datasets.

Table 2 shows the performance comparison of ELM, DBN, and DS-ELM for classification problems. It can be seen from binary classification tests that (1) DS-ELM tends to obtain the lowest standard deviation for five out of six datasets; (2) although ELM achieved best testing rate for four out of six datasets, DS-ELM has the best testing rate for Mushroom and Gisette datasets which contain the biggest number of training/testing samples. Observed from multi-category classification simulations, we found that (1) the performance of DS-ELM is more stable (with the smallest "Dev" value) than the other two approaches in all tested datasets; (2) for DNA (high dimension and medium size) and Protein (high dimension and large size) datasets, DS-ELM method achieved better testing rate compared to the other two methods.

[1] ELM: `http://www.ntu.edu.sg/home/egbhuang/elm_random_hidden_nodes.html`
[2] DeepLearnToolbox: `https://github.com/rasmusbergpalm/DeepLearnToolbox`

Table 2. Performance comparison of ELM (random hidden nodes), DBN, and DS-ELM approaches: classification problems

Datasets	ELM			DBN			DS-ELM		
	Tesing Rate(%)	Training Time(s)	Dev (%)	Tesing Rate(%)	Training Time(s)	Dev (%)	Tesing Rate(%)	Training Time(s)	Dev (%)
Diabetes	**76.85**	0.14	2.16	73.33	1.2436	6.83	75.88	0.369	**1.07**
Liver	**71.12**	0.189	3.28	68.06	1.739	5.42	70.1	0.2501	**1.83**
Mushroom	99.82	2.23	0.28	88.83	214.34	1.69	**99.9**	2.6515	**0.03**
Musk2	**95.34**	27.231	0.32	85.08	57.103	**0**	93.81	29.281	0.16
Leukemia	**81.25**	3.56	3.97	79.81	32.192	4.53	80.78	3.166	**1.95**
Gisette	92.5	98.199	1.12	90.6	3041.5	2.49	**93.04**	127.08	**0.32**
Iris	**100**	0.0031	**0**	97.91	0.1969	2.36	**100**	0.0093	**0**
Vowel	**90.74**	0.018	1.89	88.45	0.2093	1.9	89.7	0.0602	**0.24**
Segment	**98.07**	0.84	0.94	96.52	139.26	0.92	97.88	2.69	**0.91**
Shuttle	**99.75**	2.8303	0.03	98.99	1894.9	0.02	99.7	9.592	**0.01**
DNA	92.86	1.495	0.4	92.79	2207.8	0.36	**93.61**	13.119	**0.21**
Protein	83.13	20.26	0.65	84.03	1511.3	0.75	**84.78**	30.81	**0.2**

Table 3. Performance comparison of ELM (random hidden nodes), DBN, and DS-ELM approaches: regression problems

Datasets	ELM			DBN			DS-ELM		
	RMSE	Training Time(s)	Dev (%)	RMSE	Training Time(s)	Dev (%)	RMSE	Training Time(s)	Dev (%)
Pyrim	0.1317	0.089	**0.0332**	0.2023	1.592	0.2916	**0.1091**	0.159	0.0346
Housing	**0.0802**	0.239	**0.0103**	0.0838	11.042	0.0599	0.084	0.5753	0.0164
Abalone	**0.073**	1.4166	**0.0031**	0.0785	87.08	0.0438	0.0775	1.7391	0.0099
Crime	0.1381	0.719	0.0642	0.1418	116.37	0.1094	**0.1368**	1.7729	**0.06**
CTslice	3.3055	232.28	0.0125	3.8021	2092.8	0.2665	**3.278**	562.31	**0.0083**

For regression problems (Table 3), DS-ELM also showed a slight advantage over ELM on relatively large datasets (i.e. Crime and CTslice). However, we do not observe an obvious pattern of better standard deviation. It should also be noted that parameter tuning for DBN is a time-consuming task comparing to ELM; it is probably one of the reasons that DBN test results seems sub-optimal than the other approaches. In order to achieve good results, DBN need more careful and data-centric parameter tuning activities. All simulation results showed that ELM uses least training time while DBN tends to consume a lot more time (over hundreds of times more in many situations) to train. DS-ELM approach often has a comparable training time consumption to ELM; of course, the scale of its training time depends on the level of "quick-and-dirty"-ness (number of epochs) of the embodied deep network. Our parameter tuning activities also justify the fact that DS-ELM is not sensitive to the network structure; but it is not quantitatively measured yet in our research.

6 Conclusions

DS-ELM is a training schema that combines DBN and ELM. The essence of DS-ELM is using a quick-and-dirty DBN to generate a relatively stable feature space H which is, in turn, fed into an ELM to calculate the output weights. From our experimental results, we summarize the three key findings below:

(1). DS-ELM tends to achieve better testing rate over ELM and DBN on relatively large datasets (i.e. large number of samples and high dimension).
(2). DS-ELM is generally more stable (with smaller standard deviation value) than ELM and DBN in solving classification problems (for both binary and multiple category cases).
(3). DS-ELM approach often has a similar training-time cost as ELM, as long as its embodied deep network merely requires a few (usually one or two) epochs for unsupervised pre-training (per layer) and supervised fine-tune.

References

1. Huang, G.-B., Zhu, Q.-Y., Siew, C.-K.: Extreme learning machine: theory and applications. Neurocomputing 70(1), 489–501 (2006)
2. Huang, G.-B., Zhu, Q.-Y., Siew, C.-K.: Extreme learning machine: a new learning scheme of feedforward neural networks. In: Proceedingsof the 2004 IEEE International Joint Conference on Neural Networks, vol. 2, pp. 985–990. IEEE (2004)
3. Huang, G.-B., Chen, L., Siew, C.-K.: Universal approximation using incremental constructive feedforward networks with random hidden nodes. IEEE Transactions on Neural Networks 17(4), 879–892 (2006)
4. Huang, G.-B., Zhou, H., Ding, X., Zhang, R.: Extreme learning machine for regression and multiclass classification. IEEE Transactions on Systems, Man, and Cybernetics, Part B: Cybernetics 42(2), 513–529 (2012)
5. Ribeiro, B., Lopes, N.: Extreme Learning Classifier with Deep Concepts. In: Ruiz-Shulcloper, J., Sanniti di Baja, G. (eds.) CIARP 2013, Part I. LNCS, vol. 8258, pp. 182–189. Springer, Heidelberg (2013)
6. A.-r. Mohamed, G., Hinton, G.: Understanding how deep belief networks perform acoustic modelling. In: 2012 IEEE Int'l Conf. on Acoustics, Speech and Signal Processing (ICASSP), pp. 4273–4276. IEEE (2012)
7. Lopes, N., Ribeiro, B.: Gpumlib: An efficient open-source gpu machine learning library. International Journal of Computer Information Systems and Industrial Management Applications 3, 355–362 (2011)
8. Kasun, L.L.C., Zhou, H., Huang, G.-B., Vong, C.M.: Representational learning with extreme learning machine for big data. IEEE Intelligent Systems (2013)
9. Huang, G.-B., Chen, L.: Convex incremental extreme earning machine. Neurocomputing 70(16), 3056–3062 (2007)
10. Huang, G.-B., Chen, L.: Enhanced random search based incremental extreme learning machine. Neurocomputing 71(16), 3460–3468 (2008)
11. Huang, G.-B., Ding, X., Zhou, H.: Optimization method based extreme learning machine for classification. Neurocomputing 74(1), 155–163 (2010)
12. Huang, G.-B., Wang, D.H., Lan, Y.: Extreme learning machines: a survey. International Journal of Machine Learning and Cybernetics 2(2), 107–122 (2011)

13. Li, M.-B., Huang, G.-B., Saratchandran, P., Sundararajan, N.: Fully complex extreme learning machine. Neurocomputing 68, 306–314 (2005)
14. Hinton, G.E., Osindero, S., Teh, Y.-W.: A fast learning algorithm for deep belief nets. Neural computation 18(7), 1527–1554 (2006)
15. Arel, I., Rose, D.C., Karnowski, T.P.: Deep machine learning-a new frontier in artificial intelligence research [research frontier]. IEEE Computational Intelligence Magazine 5(4), 13–18 (2010)
16. Lee, T.S., Mumford, D.: Hierarchical bayesian inference in the visual cortex. JOSA A 20(7), 1434–1448 (2003)
17. Bengio, Y., Lamblin, P., Popovici, D., Larochelle, H., et al.: Greedy layer-wise training of deep networks. Greedy layer-wise training of deep networks 19, 153 (2007)
18. Hinton, G.: A practical guide to training restricted boltzmann machines. Momentum 9(1), 926 (2010)
19. Wang, G., Hoiem, D., Forsyth, D.: Building text features for object image classification. In: IEEE Conference on Computer Vision and Pattern Recognition, CVPR 2009, pp. 1367–1374. IEEE (2009)
20. Bengio, Y.: Learning deep architectures for AI. Foundations and trends in Machine Learning 2(1), 1–127 (2009)
21. Bache, K., Lichman, M.: UCI repository of machine learning repository. University of California, Irvine, School of Information and Computer Sciences (2013), `http://archive.ics.uci.edu/ml`
22. Shevade, S.K., Keerthi, S.: A simple and efficient algorithm for gene selection using sparse logistic regression. Bioinformatics 19(17), 2246–2253 (2003)
23. Duarte, M.F., Hen Hu, Y.: Vehicle classification in distributed sensor networks. Journal of Parallel and Distributed Computing 64(7), 826–838 (2004)
24. Xing, E.P., Jordan, M.I., Karp, R.M., et al.: Feature selection for high-dimensional genomic microarray data. ICML 1, 601–608 (2001)
25. Guyon, I., Gunn, S.R., Ben-Hur, A., Dror, G.: Result analysis of the nips 2003 feature selection challenge. In: NIPS, vol. 4, pp. 545–552 (2004)
26. Redmond, M., Baveja, A.: A data-driven software tool for enabling cooperative information sharing among police departments. European Journal of Operational Research 141(3), 660–678 (2002)
27. Palm, R.B.: Prediction as a candidate for learning deep hierarchical models of data. Technical University of Denmark, Palm (2012)

Extreme Learning Machine Ensemble Classifier for Large-Scale Data

Haocheng Wang[1,2], Qing He[1], Tianfeng Shang[3], Fuzhen Zhuang[1], and Zhongzhi Shi[1]

[1] Key Lab of Intelligent Information Processing of Chinese Academy of Sciences (CAS), Institute of Computing Technology, CAS, Beijing 100190, China
[2] University of Chinese Academy of Sciences, Beijing 100049, China
{wanghc,heq,zhuangfz,shizz}@ics.ict.ac.cn
[3] School of Information Systems, Singapore Management University, Singapore
tfshang@smu.edu.sg

Abstract. For classification problem, extreme learning machine (ELM) can get better generalization performance at a much faster learning speed. Nevertheless, a single ELM is unstable in data classification. The Bagging-based ensemble classifier, i.e., Bagging-ELM has been studied popularly and proved to improve the performance of ELM significantly in terms of accuracy, however, it is inappropriate to deal with large-scale datasets due to the highly intensive computation. In this study, we propose a novel ELM ensemble classifier, namely b-ELM, which leverages the Bag of Little Bootstraps technique to obtain a scalable, efficient means of classification for large-scale data. Efficiency of classification is achieved as it only requires repeated training under consideration on quantities of data that can be much smaller than the original training data. Furthermore, b-ELM is suited to implementation on modern parallel and distributed computing platforms. The experimental results demonstrate that b-ELM can efficiently handle large-scale data with a good performance on prediction accuracy.

Keywords: extreme learning machine, bag of little bootstraps, large-scale data, classification.

1 Introduction

Extreme learning machine (ELM) was proposed as a powerful machine learning technique for single-hidden-layer feedforward neural networks (SLFNs) [8,9,6], and has been studied popularly for its fast learning speed and good generalization performance [3,5,11,4,7]. The essence of ELM is that the hidden layer of SLFNs need not be tuned. Concretely, the hidden neuron parameters are randomly generated which may be independent of the training data, and the output weights are analytically resolved by using Moore–Penrose generalized inverse. Therefore, ELM can overcome the difficulties that the traditional classic gradient-based learning algorithms have to face, such as local minima, learning rate, stopping criteria, and overfitting, etc. Additionally, a wide range of activation functions

J. Cao et al. (eds.), *Proceedings of ELM-2014 Volume 1*,
Proceedings in Adaptation, Learning and Optimization 3, DOI: 10.1007/978-3-319-14063-6_14

including all piecewise continuous functions can be used as activation functions in ELM.

However, a single ELM is unstable in data classification. To improve the stability and boost the accuracy, more and more researchers consider using ensemble of ELMs. One popular ensemble learning method is Bagging [1], which takes different bootstrap resamples from the original training set and trains a classifier or predictor on each resample to build its constituent members. Several studies have demonstrated that Bagging-ELM is generally more accurate than the individual members [12,14,13].

Bagging-ELM, despite its favorable accuracy of predictability, has high or even prohibitive computational costs. Therefore, its usefulness is severely blunted by the large-scale datasets increasingly encountered in practice. In Bagging-ELM, training in question is repeatedly applied to the bootstrap resamples of the original training set. Because these resamples have size on the order of that of the original training data, with approximately 63.2% of data points appearing at least once in each resample [2], classification on large-scale data can be prohibitively costly. To reduce the computational complexity, one might spontaneously attempt to employ the modern trend toward parallel and distributed computing, i.e., different processors or compute nodes are used to process different bootstrap resamples independently in parallel. However, the large size of bootstrap resamples in the large-scale data setting renders this approach problematic.

For the sake of alleviating the aforementioned problem, we present b-ELM, a novel ELM ensemble classifier which utilizes the Bag of Little Bootstraps (BLB) [10] technique to obtain a scalable, efficient means of classification for large-scale data. The b-ELM algorithm constructs an ensemble of the predictors of bootstrapping multiple small subsets of a larger original training set, and then makes decisions for testing samples through majority voting with the ensemble. It is worth noting that b-ELM has a more favorable computational profile than Bagging-ELM, as it only requires repeated training under consideration on quantities of data that can be much smaller than the original training dataset. Moreover, b-ELM is suited to implementation on modern parallel and distributed computing architectures which are often used to process large datasets. As we show empirically, our procedure possesses superior ability to scale computationally to large-scale datasets, typically incurring less total computation to reach comparably high accuracy.

The remainder of this paper is arranged as follows. In Section 2, preliminary knowledge is described. Subsequently, we introduce b-ELM in full detail in Section 3. Our experimental results to demonstrate the efficiency and effectiveness of b-ELM are given in Section 4. Finally, we conclude in Section 5.

2 Preliminaries

This section will briefly introduce the techniques related to b-ELM.

2.1 Extreme Learning Machine

ELM was originally developed for the single-hidden-layer feedforward neural networks (SLFNs) [8,9,6] and then extended to the "generalized" SLFNs where the hidden layer need not be neuron alike. ELM typically applies random computational nodes in the hidden layer, which may be independent of the training data. It increases learning speed by means of randomly generating weights and biases for hidden nodes rather than iteratively adjusting network parameters which is commonly adopted by gradient-based methods. Different from traditional learning algorithms, ELM tends to reach not only the smallest training error but also the smallest norm of output weights.

The output function of ELM with L hidden nodes for generalized SLFNs is

$$f_L(\boldsymbol{x}) = \sum_{i=1}^{L} \boldsymbol{\beta}_i g_i(\boldsymbol{x}) = \sum_{i=1}^{L} \boldsymbol{\beta}_i G(\boldsymbol{a}_i, b_i, \boldsymbol{x}), \ \boldsymbol{x} \in \mathbb{R}^d, \ \boldsymbol{\beta}_i \in \mathbb{R}^m \tag{1}$$

where $\boldsymbol{a}_i$ is the weight vector connecting the input nodes to the ith hidden node, b_i is the bias of the ith hidden node, g_i denotes the output function i.e. activation function $G(\boldsymbol{a}_i, b_i, \boldsymbol{x})$ of the ith hidden node, and $\boldsymbol{\beta}_i$ is the weight vector linking the ith hidden node to the output nodes. For N arbitrary distinct samples $(\boldsymbol{x}_j, \boldsymbol{t}_j) \in \mathbb{R}^d \times \mathbb{R}^m$, SLFNs with L hidden nodes can approximate these N samples with zero error means that there exist $(\boldsymbol{a}_i, b_i)$ and $\boldsymbol{\beta}_i$ such that

$$\sum_{i=1}^{L} \boldsymbol{\beta}_i G(\boldsymbol{a}_i, b_i, \boldsymbol{x}_j) = \boldsymbol{t}_j, \ j = 1, \cdots, N. \tag{2}$$

The above N equations can be written compactly as

$$\boldsymbol{H}\boldsymbol{\beta} = \boldsymbol{T} \tag{3}$$

where

$$\boldsymbol{H} = \begin{bmatrix} \boldsymbol{h}(\boldsymbol{x}_1) \\ \vdots \\ \boldsymbol{h}(\boldsymbol{x}_N) \end{bmatrix} = \begin{bmatrix} G(\boldsymbol{a}_1, b_1, \boldsymbol{x}_1) \cdots G(\boldsymbol{a}_L, b_L, \boldsymbol{x}_1) \\ \vdots \quad \ddots \quad \vdots \\ G(\boldsymbol{a}_1, b_1, \boldsymbol{x}_N) \cdots G(\boldsymbol{a}_L, b_L, \boldsymbol{x}_N) \end{bmatrix}_{N \times L}$$

$$\boldsymbol{\beta} = \begin{bmatrix} \boldsymbol{\beta}_1^\top \\ \vdots \\ \boldsymbol{\beta}_L^\top \end{bmatrix}_{L \times m}, \ \boldsymbol{T} = \begin{bmatrix} \boldsymbol{t}_1^\top \\ \vdots \\ \boldsymbol{t}_N^\top \end{bmatrix}_{N \times m}$$

$\boldsymbol{H}$ is the hidden layer output matrix of the SLFN, and the ith column of $\boldsymbol{H}$ is the ith hidden node output with respect to inputs $\boldsymbol{x}_1, \boldsymbol{x}_2, \cdots, \boldsymbol{x}_N$. While the jth row of $\boldsymbol{H}$, i.e., $\boldsymbol{h}(\boldsymbol{x}_j)$ is the hidden layer feature mapping with respect to the jth input $\boldsymbol{x}_j$. As the hidden node parameters $(\boldsymbol{a}_i, b_i)$ can be randomly generated and remain fixed, the only unknown parameters in ELM are the output weights vectors $\boldsymbol{\beta}_i$ between the hidden layer and the output layer, which can simply be resolved by ordinary least-square directly. Since ELM aims to minimize the training error $\| \boldsymbol{H}\boldsymbol{\beta} - \boldsymbol{T} \|$ and the norm of weights $\| \boldsymbol{\beta} \|$, the smallest norm least-squares solution of the above linear system is

$$\widehat{\boldsymbol{\beta}} = \boldsymbol{H}^{\dagger}\boldsymbol{T} \tag{4}$$

where $\boldsymbol{H}^{\dagger}$ is the Moore–Penrose generalized inverse of matrix $\boldsymbol{H}$ [6]. Hence, the prediction value matrix $\boldsymbol{Y}$ can be expressed by

$$\boldsymbol{Y} = \boldsymbol{H}\widehat{\boldsymbol{\beta}} = \boldsymbol{H}\boldsymbol{H}^{\dagger}\boldsymbol{T} \tag{5}$$

The error matrix can be described as

$$e = \| \boldsymbol{Y} - \boldsymbol{T} \|^2 = \| \boldsymbol{H}\boldsymbol{H}^{\dagger}\boldsymbol{T} - \boldsymbol{T} \|^2 \tag{6}$$

2.2 Bag of Little Bootstraps

The Bag of Little Bootstraps (BLB), similar to the classical bootstrap, quantifies uncertainty on a statistical estimate, but is better suited for implementation on modern parallel and distributed computing platforms due to its structure. From the input data of size n, it samples without replacement subsamples of size $b = n^{\gamma}$ for typically $0.5 < \gamma \leq 0.9$, which results in a relatively small number of distinct points per subsample compared to the total input size n. From each subsample, resamples of size n are sampled with replacement, and the statistical estimator function is computed on each resample. The differences between the estimates made on the resamples for each subsample are quantified, typically with a standard deviation or variance calculation. The algorithm's output is the average of the error measurements on each of the subsamples.

Obviously, BLB only requires repeated computation on small subsets of the original dataset and avoids the bootstrap's problematic need for repeated computation of the estimate on resamples having size comparable to that of the original dataset. A standard and straightforward calculation reveals that each bootstrap resample contains approximately $0.632n$ distinct data points, which is large if n is large [2]. On the contrary, as previously mentioned, each BLB resample contains at most b distinct data points, and b can be chosen to be much smaller than n or $0.632n$. As a result, the cost of computing the estimate on each BLB resample is commonly substantially lower than the cost of computing the estimate on each bootstrap resample, or on the full dataset. Moreover, BLB typically requires less total computation (across multiple data subsets and resamples) than the bootstrap to reach comparably high accuracy.

3 b-ELM Classifier

In this section we introduce our ELM ensemble classifier, i.e., b-ELM. Our goal is to obtain a scalable, efficient means of classification for large-scale data. b-ELM employs the Bag of Little Bootstraps technique owing to its significantly

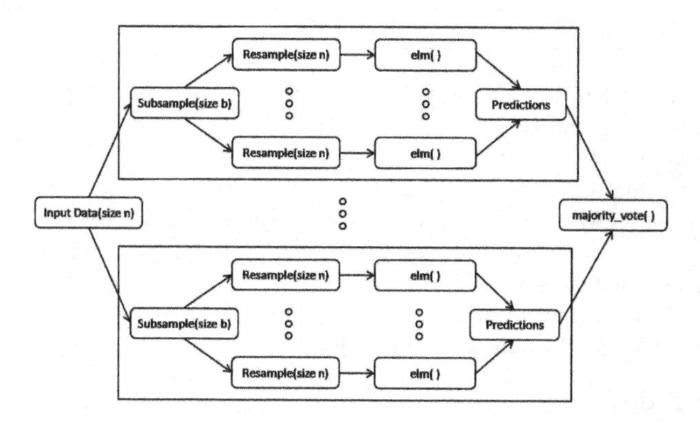

Fig. 1. The workflow of b-ELM. Subsamples are subsampled without replacement, while resamples are drawn with replacement.

computational gains and effective scalability. BLB can capture the diversities of base classifiers from relatively small subsets of data. More concretely, b-ELM is a method of generating training sets for ELM base classifiers by bootstrapping multiple small subsets of a larger original training dataset, subsequently constructing an ensemble of the predictors trained on those training sets, and then making decisions for testing samples through majority voting with the ensemble, i.e., the class that obtains the highest votes is considered as the predicted label. The workflow of b-ELM can be seen in Fig. 1.

The pseudo-code of b-ELM is described in Algorithm 1. There are two pairs of nested loops in the algorithm: the first pair of nested loops is to find the best parameters (e.g., the hidden node number L) for those base classifiers based on a k-fold cross-validation (CV); the other pair is to train the base classifiers and get their predictions for testing dataset. The aggregation phase includes a majority vote to obtain the final decision. The parameter $b = n^\gamma$ in the algorithm is the size of subsamples sampled without replacement from the entire original training dataset. According to prior knowledge, we might take $b = n^\gamma$ for typically $0.5 < \gamma \leq 0.9$. Obviously, b-ELM has a more favorable space profile than Bagging-ELM. The parameter s indicates the number of subsamples, while r represents the number of resamples bootstrapped from each subsample. In addition, both s and r determine the total number of predictions for the testing dataset.

Owing to its much smaller subsample and resample sizes, b-ELM is notably more amenable than Bagging-ELM to distribution of different subsamples and resamples and their associated computations to independent compute nodes; thus, b-ELM allows for simple parallel and distributed implementations, enabling additional large computational gains.

In the large-scale data setting, training on the entire training dataset often requires simultaneous distributed training across multiple compute nodes, among which the observed dataset is partitioned. Given the large size of each bootstrap resample, training on even a single such resample in turn also requires the use of a comparably large cluster of compute nodes; Bagging-ELM requires repetition

Algorithm 1. b-ELM

Input:
$\boldsymbol{T}$: training dataset $\{x_1, x_2, \ldots, x_n\}$
$\boldsymbol{y_T}$: labels of training dataset
$\boldsymbol{S}$: testing dataset
$\boldsymbol{y_S}$: labels of testing dataset
$\boldsymbol{P}$: parameters of ELM
b: subset size
s: number of sampled subsets
r: number of Monte Carlo iterations
Output: the aggregated prediction results $\widehat{\boldsymbol{y_S}}$

1. **for** each $\boldsymbol{P}$ **do**
2.   **for** each fold in k-fold CV **do**
3. $\boldsymbol{M} = elm\_train(\boldsymbol{T}, \boldsymbol{y_T}, \boldsymbol{P})$
4.   **end for**
5. **end for**
6. Save the best-performed parameters $\boldsymbol{P}_{best}$
7. **for** $i = 1$ to s **do**
8.   //Subsample the data
9.   Randomly sample a subset $\boldsymbol{B}_i$ of b indices from $\{1, 2, \ldots, n\}$ without replacement
10.   **for** $j = 1$ to r **do**
11.   Draw a random pseudo-sample $\boldsymbol{D}_j$ of size n from $\boldsymbol{B}_i$ with replacement
12.   //Training
13. $\boldsymbol{M}^{(i-1)\times r+j} = elm\_train(\boldsymbol{D}_j, \boldsymbol{y_{D_j}}, \boldsymbol{P}_{best})$
14.   //Testing
15. $\widehat{\boldsymbol{y_S}}^{(i-1)\times r+j} = elm\_test(\boldsymbol{M}^{(i-1)\times r+j}, \boldsymbol{S}, \boldsymbol{y_S})$
16.   Store prediction $\widehat{\boldsymbol{y_S}}^{(i-1)\times r+j}$
17.   **end for**
18. **end for**
19. //Aggregation
20. Majority Vote $\widehat{\boldsymbol{y_S}}^{k} \Rightarrow \widehat{\boldsymbol{y_S}},\ k = 1, 2, \ldots, s \times r$
21. **return** $\widehat{\boldsymbol{y_S}}$

of this training process for multiple resamples. Each training process is thus quite costly, and the aggregate computational costs of this repeated distributed training process are quite high. Indeed, the training for each bootstrap resample requires use of an entire cluster of compute nodes and incurs the associated overhead.

Conversely, b-ELM straightforwardly allows to train on multiple (or even all) subsamples and resamples simultaneously in parallel: because subsamples and resamples of b-ELM can be significantly smaller than the original training dataset, they can be transferred to, stored by, and processed on individual (or very small sets of) compute nodes. For instance, we could naturally utilize modern hierarchical distributed architectures by distributing subsamples to different compute nodes and subsequently using intra-node parallelism to train across different resamples generated from the same subsample. Hence, compared with Bagging-ELM, b-ELM both decreases the total computational cost of classification and allows more natural use of parallel and distributed computational resources.

4 Experiments

In this section, the performance of the proposed b-ELM is compared with the single ELM and Bagging-ELM. Simulations are conducted on simulated data and real data. All experiments are implemented and executed using MATLAB on a single processor. More specifically, the current version of basic ELM is employed here, and the sigmoid function $g(x) = 1/(1 + e^{-\lambda x})$ is selected as the activation function. Motivated by the need for good performance, 5-fold cross-validation is performed to select meta-parameters of basic ELM on each training set, i.e., one fold is used as testing set for a classifier built on the remaining four in each cross. The grid search is implemented once for each dataset in advance to avoid overfitting, and the parameter tuning time is not considered here. For each dataset, fifty trials are conducted for all the algorithms and the average results are reported in this paper.

4.1 Datasets

The experiments are performed on several datasets, including a simulated dataset ("3-Covers"), and nine other classical datasets from the UCI Machine Learning repository which has been extensively used in testing the performance of different kinds of classifiers. The basic information of the datasets is given in Table 1. The training and testing data of these 10 datasets are fixed for all trials of simulations. Moreover, all the attributes have been normalized into the range [0, 1].

Table 1. Specifications of classification datasets

Datasets	# Classes	# Attributes	# Training data	# Testing data
Balance	3	4	400	225
Car	4	6	1,200	528
Waveform	3	21	3,000	2,000
Mushroom	2	22	4,000	4,124
Digits	10	16	7,494	3,498
Letter	26	16	16,000	4,000
3-Covers	3	7	10,000	7,895
Adult	2	14	10,000	38,842
Covertype	7	54	15,120	565,892
Poker	10	10	25,010	1,000,000

4.2 Evaluation Metrics

Considering that the *accuracy* measure may be vulnerable to the class unbalance, we employ both the standard F_1 and *accuracy* metrics to evaluate the prediction performance of different classifiers on each dataset. *Accuracy* is the proportion of true results in the population. For each known class we calculate F_1^i as follows,

$$F_1^i = \frac{2 \times precison_i \times recall_i}{precison_i + recall_i}, \quad i \in \{1, \cdots, k\}, \tag{7}$$

where $precison_i$ and $recall_i$ are the precision and recall on the i-th known class. Then,

$$F_1 = \sum_{i=1}^{k} F_1^i / k, \tag{8}$$

where k is the number of known classes.

4.3 Simulated Data

We first utilize the simulated 3-Covers dataset to evaluate the performance characteristics of b-ELM. More concretely, we study the prediction and computational properties of b-ELM as well as Bagging-ELM. To maintain consistency of notation, we henceforth refer to the basic ELM as Single-ELM. Moreover, we consider $b = n^\gamma$ where $\gamma \in \{0.6, 0.7, 0.8, 0.9\}$ in runs of b-ELM. For both b-ELM and Bagging-ELM, identical evaluation criterions have been used to evaluate the classification quality in the experiments.

From Fig. 2, we can see that b-ELM succeeds in converging to high *accuracy* value significantly more quickly than Bagging-ELM for all values of b considered. When computing on a single processor, b-ELM generally requires less time, and hence less total computation, than Bagging-ELM to attain comparably high classification accuracy. Those results only hint at b-ELM's superior ability to scale computationally to large-scale data. As seen in Fig. 3, b-ELM also outperforms Bagging-ELM in terms of achieving higher F_1 value with less time on 3-Covers dataset.

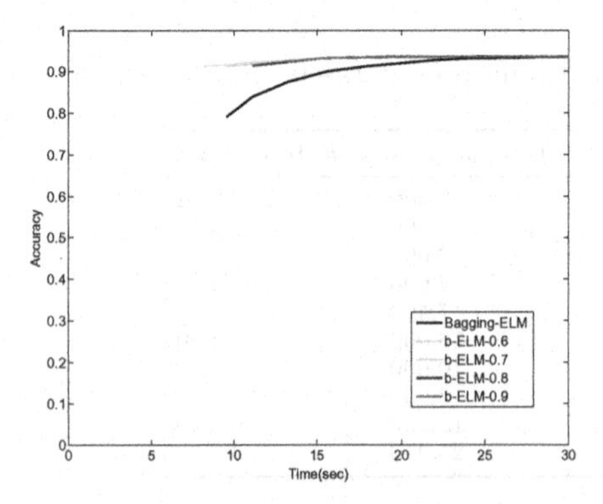

Fig. 2. *Accuracy* vs. *processingtime* for both b-ELM and Bagging-ELM classification on 3-Covers dataset. For b-ELM, $b = n^\gamma$ with the value of γ for each trajectory are given in the legend.

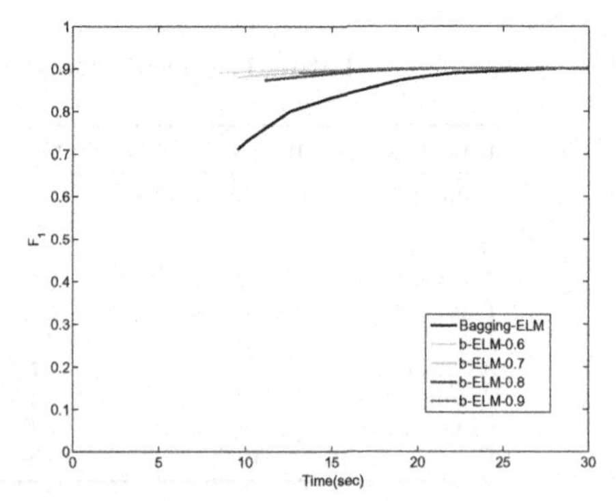

Fig. 3. F_1 vs. *processingtime* for both b-ELM and Bagging-ELM classification on 3-Covers dataset. For b-ELM, $b = n^\gamma$ with the value of γ for each trajectory are given in the legend.

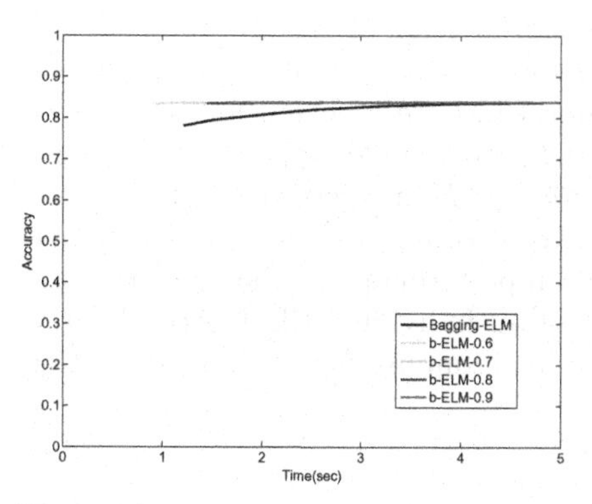

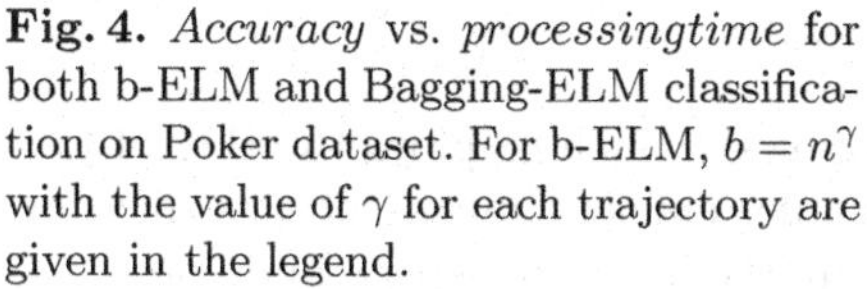

Fig. 4. *Accuracy* vs. *processingtime* for both b-ELM and Bagging-ELM classification on Poker dataset. For b-ELM, $b = n^{\gamma}$ with the value of γ for each trajectory are given in the legend.

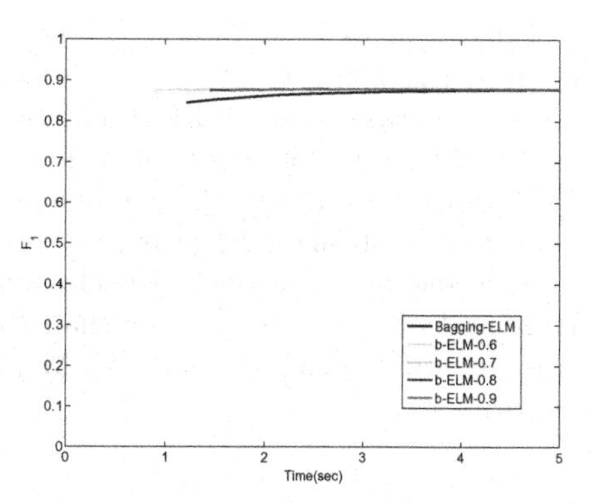

Fig. 5. F_1 vs. *processingtime* for both b-ELM and Bagging-ELM classification on Poker dataset. For b-ELM, $b = n^{\gamma}$ with the value of γ for each trajectory are given in the legend.

4.4 Real Data

We now present the results of applying b-ELM along with Bagging-ELM to several different real datasets from the UCI Machine Learning repository. As expected, the performances of b-ELM, remain substantially better than those of Bagging-ELM. Figs. 4 and 5 show *accuracy* or F_1 vs. *processingtime* for both b-ELM and Bagging-ELM classifications on Poker dataset, respectively. For b-ELM, $b = n^{\gamma}$ with the value of $\gamma \in \{0.6, 0.7, 0.8, 0.9\}$ for each trajectory are given in the legend. Notably, the outputs of b-ELM for all values of b considered, and the output of Bagging-ELM, are tightly clustered around the same value; additionally, as expected, b-ELM converges more quickly than Bagging-ELM. Furthermore, We have obtained qualitatively similar results on other additional datasets from the UCI dataset repository.

Table 2. Prediction performance comparisons among b-ELM, Bagging-ELM, and Single-ELM

Datasets	*Accuracy* (%)			F_1 (%)		
	Single-ELM	Bagging-ELM	b-ELM	Single-ELM	Bagging-ELM	b-ELM
Balance	90.24	91.26	91.27	88.93	89.76	89.83
Car	91.45	94.23	94.61	85.31	88.14	88.17
Waveform	84.79	86.73	86.75	82.27	85.13	85.17
Mushroom	88.82	89.31	89.38	87.93	88.78	88.81
Digits	95.31	97.18	97.21	94.14	96.87	96.86
Letter	93.54	97.18	97.24	92.83	95.75	95.79
3-Covers	87.94	93.54	93.61	86.67	89.91	89.97
Adult	84.12	85.23	85.24	79.57	80.87	80.91
Covertype	79.23	85.37	85.41	69.92	78.53	78.55
Poker	50.77	83.87	84.23	52.31	86.82	87.93

Table 2 gives a summary of prediction properties of b-ELM compared to Bagging-ELM and Single-ELM. Here, we use $b = n^{0.7}$ in all runs of b-ELM. For each dataset, both b-ELM and Bagging-ELM are repeated for 50 times as well as reaching relatively high *accuracy* or F_1 value under appropriate settings. Then, for each dataset, the average *accuracy* or F_1 are recorded. The average performance of Single-ELM is also recorded for comparison with the same parameter settings. Obviously, the classification performances of both b-ELM and Bagging-ELM are much better than that of an individual ELM. Also, b-ELM is more scalable than Bagging-ELM while maintaining favorable classification performance.

5 Conclusions

In this paper, the b-ELM algorithm is proposed, which employs the Bag of Little Bootstraps technique to attain a scalable, efficient means of classification for large-scale data. b-ELM typically has markedly better space and computational profiles than Bagging-ELM. Experimental results confirm the efficiency and effectiveness of the proposed algorithm. Moreover, b-ELM is suited to implementation on modern parallel and distributed computing platforms. We will deploy and demonstrate our procedure over a cluster of compute nodes in the future.

Acknowledgments. This work is supported by the National Natural Science Foundation of China (No. 61035003, 61175052, 61203297), National High-tech R&D Program of China (863 Program) (No. 2012AA011003, 2013AA01A606, 2014AA015105).

References

1. Breiman, L.: Bagging predictors. Machine Learning 24(2), 123–140 (1996)
2. Efron, B., Tibshirani, R.J.: An introduction to the bootstrap, vol. 57. Chapman and Hall (1993)
3. Huang, G.-B., Chen, L.: Convex incremental extreme learning machine. Neurocomputing 70(16), 3056–3062 (2007)
4. Huang, G.-B., Ding, X., Zhou, H.: Optimization method based extreme learning machine for classification. Neurocomputing 74(1), 155–163 (2010)
5. Huang, G.-B., Li, M.-B., Chen, L., Siew, C.-K.: Incremental extreme learning machine with fully complex hidden nodes. Neurocomputing 71(4), 576–583 (2008)
6. Huang, G.-B., Wang, D.H., Lan, Y.: Extreme learning machines: a survey. International Journal of Machine Learning and Cybernetics 2(2), 107–122 (2011)
7. Huang, G.-B., Zhou, H., Ding, X., Zhang, R.: Extreme learning machine for regression and multiclass classification. IEEE Transactions on Systems, Man, and Cybernetics–Part B: Cybernetics 42(2), 513–529 (2012)
8. Huang, G.-B., Zhu, Q.-Y., Siew, C.-K.: Extreme learning machine: a new learning scheme of feedforward neural networks. In: Proceedings of the International Joint Conference on Neural Networks (IJCNN 2004), vol. 2, pp. 985–990. IEEE (2004)

9. Huang, G.-B., Zhu, Q.-Y., Siew, C.-K.: Extreme learning machine: theory and applications. Neurocomputing 70(1), 489–501 (2006)
10. Kleiner, A., Talwalkar, A., Sarkar, P., Jordan, M.I.: The big data bootstrap. In: Proceedings of the 29th International Conference on Machine Learning (ICML-2012), pp. 1759–1766 (2012)
11. Lan, Y., Soh, Y.C., Huang, G.-B.: Ensemble of online sequential extreme learning machine. Neurocomputing 72(13), 3391–3395 (2009)
12. Tian, H., Meng, B.: A new modeling method based on bagging elm for day-ahead electricity price prediction. In: IEEE Fifth International Conference on Bio-Inspired Computing: Theories and Applications (BIC-TA), pp. 1076–1079. IEEE (2010)
13. Xu, R.-z., Geng, X.-f., Zhou, F.-y.: A short term load forecasting based on bagging-elm algorithm. In: Lu, W., Cai, G., Liu, W., Xing, W. (eds.) Proceedings of the 2012 International Conference on Information Technology and Software Engineering. LNEE, vol. 211, pp. 507–514. Springer, Heidelberg (2013)
14. Ye, R., Suganthan, P.N.: Empirical comparison of bagging-based ensemble classifiers. In: 15th International Conference on Information Fusion, pp. 917–924. IEEE (2012)

Pruned Annular Extreme Learning Machine Optimization Based on RANSAC Multi Model Response Regularization

Lavneet Singh and Girija Chetty

Faculty of ESTEM, University of Canberra, Australia

Abstract. The accuracy and performance of machine learning and statistical models are still based on tuning some parameters and optimization for generating better predictive models of learning is based on training data. Larger datasets and samples are also problematic, due to increase in computational times, complexity and bad generalization due to outliers. Using the motivation from extreme learning machine (ELM), we proposed annular ELM based on RANSAC multi model response regularization to prune the large number of hidden nodes to acquire better optimality, generalization and classification accuracy of the network in ELM. Experimental results on different benchmark datasets showed that proposed algorithm optimally prunes the hidden nodes, better generalization and higher classification accuracy compared to other algorithms, including SVM, OP-ELM for binary and multi-class classification and regression problems.

Keywords: Extreme Learning Machine, RANSAC, Regularization, Classification, Regression.

1 Introduction

1.1 Extreme Learning Machine (ELM)

Fortunately, due to exponential expansion in the technology, the improvement in machine learning and optimizing the parameters of statistical models improved by availability of large datasets and abundance of information of a studied phenomenon with maximum number of variables and samples. But, on the other hand, increase in number of variables with respect to samples in large dataset increases the redundancy and create ill posed problems. Currently, most of the machines learning models are based on deterministic learning algorithms rather than non-deterministic approach which narrow down its learning applications in real time datasets.

(Huang, Zhu, & Siew, 2006) proposed a new novel algorithm as Extreme Machine Learning (ELM) for single hidden layer feed forward neural network which has less computational time and faster speed even on large datasets. The main working core of ELM is random initialization of weights rather than learning through slow process via iteratively gradient based learning as back-propagation (Abid, Fnaiech, & Najim, 2001). In Extreme machine learning, the number of hidden nodes and their weights are randomly assigned, which distinguishes the linear differentiable between the

J. Cao et al. (eds.), *Proceedings of ELM-2014 Volume 1*,
Proceedings in Adaptation, Learning and Optimization 3, DOI: 10.1007/978-3-319-14063-6_15

output of hidden layer and output layer. The output weights can be determined by linear least square solution of hidden layer output through activation function and the data samples targets. For N arbitrary distinct samples (x_i, t_i), where $x_i= [x_{i1}\ x_{i2},..., x_{in}]^T \in R^n$ and $t_i = [t_{i1}, t_{i2},...,t_{im}]^T \in R^m$, standard SLFNs with N hidden nodes and activation function $g(x)$ are mathematically modelled as

$$\sum_{i=1}^{\tilde{N}} \beta_i g_i(x_j) = \sum_{i=1}^{\tilde{N}} \beta_i g(w_i \cdot x_j + b_i) = 0, (1), \quad j = 1, ..., N, \tag{1}$$

where $w_i = [w_{i1}, w_{i2},..., w_{in}]^T$ is the weight vector connecting the fth hidden node and the input nodes, $\beta_i = [\beta_{i1}, \beta_{i2},..., \beta_{in}]^T$ is the weight vector connecting the fth hidden node and the output nodes, and b_i is the threshold of the ith hidden node. $w_i \cdot x_j$ denotes the inner product of w_i and x_j. The output nodes are chosen linear in this paper.

That standard SLFNs with N hidden nodes with activation function $g(x)$ can approximate these N samples with zero error means that $\sum_{j=1}^{\tilde{N}} \left\| o_j - t_j \right\| = 0$, i.e., there exist β_i, w_i and b_i such that

$$\sum_{i=1}^{\tilde{N}} \beta_i g_i(w_i \cdot x_j + b_i) = t_j, \quad j = 1,..., N. \tag{2}$$

The above N equations can be written compactly as

$$H\beta = T,$$

Where

$$H(w_1,..., w_{\tilde{N}}, b_1..., b_{\tilde{N}}, x_1,..., x_N)$$

$$= \begin{bmatrix} g(w_1 \cdot x_1 + b_1) & \cdots & g(w_{\tilde{N}} \cdot x_1 + b_{\tilde{N}}) \\ \vdots & \cdots & \vdots \\ g(w_1 \cdot x_N + b_1) & \cdots & g(w_{\tilde{N}} \cdot x_N + b_{\tilde{N}}) \end{bmatrix}_{N \times \tilde{N}} \tag{3}$$

$$\beta = \begin{bmatrix} \beta_1^T \\ \vdots \\ \beta_{\tilde{N}}^T \end{bmatrix}_{\tilde{N} \times m} \text{ and } T = \begin{bmatrix} t_1^T \\ \vdots \\ t_N^T \end{bmatrix}_{N \times m} \tag{4}$$

H is called the hidden layer output matrix of the neural network; the ith column of H is the ith hidden node output with respect to inputs $x_1, x_2..., x_N$..

As named in (Huang et al., 2006), H is called the hidden layer output matrix of the neural network; the ith column of H is the ith hidden node output with respect to inputs. (Huang, Wang, & Lan, 2011) presented a comprehensive survey on extreme learning machines and it applications. Optimally pruned extreme leaning machine (OP-ELM) algorithm which is an extension of original ELM algorithm with pruning of neurons using ranking multi-response sparse regression (MRSR) method to design optimal neural architecture removing irrelevant variables was proposed by (Yoan et al., 2010). (Martínez-Martínez et al., 2011) proposed a new strategy to prune the ELM networks using regularized regression methods to acquire optimal tuned parameters. The algorithm can acquire optimal tuned parameters by identifying the degree of relevance of the weights that connects the k-th hidden element with the output layer using lasso and ridge regression. Regularized version of least squares regression with several penalties on coefficient vector are used to remove the irrelevant or low relevance hidden nodes to achieve compact neural networks.(Singh & Chetty, 2012) proposed LDA-ELM for classification of brain abnormalities in magnetic resonance images using pattern recognition and machine learning. (Lavneet Singh, 2012) proposed a Novel Approach for protein Structure prediction Using Pattern Recognition and Extreme Machine Learning.

To overcome the drawbacks of regularization or penalty method, using sparse model and removing reductant variables for better generalization an prediction accuracy, we proposed RANSAC multi model response regularization which implements a L1 penalty or the output weights by performing RANSAC multi model response regression between the hidden and output layer.

1.2 **Regularized Extreme Learning Machine.**

To resolve these limitations of ELM, constructive and heuristic approaches have proposed in the literature. In most recent years, regularization or penalty approach seems to be significant in resolving the ELM limitations. As in extreme machine learning, there is linear behaviour between hidden layer and output layer, thus as a problem of linear regression, regularization helps to reduce the number of predictors in hidden layer by using sparse model.

Significant work have been done in past for better generalization, faster learning and rate of convergence. But, unfortunately, ELM also suffers with some limitations as outliners, irrelevant variables and number of hidden nodes. To resolve these limitations of ELM, constructive and heuristic approaches have proposed in the literature.

In most recent years, regularization or penalty approach seems to be significant in resolving the ELM limitations. As in extreme machine learning, there is linear behavior between hidden layer and output layer, thus as a problem of linear regression, regularization helps to reduce the number of predictors in hidden layer by using sparse model. Least square solution with regularization is fitted to the model to find the nonzero coefficients as output weight of output layer. Using regularized sparse model, most of the predictors are move to zero with increase in lambda. Thus, it creates a sparse model of output of hidden layer of finding the beta coefficients with respect to lambda with minimum deviance or minimum convergence with respect to Mean square error. Regularization is applied to regression problems to select the relevant hidden units, by addressing over fitting trade off with respect to network size. The big architectures are selected for the network because regularization approach prunes the network with optimal hidden neurons. In regularized based ELM, the weights of the input layer connected to the hidden layer are chosen randomly. The output weights of the output layer are determined through regularized regression removing hidden units with a optimized size network. The approach used for regularization for ELM stated as

1. Lasso regularization as L_1 penalty
2. Ridge Regression or L_2 penalty
3. Elastic net combining both L_1 and L_2 penalty

To define the general case of regularization, as a single output regression represented as –

$$Y = Xw + \varepsilon \tag{5}$$

with $X = (X_1, X_2, \ldots\ldots X_n)^T$ are the inputs of a dataset and $Y = (Y_1, Y_2, \ldots\ldots Y_n)^T$ are the output and $W = (w_1, w_2, \ldots\ldots w_p)^T$ are the regression weights of the hidden layer. As discussed, the model possess a linear regression between input layer and output layer, thus the simple least square solution (OLS) is a heuristic approach to solve single output regression formulated as

$$\min_{\hat{W}} (y - X\hat{w})^T (y - X\hat{w}) \tag{6}$$

Or in least square form

$$\min_{\hat{W}} \sum_{i=1}^{n} (y_i - x_i\hat{w})^2 \tag{7}$$

with $\hat{w} = (\hat{w}_1, \ldots\ldots \hat{w}_n)^T$ the estimated regression weights.

The solution of equation (7) is then obtained by a pseudo inverse (Moore- Penrose) as

$$\hat{w}_{OLS} = (X^T X)^{-1} X^T y \tag{8}$$

Moore Penrose is not useful in every numerical problem if X in equation 8 is not full rank.

To improve better generalization and prediction accuracy, OLS doesn't provide a complete solution to remove irrelevant variables. OLS doesn't use sparse models to get related variables with respect to output.to resolve the mentioned issues with simple OLS, regularization factors or penalty approach added to minimization cost function in equation 8 to get sparse model of OLS to acquire sparse model which try to shift most of the irrelevant variables to zero. Regularization or penalty term lambda with its weights added to minimization problem with its nonzero coefficients to get beta coefficients of particular model.

1.3 The L1 Penalty : LASSO

The minimization problem with L1 penalty is formulated as

$$\underset{\lambda,\hat{w}}{\min}\left[\sum_{i=1}^{n}(y_i - x_i\hat{w})^2 + \lambda\sum_{j=1}^{p}\left|\hat{w}_j\right|\right] \tag{9}$$

Taking an example of simple linear regression with a output as dependent variable y Ɛ R and independent vector S Ɛ RQ to approximate the regression function. As $\lambda \in [0,\infty]$ increases from minimum to maximum λ, more SW predictors moves towards becoming zero tracing out the path with respect to beta coefficients exhibiting sparsity. The computation of equation 9 is a classic quadratic programming problem intensively finding the nonzero coefficients of variable correlated to output exhibiting sparsity to less correlated variables. The number of predictors in a regression model reduces using lasso regularization. Lasso regularization identifies the important predictors among the redundant predictors and produce shrinkage estimates with potentially lower predictive errors than ordinary least squares.

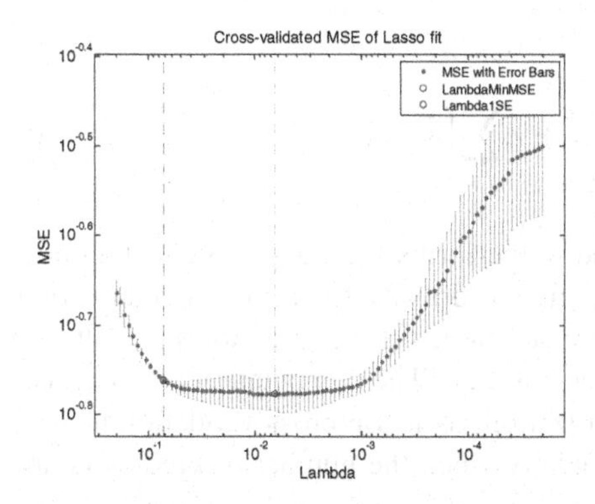

Fig. 1. Lasso plot of lambda on diabetes dataset with respect to its Mean Square Error(MSE) using regularization approach of L_1 penalty on output of Hidden Layer to calculate shrinkage estimates with potentially lower predictive errors than ordinary least squares with least predictors as hidden units.

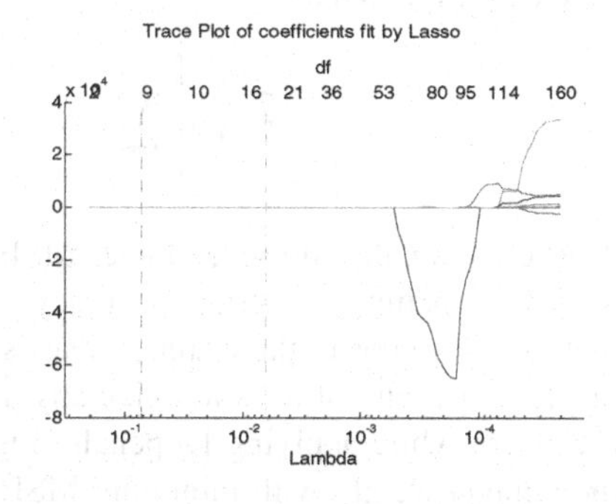

Fig. 2. lasso plot of lambda parameter using regularization approach of L_1 penalty on output of Hidden Layer for all predictors as hidden units.

Fig 1 and 2 shows Lasso plot of lambda on diabetes dataset with respect to its Mean Square Error(MSE) and lambda values for each predictors as hidden units using regularization approach of L_1 penalty on output of Hidden Layer. Lasso Regularization calculates shrinkage estimates with potentially lower predictive errors than ordinary least squares with least predictors as hidden units with the nonzero coefficients in the regression for various values of the Lambda regularization parameter. Larger values of Lambda appear on the left side of the graph, meaning more regularization, resulting in fewer nonzero regression coefficients.

The dashed vertical lines represent the Lambda value with minimal mean squared error (on the right), and the Lambda value with minimal mean squared error plus one standard deviation. This optimal value of Lambda is estimated by performing cross validation up to 5 folds. The upper part of the plot shows the degrees of freedom (df), meaning the number of nonzero coefficients in the regression, as a function of Lambda. On the left, the large value of Lambda causes most of the coefficient to be 0 producing sparsity. On the right are all five coefficients are nonzero, though the plot shows only four clearly. For small values of Lambda (toward the right in the plot), the coefficient values are close to the least-squares estimate.

A common drawback of the L_1 penalty as Lasso regularization is that it tends to be produce more sparse model with respect to increasing lambda, where $w_j = 0$. Thus, using lasso regularization, reduce the number of predictors much more with increasing sparse estimates which quite challenging to tune the model.

1.4 The L2 Penalty:- Tikhonov Regularization

Tikhonov regularization is same as almost same as Lasso regularization only the difference is in minimization problem, it involve a penalty of square of regression coefficients formulated as

$$\overset{\min}{\lambda, \hat{w}} \left[\sum_{i=1}^{n} (y_i - x_i \hat{w})^2 + \lambda \sum_{j=1}^{p} \left| \hat{w}_j^{\,2} \right| \right] \tag{10}$$

The L_2 venality provides lower MSE and outperforms the Lasso regularization in achieving better prediction accuracy. The major drawback of this regularization method is similar to the ordinary least square solution, since all variables are filtered due to L_2 penalty, it does not give any unique solution. The computation times is also a concern while applying L_2 penalty regularization, as it use cross validation to find the lambda weight with minimum MSE specially when the number of predictors are large enough.

To resolve the limitation and overcomes the drawbacks of both two regularization approaches, hybrid penalties are proposed using both L_1 and L_2 penalty in same minimization problem as

$$\overset{\min}{\hat{w}} \left[\sum_{i=1}^{n+p} (y_i - x_i \hat{w})^2 + \frac{\lambda_1}{\sqrt{1+\lambda^2}} \sum_{j=1}^{p} \left| \hat{w}_j \right| \right] \tag{11}$$

The above minimization equation is also termed as elastic net which creates great shrinkage effect on the regression coefficients than original lasso regularization. Since elastic net are effective way of implementing shrinkages estimates on regression coefficients, but still two parameter λ_1 and λ_2 need to be optimize and tuned using cross validation (CV) which is costly search in two dimensional matrix. This is hardly feasible if computation times of ELM are taken under consideration. To overcome the drawbacks of regularization or penalty method, using sparse model and removing reductant variables for better generalization an prediction accuracy, we proposed multi Ransac based regularization which implements a L_1 penalty or the output weights by performing multi Ransac response regression between the hidden and output layer.

2 Annular ELM

Circular Back Propagation (CBP) networks [12] improve over the basic formulation of MLP; the CBP model augments the input vector by one additional dimension, which is computed as the norm of the input vector itself. In a classic set-up involving a single layer network, the estimation process supported by the enhanced, CBP network is expressed as

$$y_{CBP}(x) = \xi\left(b + \sum_{j=1}^{N_h}\left[wj\xi\left(\hat{b}_{j,0} + \sum_{k=1}^{M}\hat{w}_{j,k}x_k + \hat{w}_{j,M} + 1\|x\|^2\right)\right]\right) \tag{12}$$

Based on circular back-propagation network, we proposed a new architecture network for extreme learning machines known as Annular ELM. The first formation of the annular ELM augments by adding one more dimension in both training and testing data which is computed as

Training data_{ij} = Training data_{ij+1}

Where i = {1……m} as number of observation and j = (1…….n) as number of features. Thus,

Similarly

Testing data_{pj} = Testing data_{pj+1}

As the annular topology is implemented to the input layer and feeded into the hidden layer in ELM network, the random weights and of the hidden layer and the bias is redefined as

$$Aj(x, \hat{w}_j, \hat{b}_j) = \left(\xi(\hat{w}_j.\overline{x} + \hat{b}_j)\right) \tag{13}$$

which could be rewritten as

$$Aj(x, \hat{w}_j, \hat{b}j) = \xi(z_j.\|x - c_j\|^2 + \overline{b}_j) \tag{14}$$

Where

$$Z_j = \hat{w}_{j,M+1} \tag{15}$$

$$cj = \left[\frac{\hat{w}_{j,1}}{2\hat{w}_{j,M}+1}, \ldots\ldots\ldots\ldots, -\frac{\hat{w}_{j,M}}{2\hat{w}_{j,M}+1}\right] \tag{16}$$

$$\bar{b}_j = \frac{1}{\hat{w}_{j,M+1}} \left(\sum_{k=1}^{M} \frac{\hat{w}_{j,k}^{2}}{4\hat{w}_{j,M+1}} - \hat{b}_j \right) \tag{17}$$

Input data $(X_{ij}) = [X_{ij}, \ldots\ldots\ldots X_{ij+1}]$

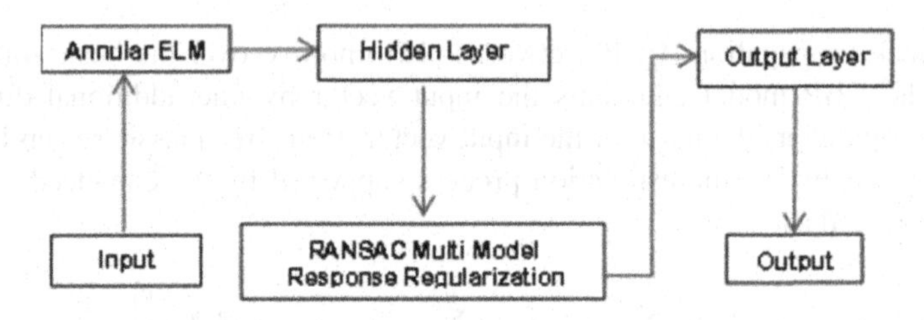

Fig. 3. Architectural Design of Annular ELM based on RANSAC Multi Model Response Regularization

The annular based ELM able to map both linear and circular separation boundaries by boosting the ability of ELM network. In annular based ELM, the initial hidden weights are chosen randomly only along with the bias too as similar like ELM. But, later new random weights and bias are approximated by applying proposed annular functions to estimate the output of hidden layer known as hidden matrix.

Later, proposed RANSAC multi response regularization is applied to the output of hidden layer using annular ELM topology to prune the hidden units for better generalization and higher accuracy.

3 RANSAC

The RANSAC (Random Sample Consensus) algorithm was proposed by (Fischler & Bolles, 1981) to estimate the parameters of a certain model from a set of data with large number of outliners. RANSAC take out the outliners from a data if it doesn't fit with a set of parameters within the error threshold with respect to maximum deviation. RANSAC can handle outlier's greater than 50% of the entire dataset known as breakdown point.

RANSAC first hypothesize minimum sample sets MSSs) randomly selected from the input dataset and the parameters of the model are estimated using MSS. To test the estimated parameters of the model using MSS, RANSAC checks which element of the entire dataset are consisted with the defined model known as consensus set (CS). RANSAC rank the consensus with a set of iterations with respect to estimated probability at certain threshold. RANSAC is extensively used in computer vision, motion detection and features matching of images and is optimized using different parameters (Raguram, Frahm, & Pollefeys, 2008),(Nistér, 2005),(Chum, Matas, & Kittler, 2003).

3.1 RANSAC Multi Model Response Regularization

3.2 RANSAC Multi Model Response Regularization for Regression Problems

To implement the RANSAC on regression problems, we proposed a RANSAC multi model response regularization which implements the sequential RANSAC on multiple models. To implement RANSAC, which in our case, are the irrelevant hidden nodes as predictor variables, and H is the hidden matrix as input from equation. In our case, the output weights follow a linear regression between hidden and output layer defined as

$$Y = mx + \epsilon \tag{18}$$

Where Y is the output of instances of data, m is the predictor's weights or slope and x is the input data and c is the constant. The output weights follow a linear regression between hidden and output layer defined as

$$Y = \text{Output weights}*H + \epsilon$$

Where

$$H = \begin{bmatrix} g(w_1 \cdot x_1 + b_1) & \cdots & g(w_{\tilde{N}} \cdot x_1 + b_{\tilde{N}}) \\ \vdots & \cdots & \vdots \\ g(w_1 \cdot x_N + b_1) & \cdots & g(w_{\tilde{N}} \cdot x_N + b_{\tilde{N}}) \end{bmatrix}_{N \times \tilde{N}} \tag{19}$$

$$OutputWeights = Multi\_RANSAC \begin{bmatrix} g(w_1 \cdot x_1 + b_1) & \cdots & g(w_{\tilde{N}} \cdot x_1 + b_{\tilde{N}}) \\ \vdots & \cdots & \vdots \\ g(w_1 \cdot x_N + b_1) & \cdots & g(w_{\tilde{N}} \cdot x_N + b_{\tilde{N}}) \end{bmatrix}_{N \times \tilde{N}} *T \tag{20}$$

$$CS = RANSAC\left(\sum_{W=1}^{m}\left(\sum_{j=1}^{n} D_{Wj}\right)\right) \tag{21}$$

$Dx_{wj} = \{x_{11}, x_{12}\ldots\ldots x_{wj}\}$ be the sets of data H with wth observations in rows and j^{th} as hidden nodes predictors in columns of D matrix. For regression problems, sequential RANSAC implements on the set of all inliners, D that are generated by W different models where $W_m = \text{rand}(H_M)$. The numbers of models are randomly generated using 20 % of the input data.

Multi Models are selected from hidden matrix and RANSAC is implemented by bootstrapping so as to get maximum sparse coefficients of CS for hidden nodes. Multi models from hidden matrix are randomly selected as 20 % of data so as to get maximum sparse CS coefficients to prune the hidden nodes. CS ranking help to determine zero and non-zero coefficients of hidden nodes which is efficiently determine by bootstrapping of RANSAC on selected multi models.

To estimate the parameters of W models, each one is represented by k dimensional parameter vector θ_w at each iteration iter. CS is estimated using MSS of each W model. The iteration run M times which is calculated before after removing the inliners from data D. the total number of inliners at iteration iter is less than total number of inliners at iteration iter-1. The whole formulation of multiple RANSAC response is defined in equation 20

The set of all inliners D is generated by W different models has cardinality CS as

$$N_I = (N_{I,1} + N_{I,2} + \dots\dots\dots N_{I,W})$$

As RANSAC is a parametric model, thus, set of parameters need to be defined before implementing RANSAC. These set of parameters are defined as

- Epsilon = False alarming rate as the probability of the algorithm throughout all the iterations will never sample a MSS containing only inliers
- Probability of inlier = the probability that a point whose fitting error is less or equal than is actually an inlier
- Sigma = Gaussian noise
- Estimate function = function that returns the estimate of the parameter vector starting from a set of data.
- Mean square function = function that returns the fitting error of the data.
- CS ranking Algorithm = Algorithm to rank the CS of data
- Minimum number of iterations
- Maximum number of iterations
- W = 20% of the training data

Let $M_w(\theta_w)$ defines the manifold of dimension k_w of all points with respect to parameter $\theta_w \in R^{kw}$ for the specified model for $1 \leq w \leq W$ with a subset S_w from D of k_w elements at iteration i called minimal sample set (MSS). To estimate the parameters of W models, each one represented parameter vector θ_w. At each i iterations, MSS for each W model is defined and CS is estimated removing all outliners.

To combine and fuse the estimated CS computed from i(W) iterations, the whole RANSAC multi model response algorithm can be summarize as follows

3.3 RANSAC Multi Model Response Regularization Algorithm

$S_w^{(i)} = S(_i)\,(\theta_w)$ be the CS of wth model at i^{th} iteration.
The all combined updated CS of $S\,(_i)\,(\theta_w)$ is updated as
$M_{(iter)} = 100;\ i = 0$
For $i \leq M_{iter}$(maximum number of iterations)
Do $i = i + 1$
$\{S_w^{(i-1)}\},\{S_w^{(i)}\}$
While $1 \leq w \leq W$

$$S^i(\theta) = S^{(i)}\theta \cup S^{(i-1)}\theta$$

W=w+1
Return $\{S_w^{(i)}\}$

To reduce the number of hidden units with respect to ranked CS estimated using RANSAC multi-response algorithm is calculated as

$$\text{Hidden Layer Output } (H_1) = S^{(i)}\theta * H \tag{22}$$

which reduces the hidden matrix as hidden layer output H_1 into zero and nonzero coefficients of ranked elements with respect to estimated CS. Nonzero coefficients are extracted from sparse matrix which reduces the no of hidden units to which are ranked less giving the hidden units coefficients highly correlated. Thus

$$\text{Output weights} = (H_1^T H_1)^{-1} H_1^T T \tag{23}$$

3.4 RANSAC Multi Model Response Regularization for Binary and Multiclass Problems

The proposed RANSAC multi model response regularization for binary and multi-class problems for ELM is implemented using one against all (OAA) method. As in OAA method, j binary classifiers will be constructed in which all the training examples will be used a teach time of training. The training data having the original class label $j_n = (1.........n)$ have each j_n elements of positive one class and the remaining training data will be of zero class, creating j_n models implementing proposed RANSAC multi model response regularization on (j_n) binary classes. Finally, CS defined as $S^i(\theta)$ of j(n) classes is computed as

$$S^i(\theta)_j = \sum_{j=1}^{n} S^{(i)}\theta_j \cup S^{(i-1)}\theta_j \tag{24}$$

$$S^i(\theta) = \sum_{j=1}^{n} S^{(i)}\theta_j \cup S^{(i)}\theta_n \tag{25}$$

For this, consider the ELM for multi-class classification problem, formulated as k binary ELM classification problem with the following form:

$$Hw_1 = y_1 \ldots Hw_j = y_j;$$

Where for each j, w_j is the output weights from the hidden layer to the output layer with output vector $y_j = (y_{1j}, \ldots\ldots y_{mj})^t \in R_m$. Thus the output of the hidden layer as H hidden matrix defines with respect to multiclass binary classifiers as

$$H_j = \sum_{j=1}^{n} H * Y \begin{pmatrix} Y_j = 1 \\ o \end{pmatrix} \tag{26}$$

where H is the hidden layer output matrix and Y is the j binary classes with m^{th} observations of training data and n binary classes as columns vectors. Thus, we get H_j

hidden matrix where each H_j belong to each binary class and RANSAC multi response regularization is implemented to acquire CS for each binary class as $S^i(\theta)_j$. It can be concluded that RANSAC multi response regularization for binary and multiclass problems work in similar fashion as OAA-ELM with j binary classes with a difference of jth label with positive class and rest other classes with -1 class.

To improve better generalization and prediction accuracy, OLS doesn't provide a complete solution to remove irrelevant variables. OLS doesn't use sparse models to get related variables with respect to output.to resolve the mentioned issues with simple OLS, regularization factors or penalty approach added to minimization cost function in equation 20 to get sparse model of OLS to acquire sparse model which try to shift most of the irrelevant variables to zero.

X = LSQR (A, B) attempts to solve the system of linear equations A*X=B for X if A is consistent, otherwise it attempts to solve the least squares solution X that minimizes norm (B-A*X). LSQR is an iterative method to find Ordinary Least Squares solution for large sparse matrix which is analytically equivalent to the standard equation of conjugate gradients, but possess more favourable numerical properties.

Section 4 defines the experimental results comparative analysis between proposed algorithm and with different other algorithms on benchmark datasets in term of testing accuracy, RMSE and number of hidden nodes.

4 Experimental Results

To justify the proposed algorithm, the performance of proposed algorithm will be investigated by comparing it with original ELM, SVM and other machine learning methods using public available benchmarks datasets of regression, binary and multiclass classification.

In the proposed algorithm, different activation functions were used as G(a,b,x) along with existing algorithms like ELM< MLP and SVM. The different activation functions of hidden layer is defined as

1. Sigmoid Function

$$G(a,b,x) = \frac{1}{1 + \exp(-(a^t x+b))}$$

2. Radial Basis Function

$$G(a,b,x) = (\|x-a\|_2^2 + b^2)^{1/2}$$

For simulations of experimental results, the input weights and the hidden nodes are chosen randomly at the beginning of the iterations and were fixed in rest of the iterations. The optimal values average training and testing accuracy, number of hidden nodes of different algorithms with proposed algorithm were determined by using 10-fold- cross validation on both training and testing datasets.

All experiments were conducted in MATLAB R2010 platform running on windows 7-64 bit operating system with 3.0 GHz Intel® core 2 i5 processor having 8 GB of RAM. LIBSVM is used for the implementation of SVM in matlab and weka platform. OP-ELM toolbox is used for implementation of OP-ELM(Miche, Sorjamaa, & Lendasse, 2008). RANSAC toolbox is used for implementation and making proposed changes in sequential RANSAC(Zuliani, 2008). In our experiments, several benchmark datasets were chosen. The data sets were collected from the University of California at Irvine (UCI) Machine Learning Repository (Blake & Merz, 1998) and they all been processed using 10 different random permutations are taken with- out replacement; for each permutation, two thirds are taken for the training set, and the remaining third for the test set by using cross validation function in MATLAB (crossvalind). Training sets are then normalized (zero-mean and unit variance) and test sets are also normalized using the very same normalization factors than for the corresponding training set.

4.1 Datasets

In our experiments, several benchmark problems were chosen. The data sets were collected from the University of California at Irvine (UCI) Machine Learning Repository [11] and their different attributes for the data sets are summarized in Table 1 and 2. The data sets have all been processed using 10 different random permutations are taken with- out replacement; for each permutation, two thirds are taken for the training set, and the remaining third for the test set (see Table 1) by using cross validation function in MATLAB (crossvalind). Training sets are then normalized (zero-mean and unit variance) and test sets are also normalized using the very same normalization factors than for the corresponding training set. The 10 fold cross validation also enables to obtain an estimate of the standard deviation of the results presented (see Table 2).

Table 1. Classification datasets attributes and classes

Dataset	Attributes	Classes	Training data size	Testing data size
WDBC	30	2	427	142
Wincosin_BC	10	2	525	174
Cleveland	13	2	228	75
Australian Credit	14	2	518	172
Ionosphere	34	2	264	87
diabetes	8	2	576	192
Liver Disorders	6	2	259	86
Iris	4	3	113	37
Wine	13	3	134	44
Glass	9	6	161	53
Auto Vehicle	18	4	635	211
Page Blocks	10	5	4105	1368
Image Seg	19	7	1733	577
Satellite	36	6	4827	1608

Table 2. Regression datasets attributes and classes

Dataset	Attrib-utes	Training data size	Testing data size
Auto-MPG	8	294	98
Machine-CPU	7	157	52
Servo	5	126	41
Forest-Fires	13	388	129
Boston	14	380	126
Concrete-CS	9	773	257
Abalone	8	3133	1044
Wine-(white)	12	3674	1224
Wine-(Red)	12	1200	399
Parkinson	22	4407	1468
Kin-8	9	6144	2048
Demo	5	1536	512
Ailerons	40	5366	1788

5 Results

5.1 Classification

Table 3 depicts the comparative analysis of proposed RANSAC multi model response regularized ELM with support Vector Machine (SVM) and other ELM variants. For each dataset, training data is trained with higher number of hidden nodes so as to prune the network with optimal hidden nodes with better classification accuracy. For binary classification, it can be seen from table 3 using sigmoid function; most of the binary datasets shows the higher testing accuracy compared to SVM, ELM and OP-ELM. Using the RBF function in table 4, the proposed model didn't showed better testing accuracy results compared to other algorithms. But in both activation function, RANSAC multi model response regularized ELM prune the number of hidden nodes improving the optimality of the ELM network.

Table 5 depicts the comparative analysis of proposed algorithm with SVM and other variants for multi-class classification. As can be seen from table 5 using sigmoid

Table 3. Experimental results in terms of testing accuracy for binary classification using Sigmoid kernel

Datasets	HN	Testing(sigmoid)			
		SVM	OP-ELM	ELM	RANSAC-ELM`
WDBC	200	95.77	90.85	95.39	**95.81**
Win-BC	200	94.82	89.66	96.53	**96.97**
Cleveland	200	76.00	78.67	**90.40**	83.31
Aus-credit	200	83.72	84.88	81.40	**86.03**
Ionosphere	200	78.16	79.31	79.45	**88.92**
Diabetes	200	69.79	77.60	71.26	**76.21**
Liver Disorders	200	58.13	54.65	57.81	**65.88**

and RBF kernel,, RANSAC multi model response ELM shows the significant higher testing accuracy results compared to other algorithms for wine, glass, auto and segmentation datasets. Table 6 defines the number of hidden nodes pruned using RANSAC multi model response ELM with optimal higher testing accuracy. As can be seen from table 6, for binary and multi-class datasets, the RANSAC multi model response ELM significantly prune the number of hidden nodes from higher number of hidden nodes maintaining the higher testing accuracy, faster implementation and better generalization performance for most of the binary and multiclass classification.

Table 4. Experimental results in terms of testing accuracy for binary classification using RBF kernel

Datasets	HN	Testing(RBF)			
		SVM	OP-ELM	ELM	RANSAC-ELM`
WDBC	200	**97.18**	90.14	83.20	93.20
Win-BC	200	96.55	**97.13**	90.74	95.78
Cleveland	200	76.00	78.61	**90.55**	81.25
Aus-credit	200	83.72	**86.63**	73.84	84.72
Ionosphere	200	89.65	**95.40**	78.41	92.90
Diabetes	200	75.00	**79.69**	69.20	77.30
Liver Disorders	200	**74.41**	68.60	57.47	67.30

Table 5. Experimental results in terms of testing accuracy for multiclass classification

Datasets	HN	Testing(sigmoid)			Testing(RBF)		
		SVM	ELM	RANSAC-ELM`	SVM	ELM	RANSAC-ELM
Iris	200	**91.89**	78.32	83.89	**97.29**	85.24	89.46
Wine	200	93.18	**92.77**	**95.45**	**97.72**	84.85	94.36
Glass	500	39.62	41.85	**51.81**	41.50	**50.60**	48.79
Auto	500	43.60	67.39	**79.68**	56.39	58.23	**79.94**
Page	500	91.30	**94.85**	92.96	**92.17**	89.47	91.43
Segment	500	83.88	94.55	**95.33**	87.17	94.67	**94.75**
Satellite	500	83.64	89.60	88.66	83.70	90.25	88.20

Table 6. Experimental results in terms of number of hidden nodes pruned by proposed method using sigmoid and radial basis functions for binary and multi-class classification

Dataset	ELM Hidden Nodes	RANSAC-ELM hidden nodes(sig)	RANSAC-ELM hidden nodes(RBF)
WDBC	200	40	48
Wincosin_Breast_cancer	200	34	30
Cleveland	200	52	53
Australian Credit	200	33	36
Ionosphere	200	90	92
diabetes	200	29	30
Liver Disorders	200	61	70
Iris	200	94	71
Wine	200	60	58
Glass	200	65	123
Auto Vehicle	500	90	118
Page Blocks	500	97	230
Image Segmentation	500	232	245
Satellite	500	250	252

5.2 Regression

Table 7 depicts the training accracy for differnt regression problem datasets using ELM and proposed algorithm with respect to different kernel function. In both activation function, for regression problems, RANSAC multi model response regularized ELM prune the number of hidden nodes improving the optimality of the ELM network and least RMSE compared to other algorithms.

For regression problems, table 8 and 9 depicts the comparative analysis of proposed RANSAC multi model response regularized ELM with ELM and other ELM variants using various kernels.. As can be seen from table 8 and 9, RANSAC multi model response annular ELM shows the significant higher testing accuracy results compared to other algorithms on different datasets. Table 10 defines the number of pruned hidden nodes using RANSAC multi model response annular ELM with optimal higher testing accuracy.

Table 7. Experimental results in terms of training root mean square accuracy for regression using sigmoid kernel

		ELM	RANSAC-ELM	ELM	RANSAC-ELM
AutoMPG	200	**1.4316**	1.4339	**1.5571**	1.6019
CPU	200	72.1778	**48.2834**	75.2482	**34.5921**
Servo	200	1.7221e-23	4.5645e-05	7.0477e-28	**0.0172**
Forest	200	0.9071	1.2545	0.9449	1.2358
Boston	200	**2.7443**	3.0085	**8.6816**	3.6345
Concrete	200	20.3213	**14.9237**	40.3428	**16.3896**
Abalone	200	**3.8052**	3.8399	3.9294	**3.8636**
Wine-W	200	0.4576	**0.4440**	1.3107	**0.4724**
Wine-R	200	**0.2943**	0.3011	0.9455	**0.2865**
Parkinson	200	0.0012	**0.0011**	0.0061	**0.0011**
Kin-8	200	**0.0176**	0.0211	0.0337	**0.0225**
Demo	200	**1.1882**	1.1902	**1.1867**	1.2408
Ailerons	200	3.4750e-08	**2.3821e-08**	**1.8721e-07**	2.4917e-08

Table 8. Experimental results in terms of testing root mean square accuracy for regression using Sigmoid kernel

Datasets	HN	Testing(sigmoid)		
		OP-ELM	ELM	RANSAC-ELM`
Auto-MPG	200	**11.7282**	91.2388	**11.9512**
Machine-CPU	200	2.0269e+04	6.2133e+08	**1.4293e+03**
Servo	200	0.5501	11.4059	**0.2861**
Forest-Fires	200	4.6637	5.5861	**1.9943**
Boston	200	29.8807	36.5606	**8.7404**
Concrete-CS	200	223.4046	58.5347	**28.3328**
Abalone	200	6.1798	10.6831	**4.1517**
Wine-(white)	200	0.5494	0.5251	**0.4637**
Wine-(Red)	200	0.4698	0.4738	**0.3627**
Parkinson	200	0.0025	0.0014	**0.0012**
Kin-8	200	0.0481	**0.0189**	0.0219
Demo	200	1.7995	54.3682	**1.4459**
Ailerons	200	0.7569	3.9234e-08	**2.5698e-08**

Table 9. Experimental results in terms of testing root mean square accuracy for regression using RBF kernel

Datasets	HN	Testing(RBF)		
		OP-ELM	ELM	RANSAC-ELM`
Auto-MPG	200	41.4800	49.6352	**8.0027**
Machine-CPU	200	9.5348e+03	6.8199e+07	**1.4612e+03**
Servo	200	0.5550	2.1920	**0.3258**
Forest-Fires	200	2.4981	4.0079	**1.8157**
Boston	200	33.7665	60.3066	**10.3812**
Concrete-CS	200	227.5232	95.1963	**31.114**
Abalone	200	5.9829	4.9659	**3.9806**
Wine-(white)	200	0.6074	1.4766	**0.4808**
Wine-(Red)	200	0.5203	1.4369	**0.3837**
Parkinson	200	0.0031	0.0065	**0.0012**
Kin-8	200	0.0426	0.0361	**0.0224**
Demo	200	1.7991	2.5114	**1.3197**
Ailerons	200	0.8324	2.0346e-07	**2.6605e-08**

Table 10. Experimental results in terms of number of hidden nodes pruned by proposed method using sigmoid and radial basis functions for regression

Dataset	ELM Hidden Nodes	RANSAC-ELM hidden nodes(sig)	RANSAC-ELM hidden nodes(RBF)
Auto-MPG	200	164	132
Machine-CPU	200	96	102
Servo	200	110	87
Forest-Fires	200	93	81
Boston	200	78	113
Concrete-CS	200	139	138
Abalone	200	118	97
Wine-(white)	200	96	79
Wine-(Red)	200	113	97
Parkinson	200	148	127
Kin-8	200	62	147
Demo	200	133	100
Ailerons	200	87	125

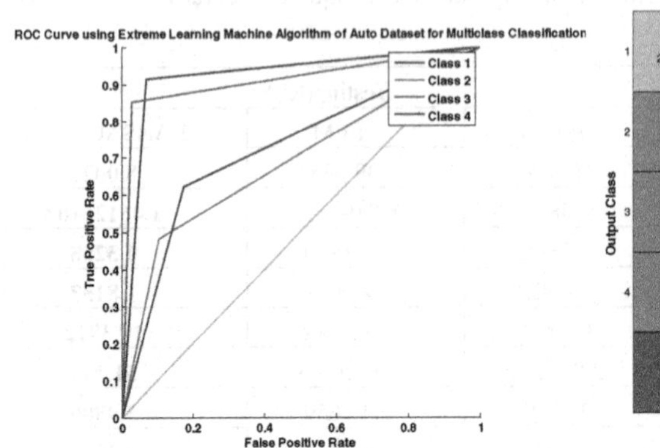

Fig. 4. ROC Curve using Extreme Learning Machine Algorithm of Auto Dataset for Multiclass Classification

Fig. 5. Confusion Matrix depicting true positive and false positive rate using Extreme Learning Machine Algorithm of Auto Dataset for Multiclass Classification

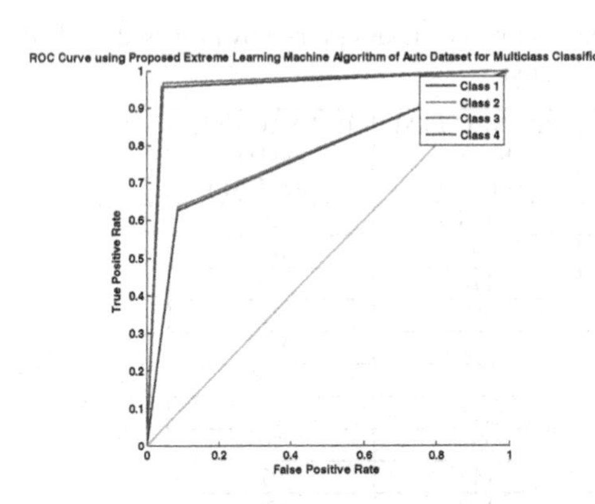

Fig. 6. ROC Curve using Proposed Extreme Learning Machine Algorithm of Auto Dataset for Multiclass Classification

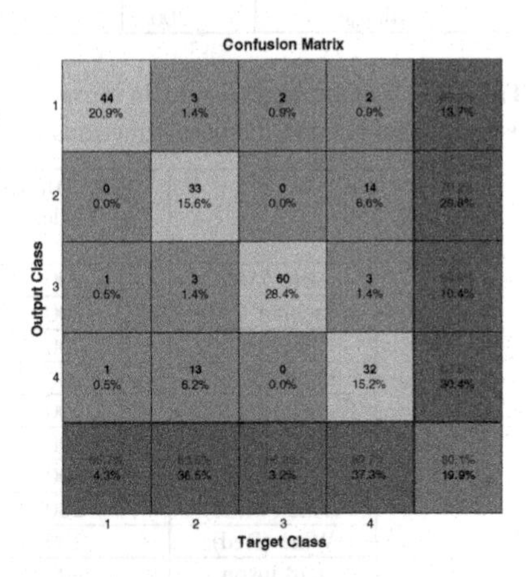

Fig. 7. Confusion Matrix depicting true positive and false positive rate using Extreme Learning Machine Algorithm of Auto Dataset for Multiclass Classification

Figure 4 represents the ROC curve using extreme learning machine algorithm of auto dataset for multiclass classification. As can be seen in the figure, the ROC curve between true positive and false positive rate is linear at least true positive rate for different classes. Figure 5 depicts the confusion matrix using ELM of auto dataset for multiclass classification. As we can see from the figure 5, the false positive rate is higher of output class with respect to target class. Compared with figure 6 showing ROC curve using proposed ELM algorithm of auto dataset for multiclass classification, the true positive rate is higher with respect to false positive rate for multi classes. As can be seen in figure 7, the true positive rate is higher of output class with respect to target class.

Thus, from above experiments, it can be stated that proposed method based on RANSAC regularization performs very well in terms of training and testing accuracy compared to traditional ELM and its variants. The above experiment also reduces the size of network to optimal decreasing computation time and faster performance of the network.

6 Conclusions

In this paper, we proposed an annular ELM based on RANSAC multi model response regularization to optimally prune the hidden nodes in a network and improve better generalization and classification accuracy. Experimental results were conducted using comparative analysis of proposed RANSAC multi model response regularization based annular ELM network on different benchmark datasets for binary and multiclass classification and regression problems.

It can be concluded that from experimental results that proposed RANSAC multi model response regularized based annular ELM works significantly well in higher classification accuracy with optimally pruned hidden units. The proposed algorithm implements faster compared to other algorithms in the study as it implements the ELM with less pruned hidden units without scarifying the higher generalization capability of ELM network. Further work will be conducted by testing the proposed algorithm in problems and datasets with images datasets such as bio-medical images, and images from videos.

References

1. Abid, S., Fnaiech, F., Najim, M.: A fast feedforward training algorithm using a modified form of the standard backpropagation algorithm. IEEE Transactions on Neural Networks 12(2), 424–430 (2001), doi:10.1109/72.914537
2. Blake, C., Merz, C.J.: UCI Repository of machine learning databases. In: Department of Information and Computer Science, vol. 55, University of California, Irvine (1998), `http://www.ics.uci.edu/mlearn/MLRepository.html`
3. Chum, O., Matas, J., Kittler, J.: Locally optimized RANSAC. In: Michaelis, B., Krell, G. (eds.) DAGM 2003. LNCS, vol. 2781, pp. 236–243. Springer, Heidelberg (2003)
4. Fischler, M.A., Bolles, R.C.: Random sample consensus: a paradigm for model fitting with applications to image analysis and automated cartography. Commun. ACM 24(6), 381–395 (1981), doi:10.1145/358669.358692
5. Huang, G.-B., Wang, D., Lan, Y.: Extreme learning machines: a survey. International Journal of Machine Learning and Cybernetics 2(2), 107–122 (2011), doi:10.1007/s13042-011-0019-y
6. Huang, G.-B., Zhu, Q.-Y., Siew, C.-K.: Extreme learning machine: Theory and applications. Neurocomputing 70(1–3), 489–501 (2006), `http://dx.doi.org/10.1016/j.neucom.2005.12.126`
7. Lavneet Singh, G.: A Novel Approach for protein Structure prediction Using Pattern Recognition and Extreme Machine Learning. In: Proceedings of International Conference of Neuro Computing and Evolving Intelligence, NCEI (2012)

8. Martínez-Martínez, J.M., Escandell-Montero, P., Soria-Olivas, E., Martín-Guerrero, J.D., Magdalena-Benedito, R., Gómez-Sanchis, J.: Regularized extreme learning machine for regression problems. Neurocomputing 74(17), 3716–3721 (2011), doi: http://dx.doi.org/10.1016/j.neucom.2011.06.013
9. Miche, Y., Sorjamaa, A., Lendasse, A.: OP-ELM: Theory, Experiments and a Toolbox. In: Kůrková, V., Neruda, R., Koutník, J. (eds.) ICANN 2008, Part I. LNCS, vol. 5163, pp. 145–154. Springer, Heidelberg (2008), http://dx.doi.org/10.1007/978-3-540-87536-9_16
10. Nistér, D.: Preemptive RANSAC for live structure and motion estimation. Machine Vision and Applications 16(5), 321–329 (2005), doi:10.1007/s00138-005-0006-y
11. Raguram, R., Frahm, J.-M., Pollefeys, M.: A Comparative Analysis of RANSAC Techniques Leading to Adaptive Real-Time Random Sample Consensus. In: Forsyth, D., Torr, P., Zisserman, A. (eds.) ECCV 2008, Part II. LNCS, vol. 5303, pp. 500–513. Springer, Heidelberg (2008)
12. Singh, L., Chetty, G.: Review of classification of brain abnormalities in magnetic resonance images using pattern recognition and machine learning. In: Proceedings of International Conference of Neuro Computing and Evolving Intelligence, NCEI (2012)
13. Yoan, M., Sorjamaa, A., Bas, P., Simula, O., Jutten, C., Lendasse, A.: OP-ELM: Optimally Pruned Extreme Learning Machine. IEEE Transactions on Neural Networks 21(1), 158–162 (2010), doi:10.1109/TNN.2009.2036259
14. Zuliani, M.: RANSAC toolbox for Matlab (2008)

Learning ELM Network Weights Using Linear Discriminant Analysis

Philip de Chazal, Jonathan Tapson, and André van Schaik

The MARCS Institute, University of Western Sydney, Penrith NSW 2751, Australia
{p.dechazal,j.tapson,a.vanschaik}@uws.edu.au

Abstract. We present an alternative to the pseudo-inverse method for determining the hidden to output weight values for Extreme Learning Machines performing classification tasks. The method is based on linear discriminant analysis and provides Bayes optimal single point estimates for the weight values.

Keywords: Extreme learning machine, Linear discriminant analysis, Hidden to output weight optimisation, MNIST database.

1 Introduction

The Extreme Learning Machine (ELM) is a multi-layer feedforward neural network that offers fast training and flexible non-linearity for function and classification tasks. Its principal benefit is that the network parameters are calculated in a single pass during the training process [1]. In its standard form it has an input layer that is fully connected to a hidden layer with non-linear activation functions. The hidden layer is fully connected to an output layer with linear activation functions. The number of hidden units is often much greater than the input layer with a fan-out of 5 to 20 hidden units per input frequently used. A key feature of ELMs is that the weights connecting the input layer to the hidden layer are set to random values. This simplifies the requirements for training to one of determining the hidden to output unit weights, which can be achieved in a single pass. By randomly projecting the inputs to a much higher dimensionality, it is possible to find a hyperplane which approximates a desired regression function, or represents a linear separable classification problem [2].

A common way of calculating the hidden to output weights is to use the Moore-Penrose pseudo-inverse applied to the hidden layer outputs using labelled training data. In this paper we present an alternative method for hidden to output weight calculation for networks performing classification tasks. The advantage of our method over the pseudo-inverse method is that the weights are the best single point estimates from a Bayesian perspective for a linear output stage. Using the same network architecture and same random values for the input to hidden layer weights, we applied the two weight calculation methods to the MNIST database and demonstrated that our method offers a performance advantage.

J. Cao et al. (eds.), *Proceedings of ELM-2014 Volume 1*,
Proceedings in Adaptation, Learning and Optimization 3, DOI: 10.1007/978-3-319-14063-6_16

2 Methods

If we consider a particular sample of input data $\mathbf{x}_k \in \mathbb{R}^{L\times 1}$ where k is a series index and K is the length of the series, then the forward propagation of the local signals through the network can be described by:

$$y_{n,k} = \sum_{m=1}^{M} w_{nm}^{(2)} g\left(\sum_{l=1}^{L} w_{ml}^{(1)} x_{l,k}\right) \tag{1}$$

Where $\mathbf{y}_k \in \mathbb{R}^{N\times 1}$ is the output vector corresponding to the input vector $\mathbf{x}_k$, l and $L$are the input layer index and number of input features respectively, m and M are the hidden layer index and number of hidden units respectively, and n and N are the output layer index and number of output units respectively. $w^{(1)}$ and $w^{(2)}$ are the weights associated with the input to hidden layer and the hidden to output layer linear sums respectively. $g(\ )$ is the hidden layer non-linear activation function.

With ELM, $w^{(1)}$ are assigned randomly which simplifies the training requirements to task of optimisation of the $w^{(2)}$ only. The choice of linear output neurons further simplifies the optimisation problem of $w^{(2)}$ to a single pass algorithm.

The weight optimisation problem for $w^{(2)}$ can be stated as

$$y_{n,k} = \sum_{m=1}^{M} w_{nm}^{(2)} a_{m,k} \text{ where } a_{m,k} = g\left(\sum_{l=1}^{L} w_{ml}^{(1)} x_{l,k}\right). \tag{2}$$

We can restate this as matrix equation by using $\mathbf{W} \in \mathbb{R}^{N\times M}$ with elements $w_{nm}^{(2)}$, and $\mathbf{A} \in \mathbb{R}^{M\times K}$ in which each column contains outputs of the hidden unit at one instant in the series $\mathbf{a}_k \in \mathbb{R}^{M\times 1}$, and the output $\mathbf{Y} \in \mathbb{R}^{N\times K}$ where each column contains output of the network at one instance in the series as follows:

$$\mathbf{Y} = \mathbf{WA}. \tag{3}$$

The optimisation problem involves determining the matrix $\mathbf{W}$ given a series of desired outputs for $\mathbf{Y}$ and a series of hidden layer outputs $\mathbf{A}$.

We represent the desired outputs for $\mathbf{y}_k$ using the target vectors $\mathbf{t}_k \in \mathbb{R}^{N\times 1}$ where $t_{n,k}$ has value 1 in the row corresponding to the desired class and 0 for the other N-1 elements. For example $\mathbf{t}_k = [0,1,0,0]^T$ indicates the desired target is class 2 (of four classes). As above we can restate the desired targets using a matrix $\mathbf{T} \in \mathbb{R}^{N\times K}$ where each column contains the desired targets of the network at one instance in the series. Substituting $\mathbf{T}$ in for the desired outputs for $\mathbf{Y}$, the optimisation problem involves solving the following linear equation for $\mathbf{W}$:

$$\mathbf{T} = \mathbf{WA}. \tag{4}$$

2.1 Output Weight Calculation Using the Pseudo-inverse

In ELM literature $\mathbf{W}$ is often determined by the taking the Moore-Penrose pseudo-inverse $\mathbf{A}^{+} \in \mathbb{R}^{K \times M}$ of $\mathbf{A}$ [3]. If the rows of are $\mathbf{A}$ are linearly independent (which normally true if K>M) then $\mathbf{W}$ maybe calculated using

$$\mathbf{W} = \mathbf{T}\mathbf{A}^{+} \text{ where } \mathbf{A}^{+} = \mathbf{A}^{T}\left(\mathbf{A}\mathbf{A}^{T}\right)^{-1}. \tag{5}$$

This minimises the sum of square error between networks outputs $\mathbf{Y}$ and the desired outputs $\mathbf{T}$, i.e.

$$\mathbf{A}^{+} \text{ minimises } \left\|\mathbf{Y}-\mathbf{T}\right\|_{2} = \sum_{k=1}^{K}\sum_{n=1}^{N}\left(y_{n,k}-t_{n,k}\right)^{2} \tag{6}$$

We refer to the pseudo-inverse method for output weight calculation as PI-ELM. We note that in cases where the classification problem is ill-posed it may be necessary to regularise this solution, using standard methods such as Tikhonov regularisation (ridge regression).

2.2 Output Weight Calculation Using Linear Discriminant Analysis

In this paper we develop an alternative approach to estimating $\mathbf{W}$ based on a maximum likelihood estimator assuming a linear model. We refer to it as the LDA-ELM method as it is equivalent to applying linear discriminant analysis to the hidden layer outputs. Our presentation is based on the notation of Ripley [4].

For an *N*-class problem Bayes' rule states that the posterior probability of the *n*th class p_n is related to its prior probability π_k and its class density function $f_n\left(\mathbf{d},\boldsymbol{\theta}_n\right)$ by

$$p_n = \frac{\pi_n f_n\left(\mathbf{d},\boldsymbol{\theta}_n\right)}{\sum_{z=1}^{N}\pi_z f_z\left(\mathbf{d},\boldsymbol{\theta}_z\right)} \tag{7}$$

where $\mathbf{d}$ is the input data vector (in our case the hidden layer output), and $\boldsymbol{\theta}_n$ are the parameters of the class density function.

The class densities are modelled with a multi-variate Gaussian model with common covariance $\boldsymbol{\Sigma}$ and class dependent mean vectors $\boldsymbol{\mu}_k$. Given an input vector $\mathbf{a}_k$ the class density is

$$f_n\left(\mathbf{a}_k,\boldsymbol{\theta}_n=\boldsymbol{\mu}_n,\boldsymbol{\Sigma}\right) = (2\pi)^{-\frac{M}{2}}\left|\boldsymbol{\Sigma}\right|^{-\frac{1}{2}}\exp\left[-\tfrac{1}{2}\left(\mathbf{a}_k-\boldsymbol{\mu}_n\right)^{T}\boldsymbol{\Sigma}^{-1}\left(\mathbf{a}_k-\boldsymbol{\mu}_n\right)\right] \tag{8}$$

We set the dimension of the Gaussian model equal to the number of hidden units so that $\mathbf{a}_k$ is as defined above for the hidden unit output and hence $\boldsymbol{\Sigma} \in \mathbb{R}^{M \times M}$ and $\boldsymbol{\mu}_n \in \mathbb{R}^{M \times 1}$.

To begin with, the training data is partitioned according to the class membership so that we have $K=\sum_{n=1}^{N} K_n$ labelled data vectors of hidden unit outputs, $\mathbf{a}_k^{(n)} \in \mathbb{R}^{M\times 1}, k=1..K_n$ where all members $\mathbf{a}^{(n)}$ belong to class n.

For a given set of hidden unit output data and class membership a likelihood function $l(\boldsymbol{\theta})$ is formed using

$$l(\boldsymbol{\theta}) = l(\boldsymbol{\theta}_1,\boldsymbol{\theta}_2,\ldots,\boldsymbol{\theta}_N) = \prod_{n=1}^{N}\prod_{k=1}^{K_n} \pi_n f_n\left(\mathbf{a}_k^{(n)},\boldsymbol{\theta}_n\right) \tag{9}$$

Our aim is to find values of $\boldsymbol{\theta}_n$ that maximise $l(\boldsymbol{\theta})$ for given set of training data. Equivalently we can maximise the value of the log-likelihood:

$$L(\boldsymbol{\theta}_1,\boldsymbol{\theta}_2,\ldots,\boldsymbol{\theta}_N) = \log\left(l(\boldsymbol{\theta}_1,\boldsymbol{\theta}_2,\ldots,\boldsymbol{\theta}_N)\right) = \sum_{n=1}^{N}\sum_{k=1}^{K_n} \log\left(f_n\left(\mathbf{a}_k^{(n)},\boldsymbol{\theta}_n\right)\right) + \sum_{n=1}^{N} K_n \log(\pi_n) \tag{10}$$

Substituting our multi-variate Gaussian model for $f_n(\mathbf{a}_k,\boldsymbol{\theta}_n)$ we get

$$\begin{aligned} &L(\boldsymbol{\theta}_1,\boldsymbol{\theta}_2,\ldots,\boldsymbol{\theta}_N) = L(\boldsymbol{\mu}_1,\ldots,\boldsymbol{\mu}_N,\boldsymbol{\Sigma}) = \\ &\sum_{n=1}^{N}\sum_{k=1}^{K_n}\left(-\tfrac{M}{2}\log(2\pi) - \tfrac{1}{2}\log(|\boldsymbol{\Sigma}|) - \tfrac{1}{2}\left(\mathbf{a}_k^{(n)}-\boldsymbol{\mu}_n\right)^T \boldsymbol{\Sigma}^{-1}\left(\mathbf{a}_k^{(n)}-\boldsymbol{\mu}_n\right)\right) + \sum_{n=1}^{N} K_n \log(\pi_n). \end{aligned} \tag{11}$$

This is maximised when

$$\boldsymbol{\mu}_n = \sum_{k=1}^{K_n} \mathbf{a}_k^{(n)} \Big/ K_n \text{ , and } \boldsymbol{\Sigma} = \sum_{n=1}^{N}\sum_{k=1}^{K_n}\left(\mathbf{a}_k^{(n)}-\boldsymbol{\mu}_k\right)\left(\mathbf{a}_k^{(n)}-\boldsymbol{\mu}_k\right)^T \Big/ K. \tag{12}$$

Having determined the $\boldsymbol{\mu}_n$'s and $\boldsymbol{\Sigma}$ from the training data we now need to find the values for $\mathbf{W}$. We begin by substituting (8) into (7), bringing the π_n into exponential function and removing the common numerator and denominator term $(2\pi)^{-\frac{M}{2}}|\boldsymbol{\Sigma}|^{-\frac{1}{2}}$, giving us

$$p_n = \frac{\exp\left[-\tfrac{1}{2}(\mathbf{a}_k-\boldsymbol{\mu}_n)^T \boldsymbol{\Sigma}^{-1}(\mathbf{a}_k-\boldsymbol{\mu}_n) + \log(\pi_n)\right]}{\sum_{z=1}^{N}\exp\left[-\tfrac{1}{2}(\mathbf{a}_k-\boldsymbol{\mu}_z)^T \boldsymbol{\Sigma}^{-1}(\mathbf{a}_k-\boldsymbol{\mu}_z) + \log(\pi_z)\right]}. \tag{13}$$

After expanding the $-\tfrac{1}{2}(\mathbf{a}_k-\boldsymbol{\mu}_n)^T \boldsymbol{\Sigma}^{-1}(\mathbf{a}_k-\boldsymbol{\mu}_n)$ terms and removing the $-\tfrac{1}{2}\mathbf{a}_k^T\boldsymbol{\Sigma}^{-1}\mathbf{a}_k$ from the numerator and denominator we get

$$p_n = \frac{\exp(y_n)}{\sum_{a=1}^{N}\exp(y_a)} \tag{14}$$

where

$$y_n = \log(\pi_n) + \boldsymbol{\mu}_n^T \boldsymbol{\Sigma}^{-1} \mathbf{a}_k - \tfrac{1}{2} \boldsymbol{\mu}_n^T \boldsymbol{\Sigma}^{-1} \boldsymbol{\mu}_n . \tag{15}$$

Classification is performed by choosing the class with the highest value of p_n. As p_n in (14) is a monotonic function of y_n in (15) we can use either function when deciding our final class. We choose to use y_n defined in (15) as it is a linear function of the input data vector $\mathbf{a}_k$ and it can be used to determine $\mathbf{W}$ for our network as follows:

$$\mathbf{W} = \begin{bmatrix} \log(\pi_1) - \tfrac{1}{2}\boldsymbol{\mu}_1^T \boldsymbol{\Sigma}^{-1} \boldsymbol{\mu}_1 & \boldsymbol{\mu}_1^T \boldsymbol{\Sigma}^{-1} \\ \log(\pi_2) - \tfrac{1}{2}\boldsymbol{\mu}_2^T \boldsymbol{\Sigma}^{-1} \boldsymbol{\mu}_2 & \boldsymbol{\mu}_2^T \boldsymbol{\Sigma}^{-1} \\ \vdots & \vdots \\ \log(\pi_N) - \tfrac{1}{2}\boldsymbol{\mu}_N^T \boldsymbol{\Sigma}^{-1} \boldsymbol{\mu}_N & \boldsymbol{\mu}_N^T \boldsymbol{\Sigma}^{-1} \end{bmatrix}. \tag{16}$$

Note that $\mathbf{W} \in \mathbb{R}^{N \times M+1}$, as a constant term has been introduced into the hidden to output layer weights (the first column of $\mathbf{W}$). If we want to determine the posterior probabilities then we use (14) applied to the network outputs.

Summary of Method

In summary calculating $\mathbf{W}$ proceeds as follows

(i) Partitioned the hidden unit output data according to the class membership so that we have $K = \sum_{n=1}^{N} K_n$ labelled data vectors, $\mathbf{a}_k^{(n)} \in \mathbb{R}^{M \times 1}, k = 1..K_n$ where all members $\mathbf{a}^{(n)}$ belong to class n.

(ii) Calculate $\boldsymbol{\mu}_n = \sum_{k=1}^{K_n} \mathbf{a}_k^{(n)} \Big/ K_n$ and $\boldsymbol{\Sigma} = \sum_{n=1}^{N} \sum_{k=1}^{K_n} \left(\mathbf{a}_k^{(n)} - \boldsymbol{\mu}_k\right)\left(\mathbf{a}_k^{(n)} - \boldsymbol{\mu}_k\right)^T \Big/ K$.

(iii) Set the prior probabilities π_n.

(iv) Calculate $\mathbf{W} = \begin{bmatrix} \log(\pi_1) - \tfrac{1}{2}\boldsymbol{\mu}_1^T \boldsymbol{\Sigma}^{-1} \boldsymbol{\mu}_1 & \boldsymbol{\mu}_1^T \boldsymbol{\Sigma}^{-1} \\ \log(\pi_2) - \tfrac{1}{2}\boldsymbol{\mu}_2^T \boldsymbol{\Sigma}^{-1} \boldsymbol{\mu}_2 & \boldsymbol{\mu}_2^T \boldsymbol{\Sigma}^{-1} \\ \vdots & \vdots \\ \log(\pi_N) - \tfrac{1}{2}\boldsymbol{\mu}_N^T \boldsymbol{\Sigma}^{-1} \boldsymbol{\mu}_N & \boldsymbol{\mu}_N^T \boldsymbol{\Sigma}^{-1} \end{bmatrix}$

To classify new data we

(i) Calculate the network output $\mathbf{y}$ in response to the hidden layer output $\mathbf{a}$ is

$$\mathbf{y} = \mathbf{W} \begin{bmatrix} 1 \\ \mathbf{a} \end{bmatrix}.$$

(ii) (Optional) Calculate the posterior probabilities

$$p_n = \frac{\exp(y_n)}{\sum_{a=1}^{N} \exp(y_a)}.$$

(iii) The final decision of the network is the output with the highest value of y_n or, equivalently, p_n.

Combining Classifiers

Equation (14) provides an easy way to combine the outputs of multiple classifiers. Once the posterior probabilities are calculated for each class for each classifier we can form a combined posterior probability and choose the class with the highest combined posterior probability. There are many schemes for doing this [5] with unweighted averaging across the posterior probability outputs being one of the most simple schemes.

3 Experiments

We applied the LDA-ELM and PI-ELM weight calculation method to the MNIST handwritten digit recognition problem [6]. Authors JT and AvS have previously reported good classification results using ELM on this database [2]. The database has 60,000 training and 10,000 testing examples. Each example is a 28*28 pixel 256 level grayscale image of a handwritten digit between 0 and 9. The 10 classes are approximately equally distributed in the training and testing sets.

The ELM algorithms were applied directly to the unprocessed images and we trained the networks by providing all data in batch mode. The random values for the input layer weights were uniformly distributed between -0.5 and 0.5. The prior probabilities for the 10 classes for LDA-ELM were each set to 0.1.

In order to perform a direct comparison of the two methods we used the following protocol:

For fan-out of 1 to 20 hidden units per input, repeat 200 times

(i) Assign random values to the input layer weights and determine the hidden layer outputs for the 60,000 training data examples.
(ii) Determine PI-ELM network weights using data from (i).
(iii) Determine LDA-ELM network weights using data from (i).
(iv) Evaluate both networks on the 10,000 test data examples and store results.

We averaged the results for the 200 repeats of the experiment for each fan-out and compared the misclassification rates. These results are shown in Fig. 1 and Table 1.

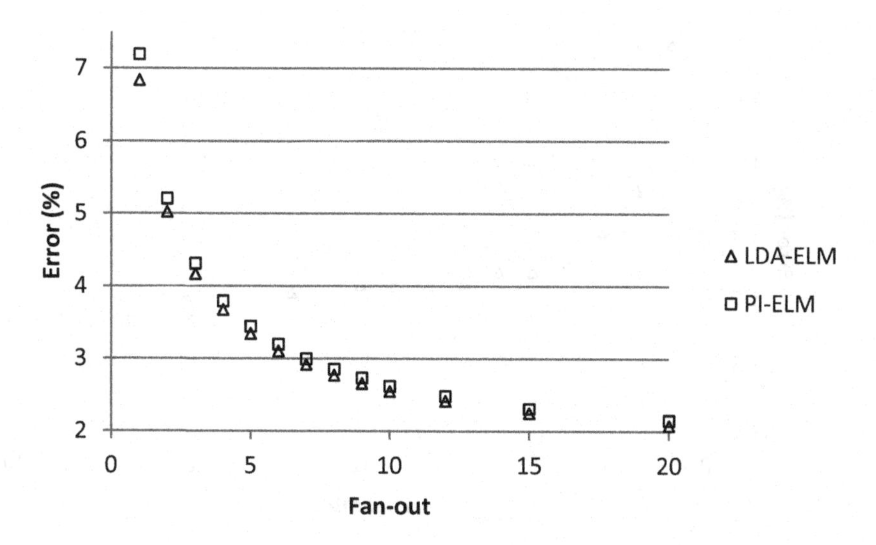

Fig. 1. The error rate of the LDA-ELM and the PI-ELM on the MNIST database for fan-out varying between 1 and 20. All results at each fan-out are averaged from 200 repeats of the experiment.

The results show that the LDA-ELM outperforms the PI-ELM at every fan-out value. The average performance benefit was a 3.1% decrease in the error rate of LDA-ELM with a larger benefit at smaller fan-out values. Table 2 below shows that there is little extra computational requirement for the LDA-ELM method.

Table 1. The error rate (%) and percentage improvement of the LDA-ELM over the PI-ELM on the MNIST database. Results averaged from 200 repeats.

	Fanout												
	1	2	3	4	5	6	7	8	9	10	12	15	20
PI-ELM	7.20	5.21	4.32	3.80	3.45	3.20	3.00	2.86	2.74	2.63	2.49	2.31	2.15
LDA-ELM	6.84	5.03	4.17	3.68	3.35	3.11	2.92	2.78	2.66	2.55	2.42	2.25	2.08
% improvement	4.9	3.5	3.3	3.1	2.9	3.0	2.6	2.9	2.8	2.7	2.6	2.6	3.3

Table 2. Computation times (in seconds). The elapsed time is shown for training the PI-ELM and LDA-ELM networks on the 60,000 images from MNIST database and testing on the 10,000 images using MATLAB R2012a code running on 2012 Sony Vaio Z series laptop with an Intel i7-2640M 2.8GHz processor and 8GB RAM.

	Fan-out													
	1	2	3	4	5	6	7	8	9	10	12	20	15	20
PI-ELM	6.2	13.9	24.9	37.8	53.3	68.5	88.2	111	136	162	228	630	339	630
LDA-ELM	6.2	13.9	25.2	38.1	54.0	69.5	90.6	113	140	167	238	702	357	702

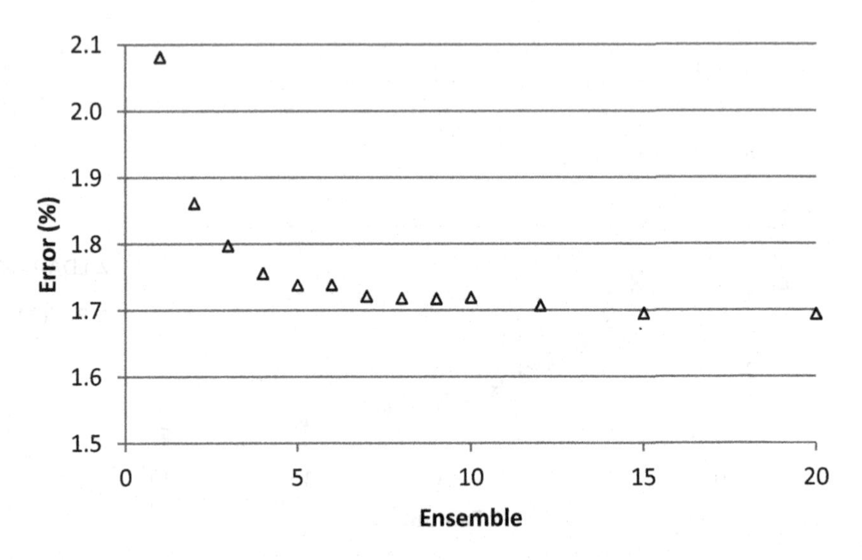

Fig. 2. The error rate of the LDA-ELM on the MNIST database at a fan-out of 20 with the ensemble number varying between 1 and 20. The result at each ensemble number is averaged from 10 repeats of the experiment.

The last experiment we performed investigated combining multiple networks using the LDA-ELM by averaging posterior probabilities. We investigated using an ensemble number between 1 and 20 and repeated the training and testing 10 times at each ensemble number. The following protocol was adopted.

For a fan-out of 20 hidden units per input, repeat 10 times

i. With an ensemble number between 1 and 20 then for each classifier of the ensemble we performed the following:
   a. Assign random values to the input layer weights and determine the hidden layer outputs for the 60,000 training data examples.
   b. Determine LDA-ELM network weights using data from (a).
   c. Evaluate the network on the 10,000 test data examples and store the output probabilities.
ii. Determine the combined network output by averaging the posterior probabilities of the individual networks.
iii. Evaluate network performance of the combined network.

The final results are shown in Fig. 2.

The results shown in Fig. 2 demonstrate the benefit of combining multiple LDA-ELM networks on the MNIST database. Combining two networks reduced the error rate from 2.08% to 1.86% and adding more networks further reduced the error. The best error rate was 1.69% achieved when 20 networks were combined.

4 Discussion

The results on the MNIST database shown in Fig. 1 suggest that there is a performance benefit to be gained by using the LDA-ELM output weight calculation over the PI-ELM method. As there is only a small extra computation overhead we believe it is a viable alternative to the pseudo-inverse method especially at small fan-out values.

Another benefit of the LDA-ELM is the ability to combine outputs from networks by combining the posterior probabilities estimates of the individual networks. When we applied this to the MNIST database we were able to reduce the error rate to 1.7%. This result is comparable to the best performance of most other 2 and 3 layer neural networks processing the raw data [7]. Further work will include comparing the two weight calculation methods on other publicly available databases such as abalone and iris data sets [8].

5 Conclusion

We have presented a new method for weight calculation for hidden to output weights for ELM networks performing classification tasks. The method is based on linear discriminant analysis and requires a modest amount of extra calculation time compared to the pseudo-inverse method (<12% for a fan-out ≤ 20). When applied to the MNIST database the average misclassification rate improvement was 3.1% in comparison to the pseudo-inverse method for identically configured and initialised networks.

References

1. Huang, G.B., Zhu, Q.Y., Siew, C.K.: Extreme Learning Machine: Theory and Applications. Neurocomputing 70, 489–501 (2006)
2. Tapson, J., van Schaik, A.: Learning the pseudoinverse solution to network weights. Neural Networks 45, 94–100 (2013)
3. Penrose, R., Todd, J.A.: On best approximate solutions of linear matrix equations. Mathematical Proceedings of the Cambridge Philosophical Society 52, 17–19 (1956)
4. Ripley, B.D.: Pattern Recognition and Neural Networks. Cambridge University Press (1996)
5. Kuncheva, L.I.: Combining Pattern Classifiers: Methods and Algorithms. Wiley Press (2004)
6. LeCun, Y., Cortes, C.: The MNIST database of handwritten digits, http://yann.lecun.com/exdb/mnist
7. LeCun, Y., Bottou, L., Bengio, Y., Haffner, P.: Gradient-Based Learning Applied to Document Recognition. Proceedings of the IEEE 86(11), 2278–2324 (1998)
8. Data sets from the UCI Machine Learning Repository, http://archive.ics.uci.edu/ml/

An Algorithm for Classification over Uncertain Data Based on Extreme Learning Machine*

Ke-yan Cao, Guoren Wang, and Donghong Han

Key Laboratory of Medical Image Computing
(Northeastern University), Ministry of Education, China
College of Information Science and Engineering,
Northeastern University, Liaoning, Shenyang 110819

Abstract. In recent years, along with the generation of uncertain data, more and more attention is paid to the mining of uncertain data. In this paper, we study the problem of classifying uncertain data using Extreme Learning Machine(ELM). We propose UU-ELM algorithm for classifying uncertain data which are uniform distribution. Finally, the performances of our methods are verified through a large number of simulation experiments. The experimental results show that our method is effective way to solve the problem of uncertain data classification.

Keywords: Extreme Learning Machine, classification, uncertain data, single hidden layer feedforward neural networks, weighted.

1 Introduction

In recent years, a large amount of uncertain data are generated and collected due to new techniques of data acquisition, which are widely used in many real-world applications, such as wireless sensor networks[1, 7], moving object detection [4, 5, 20], meteorology and mobile telecommunication. However, since the intrinsic differences between uncertain and deterministic data, it is difficult to deal uncertain data with traditional data mining algorithms for deterministic data. Therefore, many researchers put efforts in developing new techniques of data processing and mining on uncertain data [2, 3, 18, 19, 22].

Classification is one of the key problems in data mining area which can find interesting patterns, and has signification application merits in many fields. There are many published works on classification [9]. Inclusion of uncertainty to the data makes the problem far more difficult to tackle, as this will further limit the accuracy of subsequent classification. Therefore, how to effectively classify over uncertain data is greatly importance. There are many challenges raised ahead to affect uncertain data classification.

* The work is supported by the National Natural Science Foundation of China under Grant Nos. 61025007, 61328202, 61173029, 61100024, 61332006, and 61073063, the National High Technology Research and Development 863 Program of China under Grant No. 2012AA011004, and the National Basic Research 973 Program of China under Grant No. 2011CB302200-G.

J. Cao et al. (eds.), *Proceedings of ELM-2014 Volume 1*,
Proceedings in Adaptation, Learning and Optimization 3, DOI: 10.1007/978-3-319-14063-6_17

In the remainder of this paper, we first introduce the ELM in Section 2. After that, we formally define our problem in section 3. We analyze the challenge of classification over uncertain data, and develop the algorithm respectively in section 4. Section 5 presents experimental results of our method. In Section 6, we conclude this paper with directions for future work.

2 Brief of Extreme Learning Machine

In this section, we present a brief overview of ELM, developed by Huang et .al.[10, 11, 14–16]. In ELM, the hidden-layer node parameters are mathematically calculated instead of being iteratively tuned, providing good generalization performance at thousands of times higher speeds than traditional popular learning algorithms for feedforward neural networks [12]. The output function can be represented by

$$f_L(x) = \sum_{i=1}^{L} \beta_i g_i(x) = \sum_{i=1}^{L} \beta_i G(a_i, b_i, x), x \in R^d, \beta_i \in R^m$$

where g_i denotes the output function $G(a_i, b_i, x)$ of the ith hidden node [17]. For additive nodes with activation function g, g_i is defined as

$$g_i = G(a_i, b_i, x) = g(a_i \cdot x + b_i), a_i \in R^d, b_i \in R$$

For Radial Basis Function (RBF) nodes with activation function g, g_i is defined as

$$g_i = G(a_i, b_i, x) = g(b_i \|x - a_i\|), a_i \in R^d, b_i \in R^+$$

For the multiclass classifier with single output, ELM can approximate any target continuous function and the output of the ELM classifier $h(x)\beta$ can be as close as possible to the class labels in the corresponding regions [13]. The classification problem for the proposed constrained-optimization-based ELM with a single-output node can be formulated as

$$\begin{aligned} &Minimize: L_{P_{ELM}} = \tfrac{1}{2} \|\beta\|^2 + C\tfrac{1}{2} \sum_{i=1}^{N} \xi_i^2 \\ &Subject\ to: h(x_i)\beta = t_i - \xi_i, \quad i = 1, \cdots, N \end{aligned}$$

where C is user-specified parameter. Based on the KKT theorem [8], training ELM is equivalent to solve the following dual optimization problem:

$$L_{D_{ELM}} = \frac{1}{2} \|\beta\|^2 + C\frac{1}{2} \sum_{i=1}^{N} \xi_i^2 - \sum_{i=1}^{N} \alpha_i (h(x_i)\beta - t_i + \xi_i)$$

3 Problem Definition

Uncertain Data Model. Let us assume that uncertain data set consists of uncertain objects $x_1, x_2, \cdots, x_i, \cdots, x_n$. Each object is described by a continuous probability density function (pdf), x_i denotes the i-th object in data set. Probability density function of the x_i is denoted by $f(x_i)$. Meanwhile, x_i^j is any one instance of object x_i. Let $S(x_i)$ denote the set of all instances of x_i. $f(x_i^j) > 0$ for any instances x_i^j in set $S(x_i)$, and $\int_{x \in S(x_i)} f(x_i)dx = 1$.

As an example, x_a, x_b and x_c are three uncertain objects, each object is described by a continuous probability density function. Figure 1 maps some uncertain objects into a 2D coordinate. For simplicity, we only consider two attributes in the example in Figure 1. Let the shaded area and the white area represent two classes respectively, and the range of uncertain object is to be classified into the corresponding category. When the range of the uncertain objects is classified into the same class, then the uncertain object should belong to that particular class, as x_a, x_b. However, when the instances of uncertain object are classified into different classes, then we will consider probability of each instance, as x_c. For an uncertain dataset, we need to consider all possible classes which the uncertain object belongs to, the class with max probability is chosen.

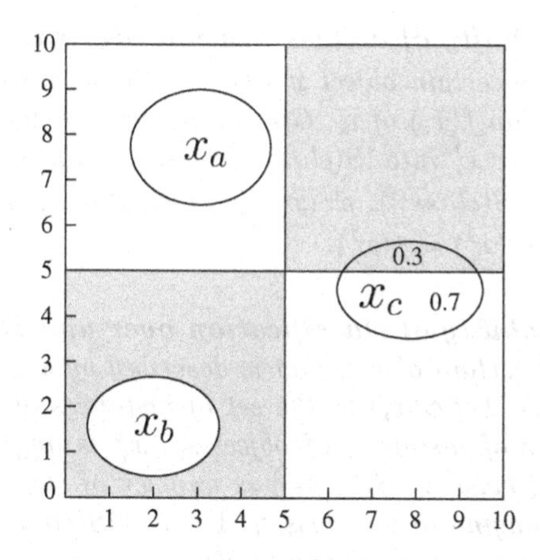

Fig. 1. Example of classification over uncertain data

The decision function of multiclass classification of ELM is as follows:

$$f(x) = sign(h(x)H^T(\frac{1}{C} + HH^T)^{-1}T) \tag{1}$$

The expected output vector is:

$$T = \begin{bmatrix} t_1 \\ \vdots \\ t_L \end{bmatrix}$$

For multiclass cases, $S(class)$ denotes the set of classes, $|S(class)|$ is the number of classes and output nodes of hidden layer.

The predicted class label of a given testing sample is the index number of the output node which has the highest output value for the given testing sample. Let $f_k(x_i^j)$ denote the output function of the k-th $(1 \leq k \leq |S(class)|)$ output node, i.e., $f(x_i^j) = [f_1(x_i^j), \cdots, f_{|S(class)|}(x_i^j)]$, then the predicted class label of sample x_i^j is:

$$label(x_i^j) = arg \max_{\tau \in (1, \cdots, |S(class)|)} f_\tau(x_i^j)$$

Due to the nature of classification over uncertain data [17], the Theorem 1 cannot be applied to uncertain multiclass classification. If the uncertain object is classified into three categories, and all probabilities in three classes are smaller than 0.5, then Theorem 1 cannot be used for multiclass classification. For this case, Theorem 1 is modified as follows:

Definition 1. ***Probability of a classification over instance.*** *Let x_i^j denote any one instance of uncertain object x_i, $f(x_i^j)$ is probability associated with x_i^j based on the pdf function $f(x_i)$ of x_i. Given a number of classes $|S(class)|$, ELM can classify the instance x_i^j into $|S(class)|$ classes. If the instance x_i^j belongs to the class c_l, $(1 \leq l \leq |S(class)|)$, $p^{c_l}(x_i^j)$ is probability of instance x_i^j belonging to the class c_l, then $p^{c_l}(x_i^j) = f(x_i^j)$.*

Definition 2. ***Probability of classification over uncertain object.*** *We assume that x_i is an uncertain object, that is described by a continuous probability density function $f(x_i)$. Let $S(x_i)$ be the set of instances of uncertain object x_i, $|S(x_i)|$ is the number of instances of object x_i, x_i^j is any one instance of x_i. $S(class)$ is the set of classes, $|S(class)|$ is number of classes. $p^{c_l}(x_i)$ is probability of object x_i belonging to the class c_l, $1 \leq l \leq |S(class)|$, $p^{c_l}(x_i)$ is sum of probabilities the instances which belong to class c_l, then*

$$p^{c_l}(x_i) = \sum_{x_i^j \in c_l} f(x_i^j)$$

Definition 3. ***Classification of uncertain data.*** *Let $S(class)$ be the set of classes, $|S(class)|$ is number of classes. c_l is the class label, $(1 \leq l \leq |S(class)|)$. x_i is an uncertain object,that is described by a continuous probability density function $f(x_i)$. $p^{c_l}(x_i)$ is probability of objects x_i belonging to the class c_l, $1 \leq l \leq |S(class)|$. If $p^{c_l}(x_i) \geq \forall p^{c_j}(x_i)$ $(1 \leq j \leq |S(class)|, j \neq l)$, then uncertain object x_i belongs to class c_l.*

4 Classification Algorithm

We proposed a grid-based method, named UU-ELM (uniform distribution uncertain data classification based on ELM) to classify uncertain data which are uniform distribution by using grid framework. We first describe UU-ELM algorithm for classifying uncertain data. In order to present the pruning approach for uncertain data, we first review the structure of the grid. For simplicity and limitation of space, we only discuss the case that uncertain objects in 2-dimensional space.

Given an uncertain database U, we divide its domain by grid G, which can be viewed as a 2-dimensional array of cells. the area of any one grid is s. Let $g(m,n)$ be a cell in G, as shown in Figure 2. Each cell is square. Let $SS(x_i)$ be the set of cells that are covered by object x_i, area of x_i is denoted by $area(x_i)$. The cells which are covered by x_i are divided two categories: complete coverage cells and partially covered cells, denoted by *c*-cell and *p*-cell respectively. Let $S_i(c)$ denote the set of all *c*-cells and $S_i(p)$ denote the set of all partially covered cells of x_i, then $S_i(c) \cup S_i(p) = SS(x_i)$ and $S_i(c) \cap S_i(p) = \emptyset$.

All cells are classified by ELM, and the results are available. Let $p^{c_l}_{g(m,n)}$ denote the probability of cell $g(m,n)$ belonging to class c_l.

Theorem 1. *Let $|c|$-cells denote the number of c-cells that are covered by object x_i, $S_i(c)$ denote the set of all c-cells of x_i, the probability of any one c-cell belonging to class c_l is $p^{c_l}_c$, the probability of all c-cells of x_i belong to class c_l is as follows:*

$$p^{c_l}_{c-cell}(x_i) = \frac{\sum_{g(m,n)\in S_i(c)} p^{c_l}_{g(m,n)}}{|c| - cell} \times s \times |c| - cell = \sum_{g(m,n)\in S_i(c)} p^{c_l}_{g(m,n)} \times s$$

if we can get:

$$p^{c_l}_{c-cell}(x_i) \geq \frac{1}{2} area(x_i)$$

then object x_i belongs to class c_l

If we can not judge which class that object x_i belongs to, based on Theorem 1, we need consider the *p*-cells of x_i. Partially covered cells are divided into four categories, as shown in Figure 3. In figure 3(a), cell $g(m,n)$ is a partially cell of object x_i, x_i and cell $g(m,n)$ in a, b and c points. We define a minimum bounding rectangle (MBR) outside which the object has zero (or negligible) probability of occurrence,denoted by $s(a,b,c,d)$, i.e. $s(a,b,c) \leq s(x_i \cap g(m,n)) \leq s(a,b,c,d)$. Let s_{min} and s_{max} denote the minimum and maximum value of $s(x_i \cap g(m,n))$ respectively. In figure 3(b), art tangent to the grid at point e, f point, makes the area of s(a,b,e,f) be minimum value, then $s_{min} = s(a,b,c,d)$, $s_{max} = s(a,b,e,f)$. According to the same principle, in figure 3(c), we can know that $s_{min} = s(a,b,c,d)$, $s_{max} = s(a,b,e,f)$. In figure 3(d), $s_{min} = s(a,b,c,d,e)$, $s_{max} = s(a,b,f,g,e)$.

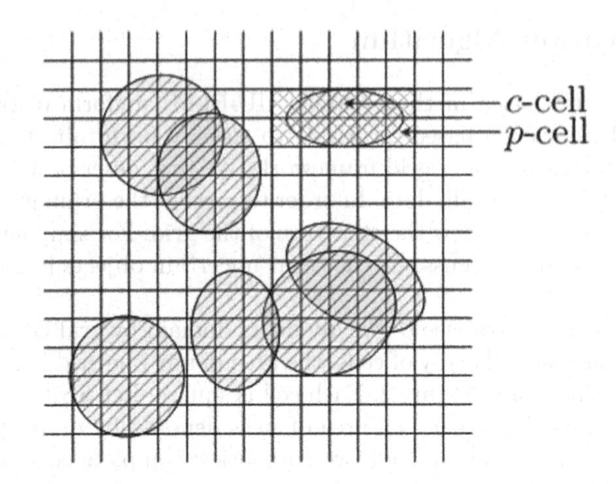

Fig. 2. Uncertain objects on grid

Theorem 2. *Let s_{min} and s_{max} denote the minimum and maximum value of area that object x_i and $g(m,n)$ interssection, $g(m,n)$ is a p-cell of object x_i, the area of $g(m,n)$ is denoted by s, $p^{c_l}_{g(m,n)}$ denote the probability of cell $g(m,n)$ belonging to class c_l. $\overline{p^{c_l}_{s_{min}}}$ denote the upper bound and lower bound of probability of instances in range S_{min} belonging to class c_l respectively. if*

$$\frac{s_{min}}{s} \geq p^{c_l}_{g(m,n)}$$

then

$$\overline{p^{c_l}_{s_{min}}} = p^{c_l}_{g(m,n)}$$

we need to consider the relationship between $\frac{s_{min}}{s}$ and $p^{c_l}_{g(m,n)}$ to determine the lower bound of $p^{c_l}_{s_{min}}$: if

$$\frac{s_{min}}{s} \geq 1 - p^{c_l}_{g(m,n)}$$

then

$$\underline{p^{c_l}_{s(min)}} = \frac{s_{min}}{s} - (1 - p^{c_l}_{g(m,n)}) = \frac{s_{min}}{s} - 1 + p^{c_l}_{g(m,n)}$$

otherwise

$$\underline{p^{c_l}_{s(min)}} = 0$$

if

$$\frac{s_{min}}{s} < p^{c_l}_{g(m,n)}$$

then

$$\overline{p^{c_l}_{s_{min}}} = \frac{s_{min}}{s}$$

the lower bound of $p^{c_l}_{s_{min}}$ is based on the relationship between $\frac{s_{min}}{s}$ and $p^{c_l}_{g(m,n)}$: if

$$\frac{s_{min}}{s} \geq 1 - p^{c_l}_{g(m,n)}$$

then

$$\underline{p^{c_l}_{s(min)}} = \frac{s_{min}}{s} - 1 + p^{c_l}_{g(m,n)}$$

otherwise

$$\underline{p^{c_l}_{s(min)}} = 0$$

The upper bound and lower bound of $p^{c_l}_{s_{max}}$ are obtained based on the same method as s_{min}. The detail is not given in this section.

Theorem 3. *Let $p^{c_l}_{c-cell}$ denote the probability of all c-cells of x_i belonging to class c_l, $\overline{p^{c_l}_{x_i}}$ and $\underline{p^{c_l}_{x_i}}$ are upper bound and lower bound of $p^{c_l}_{x_i}$ respectively, then*

$$\overline{p^{c_l}(x_i)} = p^{c_l}_{c-cell} + \sum_{g(m,n)\in S_i(p)} \overline{p^{c_l}_{s_{max}}}$$

$$\underline{p^{c_l}(x_i)} = p^{c_l}_{c-cell} + \sum_{g(m,n)\in S_i(p)} \underline{p^{c_l}_{s_{min}}}$$

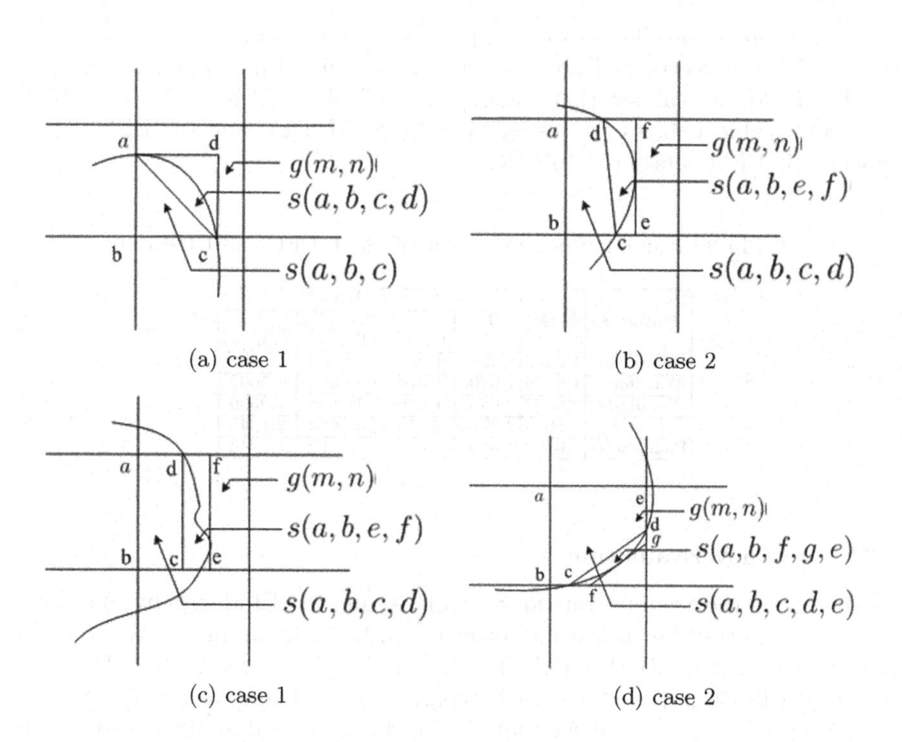

Fig. 3. Uncertain object on grid

Table 1. Example of probability of x_i

class	c_1	c_2	c_3	c_4	c_5
upper bound	0.28	0.5	0.26	0.2	0.21
lower bound	0.1	0.3	0.25	0.05	0.15

Table 2. Example of list L

L	t_1	t_2	t_3	t_4	t_5
probability	$p_{x_i}^{c2}$	$p_{x_i}^{c3}$	$p_{x_i}^{c5}$	$p_{x_i}^{c1}$	$p_{x_i}^{c4}$

5 Performance Verification

This section compares the performance of several algorithms (Support Vector Machine (SVM) [6], Dynamic Classifier Ensemble (DCE) [21] and UU-ELM,) in the real-world benchmark regression and multiclass classification data sets.

5.1 Efficiency Evaluation

In order to prove the efficiency of our proposed algorithms, we first evaluate the efficiency of our methods. Table 3 shows the training time among SVM, DEC and UU-ELM. We can see that training time of UU-ELM is shorter than SVM and DEC. Table 4 shows the testing time for SVM, DEC and UU-ELM. As we expect, UU-ELM is faster than DEC.

Table 3. Training time comparison of SVM, DEC and UU-ELM

Datasets	SVM	DEC	UU-ELM		
			sig	hardlim	sin
			Time(s)	Time(s)	Time(s)
Magic 04	1.9832	2.3245	1.5325	1.6260	1.6992
Waveform	1.6956	1.6206	1.6701	1.5603	1.5009
Pendigits	2.0692	2.0865	1.9500	1.9698	1.9506
Letter	2.8616	2.8703	2.3165	2.5682	2.1345
Pageblocks	1.6503	1.6352	1.5682	1.5103	1.5243

5.2 Accuracy Evaluation

Table 5 shows the accuracy rate for SVM, DEC and UU-ELM with five data sets. We adopt the traditional learning principle in the training phase. We compare the accuracy among SVM, DEC, and UU-ELM as shown in Table 5. It can be seen that ELM can always achieve comparable performance as SVM and DEC. Seen from Table 5, different functions of ELM can be used in different data sets in order to have similar accuracy in different size of data sets, although any output function can be used in all types of data sets. We can see that UU-ELM and DEC performs are as good as SVM.

Table 4. Testing time comparison of SVM, DEC and UU-ELM

Datasets	SVM	DEC	UU-ELM		
			sig	hardlim	sin
			Time(s)	Time(s)	Time(s)
Magic 04	2.3262	2.6591	1.9562	1.8623	1.8965
Waveform	2.0132	2.1656	1.6852	1.6503	1.5230
Pendigits	2.9538	2.8215	1.9520	1.7659	2.0291
Letter	2.9654	2.8659	2.0356	2.3016	2.1530
Pageblocks	1.7925	1.8560	1.1034	1.0362	0.3560

Table 5. Accuracy comparison for SVM, DEC and UU-ELM

Datasets	SVM	DEC	WEC-ELM		
			sig	hardlim	sin
	TR(%)	TR(%)	TR(%)	TR(%)	TR(%)
Magic 04	76.89	78.65	76.82	75.98	78.01
Waveform	73.56	73.01	72.98	74.23	74.08
Pendigits	68.52	67.98	98.29	70.08	69.86
Letter	68.02	39.56	69.31	68.95	69.55
Pageblocks	70.12	72.35	72.52	71.56	70.91

6 Conclusions

In this paper, we studied the problem of classification based on ELM over uncertain data. We proposed UU-ELM algorithm based on ELM, which is suitable for uniform distribution. Finally, experiments were conducted on real data, which showed that the algorithms proposed in this paper are efficient and are able to deal with uncertain data in a real-time fashion.

References

1. Aggarwal, C.C.: On density based transforms for uncertain data mining. In: IEEE 23rd International Conference on Data Engineering, ICDE 2007, pp. 866–875 (2007)
2. Aggarwal, C.C.: Managing and mining uncertain data. Advance in Database Systems 35 (2009)
3. Aggarwal, P.S.Y.C.C.: A survey of uncertain data algorithms and applications. IEEE Transactions on Knowledge Data Engineering (2009)
4. Chen, L., Özsu, M.T., Oria, V.: Robust and fast similarity search for moving object trajectories. In: Proceedings of the 2005 ACM SIGMOD International Conference on Management of Data (2005)
5. Cheng, R., Kalashnikov, D.V., Prabhakar, S.: Querying imprecise data in moving object environments. IEEE Transactions on Knowledge and Data Engineering 16(9), 1112–1127 (2004)
6. Cortes, C., Vapnik, V.: Support-vector networks. Machine Learning 10(3), 273–297 (1995)
7. Faradjian, A., Gehrke, J., Bonnett, P.: Gadt: A probability space adt for representing and querying the physical world. In: Proceedings of the 18th International Conference on Data Engineering (2002)

8. Fletcher, R.: Practical methods of optimization. IEEE Transactions on Systems, Man, and Cybernetics, Part B: Cybernetics (1987)
9. Liu, L.Y.H.: Toward integrating feature selectional gorithms for classification and clustering. IEEE Trans. Knowl.Data. Eng. (2005)
10. Huang, G.-B., Chen, L., Siew, C.-K.: Universal approximation using incremental constructive feedforward networks with random hidden nodes. IEEE Transactions on Neural Networks 17, 879–892 (2006)
11. Huang, G.-B., Siew, C.-K.: Extreme learning machine: RBF network case. In: Proceedings of the Eighth International Conference on Control, Automation, Robotics and Vision (ICARCV 2004), Kunming, China, December 6-9, vol. 2, pp. 1029–1036 (2004)
12. Huang, G.-B., Wang, D.H., Lan, Y.: Extreme learning machines: a survey. International Journal of Machine Learning and Cybernetics 2(2), 107–122 (2011)
13. Huang, G.-B., Zhou, H., Ding, X., Zhang, R.: Extreme learning machine for regression and multiclass classification. Transactions on Systems, Man, and Cybernetics, Part B: Cybernetics 42(2), 513–529 (2012)
14. Huang, G.-B., Zhu, Q.-Y., Mao, K.Z., Siew, C.-K., Saratchandran, P., Sundararajan, N.: Can threshold networks be trained directly? IEEE Transactions on Circuits and Systems II 53(3), 187–191 (2006)
15. Huang, G.-B., Zhu, Q.-Y., Siew, C.-K.: Extreme learning machine: Theory and applications. Neurocomputing 70, 489–501 (2006)
16. Huang, G.-B., Zhu, Q.-Y., Siew, C.-K.: Extreme learning machine: A new learning scheme of feedforward neural networks. In: Proceedings of International Joint Conference on Neural Networks (IJCNN2004), Budapest, Hungary, July 25-29, vol. 2, pp. 985–990 (2004)
17. Han, D., Ning, J., Zhang, X., Cao, K., Wang, G.: Classification of uncertain data streams based on extreme learning machine. Cognitive Computation (2014)
18. Leung, C.K.S.: Mining uncertain data. Wiley Interdiscovery Rewind: Data Mining and Knowledge Discovery (2011)
19. Liu, J., Deng, H.: Outlier detection on uncertain data based on local information. Knowledge-Based Systems (2013)
20. Ljosa, V., Singh, A.K.: Apla: Indexing arbitrary probability distributions. In: IEEE 23rd International Conference on Data Engineering, ICDE 2007 (2007)
21. Pan, S., Wu, K., Zhang, Y., Li, X.: Classifier ensemble for uncertain data stream classification. Advances in Knowledge Discovery and Data Mining (2010)
22. Zhang, C., Gao, M., Zhou, A.: Tracking high quality clusters over uncertain data streams. In: IEEE 25th International Conference on Data Engineering, ICDE 2009 (2009)

Training Generalized Feedforword Kernelized Neural Networks on Very Large Datasets for Regression Using Minimal-Enclosing-Ball Approximation

Jun Wang[1,*], Zhaohong Deng[1], Shitong Wang[1], and Qun Gao[2]

[1] School of Digital Media, Jiangnan University, Wuxi, Jiangsu, China
[2] Library and Archives of Jiangnan University, Wuxi, Jiangsu, China
wangjun_sytu@hotmail.com,
{dzh666828,wxwangst}@aliyun.com

Abstract. Training feedforward neural networks (FNNs) on very large datasets is one of the most critical issues in FNNs studies. In this work, the connection between FNNs and kernel methods is investigated and, HFSR-GCVM, a scalable learning method for GFKNN on large datasets, is proposed. In HFSR-GCVM, the parameters in the hidden nodes are generated randomly independent of the training data. Moreover, the learning of parameters in its output layer is proved equivalent to a special CCMEB problem in GFKNN hidden feature space. As most CCMEB approximation based machine learning algorithms, the proposed HFSR-GCVM training algorithm has the following merits: The maximal training time of the HFSR-GCVM training is linear with the size of training datasets and the maximal space consuming is independent of the size of training datasets. The experiments on regression tasks confirm the above conclusions.

Keywords: generalized feedforward kernel neural networks, minimal enclosing ball, scalable learning, hidden feature space learning.

1 Introduction

Neural networks have strong ability to approximate complex nonlinear functions from the input samples and can provide accurate models for a large class of natural and artificial phenomena. Out of many kinds of neural networks, feedforward neural networks (FNNs) play an important role in practical applications and have been fully investigated in recent years. Among various investigation works on FNNs, the training of the FNNs is one of the critical issues [1-8]. However, most of the training algorithms for FNNs are designed for datasets of small and middle size, and little work was addressed for the training of FNNs on large datasets, which has commonly emerged in real world applications.

In recent years, a variety of approaches for large datasets have been proposed in the context of kernel methods, whose main learning feature is that the data in the original

* Corresponding author.

J. Cao et al. (eds.), *Proceedings of ELM-2014 Volume 1*,
Proceedings in Adaptation, Learning and Optimization 3, DOI: 10.1007/978-3-319-14063-6_18

feature space are mapped nonlinearly into the kernel space in which they can be separated by a hyperplane. By applying the criterion of maximizing the separating margins of two different classes, the learning problem can be formulated as a quadratic programming (QP in abbreviation) problem which has the important computational advantage of not suffering from the problem of local minima. However, given training N patterns, a naive implementation of the QP solver takes $O(N^3)$ training time. In order to scaling up QP and find QP solutions on large datasets effectively, many methods have been proposed in recent years [9]-[17]. Among these works, the generalized core vector machine (GCVM) proposed by Tsang et al. [14][15] achieves an asymptotic time complexity that is linear in N and a space complexity that is independent of N by utilizing an approximation algorithm for the center-constrained minimum enclosing ball (CCMEB) problem in computational geometry. Experiments on large datasets for both classification and regression tasks demonstrate that GCVM is as accurate as existing kernel methods implementations, but is much faster and can handle much larger datasets than existing scale-up methods.

In order to solve the learning problems for FNNs on large datasets, in this work, we first build the connection between kernel methods and FNNs, and present generalized feedforward kernelized neural networks (GFKNN in abbreviation) as a unified framework for current kernel methods and FNNs. Then we show the virtues of GFKNN, i.e. the parameters of hidden nodes can be generated randomly, the rigorous Mercer's condition for kernel functions is not required and its learning criterion is able to take more various forms besides least squared error. As a special case of GFKNN, a novel learning algorithm called HFSR-GCVM is developed, in which the merits of generalized CVM are integrated to solve the QP problem for training GFKNN on large datasets.

2 Generalized Feedforward Kernel Neural Networks: A Unified Framework for Kernel Methods and Feedforward Neural Networks

Fig. 1 shows the structure of GFKNN, which includes input layer, hidden layers and output layer. Like traditional FNNs, all the nodes in GFKNN are connected to the nodes in the adjacent layers through unidirectional branches and the connection between nodes within one layer is not allowed. The input layer does not perform any computations, only serves to transmit input signal to the first hidden layer and the output nodes in the output layer construct the response vector of GFKNN. Each node in hidden layers is combined with a linear combiner and an activation function, whose output is the response of the nodes. Notice that in GFKNN, the activation function can takes any infinite differentiability functions such as sigmodial functions, decaying RBF functions, Mexican Hat wavelet function, Morlet wavelet function and fuzzy basis functions. Besides, it can also take any nonlinear mapping function $\varphi(\mathbf{x})$ that maps the data x from the original feature space to the Reproducing Kernel Hilbert Space. In this way, kernel methods such as SVM, PSVM and LS-SVM can be considered as specific types of single-hidden-layer GFKNN (the so-called support vector network termed by Cortes and Vapnik [18]).

In general, there can be any number of hidden layers in GFKNN. However, from a practical perspective, one hidden layer with sufficient number of nodes is enough to estimate a nonlinear function. On the other hand, a deep architecture [21] can also be developed by adding more hidden layers into GFKNN.

In order to train GFKNNs, an efficient learning mechanism is needed to adjust all the weights of the connections from the hidden layer to the output layer. Based on the structure in Fig.1, there are three main approaches in the training of GFKNNs. (1) gradient-descent based (e.g. backpropagation (BP) method for multi-layer GFKNN [1]; (2) least square error based (e.g. extreme learning machines (ELMs) [19] for the single-hidden-layer feedforward networks (SLFNs), hidden-feature-space ridge regression for the multiple-hidden layer feedforward networks (HFSR)) [8]; (3) standard optimization method based (e.g. support vector machines (SVMs) [18] for the so-called support vector network).

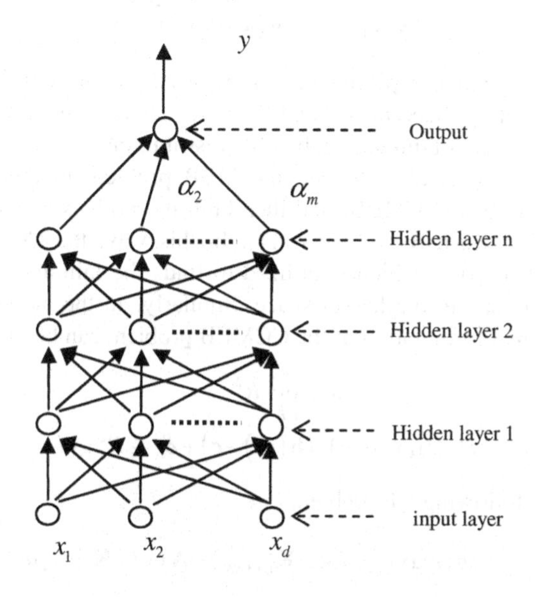

Fig. 1. Network Structure of Generalized FKNN

3 Fast Learning of Single-Hidden-Layer GFKNN on Large Datasets

Due to the simplicity of single-hidden-layer feedforward neural networks, in this work, we investigate single-hidden-layer GFKNN, in which there is only one hidden layer. The output of single-hidden-layer GFKNN can be represented by

$$f(\mathbf{x}) = \boldsymbol{\beta}^T \mathbf{h}(\mathbf{x}) \quad \mathbf{x} \in R^d \tag{1}$$

where $\boldsymbol{\beta}=[\beta_1,\ldots,\beta_L]^T$ are the weights connecting the ith hidden node to the output node and L is the nodes' number in the hidden layer, $\mathbf{h} = [g(\mathbf{x}, \theta_1), g(\mathbf{x}, \theta_2), \ldots, g(\mathbf{x}, \theta_L)]^T$, $g(\mathbf{x},\theta_i)$ is the activation function of the ith hidden node and its value denotes the response of the ith hidden node with respect to the input $\mathbf{x}$. In the following, we will develop a novel learning criterion to show that single-hidden-layer GFKNN can be scaled to handle large scaled datasets.

3.1 CCMEB in GFKNN Hidden Feature Space

By defining the mapping function explicitly, the GFKNN kernel function can be defined as

$$\begin{aligned}\mathbf{K}_{GFKNN}(\mathbf{x}_i,\mathbf{x}_j) &= \mathbf{h}(\mathbf{x}_i)\mathbf{h}^T(\mathbf{x}_j)\\ &= \left[g(\mathbf{x}_i,\theta_1) \;\; \cdots \;\; g(\mathbf{x}_i,\theta_L)\right]^T\left[g(\mathbf{x}_j,\theta_1) \;\; \cdots \;\; g(\mathbf{x}_j,\theta_L)\right]\end{aligned} \tag{2}$$

where $g(\cdot)$ is a nonlinear piecewise continuous function satisfying universal approximation capability theorems [4] and θ_is, $i=1,\ldots,L$, are randomly generated parameters according to any continuous probability distribution.

Recently, Tsang et al. extended the MEB problem to the center-constrained minimal enclosing ball (CCMEB) and then built the bridges connecting more kernel methods with the CCMEB problems [14]. In this way, the CCMEB problem was extended to kernel space, which is utilized to finds the minimal enclosing ball containing all data points in the kernel space. Similarly, in the hidden feature space of GFKNN, the primal formulation of the CCMEB problem can be expressed as

$$\begin{aligned}&\arg\min_{\mathbf{c},R} R^2\\ &s.t.\ (\mathbf{h}(\mathbf{x}_i)-\mathbf{c})^T(\mathbf{h}(\mathbf{x}_i)-\mathbf{c})+\delta_i^2 \le R^2, \forall i\end{aligned} \tag{3}$$

Its dual is the following QP problem

$$\begin{aligned}&\arg\max_{\boldsymbol{\alpha}} \boldsymbol{\alpha}^T(diag(\mathbf{K}_{GFKNN})+\boldsymbol{\Delta})-\boldsymbol{\alpha}^T\mathbf{K}_{GFKNN}\boldsymbol{\alpha}\\ &s.t.\ \boldsymbol{\alpha}^T\mathbf{1}=1, \alpha_i \ge 0, \forall i\end{aligned} \tag{4}$$

where $\boldsymbol{\Delta}=\left[\delta_1^2 \;\; \cdots \;\; \delta_N^2\right]^T \ge 0$. Because of the constraint $\boldsymbol{\alpha}^T\mathbf{1}=1$ in Eq. (4), an arbitrary multiple of $\boldsymbol{\alpha}^T\mathbf{1}$ can be added to the objective function without affecting its $\boldsymbol{\alpha}$ solution. In other words, for an arbitrary $\eta\in\mathrm{R}$, Eq. (4) yields the same optimal $\boldsymbol{\alpha}$ as

$$\begin{aligned}&\arg\max_{\boldsymbol{\alpha}} \boldsymbol{\alpha}^T(diag(\mathbf{K}_{GFKNN})+\boldsymbol{\Delta}-\eta\mathbf{1})-\boldsymbol{\alpha}^T\mathbf{K}_{GFKNN}\boldsymbol{\alpha}\\ &s.t.\ \boldsymbol{\alpha}^T\mathbf{1}=1, \alpha_i \ge 0, \forall i\end{aligned} \tag{5}$$

3.2 HFSR-GCVM and Its Relationship with CCMEB in Hidden Feature Space

Given a set of training data ($\mathbf{x}_i$, y_i), i=1, …, N, where $\mathbf{x}_i \in R^d$ and $y_i \in R$, our goal is to find a function $f(\cdot)$ in hidden feature space so that it deviates least from the training data according to the $\bar{\varepsilon}$-insensitive function

$$\left|y-f(\mathbf{x})\right|_{\bar{\varepsilon}} = \begin{cases} 0, & \left|y-f(\mathbf{x})\right| \le \bar{\varepsilon} \\ \left|y-f(\mathbf{x})\right|-\bar{\varepsilon}, & otherwise \end{cases} \tag{6}$$

while at the same time as flat as possible (i.e. $\|\boldsymbol{\beta}\|$ is as small as possible). In order to find the function, the following convex optimization problem is formulated:

$$\text{Minimize: } \|\boldsymbol{\beta}\|^2 + \frac{C}{\mu m}\sum_{i=1}^{m}\left(\xi_i^2+\xi_i^{*2}\right)+2C\bar{\varepsilon}$$

$$\text{s.t. } y_i - \boldsymbol{\beta}^T\mathbf{h}(\mathbf{x}_i) \le \bar{\varepsilon}+\xi_i$$

$$\boldsymbol{\beta}^T\mathbf{h}(\mathbf{x}_i) - y_i \le \bar{\varepsilon}+\xi_i \tag{7}$$

where C is a predefined parameter and provides a tradeoff between the flatness of the function and the training error, ξ_is and ξ_i^*s are slack variables denoting the training error corresponding to the training sample $\mathbf{x}_i$, μ is a parameter that controls the size of ε and it is analogous to the v parameter in v-SVR. By utilizing the standard method of Lagrange multipliers, we have the following dual:

$$\text{Maximize } -[\boldsymbol{\lambda}' \ \ \boldsymbol{\lambda}^*{}']\tilde{\mathbf{K}}_{HFSR-GCVM}\begin{bmatrix}\boldsymbol{\lambda} \\ \boldsymbol{\lambda}^*\end{bmatrix}+[\boldsymbol{\lambda}' \ \ \boldsymbol{\lambda}^*{}']\begin{bmatrix}\frac{2\mathbf{y}}{C} \\ -\frac{2\mathbf{y}}{C}\end{bmatrix}$$

$$\text{s.t. } [\boldsymbol{\lambda}' \ \ \boldsymbol{\lambda}^*{}']\mathbf{1}=1. \ \boldsymbol{\lambda}' \ge 0, \ \boldsymbol{\lambda}^*{}' \ge 0. \tag{8}$$

where $\mathbf{y}=[y_1,\ldots,y_m]^T$, $\boldsymbol{\lambda}=[\lambda_1, \ldots, \lambda_m]^T$, $\boldsymbol{\lambda}^*=[\lambda_1^*, \ldots, \lambda_m^*]^T$ are the $2m$ Lagrange multipliers and

$$\tilde{\mathbf{K}}_{HFSR-GCVM} = \begin{bmatrix}\left(\mathbf{K}_{GFKNN}+\frac{\mu m}{C}\mathbf{I}\right) & -\mathbf{K}_{GFKNN} \\ -\mathbf{K}_{GFKNN} & \left(\mathbf{K}_{GFKNN}+\frac{\mu m}{C}\mathbf{I}\right)\end{bmatrix}$$

is a $2m\times2m$ kernel matrix. After solving for the dual variables $\boldsymbol{\lambda}$ and $\boldsymbol{\lambda}^*$, the primal variable can be recovered as

$$\boldsymbol{\beta} = C\sum_{i=1}^{m}\left(\lambda_i-\lambda_i^*\right)\mathbf{h}(\mathbf{x}_i)$$

$$\xi_i = \lambda_i \mu m$$

$$\xi_i^* = \lambda_i^* \mu m \tag{9}$$

Then, we have the output of HFSR-GCVM as follows:

$$f(\mathbf{x})=\boldsymbol{\beta}^T\mathbf{h}(\mathbf{x})=C\sum_{i=1}^{m}(\lambda_i-\lambda_i^*)\mathbf{h}^T(\mathbf{x}_i)\mathbf{h}(\mathbf{x})$$
$$=C\sum_{i=1}^{m}(\lambda_i-\lambda_i^*)K_{GFKNN}(\mathbf{x},\mathbf{x}_i) \tag{10}$$

Now, let us investigate the relationship between HFSR-GCVM formulated in Eq. (8) and CCMEB problem in hidden feature space. For Eq.(5), let

$$\boldsymbol{\Delta}=-diag(\tilde{\mathbf{K}}_{HFSR-GCVM})+\eta\mathbf{1}+\begin{bmatrix}\frac{2\mathbf{y}}{C}\\-\frac{2\mathbf{y}}{C}\end{bmatrix}. \tag{11}$$

We can easily find that they have the same expressions. Thus, we get a significant conclusion that HFSR-GCVM formulated in Eq.(8) can now be taken as a CCMEB problem in GFKNN hidden feature space.

3.3 HFSR-GCVM Using Core-Set Based CCMEB Approximation

Given a data set $\mathbf{S}=\{\mathbf{x}_1, \ldots, \mathbf{x}_N\}\in R^d$, the minimal enclosing ball of **S**, denoted as MEB(**S**), is the smallest ball that contains all the data points in **S**. Given an $\varepsilon>0$, a ball $B(\mathbf{c}, (1+\varepsilon)r)$ is an $(1+\varepsilon)$-approximation of MEB(**S**) if $r\leq r_{MEB(S)}$ and $\mathbf{S}\subset B(\mathbf{c}, (1+\varepsilon)r)$. A subset **Q** of **S** is a core-set of **S** if an expansion by a factor $(1+\varepsilon)$ of its MEB contains **S**, i.e., $\mathbf{S}\subset B(\mathbf{c}, (1+\varepsilon)r)$, where $B(\mathbf{c}, r)$=MEB(**Q**). It is found that solving MEB problem on the core-set **Q** of data points from **S**, can often give an accurate and efficient approximation [20]. In order to achieve such an $(1+\varepsilon)$-approximation, a simple iterative scheme was proposed in [20] and further explored in [14][15][22][23]: at the tth iteration, the current estimate $B(\mathbf{c}_t,r_t)$ is expanded by including the farthest point outside the $(1+\varepsilon)$-ball $B(\mathbf{c}_t, r_t)$. This procedure is repeated until all the points in **S** are covered by $B(\mathbf{c}_t,(1+\varepsilon)r_t)$. Despite its simplicity, a surprising property is that the maximal number of iterations and the size of the final core-set depend only on ε but not on dimensional number or the size of a data set [24].

Based on the above description, we give the following general description of the training procedure of HFSR-GCVM:

```
Training procedure of HFSR-GCVM
```

Step 1: Initialize $\mathbf{Q}_0$, $\mathbf{c}_0$, and R_0. Set the iteration number t=1.

Step 2: If there is no training point $\mathbf{h}(\mathbf{x})$ which falls outside the $(1+\varepsilon)$-ball $B(\mathbf{c}_t, (1+\varepsilon)r_t)$, go to Step 6.

Step 3: Find $\mathbf{h}(\mathbf{x})$ which is farthest away from $\mathbf{c}_t$ in the hidden feature space. Set $Q_{t+1} = Q_t \cup \{\mathbf{h}(\mathbf{x})\}$.

Step 4: Find the new CCMEB, i.e., MEB $(\mathbf{Q}_{t+1})$ in the hidden feature space and then set $\mathbf{c}_{t+1}=\mathbf{c}_{MEB(Qt+1)}$ and $r_{t+1} = r_{MEB(Qt+1)}$.

Step 5: Let $t=t+1$ and go to Step 2

Step 6: Terminate the training procedure and return the obtained outputs.

HFSR-GCVM is a hidden-feature-space extension of the general core-set based CCMEB approximation algorithm based on single-hidden-layer GFKNN. Thus, the conclusions about the core-set based CCMEB approximation algorithm also hold true for HFSR-GCVM. We can directly give the following conclusions about HFSR-GCVM with the similar analysis in [14] [15][22].

Property 1: Since it has been proved that SLFNs with randomly generated hidden nodes have the universal approximation capability [4], the parameters of the hidden nodes in HFSR-GCVM for single hidden layer can be randomly generated independent of the training data.

Property 2: Given a fixed ε, the upper bound of the size of the core-set obtained by HFSR-GCVM is $O(1/\varepsilon)$, and thus the upper bound of the size of the reduced set obtained by HFSR-GCVM is not higher than $O(1/\varepsilon)$.

Property 3: Given a fixed ε, the upper bound of the time complexity of HFSR-GCVM is $O(N/\varepsilon^2 + 1/\varepsilon^4)$. i.e., it is linear with the size N of the data set. Furthermore, with use of probabilistic speedup strategy, this may be further reduced to $O(1/\varepsilon^8)$, which is independent of N.

Property 4: Given a fixed ε, the upper bound of the space complexity of GFKNN is $O(1/\varepsilon^2)$. Here, the $O(N)$ space for storing the whole data set is ignored since they can be stored outside the core memory.

Properties 2-4 give the upper bounds of algorithm HFSR-GCVM about its iteration number, its time and space complexities. In fact, we found from our experimental results that for large datasets, its real iteration number and its running time and space are far less than the theoretical upper bounds. Thereby, the proposed algorithm is very effective and efficient for large datasets.

4 Experimental Studies

In this section, the performance of HFSR-GCVM for single-hidden-layer GFKNN is compared with ELM [3]. The experiments were conducted in Matlab 2010b environment on a computer with two E5-2620 CPUs and 32 GB memory.

In the experiment, the implementation of HFSR-GCVM is based on MATLAB platform and the *quadprog()* function was utilized as the QP solver for CCMEB. We believe that HFSR-GCVM will become faster if more effective QP solver, such as SMO, is adopted. The activation function used in both ELM and HFSR-GCVM is radial basis function $g(\mathbf{x}) = \exp\left(-(a\mathbf{x}+b)^2\right)$, in which a and b were generated randomly.

To evaluate the generalization ability of the comparison algorithms, the root-mean square error (RMSE) [3] as well as the squared correlation coefficient (SCC) [13] was adopted, which were computed as follows:

$$RMSE = \sqrt{\frac{1}{n}\sum_{i=1}^{n}(f(\mathbf{x}_i) - y_i)^2} \tag{12}$$

$$SCC = \frac{\left(n\sum_{i=1}^{n} f(\mathbf{x}_i)y_i - \sum_{i=1}^{n} f(\mathbf{x}_i)\sum_{i=1}^{n} y_i\right)^2}{\left(n\sum_{i=1}^{n} y_i^2 - \left(\sum_{i=1}^{n} y_i\right)^2\right)\left(n\sum_{i=1}^{n} f^2(\mathbf{x}_i) - \left(\sum_{i=1}^{n} f(\mathbf{x}_i)\right)^2\right)} \tag{13}$$

4.1 Experimental Results on Sinc

In this section, the performance of HFSR-GCVM as well as ELM [3] was compared on *sinc*, a popular choice to illustrate machine learning algorithm for regression tasks.

$$y(x) = \begin{cases} \sin(x)/x & x \neq 0 \\ 1 & x = 0 \end{cases} \tag{14}$$

A training set $\{\mathbf{x}_i, y_i\}$ with 100,000 data and testing set $\{\mathbf{x}_i, y_i\}$ with 1000 data, respectively, were created where x is are uniformly randomly distributed on the interval [-10, 10]. In order to make the regression problem 'real', large uniform noise distributed in [-0.2, 0.2] was added to all the training samples while testing data remain noise-free.

In the first part of the experiment, we investigated the regression performance of HFSR-GCVM. Fig.1 plots the true and the approximated function of HFSR-GCVM when the training dataset size was 10,000 and the hidden nodes number was 100. As can be seen, HFSR-GCVM can obtain satisfactory results on sinc.

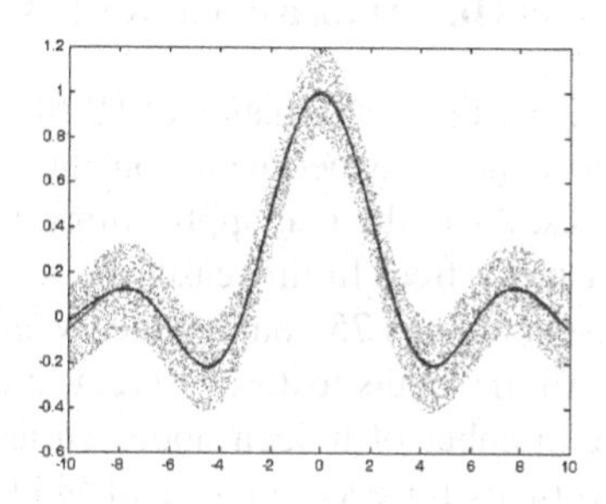

Fig. 2. Regression results of HFSR-GCVM (The true function was plot in red and the approximated one was in blue)

In the second part of the experiment, the generalization performance of ELM and HFSR-GCVM was compared when the number of hidden nodes varied from 500 to 5000. Of course, this problem needn't so many hidden nodes for training, but it is convenient for us to illustrate the scaling behavior of each algorithm. The parameter for HFSR-GCVM was set as ε=1e-10, C=1e5 and μ=1e-3. To visually show the regression performance of ELM and HFSR-GCVM, Figs.2 plots the regression results on the datasets when the number of hidden nodes varied from 500 to 5000.

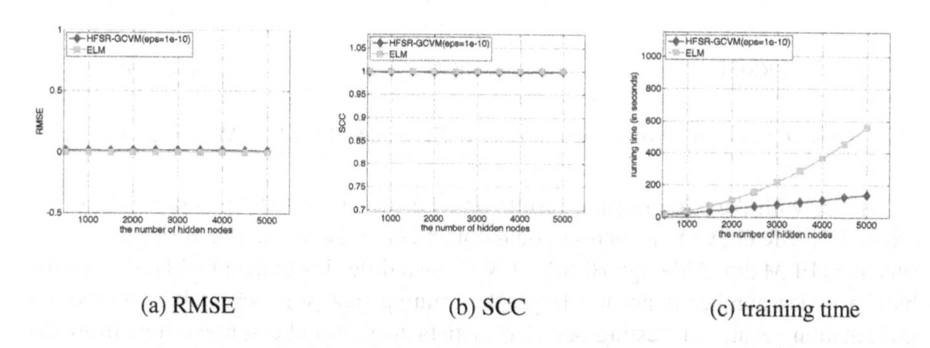

(a) RMSE (b) SCC (c) training time

Fig. 3. Comparison of the performance of ELM and HFSR-GCVM on *sinc*

As can be seen from Fig.2, both RMSE and SCC performance of HFSR-GCVM was comparable with those of ELM when the number of hidden nodes varied in a wide range from 500 to 5000. On the other hand, the running time of HFSR-GCVM increased linearly with the increase of hidden nodes number, which is mainly caused by features' increase in GFKNN hidden feature space. On the other hand, the size of the coreset was immune to the increase of the hidden nodes number, which is an ideal property for HFSR-GCVM in high dimensional hidden feature space. However, due to the inverse operation of the matrix in the matrix equation, the running time of ELM increases sharply with the increase of the number of hidden nodes, which implies that ELM will take more time in high dimensional hidden feature space.

4.2 Regression Tasks on High Dimensional Real World Datasets

In order to further investigate the performance of HFSR-GCVM on real world applications, we continued to conduct experiments on dataset *ailerons*, which contains 13750 data and 40 features. As in the real applications, the distribution of the dataset was unknown and is not noisy-free. In this case, the training dataset contains 13000 data and the testing dataset contains 750 data, which were generated randomly before each trial of simulation. All the inputs features were normalized into the range [0, 1].

In the experiment, the number of hidden nodes varied from 500 to 5000 and the training was repeated 20 times for each number of hidden nodes. The parameter for HFSR-GCVM was set as ε=1e-8, C=1 and μ=1e-5. To visually show the regression performance of ELM and HFSR-GCVM, Figs.1 plots the regression results on the datasets when the number of hidden nodes varied from 500 to 5000.

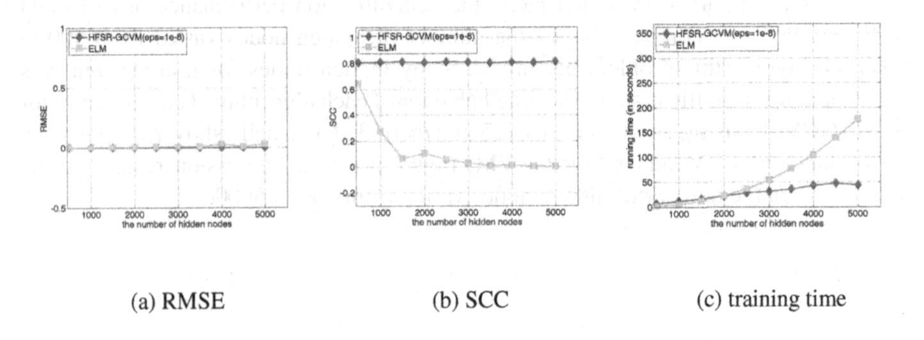

(a) RMSE (b) SCC (c) training time

Fig. 4. Comparison of the performance of ELM and HFSR-GCVM on *ailerons*

Similar with the experimental results on *sinc*, when the hidden nodes number increased, HFSR-GCVM on ailerons could obtain better approximation results in shorter time than ELM did. Although HFSR-GCVM run a little slower than ELM did when the hidden nodes number is not too large, the training time was acceptable and the approximation results on testing set were satisfactory. Another observation from the experimental results was the testing accuracy of ELM on ailerons went down dramatically when the hidden nodes number became larger and larger. This is because overfitting occurred during the learning process.

5 Conclusions

FNNs play an important role in practical applications and have been fully investigated in recent years. However, most of studies focused on FNNs performance on small or middle size datasets. In this paper, based on the connection between FNNs and kernel methods, a novel fast learning algorithm HFSR-GCVM for GFKNNs is proposed. It has a novel learning criterion which is equivalent to a special CCMEB problem in GFKNN hidden feature space while all the parameters in hidden layers can be randomly assigned. Unlike traditional kernel methods, the proposed HFSR-GCVM does not require the activation functions to be differentiable. Experimental results show that the proposed algorithm HFSR-GCVM is suitable for large datasets and has better generalization performance than ELM.

There still exists a big room for us to explore HFSR-GCVM in our future work. One challenging topic is how to determine optimal parameter combination for HFSR-GCVM. Another limitation of this work is that HFSR-GCVM was only applied for single-hidden-layer GFKNNs. When HFSR-GCVM is utilized for training multiple-hidden-layer GFKNNs, the problem of choosing an appropriate architecture, especially determining the number of hidden nodes in each layer, will become more complex. In our future work, we will further investigate these problems and report our latest advance as soon as possible.

Acknowledgements. This work was supported in part by the Hong Kong Polytechnic University under Grant 1-ZV5V, by the National Natural Science Foundation of China under Grants 61170122, 61272210, 61300151, the Fundamental Research Funds for the Central Universities (JUSRP51321B) and by the Natural Science Foundation of Jiangsu province under Grant BK2011417 and BK20130155.

References

1. Rumelhart, D.E., Hinton, G.E., Williams, R.J.: Learning representations by back- propagation errors. Nature 323, 533–536 (1986)
2. Huang, G.B., Zhu, Q.Y., Siew, C.K.: Extreme learning machine: a new learning scheme of feedforward neural networks. In: Proceedings of International Joint Conference on Neural Networks, Budapest, Hungary, July 25–29, vol. 2, pp. 985–990 (2004)
3. Huang, G.B., Zhu, Q.Y., Siew, C.K.: Extreme learning machine: theory and applications. Neurocomputing 70, 489–501 (2006)
4. Huang, G.B., Chen, L., Siew, C.K.: Universal approximation using incremental constructive feedforward networks with random hidden nodes. IEEE Transactions on Neural Networks 17(4), 879–892 (2006)
5. Huang, G.B., Chen, L.: Convex incremental extreme learning machine. Neurocomputing 70, 3056–3062 (2007)
6. Huang, G.B., Chen, L.: Enhanced random search based incremental extreme learning machine. Neurocomputing 71, 3460–3468 (2008)
7. Cortes, C., Vapnik, V.: Support vector networks. Machine Learning 20, 273–297 (1995)
8. Wang, S., Chang, F.-L., Wang, J., et al.: A Fast Learning Method for Feedforward Neural Networks. Neurocomputing (accepted)
9. Williams, C., Seeger, M., Leen, T., Dietterich, T., Tresp, V.: Using the Nyström method to speed up kernel machines. In: Advances in Neural Information Processing Systems, vol. 13, pp. 682–688. MIT Press, Cambridge (2001)
10. Smola, A., Schölkopf, B.: Sparse greedy matrix approximation for machine learning. In: Proceedings of 7th International Conference on Machine Learning, pp. 911–918. Stanford, CA (2000)

11. Achlioptas, D., McSherry, F., Schölkopf, B.: Sampling techniques for kernel methods. In: Dietterich, T., Becker, S., Ghahramani, Z. (eds.) Advances in Neural Information Processing Systems, vol. 14, pp. 335–342. MIT Press, Cambridge (2002)
12. Fine, S., Scheinberg, K.: Efficient SVM training using low-rank kernel representations. Journal of Machine Learning Research 2, 243–264 (2001)
13. Wang, S., Wang, J., Chung, F.-L.: Kernel density estimation, kernel methods, and fast learning in large datasets. IEEE Transactions on Cybernetics 44(1), 1–20 (2014)
14. Tsang, I.W., Kwok, J.T., Cheung, P.M.: Core vector machines: Fast SVM training on very large data sets. Journal of Machine Learning Research 6, 363–392 (2005)
15. Tsang, I.W., Kwok, J.T., Zurada, J.M.: Generalized core vector machines. IEEE Transactions on Neural Networks 17(5), 1126–1140 (2006)
16. Platt, J., Schölkopf, B., Burges, C., Smola, A.: Fast training of support vector machines using sequential minimal optimization. In: Advances in Kernel Methods – Support Vector Learning, pp. 185–208. MIT Press, Cambridge (1999)
17. Chu, W., Ong, C., Keerthi, S.: An improved conjugate gradient scheme to the solution of least squares SVM. IEEE Transactions on Neural Networks 16(2), 498–501 (2005)
18. Cortes, C., Vapnik, V.: Support vector networks. Machine Learning 20(3), 273–297 (1995)
19. Huang, G.B., Zhu, Q.Y., Siew, C.K.: Extreme learning machine: theory and applications. Neurocomputing 70(1-3), 489–501 (2006)
20. Badoiu, M., Clarkson, K.L.: Optimal core-sets for balls. In: DIMACS Workshop on Computational Geometry (2002)
21. Hinton, G.E., Osindero, S., Teh, Y.: A fast learning algorithm for deep belief nets. Neural Computation 18, 1527–1554 (2006)
22. Deng, Z., Chung, F.-L., Wang, S.: FRSDE: Fast reduced set density estimator using minimal enclosing ball approximation. Pattern Recognition 41(4), 1363–1372 (2008)
23. Deng, Z., Choi, K.-S., Chung, F.-L., Wang, S.: Scalable TSK Fuzzy Modeling for Very Large Datasets Using Minimal-Enclosing-Ball Approximation. IEEE Transactions on Fuzzy Systems 19(2), 210–226 (2011)
24. Kumar, P., Mitchell, J.S.B., Yildirim, A.: Approximate minimum enclosing balls in high dimensions using core-sets. ACM J. Exp. Algorithms (J/OL) 8 (2003)

An Online Multiple-Model Approach to Univariate Time-Series Prediction

Koshy George, Sachin Prabhu, and Prabhanjan Mutalik

Department of Telecommunication Engineering,
PES Institute of Technology,
100 Feet Ring Road, BSK 3rd Stage, Bangalore 560093, India
kgeorge@pes.edu, {sachinprabhuhr,prabhanjanmutalik}@gmail.com

Abstract. Predicting the future of a time-series is important in several and diverse applications. Whilst traditional methods are based on fixed linear models, techniques that use artificial neural networks are typically trained offline. We propose an online technique for time-series prediction using a single-layer feedforward neural network trained as an extreme learning machine, where the weights are updated with each new observation. The inputs to this machine not only include the actual observations of the time-series but as well as the predicted values from the machine. This ensures that feedback is incorporated into the training process. We demonstrate that such a procedure improves the prediction accuracy. Moreover, using multiple networks trained in this manner further improves the performance.

Keywords: Time-series prediction, Online Sequential ELM, Multiple Models, Switching and Tuning.

1 Introduction

Predicting the future trend of a collection of observations made sequentially in time is important in diverse fields, and a number of techniques exist for time-series prediction; see, for example, [1]. With the introduction of Yule's autoregressive technique, there was a paradigm shift from the earlier techniques which fit a trend-curve on the data, to modelling the observed phenomena as a dynamical system with specified input and output spaces. Once the structure of the model is fixed, the principal idea is to use the available data with perhaps exogenous inputs to determine the parameters of the model. Such a model is then used to extrapolate, and hence obtain an estimate of the future values of the time series. A major limitation of such techniques is that the chosen models are linear in nature to ensure mathematical tractability.

In most applications, the exogenous inputs that influence the time series are not measurable, and hence only past observations can be used for time-series prediction. Evidently, if one can detect the hidden patterns in the collection of observations, it is quite likely that one can obtain a better estimate of the future values. It is well-known that artificial neural networks (ANNs) are powerful to

J. Cao et al. (eds.), *Proceedings of ELM-2014 Volume 1*,
Proceedings in Adaptation, Learning and Optimization 3, DOI: 10.1007/978-3-319-14063-6_19

detect such patterns. Indeed it has been shown by several researchers that ANNs approximate a time-series better than the Box-Jenkins approach [14,15].

The general approach whilst using an ANN is to use the available data to tune the synaptic weights so that its output approximates better the time series. The quality of generalization, the rate of convergence, and the involved computational complexities depend on the choice of various network parameters such as the number of layers and the number and type of neurons in each layer. Furthermore, they also depend on the methodology used to tune the weights. Our interest lies in two classifications of training procedures — online and offline techniques, and sequential and batch processes. Offline techniques assume that a sufficient amount of data that accurately represents the function of interest is available a priori. On the other hand, online techniques rely on the data available at that instant. Batch processes try to analyze richness of each data point and segregate them into training, validation, and testing data. There is a need to store data-points in a batch process. Sequential processes extract the necessary information from the incoming data one-by-one.

The classical method to train the weights in ANNs is to propagate the identification error backwards through the network, and tune the weights based on the local gradients of that error. This technique is sluggish due to the large number of computations involved and may not be the best choice for fast applications. To overcome this problem, Extreme Learning Machines (ELM) was introduced in [8]. Here, in a single-layer feedforward neural network (SLFNN), the network parameters are randomly initialized, and only the weights in the linear output layer are tuned. It is an offline batch process where the weights are updated via a least-squares approach. The generalization property of such a network is comparable to a similar one trained with the back propagation algorithm (BPA); the convergence is, however, a couple of orders faster. Online Sequential ELMs (OSELM) was proposed in [9] where ELM was extended to suit sequential data processing applications.

The objective of this paper is to propose an online technique to time-series prediction that admits learning with every new observation. The adopted structure incorporates feedback during the training process to improve the prediction accuracy. The prediction-performance is further improved with the multiple models, switching, and tuning (MMST) methodology.

This paper is organized as follows: A systems perspective to time-series prediction is presented in Section 2, ANNs for this purpose are examined in Section 3, and MMST is introduced in Section 4. Simulation experiments with a variety of time-series data are presented in Section 5.

2 Systems Perspective

Let $\Sigma : \mathcal{U} \longrightarrow \mathcal{Y}$ be a given dynamical system that maps the input space $\mathcal{U}$ to the output space $\mathcal{Y}$. The general problem of time-series prediction may be stated as follows: Given $\epsilon > 0$, determine $\widehat{\Sigma}$ so that

$$\|\widehat{\Sigma}u - \Sigma u\| < \epsilon, \quad u \in \mathcal{U} \tag{1}$$

where $\|\cdot\|$ is a suitably defined norm on $\mathcal{Y}$. The most general approach amongst the class of univariate techniques is to approximate the time-series by an Autoregressive Integrated Moving Average (ARIMA) model described as follows:

$$\sum_{i=0}^{p_1} \alpha_i q^{-i}(1-q^{-1})^{p_2} y_k = \sum_{j=0}^{p_3} \beta_j q^{-j} u_k \tag{2}$$

where y_k and u_k are respectively the observation and exogenous input at instant k, q^{-1} is the unit delay operator (i.e., $q^{-1}y_k = y_{k-1}$), p_1, p_2 and p_3 are integers, and α_i and β_j are real constants. Without loss of generality, $\alpha_0 = 1$. With $p_2 = 0$, $p_1 = p_2 = 0$, and $p_2 = p_3 = 0$, we respectively obtain the Autoregressive Moving Average (ARMA), Moving Average (MA) and Autoregressive (AR) models. The goal of time-series analysis for prediction is to determine the parameters α_i and β_j so that the model $\widehat{\Sigma}$ satisfies the inequality (1) for some desired ϵ.

The issues with such an approach are three-fold: First, the model is obtained only on off-line data, and once the model is fixed it is not possible to improve it online. In other words, continuous learning is naturally precluded. Second, such an approach assumes that the exogenous input u_k is available for time-series analysis. In most applications where prediction is the objective, such an input is not measurable even if the underlying physical phenomenon admits it. Third, the models are assumed to be linear in order to guarantee the existence of a mathematically tractable solution to (1). The focus of this paper is on an online univariate technique for time-series prediction that allows a nonlinear model to learn with every new observation.

Researchers argue that if the power spectrum of the time series characterizes its relevant features, linear models are reasonably good choices [5]. However, even simple nonlinearities (such as the logistic map) cannot be approximated by a linear map. The use of multiple linear models was suggested in [16] to represent the different regions in the underlying state-space of the dynamical system generating the time-series. Evidently, estimating the required number of such linear models is an open problem. Besides, the models are linear and the procedure does not admit learning.

A candidate that admits learning is the so-called artificial neural network (ANN). It is essentially a computational architecture with massively parallel interconnections of simple processing elements. Several authors (e.g., [7]) conclusively showed ANNs are universal approximators. This result together with the fact that they have the capability to learn ensured that they were effectively used in areas such as pattern recognition, control and identification. Naturally, researchers used ANNs for time-series prediction.

A nonlinear dynamical system can be described by the differential equation:

$$y_{k+1} = f(y_k, y_{k-1}, \ldots, y_{k-n+1}, u_{k+1}, u_k, \ldots, u_{k-m+2}) \tag{3}$$

where m and n are constants that have a direct bearing on the input-output behaviour of the dynamical system. (Several special cases are described in [13].)

An ANN can be used to model (3) given the measurements of the input u_k and output y_k. Two choices of identification models [13] have been proposed:

$$\hat{y}_{k+1} = \mathcal{N}(\hat{y}_k, \hat{y}_{k-1}, \ldots, \hat{y}_{k-n+1}, u_{k+1}, u_k, \ldots, u_{k-m+2}) \tag{4}$$
$$\hat{y}_{k+1} = \mathcal{N}(y_k, y_{k-1}, \ldots, y_{k-n+1}, u_{k+1}, u_k, \ldots, u_{k-m+2}) \tag{5}$$

where $\hat{y}_k$ is the estimate of y_k given all measurements, and $\mathcal{N}(\cdot)$ denotes the ANN that approximates the nonlinear function $f(\cdot)$. Whilst model (4) mimics (3), model (5) is a function of the actual observations rather than their estimates; accordingly, they are respectively described as *parallel* and *series-parallel* models. In the context of time-series prediction, the exogenous input u_k is typically not measurable. Accordingly, (5) reduces to

$$\hat{y}_{k+1} = \mathcal{N}(y_k, y_{k-1}, \ldots, y_{k-n+1}) \tag{6}$$

Such a network is trained in open-loop and the predictions are available directly as its output. However, it has been argued in [3] that iterated predictions are better than direct predictions in the context of deterministic chaotic systems. Accordingly, we propose the following structure that uses as well the predicted values for training:

$$\hat{y}_{k+1} = \mathcal{N}(y_k, y_{k-1}, \ldots, y_{k-n+1}, \hat{y}_k, \hat{y}_{k-1}, \ldots, \hat{y}_{k-p+1}) \tag{7}$$

As demonstrated later in the sequel, the introduction of a closed-loop improves considerably the prediction accuracy for a variety of time-series.

3 Artificial Neural Networks for Time-Series Prediction

Artificial Neural Networks have been used in the past for time-series forecasting. ANNs are data-driven and no assumptions are made on the function to be approximated; accordingly, the assumption that the underlying process be linear can be eliminated. Studies show that ANNs outperform ARIMA models [15,17]. Nonetheless, the usage of ANNs is not without challenges: The choice of the network, the required number of neurons, and their interconnections, the choice of training method, and the number of data samples required for satisfactory performance are some examples. Although, Support Vector Machines (SVMs) have been shown to perform better, we use the conventional Feedforward Neural Network (FNN). The training of an SVM requires the solution to a computationally-intensive optimization problem for each data sample. This may be difficult to implement in those online applications where data is streaming in rather rapidly.

The network architecture and size are important design parameters in ANNs. Since single-layer FNN (SLFNN) are universal approximators, the only design parameter is N, the number of hidden neurons. Although several techniques have been proposed to choose N, the impact of the number of input nodes on the quality of predictions is greater [17]; i.e., the size of the regression vector in (7) is more important than N. Moreover, a recurrent network that implements an ARMA process with a SLFNN is shown to result in robust predictions in the face of non-stationary data [2]. An important consideration is that the ANN be trained in reasonable time to provide satisfactory performance. Classical training algorithms (e.g., the Back Propagation Algorithm (BPA)) adjust all the weights based on the local gradients of the error, and their slow convergence is well-known. The novel idea of adapting only the weights in the linear output layer resulted in the Extreme Learning Machine (ELM) [8]; the convergence is two orders faster and the generalization performance is comparable.

In this paper, we consider a SLFNN with N hidden neurons and one output neuron. In our context, the output at time instant k is given by $\hat{y}_k = \beta^T h_k + \beta_0$, where $\beta = \left(\beta_1\, \beta_2 \cdots \beta_N\right)^T$ is the weight vector connecting the hidden layer to the output neuron, β_0 is the bias at the output, and $h_k = \left(h_{1,k}\, h_{2,k} \cdots h_{N,k}\right)^T$ is the vector of outputs of the neurons in the hidden layer. The output of the ith hidden neuron $h_{i,k}$ is given by $h_{i,k} = g(w_i^T \phi_k + b_i)$, where w_i is the vector of weights connecting the vector of inputs ϕ_k to the ith hidden neuron with b_i the corresponding bias, and $g(\cdot)$ is the activation function. Let $\chi = \left(\phi_1\, \phi_2 \cdots \phi_L\right)$, $L \geq N$ be a chunk of data. The idea in ELM [8] is to compute the hidden layer outputs $\Omega = \left(h_1\, h_2 \cdots h_L\right)^T$ corresponding to χ and tune only β by seeking a least-squares approximation to the target sequence of values $\mathcal{Y} = \left(y_1\, y_2 \cdots y_L\right)$. We now summarize the ELM algorithm: Given a data sequence χ and an SLFNN, choose arbitrary initial values for w_i, β_i, and b_i, $1 \leq i \leq N$, and β_0. The outputs Ω of the hidden layer is calculated, and the least-squares solution to $\Omega\beta = \mathcal{Y}$ is then computed: $\hat{\beta} = \Omega^\dagger \mathcal{Y}$, where $\Omega^\dagger = (\Omega^T \Omega)^{-1} \Omega^T$ is the Moore-Penrose pseudo-inverse of Ω. To obtain a good approximation of the target vector $\mathcal{Y}$, Ω must be full ranked. In other words, the rank of Ω must be N, the number of hidden neurons.

ELM per se does not allow for sequential or online implementation of the learning process, and the network must be retrained with every new chunk of data. Thus, ELM caters to those applications where only offline learning is possible or needed. The Online Sequential Extreme Learning Machine (OSELM) is an extension of ELM [9], where β and $\Omega^\dagger$ are updated with every single, or a set of, new observation(s). In the *boosting phase*, the ELM algorithm is used to obtain initial estimates $\beta^{(0)} = \Omega^\dagger \mathcal{Y}$ and $M^{(0)} = (\Omega^T \Omega)^{-1}$, given a small set of data χ. Subsequently, in the *sequential phase*, the following are computed and updated:

$$h_{i,k} = g(w_i^T \phi_k + b_i) \tag{8}$$

$$M^{(k+1)} = M^{(k)} - \frac{M^{(k)} h_{k+1} h_{k+1}^T M^{(k)}}{1 + h_{k+1}^T M^{(k)} h_{k+1}} \tag{9}$$

$$\beta^{(k+1)} = \beta^{(k)} + M^{(k+1)} h_{k+1} (y_{k+1} - h_{k+1}^T \beta^{(k)}) \tag{10}$$

In the context of time-series prediction, the quantities h_{k+1} and $\hat{y}_{k+1}$ are computed at the kth time instant whereas y_{k+1} is available at the $(k+1)$th instant.

4 Multiple Models for Time-Series Prediction

Using multiple models for prediction is not new; e.g., linear models in [16], and offline-trained ANNs in [6]. Several researchers have combined multiple ELMs to achieve better performance. Whilst some approaches consider an ensemble of models where the outputs are combined linearly, a few others select the best one. A linear combination of ELMs (OSELMs) is essentially a single ELM (OSELM) as the input weights are never tuned. An additional drawback of multiple ELMs is that there is little scope for online learning.

In the approach adopted in this paper, diversity in the models is achieved by initializing different models with different weights. Accordingly, the predictor consists of M recurrent networks (7), $\mathcal{N}_1, \ldots, \mathcal{N}_M$, operating in parallel. The inputs to each network are the vectors $\left(y_k\, y_{k-1} \cdots y_{k-n_i+1}\right)^T$ and $\left(\hat{y}_k\, \hat{y}_{k-1} \cdots \hat{y}_{k-p_i+1}\right)^T$. The parameters n_i and p_i and the number of hidden neurons N_i for each network are chosen at random, and the networks trained independently. Unlike other approaches, we do not combine the outputs of these networks. On the contrary, we expect each network to provide an independent analysis of the time-series and provide its best estimate of $\hat{y}_{k+1}$. The best of these M estimates is selected based on the performance at the previous instant. That is, the M estimates computed at instant $k-1$ are compared with the new observation y_k, and the best network is chosen as follows:

$$j = \arg \min_{1 \le i \le M} \|y_k - \hat{y}_{i,k}\|^2 \tag{11}$$

Accordingly, the network $\mathcal{N}_j$ is chosen as the best predictor at instant k so that the estimate $\hat{y}_{j,k+1}$ is taken as the best predicted value of y_{k+1}. This process is repeated at all instants. Evidently, in our approach there are multiple models for prediction with each of them being tuned online and the best network chosen by switching. To the best knowledge of the authors, combining the individual concepts of multiple models, switching and tuning (MMST) for improving the prediction performance is novel.

Although the individual concepts of multiple models, switching, and tuning, have been developed by several researchers, the concept of using them together for improving tracking performance was first introduced in [11]. This notion has subsequently been systematically and rigorously developed into a methodology. The earlier developments together with some examples that required MMST for effective adaptive control are summarized in [12]. Example applications in other fields are treated in [4].

5 Results

In this section, we compare the prediction performance of three methods: We first consider the conventional open-loop structure (6) trained with the OSELM algorithm; we refer to this as Method A. The second method uses the proposed closed-loop structure (7) (Method B). Finally, we use the MMST approach to prediction (Method C). The performance of the three methods are compared with data from the Sante Fe Time Series Prediction and Analysis Competition held by the Sante Fe Institute in 1991 [5]. The volume of data is enormous so that automatic processing of data became essential relative to traditional time-series analysis [5]. In addition to this, we also consider the classical Mackey-Glass model of symptoms such as the irregular breathing patterns in adults with Cheynes-Stokes respiration. The example data considered data here are as follows: (i) Example I: The laser generated data is based on the fluctuations in a far-infrared laser which can approximately be modelled by three simultaneous nonlinear ordinary differential equations. Essentially the laser is in a chaotic state with the pulsations nearly following the Lorenz model. (ii) Example II: This is a multivariate data set of a patient with sleep apnea. There are no premature beats implying that the sudden changes in the heart rate are not artifacts. The heart rate is to be predicted as its variations provide information about the onset of sleep apnea. (iii) Example III: This data set corresponds to the observations of the time variation of the intensity of a variable white dwarf star PG1159-035 during March 1989. These intensity variations are the result of the superposition of independent spherical harmonic multiplets. (iv) Example IV: The exchange rate between the Swiss franc and the US dollar was observed for a period of about eight months from August 1990 to April 1991. (v) Example V: The primary symptom in a number of chronic and acute diseases is the change in the periodicity of some observable phenomenon. The irregular breathing patterns in adults with Cheynes-Stokes respiration and the fluctuations in the peripheral white blood cell counts in patients with chronic granulocytic leukemia are two examples. In their seminal paper, Mackey and Glass [10] models such phenomenon as bifurcations in differential-delay systems

$$\frac{dx(t)}{dt} = \frac{\alpha_1 \alpha_2^n x(t-\tau)}{\alpha_2^n + x^n(t-\tau)} - \alpha_3 x(t) \tag{12}$$

where α_1, α_2, α_3 and n are constants. This is considered in [9] as a benchmark problem for time-series prediction with the constants $\alpha_1 = 0.2$, $\alpha_2 = 1$, $\alpha_3 = 0.1$, $n = 10$, and the delay $\tau = 17$. For purposes of comparison, we consider the time-series generated in [9].

As mentioned earlier, the data points are considered one-by-one for training. For Methods A and B, the chosen number of hidden neurons is 3 for Examples I–IV and 120 for Example V. For Method C, the chosen number of models is 9 for Examples I–IV and 2 for Example V. For Example I, the graphs of root-mean-square error (RMSE) as a function of the number of hidden neurons are shown in Fig. 1(a). Here, the RMSE obtained for Methods A and B are respectively shown as a solid line and a dashed line. Evidently, for both techniques there is a decline in the RMSE with an increase in the number of neurons. However, the RMSE obtained with Method B is smaller than that obtained with Method A regardless of the number of hidden neurons. The RMSE values for both the techniques appear to reduce until a certain number of hidden neurons — eight in this case — after which there is hardly any reduction. The effect of the number of models in Method C on RMSE is depicted in Fig. 1(b). Evidently, the RMSE continuously

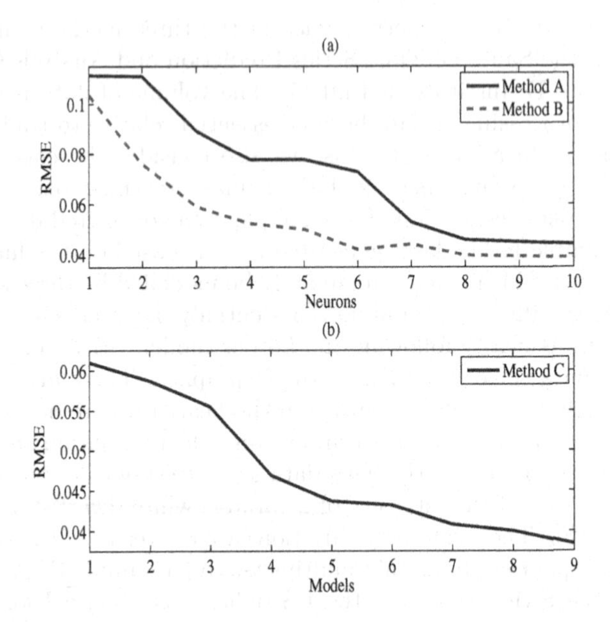

Fig. 1. Example I: (a) Effect of the number of hidden neurons on RMSE with Methods A and B. (b) Effect of the number of models on RMSE with Method C.

reduce with an increase in the number of models. For this experiment, the number of hidden neurons is fixed at ten for the first model in Method C due to which the RMSE value is approximately 0.06 with 1 model in Fig. 1(b). We randomized the number of hidden neurons in the subsequent models. A noteworthy point is a further reduction in the RMSE not achievable by increasing the number of hidden neurons in Methods A or B. On the contrary, this was made possible by increasing the number of models in Method C. This is due to the fact that all the networks are allowed to adapt at every instant, and only the best network, in the sense of (11), is chosen. Experiments with other examples reveal similar trends; these are not included here for purposes of brevity.

In order to provide a measure on the training speed of the algorithms, all experiments were conducted using MATLAB 7.14 on a general purpose computer with Intel Core i5 2.5 GHz processor. A comparison of RMSE and the training time for the different data sets with different algorithms is presented in Table 1. As discussed earlier, the RMSE obtained with Method B is observed to be smaller than that obtained with Method A. Method C outperformed both Methods A and B in terms of RMSE. On the contrary, Method A took lesser time to process the data compared to both Methods B and C. This is not surprising as at every instant k the latter techniques are required to process more inputs relative to Method A. The time taken for Method C is longer as the weights in all the models are tuned sequentially in one time instant. Evidently, a purely parallel implementation of Method C would result in lesser computational time.

Table 1. Comparison of RMSE and training times in seconds for different methods

Example	Single Model				Multiple Models	
	Method A		Method B		Method C	
	RMSE	Time	RMSE	Time	RMSE	Time
I	46.6268	0.0313	36.7056	0.0625	28.29	0.3906
II	9.905	0.2813	6.05	0.0313	2.905	0.4063
III	0.1299	0.0156	0.09	0.0313	0.0384	0.3125
IV	0.8231	0.0313	0.0403	0.0469	0.0017	0.4219
V	0.0102	4.2344	0.0083	4.5625	0.0055	4.6719

6 Conclusions

In this paper we presented a technique for online time-series prediction using single-layer feedforward neural networks. These networks are extreme learning machines where only the weights in the output layer are updated. These weights are updated with every new observation so that they continue to learn. The input to these networks consists of both the actual observations and the predicted values. The prediction performance improved with such an approach. Further when several such networks are initialized in the parametric space, the best estimate can be chosen. With this multiple models, switching and tuning approach, an additional improvement in the prediction performance is observed.

References

1. Box, G.E.P., Jenkins, G.M., Reinsel, G.C.: Time Series Analysis: Forecasting and Control. Prentice-Hall, Upper Saddle River (1994)
2. Connor, J.T., Martin, D.R., Atlas, L.E.: Recurrent neural networks and robust time series prediction. IEEE Transactions on Neural Networks 5, 240–254 (1994)
3. Farmer, J.D., Sidorowich, J.J.: Exploiting chaos to predict the future and reduce noise. In: Lee, Y.C. (ed.) Evolution, Learning, and Cognition, pp. 277–330. World Scientific, Singapore (1988)
4. George, K.: Some applications of multiple models methodology. In: Proceedings of the Fifteenth Yale Workshop on Adaptive and Learning Systems, pp. 81–86. Yale University, New Haven (June 2011)
5. Gershenfeld, N.A., Weigend, A.S.: The future of time series. In: Weigend, A.S., Gershenfeld, N.A. (eds.) Time Series Prediction: Forecasting the Future and Understanding the Past, pp. 1–70. Addison-Wesley (1993)
6. Hashem, S., Schmeiser, B.: Improving model accuracy using optimal linear combinations of trained neural networks. IEE Transactions on Neural Networks 6, 792–794 (1995)
7. Hornik, K., Stinchcombe, M., White, H.: Multilayer feedforward networks are universal approximators. Neural Networks 2, 359–366 (1989)
8. Huang, G.B., Zhu, Q.Y., Siew, C.K.: Extreme learning machine: A new learning scheme of feedforward neural networks. In: Proceedings of IEEE International Joint Conference on Neural Networks, pp. 985–990 (2004)

9. Liang, N.Y., Huang, G.B., Saratchandran, P., Sundararajan, N.: A fast and accurate online sequential learning algorithm for feedforward networks. IEEE Transactions on Neural Networks 17, 1411–1423 (2006)
10. Mackey, M.C., Glass, L.: Oscillation and chaos in physiological control systems. Science 197, 287–289 (1977)
11. Narendra, K.S., Balakrishnan, J.: Performance improvement in adaptive control systems using multiple models and switching. In: Proceedings of the Seventh Yale Workshop on Adaptive Learning Systems, pp. 27–33. Center for Systems Science, Yale University, New Haven, USA (May 1992)
12. Narendra, K.S., Driollet, O.A., Feiler, M., George, K.: Adaptive control of time-varying systems using multiple models. International Journal of Adaptive Control and Signal Processing 17(2), 87–102 (2003)
13. Narendra, K.S., Parthasarathy, K.: Identification and control of dynamical systems using neural networks. IEEE Transactions on Neural Networks 1(1), 4–27 (1990)
14. Shabri, A.: Comparison of time series forecasting methods using neural networks and Box-Jenkins model. Matematika UTM 17, 1–6 (2001)
15. Tang, Z., Fishwick, P.A.: Feed-forward neural nets as models for time series forecasting. ORSA Journal of Computing 5, 374–385 (1993)
16. Tong, H., Lim, K.S.: Threshold autoregression, limit cycles, and cyclical data. Journal of the Royal Statistical Society, B 42, 245–292 (1980)
17. Zhang, P.G., Patuwo, E.B., Hu, M.Y.: A simulation study of artificial neural networks for nonlinear time-series forecasting. Computers & Operations Research 28, 381–396 (2001)

A Self-Organizing Mixture Extreme Leaning Machine for Time Series Forecasting*

Hou Muzhou**, Chen Ming, and Zhang Yangchun

School of Mathematics and Statistics, Central South University, Changsha 410083, China
houmuzhou@sina.com, hmzw@csu.edu.cn, 383249234@qq.com

Abstract. A novel self-organizing Mixture Extreme Learning Machine (SOM-LEM) model and algorithm for time series forecasting is proposed in this paper. As the stock time series is non-stationary stochastic processes which switch their dynamics from time to time or have different models in different periods, and the ELM algorithm also has some drawbacks such as when the numbers of the samples and hidden nodes are very large, the calculation of the Moore-Penrose generalized inverse of matrix *H* will become very complicated, and the corresponding error of the elements in the matrix will become larger, and thus the generalization performance of the network will be reduced. These imply that it is not convincing and impractical for a single parametric model to capture the dynamics of the entire time series. So a SOM competitive layer is added in front of the ELM network to form the SOM-ELM model, in which, each category samples divided by SOM is then handled by a ELM model. The better generalization performance of the SOM-ELM algorithm are verified through some experiments with practical stock time series.

Keywords: SOM, ELM, time series, neural network, SOM-ELM.

1 Introduction

Stock indices and price data are always one of the most important information to investors and stock market forecasting has long been a focus of financial time series prediction that has attracted attentions of numerous scientists and financial practitioners for many years[1-5]. Unfortunately, stock indices and prices are essentially dynamic, nonlinear, nonparametric and chaotic in nature. This implies that the investors must handle the time series which are non-stationary, noisy, and have frequent structural breaks[4]. In fact, stock prices' movements are affected by many macro-economic factors such as political events, company's policies, movement of other stock market, general economic conditions, commodity prices index, bank rate, investors' expectations, institutional investors' choices and psychological factors of investors. Thus forecasting stock index and price and its movement accurately is not only

* This work was supported by the Natural Science Foundation of China under Grants 61375063, 11271378 and 61271355.

** Corresponding author.

J. Cao et al. (eds.), *Proceedings of ELM-2014 Volume 1*,
Proceedings in Adaptation, Learning and Optimization 3, DOI: 10.1007/978-3-319-14063-6_20

the most extremely challenging applications of time series prediction but also of great interest to investors.

Time series modeling and forecasting has received indispensable importance in various areas of science and engineering. Conventionally time series forecasting (TSF) has been performed predominantly using statistical-based methods, for example, the auto-regressive (AR), the moving average (MA), the auto-regressive moving average (ARMA)[6] and the autoregressive integrated moving average (ARIMA). However, over the past few decades, NNs, which exhibit superior performance on classification and regression problems in machine learning domain[7-12], have attracted tremendous attention in the TSF community. Compared to statistical-based forecasting techniques, neural network approaches have several unique characteristics, including: 1) being both nonlinear and data driven; 2) having no requirement for an explicit underlying model (nonparametric); and 3) being more flexible and universal, thus applicable to more complicated models[13].

Artificial neural networks (NNs) have been adopted extensively for time series forecasting as a powerful modeling technique with its ability to deal with high nonlinearity problems such as stock index and price prediction [3, 14], chaotic time series forecasting [15], power load forecasting [8], financial and economic forecasting [14], water resources variables forecasting [16] and sunspot series forecasting [17]. Lapedes and Farber [9] commented that NNs can model and forecast nonlinear time series with high accuracy by training feed-forward NNs to generate time series. It is evident that NNs are exhibit better forecasting performances than traditional time series approaches [5, 10].

There are many types of neural network tools[14, 18-21] that can forecasting stock time series. And recently, Guang-Bin Huang[22, 23] has proposed a simple and efficient learning algorithm for single-hidden layer feedforward neural networks (SLFNs) called extreme learning machine (ELM), which has several interesting and significant features different from traditional algorithm for feedforward neural networks. But it also has some drawbacks such as: when the numbers of samples N and hidden nodes M are very large, The calculation of matrix $H^{\dagger}_{N\times M}$ will become very complicated, then the calculation errors of the elements in $H^{\dagger}_{N\times M}$ will become larger, and thus the generalization performance of the network will be reduced.

In this paper, adaptive neural networks, in particular self-organizing maps (SOM), are explored in modeling stock time series in conjunction with extreme learning machine (ELM). This approach uses SOM to divide a time series into a number of homogeneous processes and then these processes are then modeled by the Extreme Learning Machine. The proposed network is termed self-organizing mixture extreme learning machine (SOM-ELM) model, which uses self-organizing map and extreme learning machine as component in the construction of the topological mixture model.

This paper is organized as follows. Section 2 provides the literature review about SOM, ELM and time series forecasting. In Section 3, the proposed SOME-LM methodology is described. Section 4 presents the related numerical experiments. Finally, conclusions are given in Section 5.

2 Literature Review

The main difficulty in modeling stock time series is their non-stationary. That is the mean and variance of the time series are not constant but change over time. This implies that the variables switch their dynamics from time to time or have different models in different periods. Therefore, it is not convincing and impractical for a single parametric model to capture the dynamics of the entire time series.

Due to the recent advances in computational intelligence and increased computer power, nonparametric models have been studied and used extensively in the last few years with various successes. Stock forecasting by adaptive neural networks provides strong evidence in terms of out-of-sample forecasting achievements. For example, Iebeling Kaastra and Milton Boyd[24] designed a neural network for forecasting financial and economic time series. Guilherme A. Barreto[25] reviewed the time series prediction with the Self-Organizing Map. He Ni and Hujun Yin[26] have proposed a SOMAR model for time series prediction. Mark van Heeswijk[27] present a adaptive ensemble models of Extreme Learning Machines for time series prediction.

2.1 The Self-Organizing Map

The SOM is a well-known unsupervised neural learning algorithm. The SOM learns from examples a mapping from a high-dimensional continuous input space X onto a low-dimensional discrete space (lattice) A of q neurons which are arranged in fixed topological forms, e.g., as a rectangular 2-dimensioal array. The map $i^*(x): \mathrm{X} \to \mathrm{A}$, defined by the weight matrix $W = (w_1, w_2, \cdots, w_q),\ w_i \in R^p \subset \mathrm{X}$, assigns to the current input vector $x(t) \in R^p \subset \mathrm{X}$ a neuron index

$$i^*(t) = \arg\min_{\forall i} \| x(t) - w_i(t) \|, \tag{1}$$

where $\| . \|$ denotes the Euclidean distance and t is the discrete time step associated with the iterations of the algorithm.

The weight vectors, $w_i(t)$, also called prototypes or codebook vectors, are trained according to a competitive-cooperative learning rule in which the weight vectors of a winning neuron and its neighbors in the output array are updated after the presentation of the input vector:

$$w_i(t+1) = w_i(t) + \alpha(t) h(i^*, i; t)[x(t) - w_i(t)], \tag{2}$$

where $0 < \alpha(t) < 1$ is the learning rate and $h(i^*, i; t)$ is a weighting function which limits the neighborhood of the winning neuron. A usual choice for $h(i^*, i; t)$ is given by the Gaussian function:

$$h(i^*, i; t) = \exp\left(-\frac{\| r_i(t) - r_{i^*}(t) \|^2}{2\sigma^2(t)} \right), \tag{3}$$

where $r_i(t)$ and $r_{i^*}(t)$ are, respectively, the coordinates of the neurons i and i^* in the output array A, and $\sigma(t) > 0$ defines the radius of the neighborhood function at time t. The variables $\alpha(t)$ and $\sigma(t)$ should both decay with time to guarantee convergence of the weight vectors to stable steady states.

Weight adjustment is performed until a steady state of global ordering of the weight vectors has been achieved. In this case, we say that the map has converged. The resulting map also preserves the topology of the input samples in the sense that adjacent patterns are mapped into adjacent regions on the map. Due to this topology-preserving property, the SOM is able to cluster input information and spatial relationships of the data on the map. Despite its simplicity, the SOM algorithm has been applied to a variety of complex problems and has become one of the most important ANN architectures. The SOM network structure diagram is shown in following Fig.1.

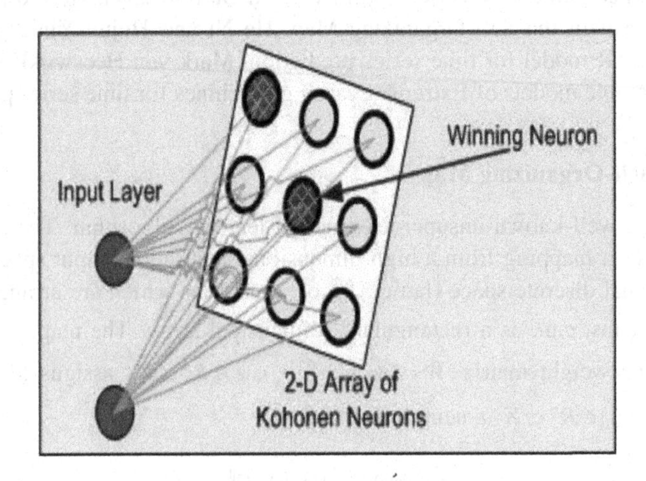

Fig. 1. Stucture model of SOM network

2.2 The Extreme Learning Machine

The Extreme Learning Machine (ELM) model has been proposed by Gung-Bin Huang et al. in [22, 23, 28, 29]. It is a type of Single-Layer Feedforward Neural Network (SLFN) and it can be used for function approximation (regression) and classification. Most traditional algorithms for training a SLFN use some learning rule that adapts all the weights based on the showing of a single training example or a batch of training examples.

Extreme Learning Machine on the other hand, rely on certain properties of the network. Namely, if the weights and biases in the input layer are randomly initialized, and the transfer functions in the hidden layer are infinitely differentiable, the optimal output weights for a given training set can be determined analytically. The obtained output weights minimize the square training error.

Since the network is trained in very few steps it is very fast to train, and it is therefore an attractive candidate model for use in a function approximation problem. A schematic overview of the structure of the ELM can be seen in Fig. 2.

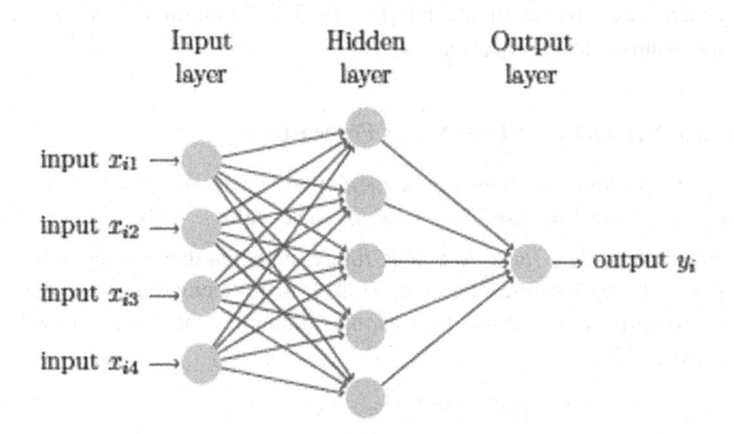

Fig. 2. A schematic overview of an ELM

Now, we consider a set of M distinct samples (X_i, y_i) with $X_i \in R^d$ and $y_i \in R$;then, a SLFN with N hidden neurons is modeled as the following sum:

$$\sum_{i=1}^{N} \beta_i f(W_i X_j + b_i),\ j \in [1, M], \tag{4}$$

where f is the activation function, W_i are the input weights to the ith neuron in the hidden layer, b_i the biases and β_i are the output weights.

In the case where the SLFN would perfectly approximate the data (meaning the error between the output $\hat{y}_i$ and the actual y_i is zero), the relation is

$$\sum_{i=1}^{N} \beta_i f(W_i X_j + b_i) = y_i,\ j \in [1, M], \tag{5}$$

which can be written compactly as

$$H\beta = Y, \tag{6}$$

where H is the hidden layer output matrix defined as:

$$H = \begin{pmatrix} f(W_1 X_1 + b_1) & \cdots & f(W_N X_N + b_N) \\ \vdots & \ddots & \vdots \\ f(W_1 X_M + b_1) & \cdots & f(W_N X_M + b_N) \end{pmatrix}, \tag{7}$$

and $\beta = (\beta_1, \cdots, \beta_N)^T$, $Y = (y_1, \cdots, y_M)^T$.

Given the randomly initialized first layer of the ELM and the training inputs $X_i \in R^d$, the hidden layer output matrix H can be computed. Now, given H and the target outputs $y_i \in R$ $(i.e. Y)$, the output weights β can be solved from the linear system defined by $H\beta = Y$. This is given by $\beta = H^{\dagger}Y$, where $H^{\dagger}$ is the Moore-Penrose generalized inverse of the matrix H. This solution for β is the unique least-squares solution to the equation $H\beta = Y$.

2.3 Neural Networks for Time Series Forecasting

Time series forecasting, or time series prediction, takes an existing series of data $X_{t-p}, \cdots, X_{t-2}, X_{t-1}$ and forecasts the $X_t, X_{t+1}, \cdots$ data values. The goal is to observe or model the existing data series to enable future unknown data values to be forecasted accurately. Throughout the literature[30, 31], many techniques have been implemented to perform time series forecasting. Examples of linear model include Auto-regressive (AR):

$$X_t = \varphi_1 X_{t-1} + \varphi_2 X_{t-2} + \cdots + \varphi_p X_{t-p} + u_t, \tag{8}$$

Moving Average (MA):

$$X_t = u_t - \theta_1 u_{t-1} - \theta_2 u_{t-2} - \cdots - \theta_q u_{t-q}, \tag{9}$$

and Auto-regressive Moving Average (ARMA)[6, 32]

$$X_t = \varphi_1 X_{t-1} + \varphi_2 X_{t-2} + \cdots + \varphi_p X_{t-p} + u_t - \theta_1 u_{t-1} - \theta_2 u_{t-2} - \cdots - \theta_q u_{t-q}. \tag{10}$$

But the time series data in practice almost exhibit high non-linearity and chaos. The above linear model cannot well solve the prediction problems of practical nonlinear time series data. And neural networks are advanced model and tool capable of extracting complex, nonlinear relationships among variables[7, 33, 34]. A three-layer feed forward NNs model is usually used to process the time series data.

The neural network forecaster can be described as follows:

$$X_t = NN(X_{t-1}, X_{t-2}, \cdots X_{t-p}). \tag{11}$$

The corresponding structure is $p \times q \times 1$, where p is the number of inputs, q is the number of neurons in the hidden layer, and one output unit. So the output of the NNs is given by

$$X_t = \sum_{j=0}^{q} c_j \varphi\left(\sum_{i=1}^{p} w_{ij} X_{t-i} + \theta_j\right), \tag{12}$$

where w_{ij} is the weight that connects the node i in the input layer neurons to the node j in the hidden layer; c_j is the weight that connects the node j in the hidden layer neurons to the node in the output layer neurons; θ_j is the threshold of neuron; and φ is the activation function.

3 Self-Organizing Mixture Extreme Leaning Machine(SOM-ELM)

The SOM-ELM neural network, just as what its name implies, is the combination of the SOM and the ELM. The SOM is a typical kind of unsupervised self-organizing neural networks, whose core concept comes from the competition and cooperation nature of biological neurons[35]. The decision of choosing the SOM as the basic model comes from the need for its special ability, that is, to build the map from high dimension input space to low dimensional output space and class the time series input data based on similarity. Then the matrix $H^{\dagger}$ of ELM model for each classification data will have smaller order. So the SOM-ELM will have smaller errors and better generalization performance.

This paper combine the characters of SOM and ELM with series. The SOM network is called the primary and ELM called secondary network. SOM network use the self-learning, needn't pre-specified the category of the training input vector during the training or learning process, it can make their own samples for cluster analysis and conduct the preliminary classification of the input data. Then for each category of SOM, we build a ELM network to forecasting the time series values. We know that the algorithm used by the SOM network is "winner takes all", namely winning neuron to its neighboring neurons from near and far and gradually becomes inhibited by the excitement. Take the position of victorious neuron of the SOM network as the input of ELM network, which will have better generalization ability because of the similarity in data of the same category and the calculation of smaller matrix $H^{\dagger}$. The combination of the two to complement each other is a novel and feasible method. The network structure diagram of SOM-ELM is shown in Fig.3.

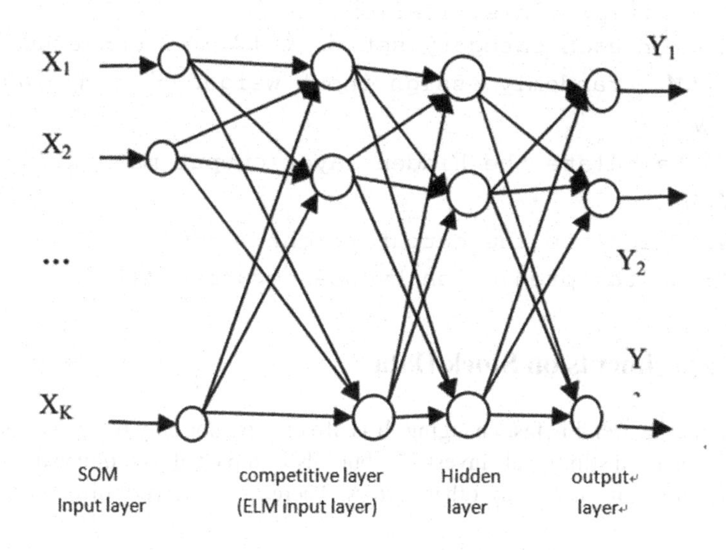

Fig. 3. Structure model of SOM-ELM

It can be seen from Fig.3, the SOM-ELM complex neural network is add a SOM competitive layer to a ELM network. First, classify the samples of input automatically by the SOM network, map the linearly inseparable samples of high-dimensional space to the linearly separable low-dimensional space, thus complete the preliminary identification of the samples in order to reduce the stress and difficulty of the prediction by the ELM network, and then through solving the $H^{\dagger}$ of ELM for supervise learning from input to output that finish the nonlinear mapping from input to output.

$$X_t = \sum_{k=1}^{S} \gamma_{(k,X)} ELM_k = \sum_{k=1}^{S} \gamma_{(k,X)} \sum_{i=1}^{N} \beta_i f(W_i X_j + b_i) = \sum_{k=1}^{S} \gamma_{(k,X)} \sum_{i=1}^{N} \beta_i f(\sum_{j}^{p} w_{ij} X_{t-j} + b_i), \quad (13)$$

where $\gamma_{(k,X)} = 0 \; or \; 1$ is a indicator or weighting factor of the $i-th$ ELM model:

$$\gamma_{(k,X)} = \begin{cases} 0 & if \; X \notin k-th \; category \\ 1 & if \; X \in k-th \; category \end{cases}, \quad (14)$$

and we have $\sum_{k=1}^{S} \gamma_{(k,X)} = 1$.

Then we can get the following SOM-ELM algorithm:

Algorithm SOM-ELM: Given the training set of time series $\aleph = \{x_1, x_2, \cdots, x_p, \cdots, x_{p+n}\}$, the number of category for SOM S, activation function $f(x)$ and the hidden neuron number N for ELM.

Step 1: Cluster the input samples preliminary by the SOM network, denoted as $A_1, A_2, \cdots, A_S$, S is the number of categories of initial classification.

Step 2: For each category set A_k $k = 1, 2, \cdots, S$, construct the model ELM_k, randomly assign input weight W_{ki} and bias b_{ki}, $i = 1, 2, \cdots, N$.

Step 3: Calculate the hidden layer output matrix H_K for each ELM_k.

Step 4: Calculate the output weight $\beta_k = H_k^{\dagger} T$

Step 5: Do the prediction by the equation (13).

4 Experiments on Stock Data

All stock market trend is fast changing. It is affected by not only the individual investors and many institutional investors, but also impacted by domestic political, economic situations and many other factors. Therefore, it is very difficult to build a classical parameter model to predict the market movement [2]. But it is easy to build a SOM-ELM model to fit the stock dataset. In this section, the approximation ability and the generalization performance of the proposed SOM-ELM compared with ELM and BP algorithms will be verified through numerical experiments.

We first collected the sample data of Chinese Shanghai Composite closing Index from internet stock database. The collection period is from 9 September 2011 to 7 November 2012 and the number of data totaled 280 (Fig. 4).

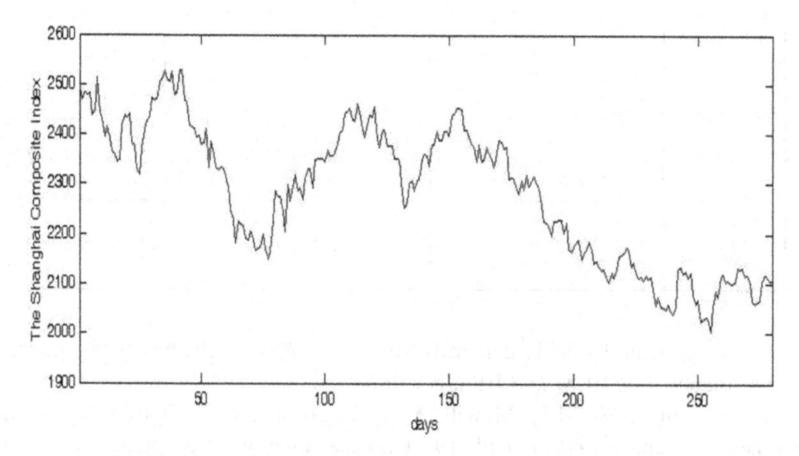

Fig. 4. Chinese Shanghai Composite closing Index in 280 days

We will use $280*85\%=238$ points of the dataset to train the neural networks and the other $280*15\%=42$ points to test the networks. And we take the memory length of time series $p=3$.

In the experiment, the training root mean square error (TRMSE) and testing root mean square error (TSMSE) are gotten as follows:

$$RMSE=\sqrt{\frac{1}{n+1}\sum_{t=0}^{n}(X_t-\hat{X}_t)}, \tag{15}$$

All the experiment are carried out in MATLAB 7.12 environment running in a Pentium ® G630 CPU with windows 7. Although there many variants of BP algorithm, a faster BP algorithms called Levenberg-Marquardt algorithm is used in our experiments. The ELM algorithm source codes are downloaded from the homepage of Extreme Learning Machine[36]. To compare the performance of the algorithms mentioned above, in each experiments 50 trials have been conducted for all the algorithms and the average TRMSE and TSMSE are shown in the following table.

Table 1. SOM-ELM Performance comparison with ELM and BP

Performance / Algorithms	average TRMSE	average TSMSE	Parameters
SOM-ELM	0	0.0037547	*S=10,p=3* *Hidden nodes=10*
ELM	0	0.0066779	*Hidden nodes=10*
BP	9.22368	0.0700886	*Hidden nodes=20*

As observed form Table 1, generally speaking, SOM-ELM has better generalization performance than ELM and BP algorithms.

Lastly, we built a SOM-ELM with 3 input neurons, 10 competitive neurons, 10 hidden neurons, and trained it with 196 Chinese Shanghai Composite closing Index validated the NNs with 42 points and predicted and tested the 42 indexes as Fig. 5.

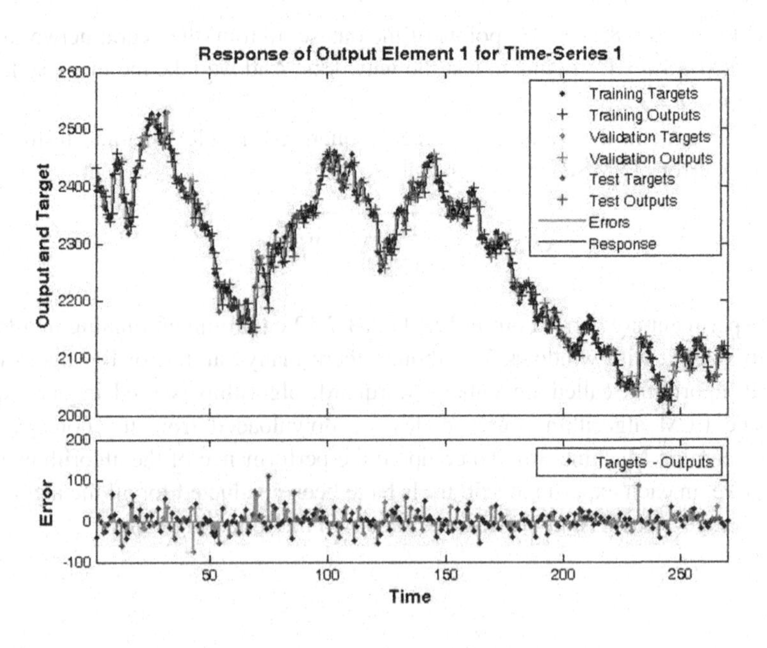

Fig. 5. The application of the SOM-ELM for prediction

5 Conclusions

In this paper, in order to improve the generalization performance, we proposed a novel SOM-ELM model and algorithm, which is built by adding a SOM competitive layer in front of the ELM model. Because the proposed SOM-ELM algorithm have used the similarity in the sample data and reduced the order and the computational complexity of the Moore-Penrose generalized inverse of matrix $H^{\dagger}$, so the error of the elements of the matrix $H^{\dagger}$ and the generalization performance of the network are improved which are verified through some experiments with practical stock time series.

References

1. Xi, L., Hou, M.Z., et al.: A new constructive neural network method for noise processing and its application on stock market prediction. Applied Soft Computing 15, 57–66 (2014)
2. Huang, C.J., Chen, P.W., et al.: Using multi-stage data mining technique to build forecast model for Taiwan stocks. Neural Computing & Applications, 1–7 (2011)
3. Liu, F., Wang, J.: Fluctuation prediction of stock market index by Legendre neural network with random time strength function. Neurocomputing (2012)
4. Oh, K.J., Kim, K.: Analyzing stock market tick data using piecewise nonlinear model. Expert Systems with Applications 22(3), 249–255 (2002)
5. Yu, S.W.: Forecasting and arbitrage of the Nikkei stock index futures: an application of backpropagation networks. Asia-Pacific Financial Markets 6(4), 341–354 (1999)
6. Flores, J.J., Graff, M., et al.: Evolutive design of ARMA and ANN models for time series forecasting. Renewable Energy 44, 225–230 (2012)
7. Adhikari, R., Agrawal, R.K.: Forecasting strong seasonal time series with artificial neural networks. Journal of Scientific & Industrial Research 71(10), 657–666 (2012)
8. Hippert, H.S., Pedreira, C.E., et al.: Neural networks for short-term load forecasting: A review and evaluation. IEEE Transactions on Power Systems 16(1), 44–55 (2001)
9. Lapedes, A., Farber, R.: Nonlinear signal processing using neural networks: Prediction and system modelling (1987)
10. Zhang, G.P.: A neural network ensemble method with jittered training data for time series forecasting. Information Sciences 177(23), 5329–5346 (2007)
11. Huang, G.B.: Learning capability and storage capacity of two-hidden-layer feedforward networks. IEEE Transactions on Neural Networks 14(2), 274–281 (2003)
12. Soyguder, S.: Intelligent control based on wavelet decomposition and neural network for predicting of human trajectories with a novel vision-based robotic. Expert Systems with Applications 38(11), 13994–14000 (2011)
13. Zhang, G., Eddy Patuwo, B., et al.: Forecasting with artificial neural networks: The state of the art. International Journal of Forecasting 14(1), 35–62 (1998)
14. Huang, W., Lai, K.K., et al.: Neural networks in finance and economics forecasting. International Journal of Information Technology & Decision Making 6(01), 113–140 (2007)
15. Han, M., Wang, Y.: Analysis and modeling of multivariate chaotic time series based on neural network. Expert Systems with Applications 36(2), 1280–1290 (2009)
16. Maier, H.R., Dandy, G.C.: Neural networks for the prediction and forecasting of water resources variables: a review of modelling issues and applications. Environmental Modelling & Software 15(1), 101–124 (2000)

17. Xie, J.X., Cheng, C.T., et al.: A hybrid adaptive time-delay neural network model for multi-step-ahead prediction of sunspot activity. International Journal of Environment and Pollution 28(3), 364–381 (2006)
18. Hou, M., Han, X.: Constructive approximation to multivariate function by decay RBF neural network. IEEE Transactions on Neural Networks 21(9), 1517–1523 (2010)
19. Muzhou, H., Xuli, H.: The multidimensional function approximation based on constructive wavelet RBF neural network. Applied Soft Computing 11(2), 2173–2177 (2011)
20. Muzhou, H., Xuli, H.: Multivariate numerical approximation using constructive $$ L^{2} (\mathbb{R}) $$ RBF neural network. Neural Computing and Applications 21(1), 25–34 (2012)
21. Muzhou, H., Xuli, H., et al.: Constructive approximation to real function by wavelet neural networks. Neural Computing & Applications 18(8), 883–889 (2009)
22. Huang, G., Zhu, Q., et al.: Extreme learning machine: Theory and applications. Neurocomputing 70(1-3), 489–501 (2006)
23. Huang, G.B., Zhu, Q.Y., et al.: Extreme learning machine: a new learning scheme of feedforward neural networks, vol. 2, pp. 985–990. IEEE (2004)
24. Kaastra, I., Boyd, M.: Designing a neural network for forecasting financial and economic time series. Neurocomputing 10(3), 215–236 (1996)
25. Barreto, G.: Time series prediction with the self-organizing map: A review. In: Hammer, B., Hitzler, P. (eds.) Perspectives of Neural-Symbolic Integration. SCI, vol. 77, pp. 135–158. Springer, Heidelberg (2007)
26. Ni, H., Yin, H.: A self-organising mixture autoregressive network for FX time series modelling and prediction. Neurocomputing 72(16), 3529–3537 (2009)
27. van Heeswijk, M., Miche, Y., Lindh-Knuutila, T., Hilbers, P.A.J., Honkela, T., Oja, E., Lendasse, A.: Adaptive ensemble models of extreme learning machines for time series prediction. In: Alippi, C., Polycarpou, M., Panayiotou, C., Ellinas, G., et al. (eds.) ICANN 2009, Part II. LNCS, vol. 5769, pp. 305–314. Springer, Heidelberg (2009)
28. Huang, G.-B., Chen, L.: Enhanced random search based incremental extreme learning machine. Neurocomputing 71(16-18), 3460–3468 (2008)
29. Huang, G.B., Chen, L.: Convex incremental extreme learning machine. Neurocomputing 70(16-18), 3056–3062 (2007)
30. Zevallos, M., Santos, B., et al.: A note on influence diagnostics in AR(1) time series models. Journal of Statistical Planning and Inference 142(11), 2999–3007 (2012)
31. Cabana, A., Cabana, E.M., et al.: Weak Convergence of Marked Empirical Processes for Focused Inference on AR(p) vs AR(p+1) Stationary Time Series. Methodology and Computing in Applied Probability 14(3), 793–810 (2012)
32. Song, P.X.K., Freeland, R.K., et al.: Statistical analysis of discrete-valued time series using categorical ARMA models. Computational Statistics & Data Analysis 57(1), 112–124 (2013)
33. Purwanto, C.E., et al.: An enhanced hybrid method for time series prediction using linear and neural network models. Applied Intelligence 37(4), 511–519 (2012)
34. Yan, W.Z.: Toward Automatic Time-Series Forecasting Using Neural Networks. IEEE Transactions on Neural Networks and Learning Systems 23(7), 1028–1039 (2012)
35. Kohonen, T.: The self-organizing map. Proceedings of the IEEE 78(9), 1464–1480 (1990)
36. `http://www3.ntu.edu.sg/home/egbhuang/`

Ensemble Extreme Learning Machine Based on a New Self-adaptive AdaBoost.RT

Pengbo Zhang and Zhixin Yang*

Department of Electromechanical Engineering
Faculty of Science and Technology
University of Macau, Macau SAR, China
{mb35515,zxyang}@umac.mo

Abstract. Extreme learning machine (ELM) has been well recognized as a new learning scheme for single-hidden layer feedforward networks (SLFNs) with extremely fast learning speed and high generalization performance. However, the stability and accuracy of ELM can be further enhanced. In this paper, a new hybrid machine learning method called robust AdaBoost.RT based ensemble ELM (RAE-ELM) for regression problems is proposed, which combined ELM with the novel self-adaptive AdaBoost.RT algorithm to achieve better approximation accuracy than using only single ELM network. To enable dynamically self-adjusting the threshold value during AdaBoost.RT computing without any empirical suggestion, the statistical parameters of applying ELM networks on the input dataset are adopted as indicators. The experiment results that applying the proposed algorithm on wide types of benchmark databases verify that RAE-ELM not only outperforms the conventional ELM but also achieves a higher stability and accuracy level than original and modified AdaBoost.RT-based ELM for regression problems.

Keywords: Extreme learning machine, Single-hidden Layer Feedforward Networks, Self-adaptive AdaBoost.RT algorithm, Regression, Ensemble.

1 Introduction

In the past decades, computational intelligence methodologies are widely adopted and have been effectively utilized in various areas of scientific research and engineering applications. Recently, Huang et al. introduced an efficient learning algorithm, named as extreme learning machine (ELM), for single-hidden layer feedforward neural networks (SLFNs) [1, 2]. Unlike conventional learning algorithms such as back-propagation (BP) methods [3] and support vector machines (SVMs) [4], ELM could randomly generate the hidden neuron parameters (the input weights and the hidden layer biases) before seeing the training data, and could analytically determine the output weights without tuning the hidden layer of SLFNs. As the random generated hidden neuron parameters are independent to the training data, ELM can reach not

* Corresponding author.

J. Cao et al. (eds.), *Proceedings of ELM-2014 Volume 1*,
Proceedings in Adaptation, Learning and Optimization 3, DOI: 10.1007/978-3-319-14063-6_21

only the smallest training error but also the smallest norm of output weights. ELM overcomes several limitations in the conventional learning algorithms, such as local minimal and slow learning speed, etc., and embodies very good generalization performance.

As a single learning machine, although ELM is quite stable comparing to other learning algorithms, its classification and regression performance may still be slightly varied among different trails on big dataset. To seek a better generalization performance, many researchers proposed to use integrated network structures which combined ELM with various ensemble methods, and verified that it performed better than using individual ELM. Lan et al. [5] proposed an ensemble of online sequential ELM (EOS-ELM), which is comprised of several OS-ELM networks. Liu and Wang [6] presented an ensemble based ELM (EN-ELM) algorithm, where the cross-validation scheme was used to create an ensemble of ELM classifiers for decision making. Besides, Xue et al. [7] proposed a genetic ensemble of extreme learning machine (GE-ELM), which adopted genetic algorithms (GAs) to produce a group of candidate networks first. More recently, Wang et al. [8] presented a parallelized ELM ensemble method based on M3-network, called as M^3-ELM.

Recently, a novel boosting algorithm, called the adaptive boosting (AdaBoost), was presented by Schapire and Freund [9]. However, many of the existing investigations on adaptive boosting algorithms focus on classification problems. These algorithms on classification problems, unfortunately, cannot be directly applied on regression problems. To solve regression problems, based on the AdaBoost algorithm on the classification problem [10-12], Schapire and Freund [9] extended AdaBoost.M2 to AdaBoost.R. Their work was followed by Shrestha and Solomatione [13, 14], who proposed a novel boosting algorithm, called as AdaBoost.RT. AdaBoost.RT projects the regression problems into the binary classification domain which could be processed by AdaBoost algorithm while filtering out those examples with the relative estimation error larger than the preset threshold value.

Combing the effective learner, ELM, with the promising ensemble method, AdaBoost.RT algorithm, could inherent their great intrinsic properties and shall be able to achieve good generalization performance for dealing with big data. Tian and Mao [15] presented an ensemble ELM based on modified AdaBoost.RT algorithm (Modified Ada-ELM) in order to predict the temperature of molten steel in ladle furnace. The novel hybrid learning algorithm combined the modified AdaBoost.RT with ELM, which possesses the advantage of ELM and overcomes the limitation of basic AdaBoost.RT by self-adaptively modifiable threshold value. However, the initial value of Φ is manually fixed to be the mean of the variation range of threshold value, which make the ensemble ELM hardly to reach a generally optimized learning effect.

This paper presents a self-adaptive and robust AdaBoost.RT based ensemble ELM (RAE-ELM) for regression problems, which combined ELM with the new self-adaptive AdaBoost.RT algorithm. The self-adaptive AdaBoost.RT algorithm not only overcomes the limitation of the original AdaBoost.RT algorithm, but also makes the threshold value of Φ be adaptive to the input dataset and ELM networks instead of manually presetting. The main idea of RAE-ELM is as follows. The ELM algorithm is selected as the 'weak' learning machines to build the hybrid ensemble model.

A new self-adaptive AdaBoost.RT algorithm is proposed to utilize the error statistics method to dynamically determine the regression threshold value rather than via manually selection which may only be ideal for very few regression cases. The proposed self-adaptive AdaBoost.RT based ensemble extreme learning machine can avoid over-fitting because of the characteristic of ELM. Moreover, as ELM is a fast learner with quite high regression performance, it contributes to the overall generalization performance of the self-adaptive AdaBoost.RT based ensemble module. The experiment results have demonstrated that the proposed self-adaptive and robust AdaBoost.RT ensemble ELM (RAE-ELM) has superior learning properties in terms of stability and accuracy for regression issues, and have better generalization performance than other algorithms.

This paper is organized as follows. Section 2 gives a brief of review of basic ELM. Section 3 introduces the proposed self-adaptive AdaBoost.RT algorithm. The hybrid self-adaptive AdaBoost.RT ensemble ELM (RAE-ELM) algorithm is then presented in Section 4. The performance evaluation of RAE-ELM and its regression ability are verified using experiments in Section 5. Finally, the conclusion is drawn in the last section.

2 Brief on ELM

Given a set of M random sample$\{(x_i, t_i) | (x_i, t_i) \in R_m \times R_n, i = 1, 2, ..., M\}$, activation function where $\boldsymbol{x}_i = [x_{i1}, x_{i2}, ..., x_{im}]^T$, and $\boldsymbol{t}_i = [\mathbf{t}_{i1}, \mathbf{t}_{i2}, ..., \mathbf{t}_{in}]^T$, the mathematical model of the standard SLFNs with $\boldsymbol{N}$ hidden nodes and activation function $\boldsymbol{f(x)}$. Extreme learning machine (ELM) can be summarized in three steps as follows:

1. Randomly assign the input weight $\boldsymbol{a}_j$ and bias b_j (j=1, 2, ..., N);
2. Calculate the hidden layer output matrix $\boldsymbol{H}$;
3. According to the equation $\hat{\beta} = \boldsymbol{H}^{\dagger}\boldsymbol{T}$,calculate the output weight β : $\beta = \boldsymbol{H}^{\dagger}\boldsymbol{T}$, where $\boldsymbol{T} = [\boldsymbol{t}_1, ..., \boldsymbol{t}_M]^T$

3 The Proposed Self-Adaptive AdaBoost.RT Algorithm

To overcome the limitations suffered by the current works on AdaBoost.RT, this paper proposes to embed statistics theory into the AdaBoost.RT algorithm. It overcomes the difficulty to optimally determine the initial threshold value and enables the intermediate threshold values to be dynamically self-adjustable according to the intrinsic property of the input data samples. The proposed self-adaptive AdaBoost.RT algorithm is described as follows.

The proposed self-adaptive AdaBoost.RT algorithm:

1) Input :
   Sequence of m sample $(x_1, y_1), \ldots, (x_m, y_m)$, where output $y \in R$;
   Weak learning algorithm (*Weak Learner*);
   Maximum number of iterations (machines) T.

---

2) Initialize:
   Iteration index t=1;
   Distribution $D_t(i)$=1/m for all i;
   The weight vector : for all i;
   Error rate ε_t=0.
3) Iterate while $t \leq T$:
   1. Call Weak Learner, WL_t, providing it with distribution:
   $$p^t = \frac{w^{(t)}}{Z_t}.$$
   where Z_t is a normalization factor chosen such that p(t) will be a distribution.
   2. Build the regression model: $f_t(x) \rightarrow y$;
   3. Calculate each error: $e_t(i) = f_t(x_i) - y_i$;
   4. Calculate the error rate: $\varepsilon_t = \sum_{i \in P} p_i^{(t)}$,
   where $P = \{i | |e_t(i) - \bar{e}_t| > \lambda\sigma_t\}$, $i \in [1, m]$. $\bar{e}_t$ stands for the expected value, $\lambda\sigma_t$ is defined as Robust Threshold (σ_t stands for the standard deviation, the relative factor λ is defined as $\lambda \in (0,1)$);
   If ε_t>1/2, then set T=t-1 and abort loop.
   5. Set $\beta_t = \varepsilon_t / 1 - \varepsilon_t$;
   6. Calculate contribution of $f_t(x)$ to the final result: $\alpha_t = -\log(\beta_t)$;
   7. Update the weight vectors:
   If $i \notin P$ then $w_i^{(t+1)} = w_i^{(t)} \beta_t$,
   else $w_i^{(t+1)} = w_i^{(t)}$
   8. Set t=t+1;
4) Normalize α_1, ..., α_T, such that $\sum_{t=1}^{T} \alpha_t = 1$;
   Output the final hypotheses: $f_{fin}(x) = \sum_{t=1}^{T} \alpha_t f_t(x)$.

---

4 A Self-Adaptive AdaBoost.RT Ensemble-Based Extreme Learning Machine

In this paper, a self-adaptive AdaBoost.RT Ensemble-based Extreme Learning Machine (RAE-ELM), which combines ELM with the self-adaptive AdaBoost.RT algorithm described in previous section, is proposed to improve the robustness and stability of ELM. In the training phase, the RAE-ELM utilizes the proposed self-adaptive AdaBoost.RT algorithm to train every ELM model and assign an ensemble weight accordingly, in order that each ELM achieves corresponding distribution based on the training output. The optimally weighted ensemble model of ELMs, is the final hypothesis output used for making prediction on testing dataset.

A. Initialization
For the first weak learner, ELM_1 is supplied with m training samples with the uniformed distribution of weights in order that each sample owns equal opportunity to be chosen during the first training process for ELM_1.

B. Distribution Updating

The relative prediction error rates are used to evaluate the performance of the current ELM networks, ELM_t. The *robust threshold* is applied to demarcate prediction errors as 'accepted' or 'rejected'. If the prediction error of one particular sample falls into the region $\mu_t \pm \lambda\sigma_t$ that is bounded by the *robust thresholds,* the prediction of this sample is regarded as 'accepted' for ELM_t, and vice versa for 'rejected' predictions. The self-adaptive AdaBoost.RT algorithm will calculate the distribution for next ELM_{t+1}.

C. Decision Making on RAE-ELM

The ELM_t with relatively superior regression performance will be granted with a larger ensemble weight. The hybrid *RAE-ELM* model combines the set of ELMs under different weights as the final hypothesis for decision making.

5 Performance Evaluation of RAE-ELM

In this section, the performance of the proposed RAE-ELM learning algorithm is compared with other ELM based algorithms on three real-world regression problems covering different domains from UCI Machine Learning Repository [16], whose specifications of benchmark data sets are shown in Table 1. The ELM based algorithms to be compared include basic ELM [2], original AdaBoost.RT based ELM (Original Ada-ELM), and the modified self-adaptive AdaBoost.RT ELM (Modified Ada-ELM) [15]. All the evaluations are conducted in Matlab 2012b environment running in a desktop with 2.20 GHz CPU and 4 GB RAM.

In our experiments, all the input attributes are normalized into the range of [-1, 1], while the outputs are normalized into [0, 1]. For each trial of simulations, the whole data set of the application is randomly partitioned into training dataset and testing dataset with the number of samples shown in Table 1, where 25% of the training data samples are used as the validation dataset.

For RAE-ELM, basic ELM, original Ada-ELM, and modified Ada-ELM algorithms, the suitable numbers of hidden nodes of them are determined using the preserved validation dataset respectively. The sigmoid function is selected as the activation function in all the algorithms. Fifty trails of simulations have been conducted for each problem, with training, validation and testing samples randomly split for each trails. The performances of the algorithms are verified using the average root mean square error (RMSE) in training and testing respectively. The significantly better results are highlighted in boldface.

Table 1. Specification of real-world regression benchmark datasets

Problems	Training data	Testing data	Attributes
Abalone	3048	1129	7
Computer hardware	100	109	7
LVST	71	54	308

In the original AdaBoost.RT-based ensemble ELM, the threshold Φ is required to be manually selected according to an empirical suggestion, which is a sensitive factor affecting the regression performance. The Original Ada-ELM with the optimal threshold values could generate satisfied results for all the three regression problems. The Modified Ada-ELM algorithm needs to select an initial value of Φ_0 to calculate the followed thresholds in the iterations. Tian and Mao suggested to set the default initial value of Φ_0 to be 0.2 [15].

The results comparisons of RAE-ELM, Basic ELM, Original Ada-ELM and Modified Ada-ELM for real-world data regressions are shown in Table 2.

Table 2. Result comparisons of training and testing RMSE of RAE-ELM, Basic ELM, Original Ada-ELM and Modified Ada-ELM for real-world data regressions

Problems	Algorithms	#Number of machines	The threshold Φ/initial value Φ_0	#Training RMSE	#Testing RMSE	#Nodes
Abalone	**RAE-ELM**	**20**		**0.0175**	**0.0168**	**30**
	Basic ELM			0.0208	0.0196	30
	Original Ada-ELM	20	0.2	0.0184	0.0177	30
	Modified Ada-ELM	20	0.2	0.0182	0.0176	30
Computer hardware	**RAE-ELM**	**20**		**0.1353**	**0.3062**	**10**
	Basic ELM			0.1522	0.3288	10
	Original Ada-ELM	20	0.35	0.1366	0.3156	10
	Modified Ada-ELM	20	0.2	0.1407	0.3094	10
LSVT	**RAE-ELM**	**20**		**0.2326**	**0.3484**	**35**
	Basic ELM			0.3004	0.3673	35
	Original Ada-ELM	20	0.25	0.2521	0.3591	35
	Modified Ada-ELM	20	0.2	0.2493	0.3563	35

The suitable number of hidden nodes of each algorithm is determined through validation process. It is easy to find that both averaged training RMSE and testing RMSE obtained by RAE-ELM for all the three cases are always the best among these four algorithms. The best performed Original Ada-ELM models for different regression problems own their correspondingly different threshold values. Therefore the manual chosen strategy is not good. The generalization performance of Modified Ada-ELM, in general, is better than Original Ada-ELM. The generalization performance of the optimized ensemble ELMs using Original Ada-ELM or Modified Ada-ELM can hardly be better than that of the proposed RAE-ELM. The three AdaBoost.RT based ensemble ELMs (RAE-ELM, Original Ada-ELM, and Modified Ada-ELM) all perform better than the Basic ELM, which verifies that an ensemble ELM using AdaBoost.RT can achieve better predication accuracy than using individual ELM as predictor.

6 Conclusion

In this paper, a self-adaptive and robust AdaBoost.RT based ensemble ELM (RAE-ELM) for regression problems is proposed, which combined ELM with the novel self-adaptive AdaBoost.RT algorithm. Combing the effective learner, ELM, with the novel ensemble method, self-adaptive AdaBoost.RT, could construct a hybrid method that inherit their intrinsic properties and achieve better predication accuracy than using only individual ELM as predictor. Therefore, selecting ELM as the 'weak' learner can avoid over-fitting. Moreover, as ELM is a fast learner with quite high regression performance, it contributes to the overall generalization performance of the ensemble ELM.

The proposed RAE-ELM is robust with respect to the difference in various regression problems and variation of approximation error rates that do not significantly affect its highly stable generalization performance. As one of the key parameter in ensemble algorithm, threshold value, do not need any human intervention; instead, is able to be self-adjusted according to the real regression effect of ELM networks on the input dataset. Such mechanism enable RAE-ELM to make sensitive and adaptive adjustment to the intrinsic properties of the given regression problem. The experimental result comparisons in terms of stability and accuracy among the four ELM based algorithms (RAE-ELM, Basic ELM, Original Ada-ELM, and Modified Ada-ELM) for regression issues verify that all the AdaBoost.RT based ensemble ELMs perform better than the Basic ELM, and more remarkably, the proposed RAE-ELM always achieves the best performance.

Acknowledgments. The authors would like to thank the funding support by the University of Macau, Grant numbers: MYRG079(Y1-L2)-FST13-YZX.

References

1. Huang, G.B., Zhu, Q.Y., Siew, C.K.: Extreme learning machine: a new learning scheme of feedforward neuralnetworks. In: Proceedings of 2004 IEEE Int. Joint Conf. on Neural Networks, vol. 2, pp. 985–990 (2004)
2. Huang, G.B., Zhu, Q.Y., Siew, C.K.: Extreme learning machine: theory and applications. Neurocomputing 70(1), 489–501 (2006)
3. Ng, S.C., Cheung, C.C., Leung, S.H.: Magnified gradient function with deterministic weight modification in adaptive learning. IEEE Transactions onNeural Networks 15(6), 1411–1423 (2004)

4. Cortes, C., Vapnik, V.: Support-vetcor networks. Machine Learning 20(3), 273–297 (1995)
5. Lan, Y., Soh, Y.C., Huang, G.B.: Ensemble of online sequential extreme learning machine. Neurocomputing 72(13), 3391–3395 (2009)
6. Liu, N., Wang, H.: Ensemble based extreme learning machine. IEEE Signal Processing Letters 17(8), 754–757 (2010)
7. Xue, X., Yao, M., Wu, Z., Yang, J.: Genetic ensemble of extreme learning machine. Neurocomputing 129(10), 175–184 (2014)
8. Wang, X.L., Chen, Y.Y., Zhao, H., Lu, B.L.: Parallelized extreme learning machine ensemble based on min–max modular network. Neurocomputing 128, 31–41 (2014)
9. Schapire, R.E., Freund, Y.: Boosting: foundations and algorithms. MIT Press, Cambridge (2012)
10. Guruswami, V., Sahai, A.: Multiclass learning, boosting, and error-corretcing codes. In: Proceedings of the Twelfth Annual Conference on Computational Learning Theory, pp. 145–155. ACM (1999)
11. Schapire, R.E., Singer, Y.: Improved boosting algorithms using confidence-rated predictions. Machine Learning 37(3), 297–336 (1999)
12. Li, L.: Multiclass boosting with repartitioning. In: Proceedings of the 23rd International Conference on Machine Learning, pp. 569–576. ACM (2006)
13. Solomatine, D.P., Shrestha, D.L.: AdaBoost. RT: a boosting algorithm for regression problems. Proceedings of 2004 IEEE Int. Joint Conf. on Neural Networks 2, 1163–1168 (2004)
14. Shrestha, D.L., Solomatine, D.: Experiments with AdaBoost. RT, an improved boosting scheme for regression. Neural Computation 18(7), 1678–1710 (2006)
15. Tian, H.X., Mao, Z.Z.: An ensemble ELM based on modified AdaBoost. RT algorithm for predicting the temperature of molten steel in ladle furnace. IEEE Transactions on Automation Science and Engineering 7(1), 73–80 (2010)
16. Bache, K., Lichman, M.: UCI Machine Learning Repository. University of California, School of Information and Computer Science, Irvine (2014), http://archive.ics.uci.edu/ml

Machine Learning Reveals Different Brain Activities in Visual Pathway during TOVA Test

Haoqi Sun[1,2,3,4], Olga Sourina[3,4], Yan Yang[1,4],
Guang-Bin Huang[2,4], Cornelia Denk[4], and Felix Klanner[4]

[1] Energy Research Institute @ NTU (ERI@N), Interdisciplinary Graduate School,
Nanyang Technological University, Singapore
[2] School of Electrical and Electronic Engineering,
Nanyang Technological University, Singapore
[3] Fraunhofer IDM @ NTU, Nanyang Technological University, Singapore
[4] Future Mobility Research Lab, A Joint Initiative of BMW Group & NTU
hsun004@e.ntu.edu.sg, {eosourina,y.yang,egbhuang}@ntu.edu.sg,
{cornelia.denk,felix.klanner}@bmw.de

Abstract. This paper explores the changes in EEG when subjects performed a modified Test of Variables of Attention (TOVA), compared to open eye resting (baseline) state. To recognize these two different brain states, two machine learning algorithms, i.e. extreme learning machine (ELM) and support vector machine (SVM), were applied and compared, using 3 statistical features and 4 power spectral density per channel. The results showed that using all 14 channels, ELM and SVM achieved similar test accuracy of 94.6% and 95.1% respectively (McNemar's test $p = 0.8 > 0.05$). Using recursive channel selection, 9 channels (ELM) and 8 channels (SVM) were selected from 14 channels. After channel selection, ELM outperformed SVM significantly (McNemar's test $p = 0.0005 < 0.01$) with average test accuracy of 95.0% and 92.5% respectively. The channel rank of each subject was weighted and merged using analytic hierarchical process to obtain a cross-subject ranking, which revealed the close correlation between TOVA and the visual pathway in brain.

Keywords: machine learning, EEG, brain activities, channel selection, TOVA, extreme learning machine.

1 Introduction

With the development of electrophysiology technologies, electroencephalography (EEG) has become a well-accepted method to discover brain activities. The EEG records the electric potential developed in the brain, using electrodes attached to the scalp [2]. In addition to its advantages of being noninvasive, easy to operate and low cost, many researches have been made possible due to its high temporal resolution at millisecond level, which is about the temporal resolution of a single spike. EEG also allows early detection of brain activity [10], which can be very important in certain applications such as driving. In this way, EEG provides a meaningful approach to inspect the activities of living and working brain.

J. Cao et al. (eds.), *Proceedings of ELM-2014 Volume 1*,
Proceedings in Adaptation, Learning and Optimization 3, DOI: 10.1007/978-3-319-14063-6_22

Most current researches on EEG focus on finding the most distinct biomarker, or feature, when performing a particular task. For example, open/closed eyes can be distinguished from lower/higher α band power (8-15Hz) respectively, and α spindle can also be used to serve the same purpose [35]. Typically a fixed or adaptive linear threshold is then set for each subject to classify different brain states. But obviously such single feature approach cannot fully grasp the complex pattern of EEG signals due to the high complexity of brain activities. As a result, multiple features with biological meanings, as well as powerful tools to classify brain states in high dimension are required to capture more complex activities such as dynamic and nonlinear changes in the brain. Machine learning is such a tool which is able to capture and approximate complex landscape of the high dimensional feature distribution, which can reveal the underlying brain states efficiently and accurately [25].

Distinguishing brain states from EEG signals, for example, on human operators' attentiveness, alertness and workload levels, are directly related to many real-world applications including the development of advanced driver state model. As a proof of concept, we designed an experiment to discover whether machine learning method can distinct the two brain states of relaxing and actively performing a task in a desktop setup, by analyzing EEG brain signals. In the next stage of the research, we will apply the developed methods to driving-related tasks, in a simulator scenario. Other data sources, for example, drivers' visual behavior will also be implemented to improve the reliability and robustness of the driver state model.

In this study, we are interested in the following 2 research questions:*i) What are the most relevant brain areas (in term of the EEG channels) when distinguishing the patterns in different brain activities? ii) How machine learning can be applied to efficiently and accurately recognize these patterns?* In order to answer these questions, we applied machine learning, i.e. extreme learning machine (ELM) with sigmoid activation function and support vector machine (SVM) with sigmoid kernel to the EEG data to recognize different brain activity during baseline (open eye resting) and task-performing (performing a modified TOVA) states. The hypothesis is: *TOVA can induce measurable changes in brain state in certain areas, compared to baseline.*

This paper is structured as follows. Section 2 outlines the related work in machine learning regarding to EEG. Section 3 describes the experimental setups. Section 4 describes the algorithm and procedures of our analysis. Results are described in Section 5. Discussion of experiment insights are presented in Section 6, followed by conclusions in Section 7.

2 Related Work

Although there is a relationship between brain activities and EEG signals, the patterns are far from straightforward. Therefore, data-driven methods, e.g. machine learning, are proposed to mine these hidden patterns in different brain states [25, 21]. The EEG analysis typically consists of feature extraction, outlier removal, feature and channel selection, model selection and classification.

Various kinds of features can be extracted from EEG signals including temporal, spectral and spatial features, from both single and multiple channels. For temporal features, Picard et al. used statistical features of the signals to classify emotions [30]. For spectral features, power spectral density of each band is commonly used and has been shown to carry biological meanings. Another biologically plausible and commonly used spectral feature is event-related spectral perturbation including event-related synchronization (ERS) and event-related desynchronization (ERD), which "allows one to model neural oscillatory dynamics" [22, 18]. EEG spatial features are less mentioned in the literature because of low spatial resolution affected by relatively large size of electrodes and the volume conduction from the cortex to electrode, but there are still methods to utilize spatial features by factoring multi-channel signals to have maximum difference in variance for a specific task, e.g. common spatial patterns (CSP) with application in motor imaginary [19]. In addition to temporal, spectral and spatial features, inter-channel features are also extracted such as mutual information [13] and phase synchronization index (PSI) [38]. Another important class of features comes from nonlinear dynamics theory, where various kinds of fractal dimension and largest Lyapunov exponent of EEG signal are evaluated with applications in the recognition of emotions and neurophysiological diseases [36, 9].

After feature extraction, outlier removal is necessary in EEG analysis if linear classifiers are used, due to the fact that the linear classifiers are more sensitive to outliers [21]. Nonlinear classifiers also suffer from outliers, but to a less extent [21]. To identify and remove outliers, it is intuitive that one can project data along the direction with maximum variance and remove outliers by setting a quantile [29]. In addition, many other advanced methods are also proposed, which can be applied to high dimensional data, such as minimum volume ellipsoid convex peeling [32].

The next step is to select and/or transform features since currently we don't have full knowledge of how brain exactly works and have to find the features which are the best to characterize the patterns. The approaches are categorized into wrapper and filter methods. Wrappers utilize the classifier and use classification accuracy as the indicator of a certain feature set. It is evident that wrappers are computationally intractable for large feature set, although for small feature space, it can find the exactly the best feature combination. Filters are methods which are independent of classifiers, instead, they use proxy variables to approximate the classification accuracy. For example, the commonly used recursive feature selection/elimination first uses filter methods such as Fisher discriminant ratio (FDR) to rank all the features to get a ranked feature list, and then a wrapper method is used to recursively evaluate the features according to classification accuracy by adding or eliminating ranked features one by one [20, 34].

After feature selection, classifiers are used to train and test the data. Classifiers are generally from two categories: linear and nonlinear. Mathematically speaking, a nonlinear classifier can be viewed as a linear one that functions in higher dimension space projected by kernels. In EEG based brain state recognition, linear discriminant analysis (LDA) is commonly used [6]. For nonlinear classifiers, support vector machine [31] and neural network, including back-propagation network [37] and extreme learning machine [39] are widely used in EEG based classification problems.

3 Experimental Setup

3.1 Subjects

10 subjects attended the experiment (3 females and 7 males), with age from 23 to 35 (mean = 26, SD = 4). All of them are right-handed and healthy with no history of neurological diseases. They were recruited from Nanyang Technological University by word of mouth and informed 5 days before the experiment. All instructions were given and all experiment data were collected by the same researcher. The experiment time was controlled between 11:00am to 12:00am before lunch without consuming coffee, tea or alcohol before and during the test to eliminate extra factors that may affect brain state.

3.2 Modified TOVA Test

A modified TOVA test [1] was used in the experiment to induce different brain state compared to open eye resting state (baseline). The original TOVA test is used in particularly to screen attention disorders such as attention deficit hyperactivity disorder (ADHD) [1]. The subject was instructed to press the space key in the keyboard as soon as the target appeared, i.e. a black square appeared at an upper position on the screen. If the square appeared at the bottom (non-target), the subject was instructed *not* to press the space key. Fig. 1 shows a target appearing. The original TOVA test contains two conditions, i.e. targets appear in high and low frequency. In the condition with low target frequency, the target-to-nontarget ratio is about 1:3.5, which is 72 targets and 248 nontargets in PEBL implementation [24], and the ratio is reversed for the high target frequency condition. The total test time in PEBL is 22.5 minutes.

The purpose of this study is to classify brain states during the task-performing and baseline. Therefore, we modified TOVA by using *only* the high target frequency condition, where the task difficulty is higher which results in more distinct brain state [23] compared to the baseline. The modified TOVA test is about 11.2 minutes.

3.3 Experiment Procedure

The whole experiment consisted of two parts, i.e. baseline and task-performing. The baseline state is defined as subject sitting in a comfortable chair in front of the computer with eyes open without large body movements for 10 minutes, during which the subject was asked to relax and *not* to pay attention to any particular objects for long time. Task-performing state is defined as the subject performing the modified TOVA test (Fig. 1). The sequence of these two parts was counterbalanced to minimize the effect of ordering, i.e. 5 randomly chosen subjects took the baseline part first and then performing the modified TOVA test, while the others took the task-performing part first and baseline second. There was 1 minute break between these two parts where the subject was allowed to have minor body movement but cannot leave the seat. Subjects' EEG data for 10 minutes of baseline and 10 minutes of task-performing were recorded for later analysis.

Fig. 1. One of the subjects doing experiment

3.4 EEG Recordings

We used Emotiv EEG headset (Emotiv Systems Inc., San Francisco, CA, USA) with 14 channels located at AF3, F7, F3, FC5, T7, P7, O1, O2, P8, T8, FC6, F4, F8 and AF4 according to the international 10-20 system. The sampling rate of EEG signals is 128Hz, notch filtered at 50Hz and 60Hz and bandpass filtered from 0.2Hz to 45Hz.

4 Methods

In Fig. 2 we show the flowchart of how EEG was analyzed from raw data to get final classification results.

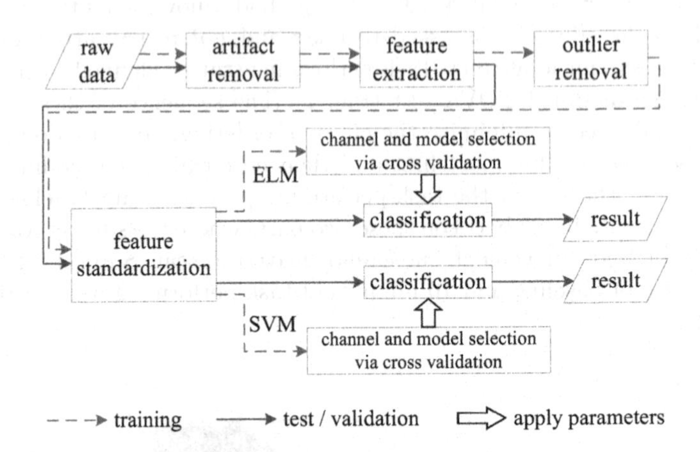

Fig. 2. EEG analysis flowchart

4.1 Artifact Removal

EEG data were first pre-processed to remove artifacts. These artifacts include blink, eyeball movement (horizontal, vertical and saccade), muscular artifacts, electrocardiogram (ECG) artifacts and loss of electrode contact with scalp. Blink artifacts usually exhibit peak-like shape in the signal, which is most dominant in frontal channels. Eyeball movement can bring staircase-like shape into the signal. Muscular artifacts resemble short burst of high frequency signals. ECG is represented as rhythmic fluctuation. Loss of electrode contact with scalp leads to dramatic change in the signal amplitude. It is expected that artifacts are statistically independent from the EEG signal from brain. Independence component analysis (ICA) was used for the whole data (including training and testing data) to get 14 components (same with the number of channels), followed by manual identification of artifact segments. Here, we did not perform a component-wise identification; instead, linear interpolation was used to replace the artifact segments contained in each component. This is because generally ICA is not guaranteed to extract all artifacts into one component. The artifact removed data was then stored and used for later processing.

4.2 Splitting Data into Training and Test Sets

We split the EEG recording of each subject into training and test sets. The first 8 minutes of each state were used for the training of classifiers and the last 2 minutes of each state were for testing (Fig. 3). 5-fold cross validation [3] was used where the optimal classifier hyperparameters and channels were selected by maximizing the average accuracy obtained on every fifth of training set as validation set. After this, the classifier was applied on test set to get test accuracy using the selected optimal hyperparameter and channels.

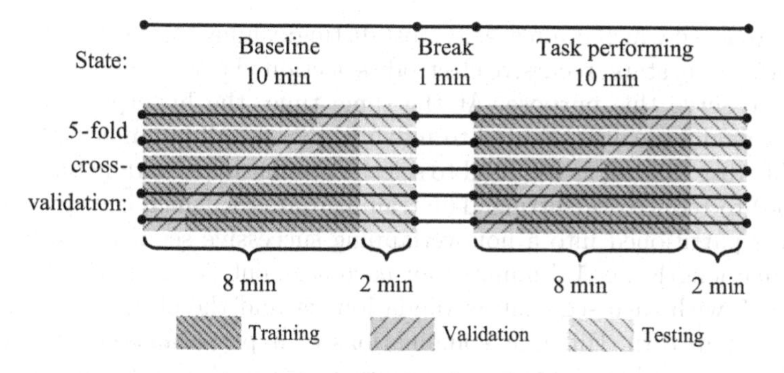

Fig. 3. Experiment procedure and data splitting

4.3 Feature Extraction

For the training set, EEG signals from each channel were first segmented to overlapping windows with length of 4 seconds and overlap of 3 seconds. The features were extracted from each segment. 3 types of statistical features were generated in this study for machine learning in next step, the signal mean, standard deviation and absolute 1st order difference (mean of absolute different between adjacent EEG amplitudes) [30].

We also used 4 power spectral density features from θ (4-7Hz), α (7-12Hz), β (12-30Hz) and γ (30-42Hz) bands, where θ band is associated with awake but relaxed or drowsy state; α band is related to idle state of brain or closing the eyes; β band is related to active thinking and alert state; γ band displays cross-modal sensory processing or short-term memory.

In total, 98 features were generated from all EEG signals (7 features per channel and 14 channels in total). In training set, features were standardized by subtracting mean and dividing standard deviation to have zero mean and unit variance *across samples*. For features in test set, we standardized in the same way by eliminating mean and dividing standard deviation from *training features*.

4.4 Outlier Removal

Before feature selection, outlier removal was applied to training set to reduce data noise and increase robustness of classification [21]. Outliers were identified and removed using iterated ellipsoidal peeling [32] due to its numerical stability and invariance to data scaling or transformation. We used Khachiyan's algorithm [17] to estimate the center of the minimum volume ellipsoid. In each peeling step, the data points with largest distance to the center were removed. We removed 2% of data points, i.e. 20 out of 954 samples are removed for each subject.

4.5 Electrode and Model Selection

In order to learn which brain area is most distinguishing during baseline state and task-performing state, recursive electrode selection via cross validation (RESCV) is used to serve this purpose. At the same time, the hyperparameters of the classifier model are also selected to maximize cross validation accuracy.

5-fold cross validation was used to select classifier hyperparameters (i.e. model selection) to reduce over-fitting. The training dataset, which lasts for 8 minutes' long, was partitioned into 5 nonoverlapping successive segments (5 folds) with equal time length, i.e. 1.6 minutes for each segment. 5 times of validation was performed, with each segment as validation set and the other four segments as training set in turn. Different combinations of hyperparameters were explored to find the optimal cross validation accuracy averaged from the 5 folds. For the ELM classifier (described in Section 4.6) with sigmoid activation function, the range of hyperparameter C was $[10^{-4}, 10^{-3}, 10^{-2}, 10^{-1}, 10^{0}, 10^{1}, 10^{2}, 10^{3}, 10^{4}]$; the range of hidden layer neuron number h was $[500, 1000, 2000]$. For the SVM classifier with sigmoid kernel, the range of hyper parameter C was the same with ELM; the range of γ is $[10^{-3}, 10^{-2}, 10^{-1}, 10^{0}]$; the range of bias term in the kernel was $[-10^{1}, -10^{0}, 0, 10^{0}, 10^{1}]$.

First, the Fisher discriminant ratio (FDR) of each feature f between two classes was calculated as follows

$$\mathrm{FDR}(f) = \frac{\text{variance between classes}}{\text{variance within classes}} = \frac{(\mu_1 - \mu_2)^2}{\sigma_1^2 + \sigma_2^2}\,, \tag{1}$$

where μ_1 and μ_2 are the means of feature f in two classes; σ_1^2 and σ_2^2 are the variances of feature f in two classes.

The FDR of a channel c was defined as the mean FDR value of the set of features F_c that belongs to c, according to Recursive Channel Elimination [20],

$$\mathrm{FDR}(c) = \frac{1}{|F_c|} \sum_{f \in F_c} \mathrm{FDR}(f)\,. \tag{2}$$

Second, we ranked the channels according to their FDR values. Then, following the final rank, we calculated the cross validation accuracy by adding one channel each time. *Selected channels and classifier hyperparameters with maximum cross validation accuracy were applied to the test set to get test accuracy.*

4.6 Classification

We used extreme learning machine (ELM) [11, 12] to classify the EEG signals into baseline and task-performing states. ELM belongs to single hidden layer feed-forward neural network (SLFN). Unlike conventional neural network using back-propagation to tune all weights, ELM uses random weight (normally distributed with zero mean and unit variance) in the input-to-hidden layer to map input data into ELM feature space with higher dimension which gives a higher possibility of linear separation in that space according to Cover's theorem [5]. According to Johnson-Lindenstrauss Lemma [14], the relative distances between data points in the transformed space are nearly preserved by normally distributed random weights, therefore the input data structure is preserved after the transformation. After the random mapping, a nonlinear activation function such as sigmoid or radius basis function is used. Formally, random mapping and nonlinear transformation are described as

$$\mathbf{H} = g(\mathbf{X}\mathbf{W} + \mathbf{1}\mathbf{b}^\top)\,, \tag{3}$$

where $\mathbf{X}$ is an $m \times n$ matrix representing input data; m is the number of samples and n is the number of features; $\mathbf{W}$ is an $n \times h$ *randomly generated* matrix representing weights from input to hidden layer; $\mathbf{b}$ is a column vector with length h to represent the bias in hidden layer; $\mathbf{1}$ is a column vector of all ones with length m in order to apply bias to each sample; the entries of $\mathbf{W}$ are random numbers sampled from any continuous probability distribution [12]; h is the number of hidden neurons; $\mathbf{H}$ is an $m \times h$ matrix containing the hidden layer output; $g(\cdot)$ is a nonlinear activation function which acts on each component of the matrix. In this paper , we used sigmoid activation function for g because of the non-local pattern of the data distribution and better data discrimination after nonlinear transformation. And we generated $\mathbf{W}$ and $\mathbf{b}$ from normal distribution $N(0,1)$.

For hidden-to-output layer, we can formulate training procedure to an unconstrained convex optimization problem, which is to minimize the error norm and weight norm simultaneously,

$$\underset{\boldsymbol{\beta}}{\text{minimize}} \quad \|\mathbf{H}\boldsymbol{\beta} - \mathbf{T}\| + C\|\boldsymbol{\beta}\|\,, \tag{4}$$

where $\mathbf{T}$ is an $m \times c$ target matrix and c is class number. Each row of $\mathbf{T}$ represents which class the corresponding input (i.e. the input in the corresponding row), by making the corresponding column as 1 and other columns as 0. $\boldsymbol{\beta}$ is an $h \times c$ matrix representing the weights from hidden to output layer. C is a hyperparameter of ELM model. $\|\cdot\|$ is a proper norm. The intention of this formulation is to have better generalization ability by limiting the norm of weight while limiting the errors at a minimum at the same time. When the errors are relatively small, the smaller weights are, the more robust the model would be. So the hyperparameter C can be understood as a trade-off between training accuracy and generalization ability, which is selected by cross validation.

For ℓ_2-norm, the optimal weight $\boldsymbol{\beta}^*$ has a closed form in terms of $\mathbf{H}$, $\mathbf{T}$ and C,

$$\boldsymbol{\beta}^* = \begin{cases} (\mathbf{H}^\top\mathbf{H} + C\mathbf{I})^{-1}\mathbf{H}^\top\mathbf{T} & \text{if } \mathbf{H}^\top\mathbf{H} \text{ is nonsingular} \\ \mathbf{H}^\top(\mathbf{H}\mathbf{H}^\top + C\mathbf{I})^{-1}\mathbf{T} & \text{if } \mathbf{H}\mathbf{H}^\top \text{ is nonsingular} \end{cases}.$$

The decision function of ELM is

$$c_{\mathbf{x}} = \text{argmax}\ g(\mathbf{x}^\top\mathbf{W} + \mathbf{b}^\top)\boldsymbol{\beta}\,, \tag{5}$$

where $\mathbf{x}$ is a feature vector of length n and $c_{\mathbf{x}}$ is the estimated class of $\mathbf{x}$. After models being built after training, we can use formula (5) to predict the class of new data.

4.7 Cross-Subject Weighted Electrode Ranking

Finally, we ranked channels by their contributions to the classifier, and to interpret the results and generate a brain activity pattern across subjects. However, since there is strong nonstationarity in EEG signal [15], the results tended to differ across subjects. We need a systematic and quantitative way to objectively evaluate the importance of each channel (representing corresponding brain area) across subjects to get an unbiased interpretation of brain activity when interacting with TOVA test.

Suppose we have a ranked channel list for every subject based on their contribution in differentiating the two brain states. An intuitive way is to look at the occurrence of each channel in the top N ranks, and the higher the ranking, the higher the channel is likely to represent brain activity related to TOVA. However, this method gives equal weights to all top N channels without reflecting their relative importance. It is obvious that the FDR value used to rank the channels could be a weighting scheme which gives the relative importance of each channel. As a result, we adopt Analytic Hierarchy Process (AHP), a structured technique which is popularly used in decision making to get a reasonable overall ranking [33] across all subjects.

To start, we need to define the objective, available alternatives and the weight of each feature. In our case, the objective is to find an objective channel ranking across all subjects, by evaluating all channels, which are the available alternatives, using the FDR value as the weight of each channel.

There are 4 steps in total to perform AHP. Firstly we applied same weight to each subject because the importance of subject was considered as being equal in our experiment. And we assigned the weights of the channels of each subject using their corresponding FDR values.

Secondly, for subject s, we formed a matrix $\mathbf{X}^{(s)}_{14\times 14}$ whose entry equals the ratio of corresponding FDR values of channels of that row and column as follows

$$\mathbf{X}^{(s)} = \begin{pmatrix} 1 & \mathrm{FDR}(c_1)/\mathrm{FDR}(c_2) & \cdots & \mathrm{FDR}(c_1)/\mathrm{FDR}(c_{14}) \\ \mathrm{FDR}(c_2)/\mathrm{FDR}(c_1) & 1 & \cdots & \mathrm{FDR}(c_2)/\mathrm{FDR}(c_{14}) \\ \vdots & \vdots & \ddots & \vdots \\ \mathrm{FDR}(c_{14})/\mathrm{FDR}(c_1) & \mathrm{FDR}(c_{14})/\mathrm{FDR}(c_2) & \cdots & 1 \end{pmatrix}. \quad (6)$$

Thirdly, we computed the eigenvector $v^{(s)}$ of $\mathbf{X}^{(s)}$ with the largest norm of eigenvalue. The ratio $r_i^{(s)}$ of each component in $v^{(s)}$ gives an objective description of how channel c_i weights for subject s.

Lastly, we repeated the step 2 and 3 for all subjects and then summed up the ratio of each channel across subjects to get an overall score r_i for channel c_i, which is an objective measure of the importance of each channel across subjects.

5 Results

In Fig. 4, we plot all cross validation and test accuracy for different numbers of channels for each subject. The dash lines show the cross validation accuracy.

The solid lines show the test accuracy. And the red and blue lines show the results from SVM and ELM respectively. The circles indicate the test accuracy where the cross validation accuracy reaches its maximum, which is the result of recursive channel selection. The values at the right end of each plot, i.e. with all 14 channels, indicate accuracy obtained without channel selection. The test accuracy was calculated using 234 samples where half (117 samples) comes from the last 2 minutes of baseline and the other half (117 samples) comes from the last 2 minutes of task-performing state.

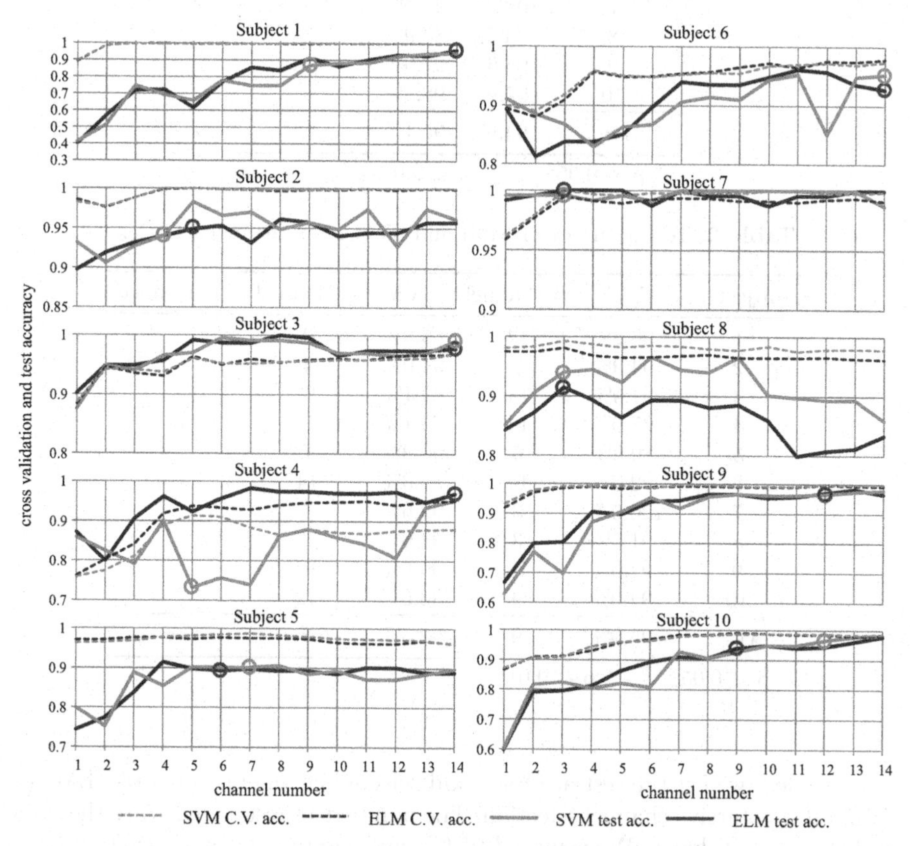

Fig. 4. Cross validation and test accuracy with increasing channel number for each subject

The test accuracy of both ELM and SVM with all 14 channels are shown in Table 1. We achieved 94.6% average test accuracy (SD = 5.0%) for ELM and 95.1% average test accuracy (SD = 4.2%) for SVM. For the average accuracy, we combined the classification results of all 10 subjects and performed McNemar's test on combined data, which contained 234 × 10 samples. On average, the difference between ELM and SVM is not statistically significant confirmed by McNemar's test ($p = 0.8 > 0.05$).

Table 1. Test accuracy of ELM and SVM using all 14 channels

subject Id	ELM	SVM	significance
1	96.2%	94.4%	
2	95.7%	96.2%	
3	97.9%	99.1%	
4	97.0%	95.3%	
5	88.9%	89.7%	
6	92.7%	95.3%	
7	100.0%	98.7%	
8	83.3%	85.9%	
9	96.6%	97.9%	
10	97.9%	98.3%	
Avg.:	94.6%	95.1%	

* $p < 0.05$ ** $p < 0.01$

Table 2. Test accuracy of ELM and SVM after channel selection

subject Id	ELM	# channel	SVM	# channel	significance
1	96.2%	14	86.8%	9	
2	94.9%	5	94.0%	4	
3	97.9%	14	99.1%	14	
4	97.0%	14	73.1%	5	**
5	89.3%	6	90.2%	7	
6	92.7%	14	95.3%	14	
7	100.0%	3	99.6%	3	
8	91.5%	3	94.0%	3	
9	97.0%	12	96.6%	12	
10	94.0%	9	96.6%	12	
Avg.:	95.0%	9	92.5%	8	**

* $p < 0.05$ ** $p < 0.01$

In Table 2 we list the test accuracy and selected channel number of ELM and SVM after recursive channel selection. The numbers of test samples are the same with those in Table 1. We achieved 95.0% average test accuracy (SD = 3.2%) for ELM and 92.5% average test accuracy (SD = 7.9%) for SVM. On average, when channel selection was considered, ELM significantly outperformed SVM (McNemar's test $p = 0.0005 < 0.01$). The average selected channel number was 9 for ELM and 8 for SVM. One may argue that the test accuracy of ELM and SVM are not comparable because of different selected channels. However, as discussed in Section 4.5, that the test accuracy was obtained using the optimal channels and hyperparameters in training. Therefore the results are comparable in terms of each classifier has reached their own optimal, although with different channels. Comparing the performance of each classifier before and after recursive channel selection, we observe that the performance of ELM is not affected by channel selection while SVM deteriorates by 2.6%, which shows ELM's better robustness against the existence of irrelevant features. We also notice that the channel number selected from training set is not always the best one for test accuracy. And the change of test accuracy with respect to channel number is less predictable than cross validation accuracy. This could be due to nonstationary in EEG signal, that the mental state in test set can be different from that in training set, so the trained model cannot fit test data well.

Table 3 shows the channel rankings of each subject. By this method, it is not conclusive, but by an overall trend that no single channel is consistently important than others across all subjects, which exhibits large subject dependence, for example, T8 ranks highest in subject 1 but ranks last in subject 2.

Table 3. Channel ranking for each subject

subject	1	2	3	4	5	6	7	8	9	10
rank										
1	T8	O2	F7	FC6	T7	P8	O2	P7	T7	AF4
2	P8	O1	AF3	F8	FC6	O2	F7	T8	F4	O1
3	O2	P8	P7	T7	P7	F3	AF4	FC5	O1	T7
4	AF4	F8	P8	AF3	O1	P7	F8	F4	AF4	T8
5	F4	P7	F8	O1	AF3	T8	F4	O1	T8	O2
6	T7	AF4	T7	O2	F4	FC5	P8	F7	F3	F3
7	O1	F4	AF4	F3	F3	FC6	FC6	AF3	P7	FC5
8	F8	AF3	O1	P7	O2	F8	T8	AF4	FC5	P7
9	P7	F3	F4	T8	AF4	O1	FC5	FC6	P8	P8
10	FC6	F7	FC5	AF4	FC5	F4	AF3	F3	O2	F7
11	FC5	FC6	O2	F4	F7	AF3	P7	P8	FC6	FC6
12	F3	FC5	T8	F7	F8	F7	T7	O2	AF3	AF3
13	AF3	T7	F3	P8	T8	AF4	F3	F8	F8	F4
14	F7	T8	FC6	FC5	P8	T7	O1	T7	F7	F8

Although by Table 3, it is not conclusive, but by an overall trend that no single channel is consistently important than other subjects, we still want to get an overall channel ranking using AHP described in Section 4.7 to gain more insights about the common pattern revealed, as shown in Table 4. The random score was computed assuming the every channel is equally important for every subject, so that it is equal to $\frac{1}{14} \times 10$. The final ranking of channels from AHP shows that T8, O1, T7, O2, P7, P8 are the most relevant channels that can distinguish different patterns in different brain activities. This suggests a major involvement of occipital and temporal lobes that is correlated with visual processing, working memory and object recognition. TOVA test is a visual intensive task involving continuous looking at the screen, memorizing the location of the fast flashing square and recognizing the associated action. So TOVA test requires visual processing, recognition and memory, which supports the "two-stream hypothesis" in visual pathway [7] where there are two streams for the visual processing pathway in brain, i.e. the ventral stream that is thought to deal with recognition of visual input and visual memory; and the dorsal stream that is thought to deal with visually guided behavior and higher layer of processing [26].

Table 4. Overall channels ranking from AHP method

channel	AHP Score	Rank	Remark
T8	0.817268	1	
O1	0.81468	2	
T7	0.796062	3	
O2	0.789472	4	
P7	0.77279	5	
P8	0.768957	6	
FC6	0.767453	7	
AF4	0.764949	8	Random score 0.714
F4	0.702232	9	
F7	0.638923	10	
F8	0.63069	11	
AF3	0.616502	12	
FC5	0.574029	13	
F3	0.545995	14	

6 Discussion

It can be seen that different brain activities can be accurately recognized by machine learning algorithms. However, from Table 3, it is not conclusive, but by an overall trend that no single channel is consistently important than others in this experiment, and the channels with the highest score (Table 4) is not far away from random guess. This phenomenon can have 3 different interpretations.

First, TOVA involves large brain areas that no single channel is significantly important than others. It requires continuous visual processing along the visual pathway, including lower level visual processing (channels O1 and O2) and higher level visual processing i.e. object recognition (channels T7 and T8, the ventral stream in visual pathway); object position recognition and visual guided behavior (channels P7 and P8, the dorsal stream in visual pathway). Vigilance (sustained attention) is also needed here, although the test itself is not difficult and long enough to induce decrement in reaction time [28]. It is known that vigilance, or attention, is a global neuronal reaction that propagate to most networks of brain regions by neuromodulators [27], as a result, most channels are incorporated (9 channels selected by ELM and 8 selected by SVM out of 14 channels).

Second, there is subject dependence in the brain activity between baseline and task-performing states. Although the test time was controlled, we cannot rule out other factors such as the biological clock of each subject and their strategy to deal with the test. During the experiment, we noticed that some subjects tended to think carefully before making the response, which leaded to less test error and slower reaction; other subjects tended to make response without too much thinking, which leaded to more test error and faster reaction. These different strategies leaded to different brain states as seen in the variance in the channel ranking of each subject. In order to deal with such subject dependence, there are basically 2 approaches. The first approach is to find the most stable or reliable features across subjects. Gudmundsson et al. found that power spectral features have the highest reliability, followed by entropy features and coherence being least reliable [8]. The second approach originates from the nonstationarity of neurological data by factorizing signal into stationary and nonstationary components. Bünau et al. introduces stationary subspace analysis (SSA) by assuming the linear superposition of stationary and nonstationary sources and nonstaitonarity are measurable in the first two moments [4].

In this paper, we looked at subject dependence from a different way using AHP to get an objective common understanding of all subjects. In this way, the channels were properly weighted according to the mean FDR value of features belonging to one channel. We can further extend this hierarchical approach to count for the signal quality of each subject to get a more objective overall channel ranking. Possible measures of signal quality include the ratio of artifact components obtained from ICA or mean/SD value of channel impedance.

Third, artifacts may affect the channel ranking of each subject. Although the EEG signals were manually screened to remove major artifacts such as blinking and electrode contact loss with scalp, we still cannot assert the signal is "clean". Small amplitude artifacts such as eyeball saccade and muscular movement can remain. And also the saline solutions could dry out gradually in the experiment, which leaded to fluctuating impedance and inaccurate signal amplitude.

For the classifiers, in this paper, we compared the performance of ELM and SVM, both of which were capable of recognizing brain activity patterns accurately. After channel selection, ELM outperformed SVM in average test accuracy very significantly (McNemar's test $p = 0.0005 < 0.01$). The advantage of ELM over SVM may be more significant in multi-class case [12] because SVM has to convert multi-class data into binary data by combining all other classes, which causes distortion in feature space. ELM does not face such distortion problem where the solution can be calculated from the same formula (4), which could bring benefits for multi-class classification. Also, the non-iterative nature of ELM leads to shorter computational time because the complexity of SVM increases with sample number. For big data problem, ELM can restrict the size of matrix under inverse $\mathbf{H}^\top\mathbf{H}$ to $h \times h$, where h is hidden layer neuron number, thus limiting the most computational demanding step $(\mathbf{H}^\top\mathbf{H} + C\mathbf{I})^{-1}$.

Another possible improvement lies in the temporal patterns of EEG [16], which is not utilized in current classifiers. Conventional classification embeds time series into static spatial feature space, but ignoring the dynamic trajectory of these feature points during the experiment. Since EEG reflects the dynamic brain activities, it is possible to learn and recognize different brain states from temporal patterns, as supported by the high temporal resolution of EEG.

7 Conclusion

This paper represented an experiment which tried to recognize brain states during open eye resting (baseline state) and performing a modified TOVA test (task-performing state) using machine learning. We applied and compared two machine learning algorithms, i.e. ELM and SVM, to classify these two states and interpreted the biological insights from the resulting accuracy and channel ranking. 3 statistical and 4 power spectral density features were used per channel. For the 234 test samples, with all 14 channels, we got 94.6% average test accuracy for ELM and 95.1% average test accuracy for SVM, which were not statistically significantly different (McNemar's test $p = 0.8 > 0.05$). Using recursive channel selection, we got 95.0% average test accuracy for ELM and 92.5% average test accuracy for SVM. ELM significantly outperformed SVM in average test accuracy (McNemar's test $p = 0.0005 < 0.01$).

We evaluated the research hypothesis that TOVA can induce measurable changes in brain state in the visual pathway, compared to the baseline. The answers to the two research questions in Section 1 are*i) The average selected channel number was 9 for ELM and 8 for SVM. No single channel was consistently important than others in this experiment due to strong nonstationarity in EEG. Based on overall channel ranking from AHP method, occipital and temporal lobes were more activated during performing the TOVA test which reflected the highly different pattern in visual pathway during TOVA test, compared to baseline. ii) We have shown that machine learning can efficiently and accurately recognize patterns in different brain states. After recursive channel selection, ELM outperformed SVM in average test accuracy very significantly (McNemar's test* $p = 0.0005 < 0.01$*) with accuracy of 95.0% and 92.5% respectively.*

Acknowledgments. Special thanks to BMW and NTU to enable the Future Mobility Research Lab. We also thank Dr. Liu Yisi for her helpful suggestions during the experiments and detailed suggestion on the analysis procedures, and Mr. Liyanaarachchi Lekamalage Chamara Kasun for his constructive discussion on extreme learning machine theories.

References

1. About the t.o.v.a, `http://www.tovatest.com/about-the-t-o-v-a/` (accessed: June 03, 2014)
2. Electroencephalography - mesh - ncbi, `http://www.ncbi.nlm.nih.gov/mesh/68004569` (accessed: June 02, 2014)
3. Breiman, L., Spector, P.: Submodel selection and evaluation in regression. the x-random case. International statistical review/revue internationale de Statistique, 291–319 (1992)

4. von Bünau, P., Meinecke, F.C., Scholler, S., Muller, K.: Finding stationary brain sources in eeg data. In: 2010 Annual International Conference of the IEEE Engineering in Medicine and Biology Society (EMBC), pp. 2810–2813. IEEE (2010)
5. Cover, T.M.: Geometrical and statistical properties of systems of linear inequalities with applications in pattern recognition. IEEE Transactions on Electronic Computers (3), 326–334 (1965)
6. Dornhege, G.: Toward brain-computer interfacing. MIT Press (2007)
7. Goodale, M.A., Milner, A.D.: Separate visual pathways for perception and action. Trends in Neurosciences 15(1), 20–25 (1992)
8. Gudmundsson, S., Runarsson, T.P., Sigurdsson, S., Eiriksdottir, G., Johnsen, K.: Reliability of quantitative eeg features. Clinical Neurophysiology 118(10), 2162–2171 (2007)
9. Güler, N.F., Übeyli, E.D., Güler, İ.: Recurrent neural networks employing lyapunov exponents for eeg signals classification. Expert Systems with Applications 29(3), 506–514 (2005)
10. Haufe, S., Treder, M.S., Gugler, M.F., Sagebaum, M., Curio, G., Blankertz, B.: Eeg potentials predict upcoming emergency brakings during simulated driving. Journal of Neural Engineering 8(5), 56001 (2011)
11. Huang, G.-B., Chen, L., Siew, C.-K.: Universal approximation using incremental constructive feedforward networks with random hidden nodes. IEEE Transactions on Neural Networks 17(4), 879–892 (2006)
12. Huang, G.-B., Zhou, H., Ding, X., Zhang, R.: Extreme learning machine for regression and multiclass classification. IEEE Transactions on Systems, Man, and Cybernetics, Part B: Cybernetics 42(2), 513–529 (2012)
13. Jin, S.-H., Lin, P., Auh, S., Hallett, M.: Abnormal functional connectivity in focal hand dystonia: mutual information analysis in eeg. Movement Disorders 26(7), 1274–1281 (2011)
14. Johnson, W.B., Lindenstrauss, J.: Extensions of lipschitz mappings into a hilbert space. Contemporary Mathematics 26(189-206) 26(189-206), 1 (1984)
15. Kaplan, A.Y., Fingelkurts, A.A., Fingelkurts, A.A., Borisov, S.V., Darkhovsky, B.S.: Nonstationary nature of the brain activity as revealed by eeg/meg: methodological, practical and conceptual challenges. Signal Processing 85(11), 2190–2212 (2005)
16. Kasabov, N., Capecci, E.: Spiking neural network methodology for modelling, classification and understanding of eeg spatio-temporal data measuring cognitive processes. Information Sciences (2014)
17. Khachiyan, L.G.: Rounding of polytopes in the real number model of computation. Mathematics of Operations Research 21(2), 307–320 (1996)
18. Knyazev, G.G., Slobodskoj-Plusnin, J.Y., Bocharov, A.V.: Event-related delta and theta synchronization during explicit and implicit emotion processing. Neuroscience 164(4), 1588–1600 (2009)
19. Koles, Z., Lazar, M., Zhou, S.: Spatial patterns underlying population differences in the background eeg. Brain Topography 2(4), 275–284 (1990), `http://dx.doi.org/10.1007/BF01129656`
20. Lal, T.N., Schroder, M., Hinterberger, T., Weston, J., Bogdan, M., Birbaumer, N., Scholkopf, B.: Support vector channel selection in bci. IEEE Transactions on Biomedical Engineering 51(6), 1003–1010 (2004)
21. Lemm, S., Blankertz, B., Dickhaus, T., Müller, K.-R.: Introduction to machine learning for brain imaging. Neuroimage 56(2), 387–399 (2011)

22. Lemm, S., Müller, K.-R., Curio, G.: A generalized framework for quantifying the dynamics of eeg event-related desynchronization. PLoS Computational Biology 5(8), e1000453 (2009)
23. Lindqvist, S., Thorell, L.B.: Brief report: Manipulation of task difficulty in inhibitory control tasks. Child Neuropsychology 15(1), 1–7 (2008)
24. Mueller, S.T., Piper, B.J.: The psychology experiment building language (pebl) and pebl test battery. Journal of Neuroscience Methods 222, 250–259 (2014)
25. Müller, K.-R., Tangermann, M., Dornhege, G., Krauledat, M., Curio, G., Blankertz, B.: Machine learning for real-time single-trial eeg-analysis: From brain-computer interfacing to mental state monitoring. Journal of Neuroscience Methods 167(1), 82–90 (2008)
26. Norman, J.: Two visual systems and two theories of perception: An attempt to reconcile the constructivist and ecological approaches. Behavioral and Brain Sciences 25(01), 73–96 (2002)
27. Noudoost, B., Moore, T.: The role of neuromodulators in selective attention. Trends in Cognitive Sciences 15(12), 585–591 (2011)
28. Nuechterlein, K.H., Parasuraman, R., Jiang, Q.: Visual sustained attention: Image degradation produces rapid sensitivity decrement over time. Science 220(4594), 327–329 (1983)
29. Parra, L., Deco, G., Miesbach, S.: Statistical independence and novelty detection with information preserving nonlinear maps. Neural Computation 8(2), 260–269 (1996)
30. Picard, R.W., Vyzas, E., Healey, J.: Toward machine emotional intelligence: Analysis of affective physiological state. IEEE Transactions on Pattern Analysis and Machine Intelligence 23(10), 1175–1191 (2001)
31. Rakotomamonjy, A., Guigue, V., Mallet, G., Alvarado, V.: Ensemble of sVMs for improving brain computer interface P300 speller performances. In: Duch, W., Kacprzyk, J., Oja, E., Zadrożny, S. (eds.) ICANN 2005. LNCS, vol. 3696, pp. 45–50. Springer, Heidelberg (2005)
32. Rousseeuw, P.J., Leroy, A.M.: Robust regression and outlier detection, vol. 589. John Wiley & Sons (2005)
33. Saaty, T.L., Peniwati, K.: Group decision making: drawing out and reconciling differences. RWS Publications (2013)
34. Shen, K.Q., Ong, C.J., Li, X.P., Hui, Z., Wilder-Smith, E.: A feature selection method for multilevel mental fatigue eeg classification. IEEE Transactions on Bio-Medical Engineering 54(7), 1231–1237 (2007)
35. Sonnleitner, A., Treder, M.S., Simon, M., Willmann, S., Ewald, A., Buchner, A., Schrauf, M.: Eeg alpha spindles and prolonged brake reaction times during auditory distraction in an on-road driving study. Accident Analysis & Prevention 62, 110–118 (2014)
36. Sourina, O., Liu, Y.: A fractal-based algorithm of emotion recognition from eeg using arousal-valence model. In: BIOSIGNALS, pp. 209–214 (2011)
37. Subasi, A., Erçelebi, E.: Classification of eeg signals using neural network and logistic regression. Computer Methods and Programs in Biomedicine 78(2), 87–99 (2005)
38. Vinck, M., Oostenveld, R., van Wingerden, M., Battaglia, F., Pennartz, C.: An improved index of phase-synchronization for electrophysiological data in the presence of volume-conduction, noise and sample-size bias. Neuroimage 55(4), 1548–1565 (2011)
39. Yuan, Q., Zhou, W., Li, S., Cai, D.: Epileptic eeg classification based on extreme learning machine and nonlinear features. Epilepsy Research 96(1), 29–38 (2011)

Online Sequential Extreme Learning Machine with New Weight-Setting Strategy for Nonstationary Time Series Prediction

Jinwan Wang[1], Wentao Mao[1,*], Liyun Wang[1], and Mei Tian[2]

[1] College of Computer and Information Engineering, Henan Normal University, Henan, Xinxiang, China 453007
[2] Management Institute, Xinxiang Medical University, Henan, Xinxiang, China 453003
maowt.mail@gmail.com

Abstract. Accurate and fast prediction of nonstationary time series is challenging and of great interest in both practical and academic areas. In this paper, an online sequential extreme learning machine with new weighted strategy is proposed for nonstationary time series prediction. First, a new leave-one-out(LOO) cross-validation error estimation for online sequential data is proposed based on inversion of block matrix. Second, a new weighted strategy based on the proposed LOO error estimation is proposed. This strategy ranks the samples' importance by means of the LOO error of each new added sample, and then assigns various weights. Performance comparisons of the proposed method with other existing algorithms are presented based on chaotic and real-world nonstationary time series data. The results show that, the proposed method outperforms the classical ELM, OS-ELM in terms of generalization performance and numerical stability.

Keywords: Extreme learning machine, Time series, Nonstationary, Leave-one-out cross-validation.

1 Introduction

Time series prediction is generally playing an important role in many engineering fields, e.g., dynamic mechanics, weather diagnostics and so on. The key goal of time series prediction is to mine the inner regular patterns in time data in order to predict future data effectively[1]. Many traditional methods such as AR, ARMA ARIMA and so on are well applied to solving stationary time series prediction. However, in practical applications, time series is almost nonstationary, which restricts the stationary methods above. From Takens' phase space delay reconstructing theory[2], this kind of data generally needs to reconstruct the phase space via delay coordinate at first. Then support vector machines(SVMs)[3], neural networks(NNs)[4], and other machine learning methods[5] are successfully introduced to approximate the spatial correlation in nonstationary time series data.

* Corresponding author.

J. Cao et al. (eds.), *Proceedings of ELM-2014 Volume 1*,
Proceedings in Adaptation, Learning and Optimization 3, DOI: 10.1007/978-3-319-14063-6_23

Generally speaking, there are two main challenges for predicting nonstationary time series effectively. One is how to choose a proper baseline algorithm which should be computationally inexpensive and enough accurate. Another is how to distinguish the importance of different samples in time series. Different with SVMs and NNs, extreme learning machine(ELM), introduced by Huang[6], have shown its very high learning speed and good generalization performance in solving many problems of regression estimate and pattern recognition[7]. As a sequential modification of ELM, online sequential ELM(OS-ELM) proposed by Liang[8] can learn data one-by-one or chunk-by-chunk. In many applications such as time-series forecasting, OS-ELMs also show good generalization at extremely fast learning speed. Therefore, OS-ELM is a proper solution for the first challenge. Many researches were devoted to solve the second challenge. As recent data usually carry more important information than the distant past data, a typical and effective method is weight-setting. Lin et al.[9] held the first sample in time series with lowest importance while the most recent sample with highest importance, and then assigned fuzzy memberships to every sample. Tay et al.[10] used exponential function to calculate every sample's importance in financial time series prediction. Very different with these stationary weight-setting strategy, Mao et al.[11] established a heurist algorithm to dynamically choose the optimal weights.

However, although ELM-based methods mentioned above work well in time series prediction, it still not yet successfully solve the second challenge, i.e., distinguish different samples' significance. Specifically speaking, as a sample is added sequentially, it seems not clear whether this sample is most importance, even it is newest. Therefore, to solve this problem, this paper firstly develops a new leave-one-out(LOO) cross-validation error estimation for OS-ELM aiming at time series prediction. Based on inversion of block matrix, this LOO estimation is enough fast for time series data. To our best knowledge, this LOO error estimation is the first attempt to evaluate the generalization performance of OS-ELM on time series data. Moreover, this paper utilizes this LOO error estimation of each new added sample to measure its importance. Obeying this weigh-setting strategy, this paper then proposes a new weighted learning method for OS-ELM. Experimental results on chaotic and real-life time series data demonstrate the proposed method outperforms the traditional ELMs in generalization performance and numerical stable.

2 Brief Review

Given a set of i.i.d. training samples $\{(\mathbf{x}_1, \mathbf{y}_1), \cdots, (\mathbf{x}_N, \mathbf{y}_N)\} \subset \mathbf{R}^d \times \mathbf{R}^n$, standard SLFNs with $\tilde{N}$ hidden nodes are mathematically formulated as[6]:

$$\sum_{i=1}^{\tilde{N}} \beta_i g_i(\mathbf{x}_j) = \sum_{i=1}^{\tilde{N}} \beta_i g_i(\mathbf{w}_i \cdot \mathbf{x}_j + b_i) = \mathbf{o}_j, \ \ j = 1, ..., N \tag{1}$$

where $g(x)$ is activation function, $\mathbf{w}_i = [w_{i1}, w_{i2}, ..., w_{id}]^T$ is input weight vector connecting input nodes and the ith hidden node, $\boldsymbol{\beta}_i = [\beta_{i1}, \beta_{i2}, ..., \beta_{in}]^T$ is the output weight vector connecting output nodes and the ith hidden node, b_i is bias of the ith hidden node. Huang[6] has rigorously proved that then for N arbitrary distinct samples and any $(\mathbf{w}_i, b_i)$ randomly chosen from $\mathbf{R}^d \times \mathbf{R}$ according to any continuous probability distribution, the hidden layer output matrix $\mathbf{H}$ of a standard SLFN with N hidden nodes and is invertible and $\|\mathbf{H}\boldsymbol{\beta} - \mathbf{T}\| = 0$ with probability one if the activation function $g : \mathbf{R} \mapsto \mathbf{R}$ is infinitely differentiable in any interval. Then given $(\mathbf{w}_i, b_i)$, training a SLFN equals finding a least-squares solution of the following equation:

$$\mathbf{H}\boldsymbol{\beta} = \mathbf{T} \tag{2}$$

where

$$\mathbf{H}(\mathbf{w}_1, ..., \mathbf{w}_{\tilde{N}}, b_1, ..., b_{\tilde{N}}, \mathbf{x}_1, ..., \mathbf{x}_{\tilde{N}}) = \begin{bmatrix} g(\mathbf{w}_1 \cdot \mathbf{x}_1 + b_1) & \cdots & g(\mathbf{w}_{\tilde{N}} \cdot \mathbf{x}_1 + b_{\tilde{N}}) \\ \vdots & \cdots & \vdots \\ g(\mathbf{w}_1 \cdot \mathbf{x}_N + b_1) & \cdots & g(\mathbf{w}_{\tilde{N}} \cdot \mathbf{x}_N + b_{\tilde{N}}) \end{bmatrix}_{N \times \tilde{N}}$$

$$\boldsymbol{\beta} = [\beta_1, ..., \beta_{\tilde{N}}]^T$$

$$\mathbf{T} = [\mathbf{y}_1, ..., \mathbf{y}_N]^T$$

Considering most cases that $\tilde{N} \ll N$, $\boldsymbol{\beta}$ cannot be computed through the direct matrix inversion. Therefore, the smallest norm least-squares solution of equation (3) is calculated as:

$$\hat{\boldsymbol{\beta}} = \mathbf{H}^{\dagger}\mathbf{T} \tag{3}$$

where $\mathbf{H}^{\dagger}$ is the Moore-Penrose generalized inverse of matrix $\mathbf{H}$. Based the above analysis, Huang[6] proposed ELM whose framework can be stated as follows:

Step 1. Randomly generate input weight and bias $(\mathbf{w}_i, b_i)$, $i = 1, \cdots, \tilde{N}$.
Step 2. Compute the hidden layer output matrix $\mathbf{H}$.
Step 3. Compute the output weight $\hat{\boldsymbol{\beta}} = \mathbf{H}^{\dagger}\mathbf{T}$.

Therefore, the output of SLFN can be calculated by $(\mathbf{w}_i, b_i)$ and $\hat{\boldsymbol{\beta}}$:

$$f(\mathbf{x}_j) = \sum_{i=1}^{\tilde{N}} \hat{\beta}_i g_i(\mathbf{w}_i \cdot \mathbf{x}_j + b_i) = \hat{\boldsymbol{\beta}} \cdot h(\mathbf{x}_j) \tag{4}$$

Like ELM, all the hidden node parameters in OS-ELM are randomly generated, and the output weights are analytically determined based on the sequentially arrived data. OS-ELM process is divided into two steps: initialization phase and sequential learning phase[8].

Step 1. Initialization phase: choose a small chunk $M_0 = \{(x_i, t_i), i = 1, 2, ..., N_0\}$ of initial training data, where $N_0 \geq \tilde{N}$.

1) Randomly generate the input weight $\mathbf{w}_i$ and bias b_i,$i = 1, 2, ..., \tilde{N}$. Calculate the initial hidden layer output matrix $\mathbf{H}_0$:

$$\mathbf{H}_0 = \begin{bmatrix} \mathbf{h}(\mathbf{x}_1) \\ \mathbf{h}(\mathbf{x}_2) \\ \cdot \\ \cdot \\ \cdot \\ \mathbf{h}(\mathbf{x}_{N_0}) \end{bmatrix} = \begin{bmatrix} g(\mathbf{w}_1 \cdot \mathbf{x}_1 + b_1) & \dots & g(\mathbf{w}_{\tilde{N}} \cdot \mathbf{x}_1 + b_{\tilde{N}}) \\ g(\mathbf{w}_1 \cdot \mathbf{x}_2 + b_1) & \dots & g(\mathbf{w}_{\tilde{N}} \cdot \mathbf{x}_2 + b_{\tilde{N}}) \\ & \dots & \\ & \dots & \\ & \dots & \\ g(\mathbf{w}_1 \cdot \mathbf{x}_{N_0} + b_1) & \dots & g(\mathbf{w}_{\tilde{N}} \cdot \mathbf{x}_{N_0} + b_{\tilde{N}}) \end{bmatrix}_{N_0 \times \tilde{N}} \tag{5}$$

2) Calculate the output weight vector:

$$\beta^0 = \mathbf{D}_0 \mathbf{H}_0{}^T \mathbf{T}_0 \tag{6}$$

where $\mathbf{D}_0 = (\mathbf{H}_0{}^T \mathbf{H}_0)^{-1}$, $\mathbf{T}_0 = [t_1, t_2, ... t_{N_0}]^T$.

3) Set $k = 0$

Step 2. Sequential learning phase

1) Learn the $(k+1)-$th training data: $d_{k+1} = (\mathbf{x}_{N_0+k+1}, t_{N_0+k+1})$

2) Calculate the partial hidden layer output matrix:

$$\mathbf{H}_{k+1} = [g(\mathbf{w}_1 \cdot \mathbf{x}_{N_0+k+1} + b_1) \quad \dots \quad g(\mathbf{w}_L \cdot \mathbf{x}_{N_0+k+1} + b_L)]_{1 \times L} \tag{7}$$

Set $\mathbf{T}_{k+1} = [t_{N_0+k+1}]^T$.

3) Calculate the output weight vector

$$\mathbf{D}_{k+1} = \mathbf{D}_k - \mathbf{D}_k \mathbf{H}_{k+1}^T (\mathbf{I} + \mathbf{H}_{k+1} \mathbf{D}_k \mathbf{H}_{k+1}{}^T)^{-1} \mathbf{H}_{k+1} \mathbf{D}_k \tag{8}$$

$$\beta^{k+1} = \beta^k + \mathbf{D}_{k+1} \mathbf{H}_{k+1}{}^T (\mathbf{T}_{k+1} - \mathbf{H}_{k+1} \beta^k) \tag{9}$$

4) Set $k = k + 1$. Go to step 2(1).

3 OS-ELM with LOO Weighted Strategy

3.1 LOO Error Estimation of ELM

The fast LOO error estimation of ELM proposed by Liu[12], derived that the generalization error in $i - th$ LOO iteration can be expressed as:

$$r_i = t_i - f_i(\mathbf{x}_i) = \frac{t_i - \mathbf{H}\mathbf{x}_i \mathbf{H}^+ \mathbf{T}}{1 - (\mathbf{H}\mathbf{x}_i \mathbf{H}^+)_i} \tag{10}$$

By the simulation experiment of artificial and real data sets, it has been verified that the LOO cross-validation algorithm based on ELM is efficient and has good generalization performance.

Obviously, equation (10) works mainly on offline learning setting rather than online sequential scenario. To avoid complex calculation and make the established model simple, we follow the idea of block matrix inversion[13], which transforms the complex calculation into linear operation, for decreasing the computation greatly in online learning stage.

3.2 Initial Stage of Training

Supposed there are N training samples $(\mathbf{x}_{i+1}, t_{i+1}), ...(\mathbf{x}_{i+N}, t_{i+N})$. The hidden layer output matrix is $\mathbf{H}_i = [\mathbf{h}_{i+1}{}^T ... \mathbf{h}_{i+N}{}^T]^T$, and the output vector is $\mathbf{T}_i = [t_{i+1} ... t_{i+N}]^T$, calculating the output weight vector

$$\boldsymbol{\beta}_i = \mathbf{H}_i{}^+\mathbf{T}_i \tag{11}$$

where

$$\mathbf{H}_i{}^+ = \mathbf{H}_i{}^T(\mathbf{H}_i\mathbf{H}_i{}^T)^{-1} \tag{12}$$

Let $\mathbf{A}_i = \mathbf{H}_i\mathbf{H}_i{}^T$, equation (12) can be rewritten as $\mathbf{H}_i{}^+ = \mathbf{H}_i{}^T\mathbf{A}_i{}^{-1}$.

3.3 Add New Sample

Add the new arrived sample $(\mathbf{x}_{i+N+1}, t_{i+N+1})$ into training set. The output vector becomes $\mathbf{T}_{i+1} = [t_{i+1} ... t_{i+N}\ t_{i+N+1}]^T = [\mathbf{T}_i{}^T\ t_{i+N+1}]^T$, and the hidden layer matrix becomes $\mathbf{H}_{i+1} = [\mathbf{H}_i{}^T\ \mathbf{h}_{i+N+1}{}^T]^T$. Then we have

$$\mathbf{H}_{i+1}{}^+ = \mathbf{H}_{i+1}{}^T(\mathbf{H}_{i+1}\mathbf{H}_{i+1}{}^T)^{-1} \tag{13}$$

Let $\mathbf{A}_{i+1} = \mathbf{H}_{i+1}\mathbf{H}_{i+1}{}^T$, then

$$\mathbf{H}_{i+1}{}^+ = \mathbf{H}_{i+1}{}^T\mathbf{A}_{i+1}{}^{-1} \tag{14}$$

Because

$$\begin{aligned}\mathbf{A}_{i+1} &= \mathbf{H}_{i+1}\mathbf{H}_{i+1}{}^T = [\mathbf{H}_i{}^T\ \mathbf{h}_{i+N+1}{}^T]^T[\mathbf{H}_i{}^T\ \mathbf{h}_{i+N+1}{}^T] \\ &= \begin{bmatrix}\mathbf{H}_i\mathbf{H}_i{}^T & \mathbf{H}_i\mathbf{h}_{i+N+1}{}^T \\ \mathbf{h}_{i+N+1}\mathbf{H}_i{}^T & \mathbf{h}_{i+N+1}\mathbf{h}_{i+N+1}{}^T\end{bmatrix} \\ &= \begin{bmatrix}\mathbf{A}_i & \mathbf{H}_i\mathbf{h}_{i+N+1}{}^T \\ \mathbf{h}_{i+N+1}\mathbf{H}_i{}^T & \mathbf{h}_{i+N+1}\mathbf{h}_{i+N+1}{}^T\end{bmatrix}\end{aligned} \tag{15}$$

For equation (15), according to Block matrix inversion, we have:

$$\mathbf{A}_{i+1}{}^{-1} = \begin{bmatrix}\mathbf{A}_i{}^{-1} \\ \mathbf{0}^T\end{bmatrix} + \frac{1}{B}\begin{bmatrix}\mathbf{A}_i{}^{-1}\mathbf{C} \\ -1\end{bmatrix}\begin{bmatrix}\mathbf{C}^T\mathbf{A}_i{}^{-1} & -1\end{bmatrix} \tag{16}$$

where: $B = \mathbf{h}_{i+N+1}\mathbf{h}_{i+N+1}{}^T - \mathbf{C}^T\mathbf{A}_i{}^{-1}\mathbf{C}$, $\mathbf{C} = \mathbf{H}_i\mathbf{h}_{i+N+1}{}^T$. So, $\mathbf{A}_{i+1}{}^{-1}$ can be calculated based on $\mathbf{A}_i{}^{-1}$, which reduces computational cost largely. Then we have $\mathbf{H}_{i+1}{}^+$ by substituting equation (16) into (14).

3.4 Calculate the LOO Error

Let $\mathbf{H}_{i+1}{}^+$ to set up the online LOO model, then the LOO error in $i-th$ LOO iteration can be expressed as:

$$r_j = t_j - f_j(\mathbf{x}_j) = \frac{t_j - \mathbf{Hx}_j\mathbf{H}_{i+1}{}^+\mathbf{T}}{1 - (\mathbf{Hx}_j\mathbf{H}_{i+1}{}^+)_j}, j = i+1, i+2, ..., i+N+1 \tag{17}$$

Then we can obtain the corresponding LOO error, r_j, of each sample from equation (17). According to the value of r_j, where $j = i+1, i+2, ..., i+N+1$, we set the relevant weight w_j of each sample, where $0.98 \leq w_j \leq 1$. Note that r_j is smaller, w_j is bigger. To emphasize the newest sample and make the decision model simple, we re-set the weight w_{i+N+1} of newest sample $(\mathbf{x}_{i+N+1}, t_{i+N+1})$ as w'_{i+N+1}. This article defines w'_{i+N+1} as 1.02. And we set the weight w_{i+1} of oldest sample $(\mathbf{x}_{i+1}, t_{i+1})$ as zero, namely we set its contribution to the model is zero.

3.5 Weighted Training

After adding the new sample $(\mathbf{x}_{i+N+1}, t_{i+N+1})$, we set the weight w_{i+1} of oldest sample $(\mathbf{x}_{i+1}, t_{i+1})$ is zero, namely excluding this sample. After excluding $(\mathbf{x}_{i+1}, t_{i+1})$, the output vector becomes $\mathbf{T}_{i+2} = [t_{i+2} \ldots t_{i+N+1}]^T$, and the hidden matrix becomes $\mathbf{H}_{i+2} = [{\mathbf{h}_{i+2}}^T \ldots {\mathbf{h}_{i+N+1}}^T]^T$. Then we have:

$$ {\mathbf{H}_{i+2}}^{+} = {\mathbf{H}_{i+2}}^T (\mathbf{H}_{i+2}{\mathbf{H}_{i+2}}^T)^{-1} \tag{18} $$

Let $\mathbf{A}_{i+2} = \mathbf{H}_{i+2}{\mathbf{H}_{i+2}}^T$, then

$$ {\mathbf{H}_{i+2}}^{+} = {\mathbf{H}_{i+2}}^T {\mathbf{A}_{i+2}}^{-1} \tag{19} $$

From equation (19), ${\mathbf{H}_{i+2}}^{+}$ contains two parts: ${\mathbf{H}_{i+2}}^T$ and ${\mathbf{A}_{i+2}}^{-1}$. Because the calculation of ${\mathbf{A}_{i+2}}^{-1}$ involves matrix inversion, we only set weight on ${\mathbf{H}_{i+2}}^T$ in order to avoid the huge computational cost in calculating LOO error. Then the hidden matrix

$$ \mathbf{H}_{i+2} = [\mathbf{h}_{i+2}; \mathbf{h}_{i+3}; \ldots; \mathbf{h}_{i+N}; \mathbf{h}_{i+N+1}] $$

becomes

$$ \mathbf{H}_{i+2} = \begin{bmatrix} \mathbf{h}_{i+2} \\ \mathbf{h}_{i+3} \\ \ldots \\ \mathbf{h}_{i+N} \\ \mathbf{h}_{i+N+1} \end{bmatrix} \bullet \begin{bmatrix} w_{i+2} \\ w_{i+3} \\ \ldots \\ w_{i+N} \\ w_{i+N+1} \end{bmatrix} = \begin{bmatrix} w_{i+2}\mathbf{h}_{i+2} \\ w_{i+3}\mathbf{h}_{i+3} \\ \ldots \\ w_{i+N}\mathbf{h}_{i+N} \\ w_{i+N+1}\mathbf{h}_{i+N+1} \end{bmatrix} $$

And ${\mathbf{H}_{i+2}}^T$ becomes

$$ {\mathbf{H}_{i+2}}^T = [w_{i+2}{\mathbf{h}_{i+2}}^T, w_{i+3}{\mathbf{h}_{i+3}}^T, \ldots, w_{i+N}{\mathbf{h}_{i+N}}^T, w_{i+N+1}{\mathbf{h}_{i+N+1}}^T] \tag{20} $$

where $w_{i+2}, w_{i+3}, \ldots, w_{i+N}, w_{i+N+1}$ are the corresponding weights, $0.98 \leq w_j \leq 1$, $j = i+2, i+3, \ldots, i+N$, $w_{i+N+1} = w'_{i+N+1} = 1.02$. Because

$$ \begin{aligned} \mathbf{A}_{i+1} = \mathbf{H}_{i+1}{\mathbf{H}_{i+1}}^T &= [{\mathbf{h}_i}^T\ {\mathbf{H}_{i+2}}^T]^T [{\mathbf{h}_i}^T\ {\mathbf{H}_{i+2}}^T] \\ &= \begin{bmatrix} \mathbf{h}_i{\mathbf{h}_i}^T & \mathbf{h}_i{\mathbf{H}_{i+2}}^T \\ \mathbf{H}_{i+2}{\mathbf{h}_i}^T & \mathbf{H}_{i+2}{\mathbf{H}_{i+2}}^T \end{bmatrix} \\ &= \begin{bmatrix} \mathbf{h}_i{\mathbf{h}_i}^T & \mathbf{h}_i{\mathbf{H}_{i+2}}^T \\ \mathbf{H}_{i+2}{\mathbf{h}_i}^T & \mathbf{A}_{i+2} \end{bmatrix} \end{aligned} \tag{21} $$

From equation (21), there is a relationship between $\mathbf{A}_{i+2}$ and $\mathbf{A}_{i+1}$. So ${\mathbf{A}_{i+2}}^{-1}$ can be calculated on the base of ${\mathbf{A}_{i+1}}^{-1}$ to simplify the calculation. We follow the process of block matrix inversion[13]. Assume that ${\mathbf{A}_{i+1}}^{-1}$ can be partitioned and expressed as:

$$ {\mathbf{A}_{i+1}}^{-1} = \begin{bmatrix} a & \mathbf{F}^T \\ \mathbf{F} & \mathbf{G} \end{bmatrix} \tag{22} $$

where $a \in \mathbf{R}, \mathbf{F} \in \mathbf{R}^N, \mathbf{G} \in \mathbf{R}^{N\times N}$.

As in equation (21), let $g = \mathbf{h}_i{\mathbf{h}_i}^T$, $\mathbf{P} = \mathbf{h}_i{\mathbf{H}_{i+2}}^T$, then equation (21) is equivalent to

$$ \mathbf{A}_{i+1} = \begin{bmatrix} g & \mathbf{P} \\ \mathbf{P}^T & \mathbf{A}_{i+2} \end{bmatrix} \tag{23} $$

By the definition of matrix inversion: $\mathbf{A}_{i+1}{\mathbf{A}_{i+1}}^{-1} = \mathbf{E}_{(N+1)\times(N+1)}$, namely

$$ \begin{bmatrix} g & \mathbf{P} \\ \mathbf{P}^T & \mathbf{A}_{i+2} \end{bmatrix} \begin{bmatrix} a & \mathbf{F}^T \\ \mathbf{F} & \mathbf{G} \end{bmatrix} = \begin{bmatrix} 0 & \mathbf{0} \\ \mathbf{0} & \mathbf{E}_{N\times N} \end{bmatrix} \tag{24} $$

Through the block matrix multiplication, we have

$$ \begin{cases} \mathbf{P}^T a + \mathbf{A}_{i+2}\mathbf{F} = 0 \\ \mathbf{P}^T\mathbf{F}^T + \mathbf{A}_{i+2}\mathbf{G} = \mathbf{E}_{N\times N} \end{cases} \tag{25} $$

Calculating equations (25), we have

$$ {\mathbf{A}_{i+2}}^{-1} = \mathbf{G} - \mathbf{F}\mathbf{F}^T/a \tag{26} $$

Thus, we have ${\mathbf{H}_{i+2}}^{+}$ by substituting equation (22) and equation (26) into equation (19).

Then we can update the network weights according to equation (27):

$$ \beta = {\mathbf{H}_{i+2}}^{+}\mathbf{T}_{i+2} \tag{27} $$

4 Experimental Results

In this section, we examine one typical nonstationary real-world data set, i.e., air pollutants forecasting in Macau[22]. The goal is to test the generalization performance on time series data as well as the running speed. In order to further test the stability and generalization of LW-OSELM, we choose suspended particulate matters (PM_{10}) to conduct experiment. Due to the limitation of the acquisition data, this paper adopts the air quality data of Macao meteorological bureau to conduct the simulation experiments[14]. Due to the limitation of paper's space, the data pre-processing procedure is omitted here.

For comparison, we choose two baselines: the classical ELM[6] and OS-ELM[8]. The proposed OS-ELM algorithm with fast LOO weighted strategy is named LW-OSELM. We also check the fixed-weighting OS-ELM (namely WELM). The WELM will set weight value as 0.98 for all old samples, and 1.02 for the latest sample. We

Table 1. Comparative results of LW-OSELM and WELM

	LW-OSELM	WELM
Training Time(s)	2.9703	0.4056
Test Time(s)	0.0094	0.001
Training Error	0.1767	0.4276
Test error	0.1794	0.431

Table 2. Comparative results of three models

	ELM	OSELM	LW-OSELM
Training Time(s)	0.0125	0.0998	3.3961
Test Time(s)	0.0031	0.0094	0.0012
Training Error	0.1706	0.1721	0.1705
Test error	0.1811	0.1808	0.1739

set hidden neurons as 25. The comparison of LW-OSELM and WELM is illustrated as Table 1. In each experiment, all results are the mean of 100 trials. RBF activation function is used in each algorithm. Each variable is linearly rescaled. A method with higher classification accuracy is better.

From Table 1, compared with WELM, the training error and test error of LW-OSELM are much smaller, which demonstrates that the dynamic weight-setting strategy, i.e., setting weights on samples according to their online LOO errors, is feasible.

We compare the performance of ELM, OSELM and LW-OSELM. Given the hidden layer activation function of RBF kernel function, the mean RMES of 100 trials on Macao meteorological time series are listed in Table 2. Here the numbers of neurons are 15.

By contrast, LW-OSELM has little smaller training error and test error while its training time is a little longer. The reason is quite likely that we didn't employ embedding dimension, i.e., reconstructing phase space like in equation (1). Here we merely use the data in the latest day as input sample, rather than using the data of past few days.

We also examine the effect of hidden neurons. Fig.1 shows the change of training error and test error with different number of hidden neurons. The training error and test error of LW-OSELM are both smaller than the others with most hidden neurons. ELM tends to be most unstable with drastic fluctuation. And LW-OSELM performs similarly stable with OSELM, which keeps pace with the results in Table 2.

Moreover, we report the generalization performance of three algorithms with different prediction step, as in Fig.2. Obviously, LW-OSELM is better than the others. For further clarification, we also compare the mean and variance of three models. The mean of ELM, OSELM and LW-OSELM is respectively 0.0726, 0.0617, 0.0605, and the variance is respectively 3.8340e-004, 3.5584e-004, 3.0809e-004. Obviously, the average error and variance of LW-OSELM is both the least, which indicates that LW-OSELM has better accuracy and stronger stability.

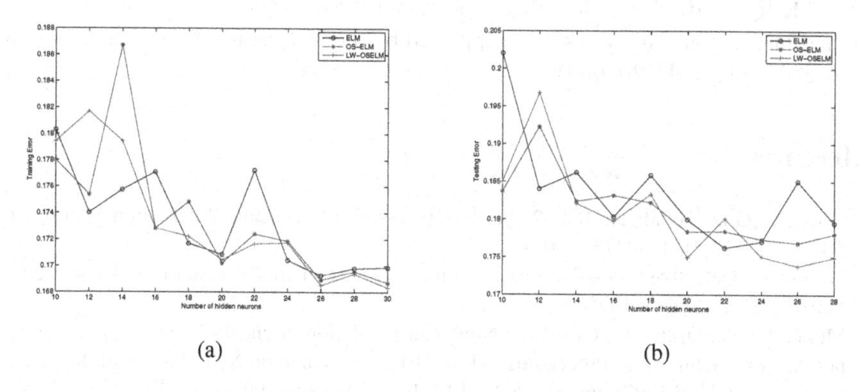

Fig. 1. Performance of three models with different number of hidden neurons in terms of (a) Training error and (b) Test error

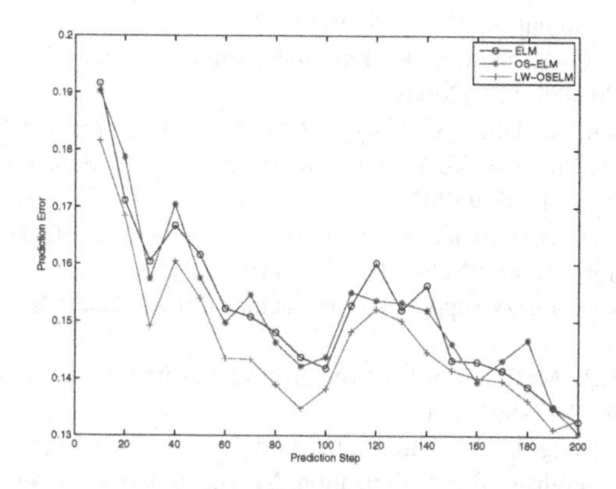

Fig. 2. Accuracy of three algorithms with different prediction step

5 Conclusion and Future Work

In this paper, nonstationary time series prediction is addressed. The key idea is distinguishing the importance of samples in time series using its LOO cross-validation error. This idea is a new attempt for weight-setting strategy. To realize this strategy, this paper utilizes OS-ELM as baseline algorithm, and proposes a new LOO error estimation which is fast and quite applied to time series prediction. Based on this estimation, this paper proposes a dynamic weight-setting algorithm for OS-ELM. The experimental results on two benchmark chaotic time series data sets and a real-world data set demonstrate the effectiveness of the proposed approach.

Acknowledgement. We wish to thank the author C.M.Vong of [14] for useful discussion and instruction. This work was supported by the National Natural Science Foundation of China(No. U1204609).

References

1. Gooijer, J.G.D., Hyndman, R.J.: 25 years of time series forecasting. International Journal of Forecasting 22(3), 443–473 (2005)
2. Takens, F.: Detecting strange attractors in turbulence. Dynamical systems and Turbulence, Warwick 1980, 366–381 (1981)
3. Mukherjee, S., Osuna, E., Girosi, F.: Nonlinear prediction of chaotic time series using support vector machines. In: Proceedings of the IEEE Workshop on Neural Network for Signal Processing (NNSP 1997), pp. 511–520. IEEE Press, Amelia (1997)
4. Du, D., Li, X., Fei, M., Irwin, G.W.: A novel locally regularized automatic construction method for RBF neural models. Neurocomputing 98, 4–11 (2012)
5. Estabrooks, A., Jo, T., Japkowicz, N.: A multiple resampling method for learning from imbalanced datasets. Comput.Intell. 20, 18–36 (2004)
6. Huang, G.-B., Zhu, Q.-Y., Siew, C.-K.: Extreme learning machine: Theory and applications. Neurocomputing 70, 489–501 (2006)
7. Huang, G.-B., Zhou, H., Ding, X., Zhang, R.: Extreme Learning Machine for Regression and Multiclass Classification. IEEE Transactions on Systems, Man, and Cybernetics - Part B: Cybernetics 42(2), 513–529 (2012)
8. Liang, N.Y., Huang, G.B.: Fast accurate online sequential learning algorithm for feedforword networks. IEEE Trans. Neural Networks 17, 1411–1423 (2006)
9. Lin, C.F., Wang, S.D.: Fuzzy support vector machines. IEEE Trans. Neural Networks 13(2), 464–471 (2002)
10. Tay, F.E.H., Cao, L.J.: Modified support vector machines in financial time series forecasting. Neurocomputing 48, 847–861 (2002)
11. Mao, W., Yan, G., Dong, L.: Weighted solution path algorithm of support vector regression based on heuristic weight-setting optimization. Neurocomputing 73, 495–505 (2009)
12. Liu, X., Li, P., Gao, C.: Fast leave-one-out cross-validation algorithm for extreme learning machine. Journal of Shanghai Jiaotong University 45(8), 6–11 (2011)
13. Zhang, X., Wang, H.: Local extreme learning machine and its application to condition on-line monitoring. Journal of Shanghai Jiaotong University 45(2), 236–240 (2011)
14. Vong, C.M., IP, W.F., Wong, P.K., Chiu, C.C.: Predicting minority class for suspended particulate matters level by extreme learning machine. Neurocomputing 128, 136–144 (2014)

RMSE-ELM: Recursive Model Based Selective Ensemble of Extreme Learning Machines for Robustness Improvement

Bo Han[1], Bo He[1,*], Mengmeng Ma[1], Tingting Sun[1], Tianhong Yan[2,*], and Amaury Lendasse[3,4]

[1] School of Information and Engineering, Ocean University of China, Shandong, Qingdao, China 266000

[2] School of mechanical and Electrical Engineering, China Jiliang University, Zhejiang, Hangzhou, China 310018

[3] Department of Mechanical and Industrial Engineering and the Iowa Informatics Initiative, The University of Iowa, Iowa City, IA 52242-1527, USA

[4] Arcada University of Applied Sciences, 00550 Helsinki, Finland

bhe@ouc.edu.cn, thyan@163.com

Abstract . For blended data, the robustness of extreme learning machine (ELM) is so weak because the coefficients (weights and biases) of hidden nodes are set randomly and the noisy data exert a negative effect. To solve this problem, a new framework called "RMSE-ELM" is proposed in this paper. It is a two-layer recursive model. In the first layer, the framework trains lots of ELMs in different ensemble groups concurrently, then employs selective ensemble approach to pick out an optimal set of ELMs in each group, which can be merged into a large group of ELMs called candidate pool. In the second layer, selective ensemble approach is recursively used on candidate pool to acquire the final ensemble. In the experiments, we apply UCI blended datasets to confirm the robustness of our new approach in two key aspects (Mean Square Error and Standard Deviation). The space complexity of our method is increased to some degree, but the results have shown that RMSE-ELM significantly improves robustness with a rapid learning speed compared to representative methods (ELM, OP-ELM, GASEN-ELM, GASEN-BP and E-GASEN). It becomes a potential framework to solve robustness issue of ELM for high-dimensional blended data in the future.

Keywords: Extreme Learning Machine, Recursive Model, Selective Ensemble, RMSE-ELM, Robustness Improvement.

1 Introduction

In recent two or three decades, neural networks are increasingly popular in machine learning community. Especially for recent five years, lots of researchers mainly have paid their attention on deep structures such as Deep Boltzmann Machine [1],

* Corresponding author.

J. Cao et al. (eds.), *Proceedings of ELM-2014 Volume 1*,
Proceedings in Adaptation, Learning and Optimization 3, DOI: 10.1007/978-3-319-14063-6_24

Convolution Neural Network [2] and so on. However, the deep networks are hardly applied into real-time area in big data era because of two reasons: First of all, there is no free lunch in any algorithms. Though the training accuracy of deep network is pretty high, the training time is so long that we can hardly bear the computational cost [3]. Secondly, the deep structures tend to fall into the pit called "over-fitting", which means that it has a bad generalization. What's more, the tuning of parameters in deep networks is very time-consuming [4]. So the shallow structure is naturally our intuition for big data analysis and real-time application.

Recently, the Extreme Learning Machine (ELM) [5] as an emerging branch of shallow networks was proposed by Guang-Bin Huang et. al. It was evolved from single hidden layer feed-forward networks (SFLNs). It has shown the excellent generalization performance and fast learning speed compared to Deep Belief Networks [6] or Deep Boltzmann Machines [7]. In essence, the algorithm of ELM has two main steps: In the first step, the input weights and biases can be assigned randomly, which will definitely reduce computational cost because they do not need to be tuned manually. In the second step, the output weights of ELM can be computed easily by the generalized inverse of hidden layer output matrix and target matrix [8]. In terms of the computational performance of ELM, it tends to reach not only the smallest training error but also the smallest norm of output weights with rapid speed. Based on above merits of ELM, a lot of researchers in machine learning community now increasingly customize their own frameworks based on ELM for specific issues. For equalization problems, ELM based complex-valued neural networks are a powerful tool. For regression or multi-label issues, the kernel based ELM proposed by Huang et. al is effective [9,10]. For generalization problem, Incremental ELM [11] outperforms many representative algorithms like SVM [12], stochastic BP [13] and so on. What's more, various extended ELMs also attract our attention. For example, online sequential ELM [14] is an efficient learning algorithm to handle both additive [15] and RBF [16,17] nodes in the unified framework. In complex dimensional space, the kernel implementation of ELM is superior to conventional SVM. From the above discussion, we can conclude that ELM is an excellent algorithm for different issues in machine learning area.

However, as the keynote given by Guang-Bin Huang indicates, the robustness analysis is still one of the open problem in ELM community [5,18]. Different researchers have different research styles to tackle with the same problem. Previously, Rong et.al presented pruning algorithm called P-ELM to improve the robustness of ELM [19]. And also Miche and Lendasse, proposed an algorithm called OP-ELM [20,21] to improve the robustness due to its variable selection mechanism, which removes the irrelevant variables from blended data efficiently [22,23]. However, for blended data (namely the raw data is blended with noisy data), they do not work very well because of two reasons. First, the mechanism of variables pruning is very time-consuming. What's more, the standard deviations of training error in above two models are relatively high, which means that these models are not the top choice for robustness improvement. If we want to improve the robustness of original ELM, we should initially clarify why the ELM is so weak for blended data. First of all, we believe ELM sets its initial weights and biases randomly, which largely reduce the computational time but cannot guarantee the suitable parameters of hidden nodes for good robustness. Second, the noisy data exert a negative effect on robustness of ELM. So for blended data, my initial intuition is that if we train a batch of different ELMs and then ensemble them averagely, we might improve the robustness because of Hansen and Salamon's theory [24]. It proved that the robustness performance of a single network can be improved by an ensemble of neural networks. Sollich and Krogh [25] confirmed it later. Thus, based on this theory, Sun et. al proposed the average weighted ELM ensemble [26], which has a better generalization than original ELM on raw data. But on blended data, the average weighted ELM ensemble does not work well because it is negatively affected by noisy data such as Gaussian noise or Uniform noise. Zhou et. al [27] proposed a new framework called GASEN, which can resist the negative effect from noisy data. In his theory, the ensemble of several optimal networks may be better than the ensemble of all networks. The GASEN is fully based on genetic algorithm and Back-Propagation (BP) neural networks. Therefore, in real-time area, we should not apply GASEN directly for robustness improvement because of high computation cost.

Inspired by above observations, for blended data [28], we hope to create a new computational framework, which not only improves the robustness largely but also keeps a rapid learning speed. So in this paper, a new approach called "RMSE-ELM" is proposed. Our tuition can be concluded into two aspects: First, selective ensemble approach is an effective tool to resist noisy data but the kernel of framework is usually the BP networks. What's more, the genetic algorithm itself is a little bit complicated. Therefore, the training process is so time-consuming [29]. So we hope to employ the advantage of ELM to speed up the selective ensemble approach. Second, in cognitive science, the information processing of human brain is constructed hierarchically, and it can extract different useful information layer by layer. However, the more layers we construct, the more parameters the algorithm will learn, which will definitely increase the computational cost. Therefore, we hope to construct a semi-shallow framework for a good compromise between robustness and computational cost. For technical details, it is a two-layer recursive model. In the first layer, we concurrently train lots of ELMs in different groups, then we employ selective ensemble approach to pick out several ELMs in each group, which can be transmitted into the second layer called candidates pool. In the second layer, we employ selective ensemble approach recursively to pick out several ELMs for the average ensemble. In the experiments, we apply UCI blended datasets [30] to confirm the robustness of new method, which is compared to that of several methods such as ELM, OP-ELM, GASEN-ELM, GASEN-BP and E-GASEN in two key aspects: Mean Square Error and Standard Deviation. Though the space complexity of our method is increased to some degree, the results have shown that the RMSE-ELM significantly improves the robustness with a rapid learning speed. We will further explore how many layers can achieve the optimal compromise between the robustness and computational cost in our framework. The extended RMSE-ELM has a great potential to be a trend framework to solve robustness issue of ELM for high-dimensional blended data in the future.

We organize the rest of the paper as follows. In Section 2, we discuss previous work on classical ELM and Selective Ensemble. In Section 3 we describe our new method called RMSE-ELM from structure to theory. In Section 4, for UCI blended datasets, several experimental results on ELM, OP-ELM, GASEN-ELM, GASEN-BP, E-GASEN are reported respectively. In Section 5, we present our discussions the motivation of benchmark selection and other facts revealed by experiments. Finally, in Section 6, conclusions are drawn and future work and direction are indicated.

2 Previous Works

2.1 Extreme Learning Machine

Extreme learning machine (ELM) has been developed to obtain a much faster learning speed and higher generalization performance both in the regression and classification problem. The essence of ELM is the hidden layers of SFLNs need not to be tuned iteratively [5,31], that is, the parameters of the hidden nodes which include input weights and biases can be randomly generated, and then it only needs to solve the output weights. The structure of ELM is shown below.

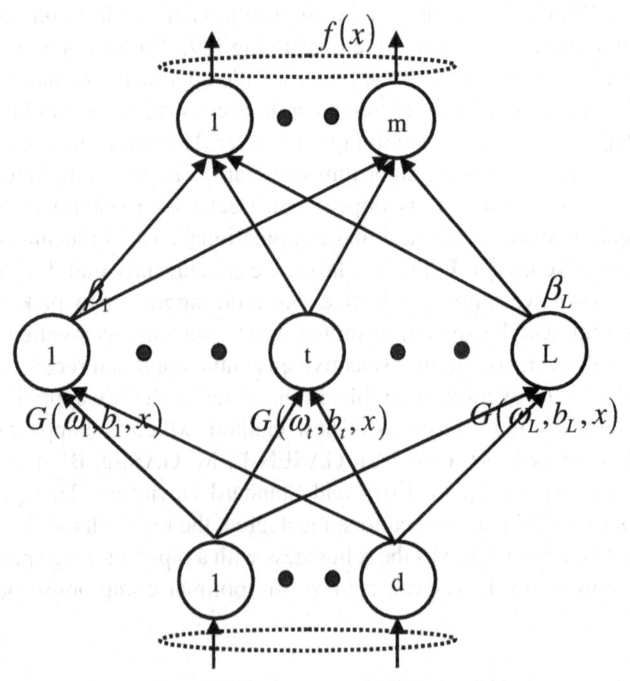

Fig. 1. The structure of ELM algorithm

For the given N learning samples $\{x_i, y_i\}_{i=1}^{N}$, where $x_i = [x_{i1}, \ldots, x_{id}]'$ and $y_i = [y_{i1}, \ldots, y_{im}]'$ the standard model of the ELM learning with L hidden neurons and activation function $G(\omega_t, b_t, x_i)$ can be written as

$$\sum_{t=1}^{L} \beta_t \, G(\omega_t, b_t, x_i) = o_i, \qquad i = 1, \cdots, N \tag{1}$$

Where $\omega_t = [\omega_{t1}, \ldots, \omega_{td}]'$ is the weight vector connecting the t_th hidden neuron and the input neurons. $\beta_t = [\beta_{t1}, \ldots, \beta_{tm}]'$ denotes the weight vector connecting the t_th hidden neuron and the output neurons. b_j is the bias of the t_th hidden neuron.

ELM can approximate these N samples with zero error means that

$$\sum_{i=1}^{N} \|o_i - y_i\| = 0 \tag{2}$$

Namely, there exist (ω_j, b_j) and β_j such that

$$\sum_{t=1}^{L} \beta_t \, G(\omega_t, b_t, x_i) = y_i, \quad i = 1, \cdots, N \tag{3}$$

The activation function $G(\omega_t, b_t, x_i)$can be arbitrarily chosen from the sigmoid function, the Hard-limit function, the Gaussian function, the Multi-quadric function and any other function which is infinitely differentiable in any interval so that the hidden layer parameters can be randomly generated. The above equation can also be written compactly as:

$$H\beta = Y \tag{4}$$

Where

$$H = \begin{bmatrix} G(\omega_1, b_1, x_1) \ldots G(\omega_L, b_L, x_1) \\ \vdots \\ G(\omega_1, b_1, x_N) \ldots G(\omega_L, b_L, x_N) \end{bmatrix}_{N \times L} \tag{5}$$

$$\beta = [\beta_1', \ldots, \beta_L']'_{L\times m} \tag{6}$$

$$Y = [y_1', \ldots, y_N']'_{N\times m} \tag{7}$$

Here H is called the hidden layer output matrix of the neural network. When the training set x_i is given and the parameters (ω_t, b_t) are randomly generated, matrix H can be obtained. And then the output weights β can be generated as:

$$\beta = H^{\dagger}Y \tag{8}$$

Where $H^{\dagger}$denotes the Moore-Penrose generalized inverse of matrix H [32,33]. In summary, the ELM algorithm can be presented as follows:

Algorithm 1 Extreme Learning Machine

Input: The N training set $\{x_i, y_i\}_{i=1}^{N}$, the activation function $G(\omega_t, b_t, x_i)$, and the number of hidden nodes L.

Steps:

1. Randomly generate input weights ω_t and biases b_t,$t = 1, \dots, L$
2. Calculate the hidden layer output matrix H.
3. Calculate the output weight vector $\beta = H^{\dagger}Y$.

2.2 Selective Ensemble

In recent years, ensemble learning has received lots of attention from machine learning community due to its potential to improve the generalization capability of a learning system [34,35]. With the increase of size, the prediction speed of an ensemble machine decreases significantly but its storage increases quickly. Z.H Zhou et. al[36] has proved that many could be better than all and proposed a new framework called selective ensemble. The aim of selective ensemble learning is to further improve the prediction accuracy of an ensemble machine, to enhance its prediction speed as well as to decrease its storage need. Selective ensemble learning mainly involves three steps [37]:

(1) Training a set of base learners individually generated from bootstrap samples of a fixed training data.

(2) Selecting right components from all the available learners and excluding the bad base learners to form an optimal ensemble. Genetic algorithm is used for components selection. The population of base learners is encoded as real chromosomes so that one bit represents the average weight of initial learner ensemble. Suppose x is randomly sampled through a distribution $p(x)$, and the expected output is y, and the output of the ith base ELM is $f_i(x)$. The optimum weight ω^* is expressed as empirical Eq(9) which minimizes the generalization error of the ensemble model.

$$\omega^* = \underset{\omega}{arg\,min}\left(\sum_{i=1}^{N}\sum_{j=1}^{N}\omega_i\,\omega_j C_{ij}\right) \tag{9}$$

C_{ij} is the correlation between the ith and the jth individual base learner. And the definition is as follows.

$$C_{ij} = \int dxp(x)\,(f_i(x) - y)(f_j(x) - y) \tag{10}$$

Therefore, the $kth\,(k = 1\ldots, N)$ of optimum weight ω^* can be solved by Lagrange multiplier, which satisfies Eq(11):

$$\omega_k^* = \frac{\sum_{j=1}^{N} C_{kj}^{-1}}{\sum_{i=1}^{N} \sum_{j=1}^{N} C_{ij}^{-1}} \tag{11}$$

Genetic algorithm based selective ensemble assigns a random weight to every base ELM first. Then, genetic algorithm is used to evolve those weights so that they can characterize the fitness of the ELM in joining the ensemble to some extent.

(3) Combining the selected base learner components to get the final predictions.

3 New Method

3.1 The Structure of RMSE-ELM

Inspired by above discussions, for blended data, we hope to create a new computational framework, which not only improves the robustness performance of ELM largely but also keeps a rapid learning speed. We naturally have two tuitions below.

First of all, Traditional selective ensemble approach like GASEN algorithm is definitely an effective tool to resist noisy data because it utilizes fewer but better individual models to ensemble, which achieves stronger generalization ability. But both genetic algorithm employed by GASEN and the training process of individual kernels (BPs) are so time-consuming, which can hardly be used in industry or real-time situation. So we hope to build our customized selective ensemble based on ELM kernels because of its rapid learning speed.

Secondly, from the point view of cognitive science, the information processing of human brain is constructed hierarchically, and it can extract different useful information layer by layer. However, if we completely construct our networks as our brain, for example a deep-layer network, we may encounter several training problems. Firstly, the training time is so long that we can rarely bear the computational cost, not to mention big data analysis. Secondly, the deep structures tend to fall into the pit called "over-fitting" which in turn means the weak generalization. Moreover, the tuning of parameters in deep networks needs large amount of time and personal experience. So the semi-shallow structure is naturally top choice for big data analysis and real-time application.

In this paper, we present a framework called "RMSE-ELM" to improve the robustness of ELM for blended data with acceptable computational cost. The figure of our framework shows in below.

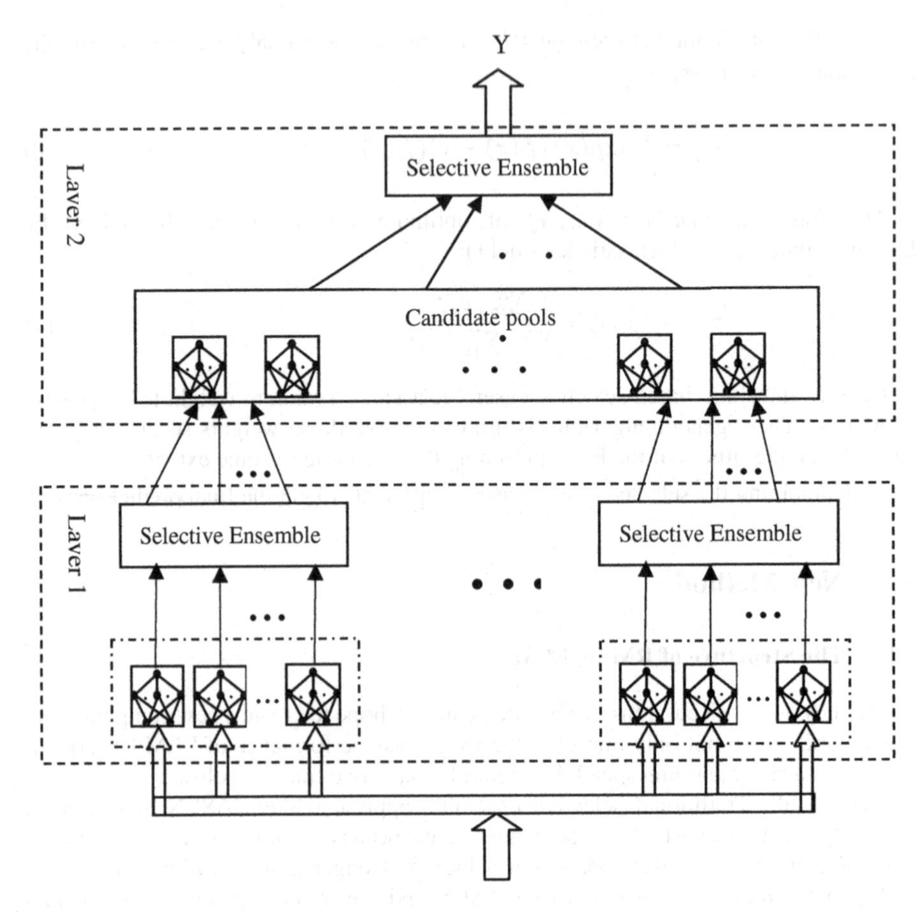

Fig. 2. The Framework of RMSE-ELM

Just as the Figure 2, it is a two-layer recursive model, which is a good compromise between shallow and deep network. In the first layer, we concurrently train lots of ELMs that belong to the different ensemble groups, then we employ selective ensemble approach to pick out several ELMs in each group, which can be transmitted into our second layer – the pool of better candidates. In the second layer, we employ selective ensemble recursively to pick from selected ELMs and then ensemble an optimal set of ELMs to acquire the final result.

Although our framework is relatively simple compared with deep structure networks, we believe that it locates in the right track to solve the robustness issues of ELM.

3.2 The Theory of RMSE-ELM

Now let's first analyze our framework in theory. From above discussion, we can clearly see our framework recursively employ selective ensemble approach. In essence, the recursive model algorithm based selective ensemble can be explained as the hierarchical model based selective ensemble. So if the selective ensemble can work well, theoretically, the recursive model based selective ensemble can work better.

So firstly we should analyze whether the selective ensemble of extreme learning machine are good enough. Please note currently the individual networks are ELMs instead of BP networks. To be honest, it is not an easy task excluding the bad ELMs from our target group. In order to generate the ensemble ELM with small size but stronger generation ability, genetic algorithm is used to select the ELM models with high fitness from a set of available ELMs. Suppose that the learning task is to approximate a function $f: R^m \rightarrow R^n$, it can be represented by an ensemble of N base ELM learners. The predictions of the base ELM learners are combined by weighted averaging, where a weight $\omega_i (i = 1 \ldots N)$ is assigned to the individual base ELM learner $f_i (i = 1 \ldots N)$, and ω_i satisfies Eq(12).

$$0 \leq \omega_i \leq 1, \quad \sum_{i=1}^{N} \omega_i = 1 \tag{12}$$

Then the output of ensemble is:

$$\bar{f}(x) = \sum_{i=1}^{N} \omega_i f_i(x) \tag{13}$$

Where f_i is the output of the ith base ELM learner.

We assume that each base ELM learner has only one output. Suppose $x \in R^m$ is randomly sampled through a distribution $p(x)$. And the target for x is $d(x)$. Then the error $E_i(x)$ of the ith base ELM learner and the error $E(x)$ of the ensemble on input x are respectively:

$$E_i(x) = \left(f_i(x) - d(x)\right)^2 \tag{14}$$

$$E(x) = \left(\bar{f}(x) - d(x)\right)^2 \tag{15}$$

Then the generalization error E_i of the ith base ELM learner and the generalization error E of the ensemble on the distribution $p(x)$ are respectively:

$$E_i = \int dx p(x) E_i(x) \tag{16}$$

$$E = \int dx p(x) E(x) \tag{17}$$

Define the correlation between the ith and the jth individual base ELM learner as:

$$C_{ij} = \int dxp(x)\left(f_i(x) - d(x)\right)\left(f_j(x) - d(x)\right) \tag{18}$$

Apparently, C_{ij} satisfies:

$$C_{ii} = E_i \; and \; C_{ij} = C_{ji} \tag{19}$$

According to Eq(13) and Eq(15):

$$E(x) = \left(\sum_{i=1}^{N} \omega_i f_i(x) - d(x)\right)\left(\sum_{j=1}^{N} \omega_j f_j(x) - d(x)\right) \tag{20}$$

Then according Eq(17), Eq(18) and Eq(20):

$$E = \sum_{i=1}^{N}\sum_{j=1}^{N} \omega_i \, \omega_j C_{ij} \tag{21}$$

When the base ELM learners are combined by the simple ensemble method, that is $\omega_i = \frac{1}{N}$ for every i, we have

$$E = \sum_{i=1}^{N}\sum_{j=1}^{N} C_{ij}/N^2 \tag{22}$$

Now, we assume that the kth base learner is omitted, the new generalization error $\widehat{E}$:

$$\hat{E} = \sum_{\substack{i=1 \\ i \neq k}}^{N}\sum_{\substack{j=1 \\ j \neq k}}^{N} C_{ij}/(N-1)^2 \tag{23}$$

According to Eq(16), the generalization error of the kth base ELM learner:

$$E_k = \int dxp(x)E_k(x) \tag{24}$$

Therefore,

$$E - \hat{E} = \frac{2\sum_{\substack{i=1 \\ i \neq k}}^{N} C_{ik} + E_k - (2N-1)E}{(N-1)^2} \tag{25}$$

So if

$$2\sum_{\substack{i=1\\i\neq k}}^{N} C_{ij} + E_k - (2N-1)E > 0 \tag{26}$$

Then,

$$E > \hat{E} \tag{27}$$

Which means new ensemble omitting the kth learner is now more robust than original ensemble.

So we can get a constraint condition from Eq(26) and Eq(27),

$$(2N-1)\hat{E} < (2N-1)E < 2\sum_{\substack{i=1\\i\neq k}}^{N} C_{ik} + E_k \tag{28}$$

If we multiply Eq(28) by $(N-1)^2$,

$$(2N-1)(N-1)^2\hat{E} < 2(N-1)^2\sum_{\substack{i=1\\i\neq k}}^{N} C_{ik} + (N-1)^2E_k \tag{29}$$

According to Eq(23) and Eq(29), the constraint condition can be deduced as follows.

$$(2N-1)\sum_{\substack{i=1\\i\neq k}}^{N}\sum_{\substack{j=1\\j\neq k}}^{N} C_{ij} < 2(N-1)^2\sum_{\substack{i=1\\i\neq k}}^{N} C_{ik} + (N-1)^2E_k \tag{30}$$

Therefore, it is proved that when using the simple ensemble method and when constraint condition Eq(30) is satisfied, then omitting the kth base learner will improve the ensemble's generalization ability.

There is a conclusion that after lots of ELMs are trained, ensembling an appropriate subset of them is superior to ensembling all of them in some cases. The individual ELMs that should be omitted satisfy Eq(30). This result implies that the ensemble does not use all the networks to achieve good performance. Therefore, the selective ensemble of ELM can work well.

According to the above proofs, the recursive model based selective ensemble of extreme learning machine might be better than the selective ensemble of extreme learning machine because of three reasons below: firstly, the best result comes from the better results more easily, so if the first layer of our framework can effectively select an optimal group of different ELMs, the second layer has a great potential to produce a better result based on an optimal group of ELMs. Secondly, from the network structure, the recursive model based selective ensemble can be explained as the hierarchical model based selective ensemble. And the RMSE-ELM is a natural extension of selective ensemble of extreme learning machine. Therefore, if each part can work well, the whole system can work well at least. Finally, lots of experiments in recent years have shown that if more neural networks are included, in some cases the generalization error of the ensemble might be further reduced.

From above theoretical discussion, we see that why the recursive model based selective ensemble of extreme learning machine can work better. However, we will further explore how many layers can achieve the optimal compromise between robustness and computational cost. The pseudo code of our current framework is organized as follows:

Algorithm 2 RMSE-ELM

Given: training set (X, Y), M(the size of ensemble groups in the first layer), N_1(the size of each ensemble in the first layer), N_2(the size of candidates pool in the second layer), $\omega^*$is defined in Eq(9), threshold λ is a pre-set value (reciprocal value of N_1 or N_2).

Steps:

1. for $group = 1 \ldots M$

    { $N_2 = 0$;

    for $element = 1 \ldots N_1$

    { Training each ELM network;

    Generating a population of weight vector;

    Using selective ensample to get the best weight vector $\omega_1{}^*$;

    Removing base ELMs that the weights less than $\lambda_1 = 1/N_1$;

    }

    Calculating the whole remained ELMs of group i are n_i;

    $$N_2 = N_2 + n_i;$$

    }

2.Training N_2 remained ELM;

3.Using selective ensemble to get the best weight vector $\omega_2{}^*$;

4.Removing base ELMs that the weights less than $\lambda_2 = 1/N_2$;

5. Getting the final prediction;

---

4 Experiments

In this section, we present some experiments on 4 UCI blended datasets to verify whether RMSE-ELM performs better in robustness than other methods such as ELM, OP-ELM, GASEN-ELM, GASEN-BP and E-GASEN for blended data. At the same time, computational cost is also a significant parameter to evaluate the usefulness of our new framework. All simulations are carried out in Matlab environment running in an Intel Corei5-3470 (3.20GHz CPU).

Table 1. Specification of the 2 tested regression data sets

Task	# variables	# training	# test	Abbr.
Boston Housing	13	400	106	BH
Abalone	8	2000	2177	Aba
Red Wine	11	1065	534	RW
Waveform	21	3000	2000	Wav

Four types of datasets are all selected from the UCI machine learning repository [39]. The first one is Boston Housing dataset which contains 506 samples. Each sample is composed of 13 input variables and 1 output variable. And this dataset is divided into a training set of 400 samples and a testing set of the rest. The second one is Abalone dataset. There are 7 continuous input variables, 1 discrete input variable and 1 categorical attribute in this dataset. It comprises 4177 samples, among which, 2000 samples are used for training and the rest 2177 samples are used for testing. The third one is Red Wine dataset which contains 1599 samples. Each sample consists of 11 input variables and 1 output variable, the dataset is divided into two sections: 1065 samples for training set and the rest samples for testing set. Finally, Waveform dataset with more number of input variables is selected. This dataset contains 21 input variables and 1 output variable. The specification of the four types of datasets is shown in table 1.

Firstly, we randomly mix several irrelevant Gaussian noises with the original UCI data, and all features of data are normalized into a similar scale. Secondly, we train the different models such as ELM, OP-ELM, GASEN-ELM, GASEN-BP, E-GASEN and RMSE-ELM on the training set of blended data. Finally, we test the different models on the testing set of blended data to acquire experimental results including Mean Square Error (MSE), Standard Deviation (STD) and Computational Cost (CC). In our experiments, the genetic algorithm employed by RMSE-ELM is implemented by the GAOT toolbox developed by Houck et al. In the toolbox, the genetic operators (selecting, crossover probability, mutation probability and stopping criterion) are set to the default values. The first group of original UCI data is blended with 7 irrelevant variables that all conform to the Gaussian distributions, such as $N(0,2)$, $N(0,1)$, $N(0,0.5)$, $N(0,0.1)$, $N(0,0.005)$, $N(0,0.001)$, $N(0,0.0005)$.To acquire the convincing result, the second group of original data is blended with 10 irrelevant Gaussian variables, such as $N(0,2)$, $N(0,1)$, $N(0,0.5)$, $N(0,0.1)$, $N(0,0.05)$, $N(0,0.01)$, $N(0,0.005)$, $N(0,0.001)$, $N(0,0.0005)$, $N(0,0.0001)$. For different ensemble frameworks (GASEN-ELM, GASEN-BP, E-GASEN and RMSE-ELM), The number of ELMs in each ensemble group is initially set to 20 [38], so the threshold λ used by selective ensemble is set to 0.05 because it is the reciprocal value of the size of each ensemble according to Zhou's experiment. For hierarchical models such as E-GASEN and RMSE-ELM, the number of ensemble groups is set to 4 according to the Zhou's experiments. In addition, the number of hidden units in each ELM is set to 50 because it can acquire the better performance at this point. Specifically speaking, the testing RMSE curve gradually decreases to a constant value and also the learning time is still less after this point [40]. For each algorithm we perform 5 runs and record the average value of MSE, STD and CC. The experimental results are shown in following tables and figures.

Table 2. MSE for UCI blended datasets(7 irrelevant variables)

Data set	ELM	OP-ELM	GASEN-ELM	GASEN-BP	E-GASEN	**RMSE-ELM**
BH	5.8564	4.9823	5.0543	4.7869	4.8822	**4.7763**
Aba	34.5586	31.4742	30.0193	29.5716	28.3969	**26.0626**
RW	0.4998	0.4946	0.4514	0.5412	0.4488	**0.4374**
Wav	0.3733	0.3412	0.3429	0.2671	0.3371	**0.3276**

Table 3. MSE for UCI blended datasets (10 irrelevant variables)

Data set	ELM	OP-ELM	GASEN-ELM	GASEN-BP	E-GASEN	**RMSE-ELM**
BH	6.3748	5.0672	5.7973	4.8495	5.6263	**5.4462**
Aba	34.7401	29.5260	29.7477	27.6825	27.5196	**26.2389**
RW	0.5069	0.4969	0.4613	0.5399	0.4512	**0.4422**
Wav	0.3750	0.3339	0.3489	0.2747	0.3449	**0.3347**

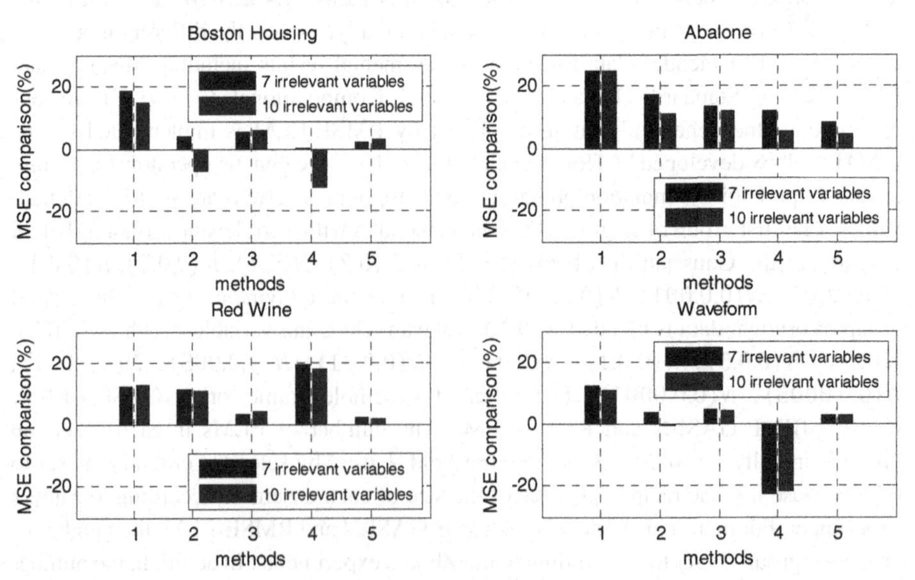

Fig. 3. MSE comparison between RMSE-ELM and other methods (x-axis 1:ELM,2:OP-ELM,3:GASEN-ELM,4:GASEN-BP,5:E-GASEN)

There are two important criteria for robustness assessment (MSE and STD). Let's first analyze the MSE among different methods on UCI blended datasets. For the evaluation of MSE, we visualize the experimental results in Table 2 and Table 3 into Figure 3. We define the difference of MSE between RMSE-ELM and other methods as MSE comparison. The formula is

$$\text{MSE comparison} = \frac{\text{MSE(other methods)} - \text{MSE(RMSE\_ELM)}}{\text{MSE(other methods)}} \times 100\% \tag{31}$$

Therefore, in Figure 3, positive percentage means the MSE of new method (RMSE-ELM) is lower than other methods, which in turn proves that the robustness of new method is better, or vice versa. In four types of UCI blended datasets, the results show that the MSE of our method is lower than that of other methods in most cases. In particular, the difference of MSE between our method and ELM is more obvious, which definitely proves that our framework improves the robustness performance of original ELM for blended data. However, in some cases, the MSE of GASEN-BP and OP-ELM is obviously lower than that of RMSE-ELM.

Table 4. STD for UCI blended datasets (7 irrelevant variables)

Data set	ELM	OP-ELM	GASEN-ELM	GASEN-BP	E-GASEN	**RMSE-ELM**
BH	0.2236	0.1416	0.1024	0.1551	0.0494	**0.1109**
Aba	3.2644	7.2611	1.3031	1.6831	0.4601	**1.3439**
RW	0.0191	0.0091	0.0092	0.0270	0.0033	**0.0110**
Wav	0.0094	0.0187	0.0031	0.0069	0.0020	**0.0041**

Table 5. STD for UCI blended datasets (10 irrelevant variables)

Data set	ELM	OP-ELM	GASEN-ELM	GASEN-BP	E-GASEN	**RMSE-ELM**
BH	0.1864	0.1807	0.0923	0.1702	0.0400	**0.1047**
Aba	3.1029	4.3826	1.7374	1.8569	0.4019	**1.4385**
RW	0.0168	0.0166	0.0086	0.0216	0.0023	**0.0085**
Wav	0.0107	0.0233	0.0039	0.0098	0.0016	**0.0026**

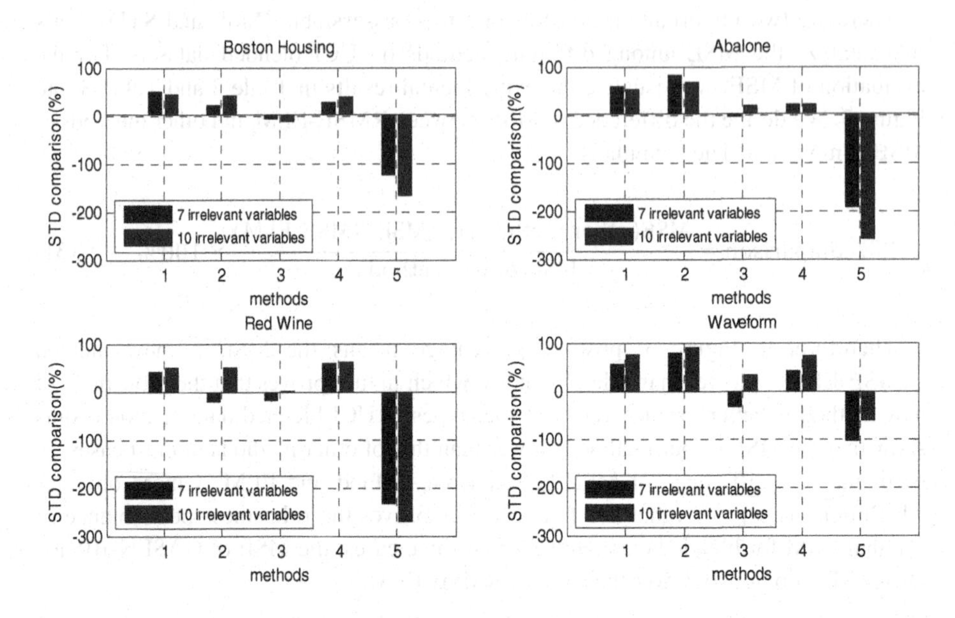

Fig. 4. STD comparison between RMSE-ELM and other methods (x-axis 1:ELM, 2:OP-ELM, 3:GASEN-ELM, 4:GASEN-BP,5:E-GASEN)

Secondly, for the evaluation of STD, we visualize the experimental results in Table 4 and Table 5 into Figure 4. We define the difference of STD between RMSE-ELM and other methods as STD comparison. The formula is

$$\text{STD comparison} = \frac{\text{STD(other methods)} - \text{STD(RMSE\_ELM)}}{\text{STD(other methods)}} \times 100\% \quad (32)$$

In Figure 4, positive percentage means the STD of our method is lower than that of other methods, which proves that the robustness of our new method is better, or vice versa. In four types of blended datasets, the results show that the STD of our method is lower than that of other methods, which confirms that our framework really improve the robustness performance for blended data. However, in some cases, the STD of E-GASEN is obviously lower than that in RMSE-ELM.

Table 6. CC for UCI blended datasets (7 irrelevant variables, unit: seconds)

Data set	ELM	OP-ELM	GASEN-ELM	GASEN-BP	E-GASEN	**RMSE-ELM**
BH	0.0920	234.5413	2.5023	574.1617	4.6832	**3.7206**
Aba	0.0250	25.7682	1.4180	205.4845	7.6893	**2.4960**
RW	0.0390	189.7191	1.8720	361.7819	3.0015	**2.9203**
Wav	0.1427	534.6310	2.8408	1534.0000	4.8984	**3.8485**

Table 7. CC for UCI blended datasets (10 irrelevant variables, unit: seconds)

Data set	ELM	OP-ELM	GASEN-ELM	GASEN-BP	E-GASEN	**RMSE-ELM**
BH	0.0952	281.5818	2.7363	634.8929	3.8517	**3.9226**
Aba	0.0250	33.0161	1.4383	229.8675	6.8874	**2.7191**
RW	0.0406	263.2673	1.7581	431.6392	2.3665	**3.0373**
Wav	0.1045	559.4664	2.7924	1995.4000	6.2244	**3.8454**

Finally, according to Table 6 and Table 7, the results show that the CC of our method is acceptable. However, the CC of GASEN-BP and OP-ELM is too long to apply in the real-time area or industry.

There are two interesting observations above, and we hope to explain further. Firstly, although in some cases, the MSE of GASEN-BP and OP-ELM is lower than that of RMSE-ELM, from the view of statistics, the MSE of RMSE-ELM is lower than that of GASEN-BP and OP-ELM on the whole. For example, we have 4 types of UCI datasets and 2 types of Gaussian noisy variants. If we run above 3 algorithms on 8 types of blended data, for MSE comparison between RMSE-ELM and GASEN-BP, the MSE of RMSE-ELM is lower on 5 types of blended data while the MSE of GASEN-BP is lower on 3 types of blended data. For MSE comparison between RMSE-ELM and OP-ELM, the MSE of RMSE-ELM is lower on 6 types of blended data while the OP-ELM is lower on only 2 types of blended data. What's more, the CC of RMSE-ELM is much shorter than that of OP-ELM and GASEN-BP. Secondly, in some cases, though the STD of E-GASEN is lower than that of RMSE-ELM, the MSE of RMSE-ELM is totally lower than that of E-GASEN. Moreover, the CC of RMSE-ELM is shorter than that of E-GASEN except RW dataset for 10 irrelevant noisy variables.

In conclusion, we believe that our new method in robustness is definitely better than ELM. We believe that our framework is a good compromise between robustness performance and learning speed. However, how many groups in the first layer of RMSE-ELM should we choose for the best robustness performances? It should be further explored.

5 Discussions

Until now, we are very clear about the structure and performance of RMSE-ELM. In the design of experiments, for added noises, the Gaussian noises are selected because they are common in real world. For comparable methods, we select OP-ELM as one of the benchmark methods because it is almost the first generation of extended ELM to probe the robustness issue. And both the GASEN-ELM and E-GASEN are also selected because they have the similar mechanism with RMSE-ELM. However, the differences in structure and mechanism among them are also obvious. For example, GASEN-ELM is a one-layer ensemble network using selective ensemble approach. Though the E-GASEN is a two-layer ensemble network like RMSE-ELM, the ensemble in the second layer is regarded as the simple ensemble instead of the selective ensemble approach employed by RMSE-ELM. According to the selection of UCI blended data and benchmark approaches, we believe that our experimental results should be fair and convincing.

In the experiments, we tested new method on four types of UCI datasets, which are blended with 7-dimensional and 10-dimensional Gaussian noises separately. It is clear that the MSE of our method is almost lower than that of other methods except for GASEN-BP in some cases. For GASEN-BP and RMSE-ELM, the CC of GASEN-BP limit its wide use in industry and real-time area compared with RMSE-ELM. And also the STD of our method is lower than that of other methods except for E-GASEN. For E-GASEN and RMSE-ELM, though the E-GASEN is lower in STD, which means that E-GASEN is more stable in fluctuation of MSE, in the rest aspects (MSE and CC), the performance of E-GASEN is totally worse than that of RMSE-ELM. In conclusion, the robustness performance of our method is than that of other methods for blended data with relatively fast speed. In essence, the ELM has a weak robustness performance for blended data mainly because of its simple structure, so the hierarchical model like recursive model inference is our natural consideration.

6 Conclusions

In this paper, we proposed a new method called RMSE-ELM. To be more specific, the structure of our framework is the two-layer ensemble architecture, which recursively employs selective ensemble to pick out several optimal ELMs from bottom to top for the final ensemble. The experiments prove that the robustness performance of RMSE-ELM is better than original ELM and representative methods for blended data. Through analysis of experiments, the reasons why our approach works are proposed as follows. Firstly, the selective ensemble extracts the optimal subset effectively from each group in the first layer and from candidate pool in the second layer. Secondly, the kernel of our framework is ELM, which has excellent generalization and rapid learning speed. Finally, the recursive model in essence is a special case of hierarchical network, which is a good compromise between shallow network and deep network. However, analyses presented in this paper are very preliminary. More experiments and principles still need to be completed in order to modify our framework further. Our future work will focus on three main directions: First, in the framework of RMSE-ELM, how many groups in the first layer should we choose to acquire the best robustness. And how many layers can achieve the optimal compromise between robustness and computational cost based on our framework. Second, whether the space complexity of our method can be largely reduced under regularized framework. For example, if the weight of our framework can be sparse enough under regularization, the complexity of our framework might be largely reduced. Third, whether the selective ensemble approach in the top layer can be replaced by other criteria for a better robustness performance. In general, it may be an interesting work to develop a combination of ensemble learning and hierarchical model to enhance the robustness performance of ELM in the future.

Acknowledgments. This work is partially supported by Natural Science Foundation of China (41176076, 31202036, 51379198 and 51075377).

References

[1] Salakhutdinov, R., Hinton, G.-E.: Deep BoltzmannMachine. In: 12th International Conference on Artificial Intelligence and Statistics Proceedings (AISTATS), vol. 5, pp. 448–455 (2009)

[2] Han, C.-C., Wang, C.-T., Jenget, B.-S., et al.: The Application of a Convolution Neural Network on Face and License Plate Detection. In: 12th International Conference on Pattern Recognition, vol. 3, pp. 552–555 (2006)

[3] Hinton, G.-E., Osindero, S., Teh, Y.-W.: A fast learning algorithm for deep belief nets. Neural Computation 18, 1527–1554 (2006)

[4] Bengio, Y.: Learning deep architectures for AI. Foundations and Trends in Machine Learning 2, 1–127 (2009)

[5] Huang, G.-B., Zhu, Q.-Y., Siew, C.-K.: Extreme learning machine: theory and applications. Neurocomputing 70, 489–501 (2006)

[6] Hinton, G.-E., Salakhutdinov, R.: Reducing the dimensionality of Data eith Niural Networks. Science 313, 504–507 (2006)

[7] Salakhutdinov, R., Larochelle, H.: Efficient learning of deep boltzmann machines. International Conference on Artificial Intelligence and Statistics (2010)

[8] Huang, G.-B., Wang, D.-H., Lan, Y.: Extreme learning machines: a survey. International Journal of Machine Learning and Cybernetics 2, 107–122 (2011)

[9] Huang, G.-B., Siew, C.-K.: Extreme learning machine with randomly assigned RBF kernels. In: Eighth International Conference on Control, Automation, Robotics and Vision Proceedings, ICARCV (2004)

[10] Frénay, B., Verleysen, M.: Parameter-insensitive kernel in extreme learning for non-linear support vector regression. Neurocomputing 74, 2526–2531 (2011)

[11] Huang, G.-B., Chen, L., Siew, C.-K.: Universal Approximation using incremental constructive feedforward networks with random hidden node. IEEE Transactions on Neural Networks 17, 879–892 (2006)

[12] Schuldt, C., Laptev, I., Caputo, B.: Recognizing human axtions: a local SVM approach. In: 17th International Conference on Pattern Recognitionproceedings, vol. 3, pp. 32–36 (2004)

[13] Rumelhart, D.-E., Hinton, G.-E., Williams, R.-J.: Learning representations by back-propagation errors. Nature 323, 533–536 (1986)

[14] Liang, N.-Y., Huang, G.-B., Saratchandran, P., et al.: A fast and accurate on-line sequential learning algorithmfor feedforward networks. IEEE Transactions on Neural Networks 17, 1411–1423 (2006)

[15] LeCun, Y.A., Bottou, L., Orr, G.B., Müller, K.-R.: Efficient backProp. In: Orr, G.B., Müller, K.-R. (eds.) NIPS-WS 1996. LNCS, vol. 1524, pp. 9–50. Springer, Heidelberg (1998)

[16] Huang, G.-B., Saratchandran, P., Sundararajan, N.: An efficient sequential learning algorithm for growing and pruning RBF (GAPP-RBF) networks. IEEE Transitions Man Cybernet 34, 2284–2292 (2004)

[17] Huang, G.-B., Saratchandran, P., Sundararajan, N.: A generalized growing and pruning RBF (GGAP-RBF) neural network for function approximation. IEEE Transitions Neural Network 16, 57–67 (2005)

[18] Huang, G.-B., Wang, D.-H., Lan, Y.: Extreme learning machines: a survey. International Journal of Machine Learning and Cybernetics 2, 107–122 (2011)

[19] Rong, H.-J., Ong, Y.-S., Tan, A.-H., et al.: A fast pruned-extreme learning machine for classification problem. Neurocomputing 72, 359–366 (2008)

[20] Miche, Y., Sorjamaa, A., Bas, P., et al.: OP-ELM: optimally pruned extreme learning machine. IEEE Transactions on Neural Networks 21, 158–162 (2010)
[21] Miche, Y., Sorjamaa, A., Lendasse, A.: OP-ELM: Theory, experiments and a toolbox. In: Kůrková, V., Neruda, R., Koutník, J. (eds.) ICANN 2008, Part I. LNCS, vol. 5163, pp. 145–154. Springer, Heidelberg (2008)
[22] Miche, Y., Sorjamaa, A., Lendasse, A.: OP-ELM: Theory, experiments and a toolbox. In: Kůrková, V., Neruda, R., Koutník, J. (eds.) ICANN 2008, Part I. LNCS, vol. 5163, pp. 145–154. Springer, Heidelberg (2008)
[23] Miche, Y., Bas, P., Jutten, C., et al.: A methodology for building regression models using extreme learning machine: OP-ELM. In: Proceedings of the European Symposium on Artificial Neural Networks (ESANN), pp. 247–252 (2008)
[24] Hansen, L.K., Salamon, P.: Neural network ensembles. IEEE Transactions on Pattern Analysis and Machine Intelligence 12, 993–1001 (1990)
[25] Krogh, A., Sollich, P.: Statistical mechanics of ensemble learning. The American Physical Society (1997)
[26] Sun, Z.-L., Choi, T.-M., Au, K.-F., et al.: Sales forecasting using extreme learning machine with applications in fashion retailing. Decision Support Systems 46, 411–419 (2008)
[27] Zhou, Z.-H., Wu, J.-X., Yuan, J., et al.: Genetic algorithm based selective neural network ensemble. In: IJCAI-2001: Proceedings of the Seventeenth International Joint Conference on Artificial Intelligence, Seattle, Washington, August 4-10, p. 797 (2001)
[28] Tang, Y., Biondi, B.: Least-squares migration/inversion of blended data. SEG Technical Program Expanded Abstracts, pp. 2859–2863 (2009)
[29] Li, N., Zhou, Z.-H.: Selective ensemble under regularization framework Multiple Classifier Systems, pp. 293–303. Springer (2009)
[30] Asuncion, A., Newman, D.J.: UCI machine learning repository (2007)
[31] Huang, G.-B., Zhu, Q.-Y., Siew, C.-K.: Extreme learning machine: a new learning scheme of feedforward neural networks. IEEE International Joint Conference on Neural Networks 2, 985–990 (2004)
[32] Serre, D.: Matrices: Theory and Applications. Springer, New York (2002)
[33] Rao, C.-R., Mitra, S.-K.: Generalized inverse of a matrix and its applications. Wiley, New York (1972)
[34] van Heeswijk, M., Miche, Y., Lindh-Knuutila, T., Hilbers, P.A.J., Honkela, T., Oja, E., Lendasse, A.: Adaptive ensemble models of extreme learning machines for time series prediction. In: Alippi, C., Polycarpou, M., Panayiotou, C., Ellinas, G. (eds.) ICANN 2009, Part II. LNCS, vol. 5769, pp. 305–314. Springer, Heidelberg (2009)
[35] Van, H.-M., Miche, Y., Oja, E., et al.: Gpuaccelerated and parallelized ELM ensembles for large-scale regression. Neurocomputing (2011)
[36] Zhou, Z.-H., Wu, J.-X., Tang, W.: Ensemble neural networks: Many could be better than one. Artificial Intelligence 137, 239–263 (2002)
[37] Zhao, L.-J., Chai, T.-Y., Yuan, D.-C.: Selective ensemble extreme learning machine modeling of effluent quality in wastewater treatment plants. International Journal of Automation and Computing 9, 627–633 (2012)
[38] Opitz, D.W., Shavlik, J.W.: Generating accurate and diverse members of a neural-network ensemble. Advances in Neural Information Processing Systems, 535–541 (1996)
[39] Arthur, A., Newman, D.: UCI machine learning repository (2007)
[40] Guang-Bin, H., Chen, L., Siew, C.-K.: Universal approximation using incremental constructive feedforward networks with random hidden nodes. IEEE Transactions on Neural Networks 17(4), 879–892 (2006)

Extreme Learning Machine for Regression and Classification Using L_1-Norm and L_2-Norm

Xiong Luo*, Xiaohui Chang, and Xiaojuan Ban

School of Computer and Communication Engineering,
University of Science and Technology Beijing (USTB), 100083 Beijing, China
xluo@ustb.edu.cn

Abstract. Extreme learning machine (ELM) has been studied extensively in recent years. It is a very simple machine learning algorithm which can achieve a good generalization performance with extremely fast speed. Thus, it has practical significance for Big Data analysis. Normally, it is implemented under the empirical risk minimization scheme and it may tend to generate a large-scale and over-fitting model. In this paper, an ELM model based on L_1-norm and L_2-norm regularizations is proposed to deal with regression and multiple class classification problems in a unified framework, and it can reduce the complexity of the network and prevent over-fitting. We test the proposed algorithm on eight benchmark data sets. Simulation results have shown that the proposed algorithm outperforms the original ELM and other advanced ELM algorithm in terms of prediction accuracy and stability.

Keywords: extreme learning machine, ridge regression, elastic net, model selection.

1 Introduction

More recently, data is being collected at an unprecedented scale. There are increasing demand of effective data analysis for making decisions to fully realize the potential of Big Data. Single-hidden layer feedforward network (SLFN) based on extreme learning machine (ELM) [1] is one of the important methods used in data analysis due to its powerful nonlinear mapping capability and extremely fast learning speed. However, original ELM solution may tend to generate an over-fitting model and are less stable in some situations [2]. Moreover, the structure of the neural network (NN) is still a question in ELM design.

To overcome the problems ELM faced, several schemes have been proposed. In [3], Rong *et al.* proposed a fast pruned ELM for classification problems. Martínez-Martínez *et al.* proposed a regularized ELM for regression problems in [4]. In [3] and [4], although those algorithms can generate a spare NN structure, they do not provide a unified NN framework for both regression and classification problems.

* This work was jointly supported by the National Natural Science Foundation of China under Grants 61174103, 61174069, and 61004021, and the Fundamental Research Funds for Central Universities under Grant FRF-TP-11-002B.

J. Cao et al. (eds.), *Proceedings of ELM-2014 Volume 1*,
Proceedings in Adaptation, Learning and Optimization 3, DOI: 10.1007/978-3-319-14063-6_25

Miche *et al.* proposed an optimally pruned ELM for regression and classification in [5], which was a regularized ELM by using the least angle regression (LARS) algorithm, i.e., a L_1 penalty, but this algorithm has its limitation while facing a group of high correlated variables.

Considering those problems in ELM design analyzed above, we propose a novel ELM algorithm based on L_1 penalty and L_2 penalty to deal with both multiple output regression tasks and multiple class classification tasks in a unified framework. Here, elastic net algorithm is used to solve this mixed penalties [6]. Then separate elastic net algorithm and the Bayesian information criterion (BIC) [7] are adopted to find the optimal model for each response variable. Thus, the proposed algorithm tends to reduce over-fitting and provide a more robust model.

This paper is organized as follows. Section 2 analyses the SLFN based on ELM and classic regularization methods. Section 3 presents the proposed ELM model based on L_1-norm and L_2-norm regularizations. Section 4 provides the simulation results and discussion. Section 5 summarizes the conclusion.

2 Model Description

2.1 SLFN Based on ELM

ELM theories claim that the hidden node learning parameters can be randomly assigned and the output weights can be determined by solving a linear system [8], thus the ELM can be implemented with few steps and low computational cost.

For P arbitrary distinct samples (x_i, t_i), where $x_i = [x_{i1}, x_{i2}, \cdots, x_{im}]^T \in \mathbb{R}^m$ and $t_i = [t_{i1}, t_{i2}, \cdots, t_{in}]^T \in \mathbb{R}^n$, a standard SLFN with L hidden nodes can be mathematically modeled as:

$$o_i = \sum_{j=1}^{L} \beta_j G(a_j, b_j, x_i), \quad i = 1, 2, \cdots, P \tag{1}$$

where a_j and b_j are the learning parameters of hidden nodes, β_j is the link connecting the j-th hidden node to the output nodes, $G(a_j, b_j, x_i)$ is the output of the j-th hidden node with respect to the input x_i, and o_i is the actual output.

The SLFN with L hidden nodes can approximate these P samples with zero error, which means that the cost function $E = \sum_{i=1}^{P} ||o_i - t_i||_2 = 0$, i.e., there exist (a_j, b_j) and β_j such that:

$$t_i = \sum_{j=1}^{L} \beta_j G(a_j, b_j, x_i), \quad i = 1, 2, \cdots, P \tag{2}$$

where $\|\cdot\|_2$ represents the L_2-norm.

The above P equations can be written compactly as :

$$H\beta = T \tag{3}$$

where

$$H = \begin{bmatrix} G(a_1, b_1, x_1) & \cdots & G(a_L, b_L, x_1) \\ \vdots & \ddots & \vdots \\ G(a_1, b_1, x_P) & \cdots & G(a_L, b_L, x_P) \end{bmatrix}_{P \times L}, \beta = \begin{bmatrix} {\beta_1}^T \\ \vdots \\ {\beta_L}^T \end{bmatrix}_{L \times n}, T = \begin{bmatrix} {t_1}^T \\ \vdots \\ {t_P}^T \end{bmatrix}_{P \times n}.$$

Here, H is called the hidden layer output matrix of the SLFN. Thus, the system (3) becomes a linear model and the output weights can be analytically determined by finding a least-square solution of this linear system as follows:

$$\beta = H^{\dagger}T \tag{4}$$

where $H^{\dagger}$ is the Moore-Penrose generalized inverse of matrix H [1].

Although ELM has been developed to work at a much faster learning speed with the higher generalization performance, it also has some drawbacks:

1) ELM is designed with the empirical risk minimization (ERM) principle and may tend to generate an over-fitting model.

2) ELM provides weak control capacity and is less stable because it is implemented by using a classical least-square method.

3) Users have to choose the number of hidden nodes through trial-and-error.

2.2 Regularization Methods

Multiple linear regression is often used to investigate the relationship between the predictor variables and the response variables. Then both the prediction accuracy and the size of the model should be considered.

Considering the general setup for a single-output regression problem:

$$y = H\beta + \varepsilon \tag{5}$$

where H is the inputs data set, and it is a $P \times L$ matrix. Here y is the actual output, β is the regression weights, and ε is the residuals. The traditional approach used to solve the above problem is the ordinary least square (OLS) estimates, which can be formulated as follows:

$$\widehat{\beta} = \arg\min_{\beta} ||y - H\beta||_2^2 \tag{6}$$

where $\widehat{\beta} = [\widehat{\beta}_1, \widehat{\beta}_2, \cdots, \widehat{\beta}_L]^T$ is the estimated regression weights. It is well known that OLS often performs not well in terms of both prediction accuracy and the model size [9]. Regularization techniques have been proposed to improve OLS.

The L_1-norm, which is also called the least absolute shrinkage and selection operator (Lasso) [10], represents the most basic augmentation of the OLS solution. The Lasso estimate $\widehat{\beta}$ is defined by:

$$\widehat{\beta} = \arg\min_{\beta} \{||y - H\beta||_2^2 + \lambda||\beta||_1\} \tag{7}$$

where λ is a positive regularization parameter, $||\cdot||_1$ represents the L_1-norm. As λ increases, the number of nonzero components of $\widehat{\beta}$ decreases.

Due to its nature of both continuous shrinkage and automatic variable selection simultaneously, the Lasso has shown success in many situations. But it has some limitations as noted by Zou and Hastie in [6].

To overcome the drawbacks of L_1-norm, both the L_1 penalty and the L_2 penalty are used in the same minimization problem. The mathematic model of this mixed penalties can be formulated as follows:

$$\widehat{\beta} = \arg\min_{\beta} \{||y - H\beta||_2^2 + \lambda||\beta||_1 + \xi||\beta||_2^2\} \tag{8}$$

where both λ and ξ are tuning parameters. In [11], the elastic net was proposed to solve (8).

The elastic net simultaneously does automatic variable selection and continuous shrinkage, and it can select group of correlated variables. It has been shown that the elastic net often outperforms the Lasso in terms of prediction accuracy, while enjoying a similar sparsity of representation.

3 L_1-L_2-ELM Model

3.1 Solution of the Elastic Net

The basic idea of solving the elastic net is to reduce the elastic net problem to an equivalent Lasso problem.

For data set (y, H) and (λ, ξ) defined in (8), an artificial data set (y^*, H^*) is generated as follows:

$$\begin{cases} H^*{}_{(P+L)\times L} = \frac{1}{\sqrt{(1+\xi)}} \begin{pmatrix} H \\ \sqrt{\xi} I \end{pmatrix} \\ y^*{}_{(P+L)\times L} = \begin{pmatrix} y \\ 0 \end{pmatrix} \end{cases} \tag{9}$$

Then the naive elastic net criterion can be written as:

$$\widehat{\beta}^* = \arg\min_{\beta^*} \{||y^* - H^*\beta^*||_2^2 + r||\beta^*||_1\} \tag{10}$$

where $r = \frac{\lambda}{\sqrt{1+\xi}}$ and $\beta^* = \sqrt{1+\xi}\beta$. Thus, $\widehat{\beta}$ can be represented as follows:

$$\widehat{\beta} = \frac{1}{\sqrt{1+\xi}} \widehat{\beta}^* \tag{11}$$

However, the above solution may incur a double shrinkage, which may introduce unnecessary extra bias. Then the elastic net estimates $\widehat{\beta}$ as follows:

$$\widehat{\beta} = \sqrt{1+\xi}\, \widehat{\beta}^* \tag{12}$$

Thus the elastic net avoids the ridge shrinkage effect by introducing a scaling factor $(1 + \xi)$, while still keep the grouping effect feature of ridge regression. Hence, the solution of elastic net has been successfully transformed into the lasso problem, and an efficient LARS-EN algorithm was implemented to solve the elastic net solution paths for any fixed ξ [6] .

For each fixed ξ, the LARS-EN algorithm will produce a set of candidate models. Then we adopt the BIC to do the model selection to balance the accuracy and the network size. The BIC was defined as follows:

$$\text{BIC}= -2\ln(Q)+M\ln(P) \tag{13}$$

where Q is the value of the likelihood function for the estimated model, M is the number of hidden nodes to be estimated, and P is the number of samples.

3.2 L_1-L_2-ELM Model

Both multiple output regression and the multiclass classification tasks can be implemented using a unified network model in the proposed algorithm. For multiclass classification problem, it can be transformed into a multiple output regression problem. Assume a set of multiclass training samples (x_i, t_i) $(i = 1, 2, \cdots, P)$, $t_i \in \{1, 2, \cdots, n\}$, each class label is expanded into a label vector of length n according to the original ELM algorithm. For example, in a training sample (x_i, t_i), if x_i is the third class, the corresponding output label vector is $t_i = [-1, -1, 1, -1, \cdots, -1]$, i.e, the output node with the largest value indicates its class label.

Then, for the regression problem with n output nodes, the proposed algorithm uses n separate elastic nets to generate the entire solution paths for each output node, then adopts the BIC to find the optimal candidate model. Thus, the output weight of ELM consists of all the optimal candidate models for each response variable. Overall, the proposed algorithm, namely, L_1-L_2-ELM, can be summarized as Algorithm 1.

4 Simulation Results and Discussion

4.1 Experimental Setup

To verify the effectiveness of the proposed algorithm L_1-L_2-ELM, eight data sets from the UCI machine learning repository [12] have been used to test this algorithm, and we compare it with the original ELM and the OP-ELM [13]. The number of hidden neurons L=100 is assigned in ELM, while the OP-ELM and the L_1-L_2-ELM use a maximum number of 100 neurons. In L_1-L_2-ELM algorithm, the fixed ξ is assigned the value of 10^{-3}. In the experiments, each data set is normalized to zero mean and unit variance, and 50 trials have been conducted for all the algorithms. Then the best performance and the standard deviations (DEV) are recorded. Nodes required by L_1-L_2-ELM in each data set can be obtained by calculating the average of the numbers of selected neurons for each response node.

Algorithm 1. L_1-L_2-ELM

Input: a training set: $\{(x_i, t_i) \mid x_i \in \mathbb{R}^m, t_i \in \mathbb{R}^n, i = 1, \cdots, P\}$;
hidden node activation function: $g(x)$;
the max hidden node number: L;
fixed L_2 penalty term: ξ .

Output: the output weight: β.

1 Assign arbitrary learning parameters of hidden nodes a_j and b_j, $1 \leqslant j \leqslant L$;
2 Calculate the hidden layer output matrix H based on (3);
3 **for** $1 \leqslant i \leqslant n$ **do**
4 $\quad \beta' =$ LARS-EN$(H, y(i), \xi)$, where $y(i) = [t_{1i}, \cdots, t_{Pi}]^T$;
5 $\quad \beta' = (1 + \xi)\beta'$;
6 $\quad$ **for** $1 \leqslant k \leqslant \text{size}(\beta')$ **do**
7 $\quad\quad$ Calculate the BIC(k) for every candidate model based on (13) and $\beta'(k)$.
8 $\quad$ **end**
9 $\quad k^* = \arg\min_k \{\text{BIC}(k)\}_{k=1}^{\text{size}(\beta')}$, where k^* is the index of the minimum value in vector BIC;
10 $\quad \beta'_{\text{optimal}} = \beta'(k^*)$;
11 $\quad \beta = [\beta \quad \beta'_{\text{optimal}}]$.
12 **end**

Table 1. Information of the regression data sets

Data sets	Attributes	Samples	
		Training	Testing
Abalone	8	2000	2177
Delta_elevators	6	6300	3217
Machine_CPU	6	139	70
Servo	4	110	57

4.2 Real-World Regression Problems

The specifications of the 4 real-world benchmark data sets [12] are listed in Table 1 while the comparison results of algorithms are provided in Table 2. As we can see from Table 2, the proposed algorithm is better than the OP-ELM and the original ELM in terms of the testing average root mean square error (RMSE) and the DEV, which means that the proposed algorithm has a better predicting accuracy and is more robust than the other two algorithms. And the L_1-L_2-ELM algorithm has a better variable selection procedure than the OP-ELM in most cases.

4.3 Real-World Classification Problems

The specifications of the 4 real-world classification data sets [12] are listed in Table 3. The comparison results are shown in Table 4. In Table 4, the L_1-L_2-

Table 2. RMSE and DEV in ELM, OP-ELM, and L_1-L_2-ELM on regression data sets

Methods	Datasets	RMSE		DEV		Nodes
		Training	Testing	Training	Testing	
ELM	Abalone	1.9807	2.1804	0.0071	0.0321	100
OP-ELM		2.0539	2.2042	0.0224	0.0287	55
L_1-L_2-ELM		2.0744	2.1087	0.0137	0.0123	37
ELM	Delta_elevators	0.0014	0.0015	2.5996e-05	3.0653e-05	100
OP-ELM		0.0014	0.0014	8.6406e-06	8.7159e-06	65
L_1-L_2-ELM		0.0014	0.0014	2.9027e-06	1.7586e-06	30
ELM	Machine_CPU	137.7941	235.7793	7.1300	4.1434e+11	100
OP-ELM		19.0124	75.8023	14.5489	27.8916	65
L_1-L_2-ELM		23.2940	45.2975	1.2069	3.5647	48
ELM	Servo	0.0424	3.5394	0.0155	1.4206	100
OP-ELM		0.2556	0.8962	0.1022	0.1219	60
L_1-L_2-ELM		0.2631	0.6979	0.0196	0.0393	63

Table 3. Information of the classification data sets

Data sets	Attributes/Classes	Samples	
		Training	Testing
Iris	4/3	100	50
Wine	13/3	120	58
Glass Identification	9/6	170	44
Landsat Satellite	36/6	4435	2000

Table 4. Success rate and DEV in ELM, OP-ELM, and L_1-L_2-ELM on classification data sets

Methods	Datasets	Success Rate		DEV		Nodes
		Training	Testing	Training	Testing	
ELM	Iris	1.0000	0.7800	0.0000	0.1017	100
OP-ELM		0.9800	0.9600	0.0167	0.0341	30
L_1-L_2-ELM		0.9900	0.9600	0.0045	0.0110	14.667
ELM	Wine	0.8333	0.8276	0.0686	0.0932	100
OP-ELM		0.9917	0.9483	0.0121	0.0294	55
L_1-L_2-ELM		0.9917	0.9828	0.0035	0.0151	34
ELM	Glass Identification	0.8412	0.6591	0.0214	0.0521	100
OP-ELM		0.8824	0.6818	0.0413	0.0503	10
L_1-L_2-ELM		0.7353	0.7045	0.0403	0.0412	6.333
ELM	Landsat Satellite	0.7445	0.7320	0.0322	0.0336	100
OP-ELM		0.8397	0.8140	0.0053	0.0066	80
L_1-L_2-ELM		0.8616	0.8420	0.0039	0.0057	29.166

ELM can achieve a higher success rate for testing samples. And the DEV is much lower than the other algorithms, which means that the proposed algorithm has more accurate and stable classification performance.

5 Conclusion

Data analysis plays a guidance role for making future plans in this Big Data era. In this paper, a novel algorithm called L_1-L_2-ELM was proposed as an effective technology in data analysis. It can deal with multiple output regression and multiple class classification problems in a unified framework. In the proposed algorithm, for W multiple output applications, W separate elastic nets need to be used to find the optimal candidate model. Simulation results have shown that the proposed algorithm has a better generalization performance and variable selection ability than the ELM and OP-ELM especially in multiple class applications. Meanwhile the L_1-L_2-ELM is more robust than the other two algorithms.

References

1. Huang, G.B., Zhu, Q.Y., Siew, C.K.: Extreme Learning Machine: A New Learning Scheme of Feedforward Neural Networks. In: IEEE International Joint Conference on Neural Networks, vol. 2, pp. 985–990. IEEE Press, New York (2004)
2. Horata, P., Chiewchanwattana, S., Sunat, K.: Robust Extreme Learning Machine. Neurocomputing 102, 31–44 (2013)
3. Rong, H.J., Ong, Y.S., Tan, A.H., Zhu, Z.: A Fast Pruned-Extreme Learning Machine for Classification Problem. Neurocomputing 2, 359–366 (2008)
4. Martínez-Martínez, J.M., Escandell-Montero, P., Soria-Olivas, E., Martn-Guerrero, J.D., Magdalena-Benedito, R., Gómez-Sanchis, J.: Regularized Extreme Learning Machine for Regression Problems. Neurocomputing 74, 3716–3721 (2011)
5. Miche, Y., Sorjamaa, A., Bas, P., Simula, O., Jutten, C., Lendasse, A.: OP-ELM: Optimally Pruned Extreme Learning Machine. IEEE Trans. Neural Networks 21, 158–162 (2010)
6. Zou, H., Hastie, T.: Regularization and Variable Selection via the Elastic Net. J. R. Statist. Soc. B 67, 301–320 (2005)
7. Burnham, K.P., Anderson, D.R.: Multimodel Inference Understanding AIC and BIC in Model Selection. Sociological Methods & Res 33, 261–304 (2004)
8. Cao, J., Lin, Z., Huang, G.B., Liu, N.: Voting Based Extreme Learning Machine. Inf. Sci. 1, 66–77 (2012)
9. Tibshirani, R.: Regression Shrinkage and Selection via the Lasso. J. R. Statist. Soc. B 58, 267–288 (1996)
10. Jacob, L., Obozinski, G., Vert, J.P.: Group Lasso with Overlap and Graph Lasso. In: 26th Annual International Conference on Machine Learning, pp. 433–440. ACM Press, New York (2009)
11. De Mol, C., De Vito, E., Rosasco, L.: Elastic-Net Regularization in Learning Theory. J. Complexity 25, 201–230 (2009)
12. Bache, K., Lichman, M.: UCI Machine Learning Repository (2013), `http://archive.ics.uci.edu/ml/`
13. Grigorievskiy, A., Miche, Y., Ventelä, A.M., Séverin, E., Lendasse, A.: Long-Term Time Series Prediction Using OP-ELM. Neural Netw. 51, 50–56 (2014)

A Semi-supervised Online Sequential Extreme Learning Machine Method

Xibin Jia[1,*], Runyuan Wang[1], Junfa Liu[2], and David M.W. Powers[1]

[1] Beijing Municipal Key Laboratory of Multimedia and Intelligent Software Technology, Beijing University of Technology, Beijing, 100124, P.R. China
[2] Institute of Computing Technology, Chinese Academy of Sciences, 100190, P.R. China
jiaxibin@bjut.edu.cn.

Abstract. Online sequential ELM (OS-ELM) provides a solution for streaming data application by only learning the newly arrived single or chunk of observations, and presents outstanding performance for learning problems. However, the algorithm relies on the labeled data, which usually involves high cost in labor and time. Moreover, manually labeled data suffers from inaccuracy caused by individual bias. Considering the semi-supervised ELM (SS-ELM) provides a way to fully utilize the easily acquired unlabeled data, the paper proposes a semi-supervised online sequential ELM, denoted as SOS-ELM. The proposed SOS-ELM not only has the advantage of learning in a sequential way, but also makes the most use of unlabeled data. Experiments have been done on benchmark problems of regression and classification and the results show that the proposed SOS-ELM outperforms OS-ELM in generalization performance with similar training speed and outperforms SS-ELM with much lower training time cost.

Keywords: Online Sequential ELM (OS-ELM), Semi-supervised ELM (SS-ELM), Semi-supervised online sequential ELM (SOS-ELM).

1 Introduction

In recent years, Extreme Learning Machine (ELM), proposed by Huang et al. [1–4], is attracting more and more attention because of its outstanding performance in training speed, predicting accuracy and generalization ability [5–8]. Especially, it is shown that ELM tends to outperform support vector machine (SVM) in both regression and classification applications with much easier implementation [9]. However, batch ELM is still a time consuming affair, although much faster than traditional learning algorithms (including SVM). Online Sequential ELM (OS-ELM) [10] learns the training data chunk-by-chunk and updates the output weight only by new training data. Thus OS-ELM not only saves storage but also decreases the computational complexity.

* Corresponding author.

J. Cao et al. (eds.), *Proceedings of ELM-2014 Volume 1*,
Proceedings in Adaptation, Learning and Optimization 3, DOI: 10.1007/978-3-319-14063-6_26

Although OS-ELM has many advantages in learning, it still can't avoid the dependency on a large amount of labeled data, which usually involves high cost in labor and time. To exploit unlabeled data, some semi-supervised ELM variants [11–14] have been proposed. As typical examples, [12] and [14] propose a kind of semi-supervised ELM based on manifold regularization, so that the learning system can balance the empirical risk and the complexity of the learned function f, where [12] is an improvement of [14] in terms of semi-supervised ELM. However, the semi-supervised ELM mentioned above learns in a batch way, so its training speed decreases rapidly as the sample size gets larger.

To solve the manual labeling cost problem and meet the demand of sequential learning for many real applications, we propose a new type of online sequential ELM, which we name as semi-supervised online sequential ELM (SOS-ELM). It inherits not only the training and testing speed of OS-ELM but also the accurate prediction of SS-ELM. The experiment results show that using the same number of labeled samples, our proposed SOS-ELM has higher accurate prediction rate than that of OS-ELM and much faster training speed than that of SS-ELM.

The detail of our proposed SOS-ELM is elaborated in the following part and the rest of paper is organized as follows. Section 2 gives a brief review of the related work. Section 3 illustrates the derivation of SOS-ELM. Section 4 presents the experiment results and discussion based on the benchmark problems in regression and classification. Conclusions based on the study are given in Section 5.

2 Related Work: ELM, SS-ELM and OS-ELM

This section briefly reviews the batch ELM, OS-ELM and SS-ELM, which are foundation of our extended algorithm: SOS-ELM.

ELM is a kind of single hidden layer feed-forward neural network (SLFN), and it is provable that ELM has the universal approximation property [3]. Compared to ELM, least square support vector machine (LS-SVM) and proximal support vector machine (PSVM) provide suboptimal solutions and require higher computational complexity [5]. In order to have SLFNs work as universal approximators, one may simply randomly choose hidden nodes and then just train the output weights linking the hidden layer and the output layer.

If an SLFN with L hidden nodes can approximate N samples $(\boldsymbol{x}_k, \boldsymbol{t}_k)$ with zero error, where $\boldsymbol{x}_k \in \boldsymbol{R}^d$ is the input signal feature vector, and $\boldsymbol{t}_k \in \boldsymbol{R}^m$ is the output target value or category label, it means that there exists $\boldsymbol{\beta}_i, \boldsymbol{a}_i, b_i$ such that

$$f_L(\boldsymbol{x}_k) = \sum_{i=1}^{L} \boldsymbol{\beta}_i G(\boldsymbol{a}_i, b_i, \boldsymbol{x}_k) = \boldsymbol{t}_k, k = 1, ..., N, \boldsymbol{a}_i \in \boldsymbol{R}^d, b_i \in R \tag{1}$$

where $\boldsymbol{\beta}_i$ is the weight vector connecting the ith hidden node to the output nodes. Equation (1) can be written in a matrix format as

$$\boldsymbol{H}\boldsymbol{\beta} = \boldsymbol{T} \tag{2}$$

where

$$\boldsymbol{H} = \begin{bmatrix} G(\boldsymbol{a}_1, b_1, \boldsymbol{x}_1) & \ldots & G(\boldsymbol{a}_L, b_L, \boldsymbol{x}_1) \\ \vdots & \ldots & \vdots \\ G(\boldsymbol{a}_1, b_1, \boldsymbol{x}_N) & \ldots & G(\boldsymbol{a}_L, b_L, \boldsymbol{x}_N) \end{bmatrix}_{N \times L}, \boldsymbol{\beta} = \begin{bmatrix} \boldsymbol{\beta}_1^T \\ \vdots \\ \boldsymbol{\beta}_L^T \end{bmatrix}_{L \times m}$$

and

$$\boldsymbol{T} = \begin{bmatrix} \boldsymbol{t}_1^T \\ \vdots \\ \boldsymbol{t}_N^T \end{bmatrix}_{N \times m} \tag{3}$$

The above equation then can be considered as a linear system and training the SLFN is simply equivalent to finding a least squares solution of the liner system. It is proved that equation (4) is the unique smallest norm least squares solution to learn and obtain output weight $\boldsymbol{\beta}$ of this linear system [16]:

$$\boldsymbol{\beta} = \boldsymbol{H}^{\dagger}\boldsymbol{T} \tag{4}$$

where $\boldsymbol{H}^{\dagger}$ is the Moore-Penrose generalized inverse of the hidden layer output matrix, in which the weight and bias parameters $\boldsymbol{a}_i$ and b_i are randomly assigned.

To enhance learning performance by utilizing unlabeled samples together with labeled ones, semi-supervised extreme learning machine (SS-ELM) [12] is extended. It assumes the change of data is generally smooth. In other word, if treat the data in training dataset as graph, there should not exist too much jump between neighbor instances. According to the learning theory on structural risk minimization (SRM) [17, 19], to get high performance on generalization, the learning system needs to remain balanced between the empirical risk and the complexity of learnt function f. To meet the requirement, according to [12] the unique smallest norm least squares solution $\boldsymbol{\beta}$ can be derived as (5).

$$\boldsymbol{\beta} = \left(\boldsymbol{I} + \boldsymbol{H}^T\boldsymbol{J}\boldsymbol{H} + \lambda\boldsymbol{H}^T\boldsymbol{L}\boldsymbol{H}\right)^{-1}\boldsymbol{H}^T\boldsymbol{J}\boldsymbol{T} \tag{5}$$

where $\boldsymbol{J}$ is a user-defined diagonal matrix of penalty coefficient on the training errors and $\boldsymbol{L}$ is the graph Laplacian.

The batch ELM described previously assumes that all the training data (N samples) is available for training at one time. However, in real applications, the training data may arrive sequentially. If the newly coming data were trained with the older ones again, it would be wasteful with recalculating old data with extra computing cost. OS-ELM is developed and learns data chunk-by-chunk or one-by-one (a special case of chunk) with fixed or varied chunk size. The output weights are refined based on sequentially arriving data [10]. And the output weight $\boldsymbol{\beta}$ under least-squares solution can be written as

$$\boldsymbol{K}_{k+1} = \boldsymbol{K}_k + \boldsymbol{H}_{k+1}^T\boldsymbol{H}_{k+1} \tag{6}$$

$$\boldsymbol{\beta}^{(k+1)} = \boldsymbol{\beta}^{(k)} + \boldsymbol{K}_{k+1}^{-1}\boldsymbol{H}_{k+1}^T\left(\boldsymbol{T}_{k+1} - \boldsymbol{H}_{k+1}\boldsymbol{\beta}^{(k)}\right) \tag{7}$$

where $\boldsymbol{\beta}^{(0)} = \boldsymbol{K}_0^{-1}\boldsymbol{H}_0^T\boldsymbol{T}_0$, and $\boldsymbol{K}_0 = \boldsymbol{H}_0^T\boldsymbol{H}_0$.

3 Proposed SOS-ELM

Both SS-ELM and OS-ELM improve the performance of basic ELM from different points. However, the SS-ELM does the training with all labeled and unlabeled data in batch way, whilst OS-ELM only utilizes the labeled data. As we have known, it is feasible in practical application to process the sequential data and take most use of unlabeled data. Therefore, this paper proposes to integrate the both advantage together by modifying SS-ELM algorithm to suit for sequential learning, which we refer the new algorithm as semi-supervised online sequential extreme learning machine, viz. SOS-ELM.

Given SS-ELM [12] can balance the empirical risk and the complexity of learnt function f, which enhances learning performance dramaticlly. We reconsidered the output weight matrix of SS-ELM shown in (5) and derived the proposed SOS-ELM. To simplify the form, we let $\boldsymbol{K} = \boldsymbol{I} + \boldsymbol{H}^T\boldsymbol{J}\boldsymbol{H} + \lambda\boldsymbol{H}^T\boldsymbol{L}\boldsymbol{H}$, then the output weight matrix of SS-ELM can be written as (8).

$$\boldsymbol{\beta} = \boldsymbol{K}^{-1}\boldsymbol{H}^T\boldsymbol{J}\boldsymbol{T} \tag{8}$$

Given a chunk of initial training set of sequential training instances $\aleph_0 = \{(\boldsymbol{x}_i, \boldsymbol{t}_i) \text{ or } \boldsymbol{x}_i^{'}\}_{i=1}^{N_0}$, in which both labeled data $(\boldsymbol{x}_i, \boldsymbol{t}_i)$ and unlabeled data $\boldsymbol{x}_i^{'}$ are included. Based on the criteria of minimizing the training error, the output weight matrix of the first chunk of training dataset is calculated as (9).

$$\boldsymbol{\beta}^{(0)} = \boldsymbol{K}_0^{-1}\boldsymbol{H}_0^T\boldsymbol{J}_0\boldsymbol{T}_0 \tag{9}$$

where

$$\boldsymbol{K}_0 = \boldsymbol{I} + \boldsymbol{H}_0^T\boldsymbol{J}_0\boldsymbol{H}_0 + \lambda\boldsymbol{H}_0^T\boldsymbol{L}_{\aleph_0}\boldsymbol{H}_0 \tag{10}$$

Suppose we get another chunk of data $\aleph_1 = \{(\boldsymbol{x}_i, \boldsymbol{t}_i) \text{ or } \boldsymbol{x}_i^{'}\}_{i=N_0+1}^{N_0+N_1}$, which also includes labeled data and unlabeled data, according to the SS-ELM, the weight matrix of joint training instances including the initial and new chunk of data $\aleph_0$ and $\aleph_1$ is calculated as (11).

$$\boldsymbol{\beta}^{(1)} = \boldsymbol{K}_1^{-1}\begin{bmatrix}\boldsymbol{H}_0\\ \boldsymbol{H}_1\end{bmatrix}^T\begin{bmatrix}\boldsymbol{J}_0 & 0\\ 0 & \boldsymbol{J}_1\end{bmatrix}\begin{bmatrix}\boldsymbol{T}_0\\ \boldsymbol{T}_1\end{bmatrix} \tag{11}$$

where

$$\boldsymbol{K}_1 = \boldsymbol{I} + \begin{bmatrix}\boldsymbol{H}_0\\ \boldsymbol{H}_1\end{bmatrix}^T\begin{bmatrix}\boldsymbol{J}_0 & 0\\ 0 & \boldsymbol{J}_1\end{bmatrix}\begin{bmatrix}\boldsymbol{H}_0\\ \boldsymbol{H}_1\end{bmatrix} + \lambda\begin{bmatrix}\boldsymbol{H}_0\\ \boldsymbol{H}_1\end{bmatrix}^T\boldsymbol{L}_{\aleph_0\cup\aleph_1}\begin{bmatrix}\boldsymbol{H}_0\\ \boldsymbol{H}_1\end{bmatrix} \tag{12}$$

and

$$\boldsymbol{L}_{\aleph_0\cup\aleph_1} = \begin{bmatrix}\boldsymbol{L}_{\aleph_0} + \boldsymbol{D}_{\aleph_0\aleph_1} & -\boldsymbol{W}_{\aleph_0\aleph_1}\\ -\boldsymbol{W}_{\aleph_0\aleph_1}^T & \boldsymbol{L}_{\aleph_1} + \boldsymbol{D}_{\aleph_1\aleph_0}\end{bmatrix} \tag{13}$$

Obviously, the structure relationship among instances from the first and second chunks separately may cause the existing of $-\boldsymbol{W}_{\aleph_0\aleph_1}$, $\boldsymbol{D}_{\aleph_0\aleph_1}$ and $\boldsymbol{D}_{\aleph_1\aleph_0}$ as in (13). This requires to do recalculation upon both the old and new chunk of data. Recalling the theory of SS-ELM in section 2.2, minimizing the complexity of learnt function is an assistant condition to enhance the smoothness of regression or classification function. So it will have little impact on results but with less computation cost when we discard counting the structural relationship among old and new coming instances. Therefore, the graph Laplacian between different chunks is ignored when calculating the graph Laplacian directly. The final graph Laplacians is obtained as (14).

$$\boldsymbol{L}_{\aleph_0 \cup \aleph_1} = \begin{bmatrix} \boldsymbol{L}_{\aleph_0} & 0 \\ 0 & \boldsymbol{L}_{\aleph_1} \end{bmatrix} \tag{14}$$

Substitute (10) and (14) into (12), we can get the recursive form of $\boldsymbol{K}_1$ as (15).

$$\boldsymbol{K}_1 = \boldsymbol{K}_0 + \boldsymbol{H}_1^T \left(\boldsymbol{J}_1 + \lambda \boldsymbol{L}_{\aleph_1}\right) \boldsymbol{H}_1 \tag{15}$$

In addition,

$$\begin{aligned}
\begin{bmatrix} \boldsymbol{H}_0 \\ \boldsymbol{H}_1 \end{bmatrix}^T \begin{bmatrix} \boldsymbol{J}_0 & 0 \\ 0 & \boldsymbol{J}_1 \end{bmatrix} \begin{bmatrix} \boldsymbol{T}_0 \\ \boldsymbol{T}_1 \end{bmatrix} &= \boldsymbol{H}_0^T \boldsymbol{J}_0 \boldsymbol{T}_0 + \boldsymbol{H}_1^T \boldsymbol{J}_1 \boldsymbol{T}_1 \\
&= \boldsymbol{K}_0 \boldsymbol{K}_0^{-1} \boldsymbol{H}_0^T \boldsymbol{J}_0 \boldsymbol{T}_0 + \boldsymbol{H}_1^T \boldsymbol{J}_1 \boldsymbol{T}_1 \\
&= \boldsymbol{K}_0 \boldsymbol{\beta}^{(0)} + \boldsymbol{H}_1^T \boldsymbol{J}_1 \boldsymbol{T}_1 \\
&= \left[\boldsymbol{K}_1 - \boldsymbol{H}_1^T \left(\boldsymbol{J}_1 + \lambda \boldsymbol{L}_{\aleph_1}\right) \boldsymbol{H}_1\right] \boldsymbol{\beta}^{(0)} + \boldsymbol{H}_1^T \boldsymbol{J}_1 \boldsymbol{T}_1 \\
&= \boldsymbol{K}_1 \boldsymbol{\beta}^{(0)} - \boldsymbol{H}_1^T \left(\boldsymbol{J}_1 + \lambda \boldsymbol{L}_{\aleph_1}\right) \boldsymbol{H}_1 \boldsymbol{\beta}^{(0)} + \boldsymbol{H}_1^T \boldsymbol{J}_1 \boldsymbol{T}_1
\end{aligned} \tag{16}$$

Substitute (16) into (11), the output weight matrix is given as in the recursive form as in (17), viz. training in the online way.

$$\boldsymbol{\beta}^{(1)} = \boldsymbol{\beta}^{(0)} + \boldsymbol{K}_1^{-1} \boldsymbol{H}_1^T \left[\boldsymbol{J}_1 \boldsymbol{T}_1 - \left(\boldsymbol{J}_1 + \lambda \boldsymbol{L}_{\aleph_1}\right) \boldsymbol{H}_1 \boldsymbol{\beta}^{(0)}\right] \tag{17}$$

The derived output weight $\boldsymbol{\beta}^{(1)}$ and relative $\boldsymbol{K}_1$ for squential problem with semi-supervised method in (15) and (17) can be extended to the general situation of the *(k+1)*th chunk of training data by updating equation like OS-ELM, so we get (18) and (19):

$$\boldsymbol{K}_{k+1} = \boldsymbol{K}_k + \boldsymbol{H}_{k+1}^T \left(\boldsymbol{J}_{k+1} + \lambda \boldsymbol{L}_{\aleph_{k+1}}\right) \boldsymbol{H}_{k+1} \tag{18}$$

$$\boldsymbol{\beta}^{(k+1)} = \boldsymbol{\beta}^{(k)} + \boldsymbol{K}_{k+1}^{-1} \boldsymbol{H}_{k+1}^T \left[\boldsymbol{J}_{k+1} \boldsymbol{T}_{k+1} - \left(\boldsymbol{J}_{k+1} + \lambda \boldsymbol{L}_{\aleph_{k+1}}\right) \boldsymbol{H}_{k+1} \boldsymbol{\beta}^{(k)}\right] \tag{19}$$

So far we have derived the formula to calculate the output weight in our proposed SOS-ELM method, which provides an effective solution of make most use of labeled and unlabeled data for the sequential problem. The algorithm flow of our proposed SOS-ELM is summarized as follows:

Given a dataset including labeled samples $\{(\boldsymbol{x}_i, \boldsymbol{t}_i) \,|\boldsymbol{x}_i \in \boldsymbol{R}^n, \boldsymbol{t}_i \in \boldsymbol{R}^m, i = 1, 2, ..., N_l\}$ and unlabeled samples $\{\boldsymbol{x}_i^{'} | \boldsymbol{x}_i^{'} \in \boldsymbol{R}^n, i = 1, 2, ..., N_u\}$, when specified the hidden nodes L, the constant λ and the activation function $g(x)$, two training steps are illustrated as follows.

Step 1: Initial phase: Given a small initial training set $\aleph_0 = \{(\boldsymbol{x}_i, \boldsymbol{t}_i) \text{ or } \boldsymbol{x}_i^{'}\}_{i=1}^{N_0}$ to calculate the the output weight matrix $\boldsymbol{\beta}^{(0)}$ and $\boldsymbol{K}_0$ through the following procedure:

(a) Assign random input weight $\boldsymbol{\omega}_i$ and bias b_i or center $\boldsymbol{\mu}_i$ and width σ_i, $i = 1, 2, ..., L$.

(b) Record the number of unlabeled instances n_u and labeled instances n_l, calculate the initial $\boldsymbol{J}_0$ and $\boldsymbol{L}_{\aleph_0}$.

(c) Estimate the initial output weight $\boldsymbol{\beta}^{(0)} = \boldsymbol{K}_0^{-1}\boldsymbol{H}_0^T\boldsymbol{J}_0\boldsymbol{T}_0$, where $\boldsymbol{K}_0 = \boldsymbol{I} + \boldsymbol{H}_0^T\boldsymbol{J}_0\boldsymbol{H}_0 + \lambda\boldsymbol{H}_0^T\boldsymbol{L}_0\boldsymbol{H}_0$.

(d) Set $k = 0$.

Step 2: Sequential Learning Phase: When a new chunk of training data comes, do

(a) Record the number of unlabeled instances n_u and labeled instances n_l, then calculate $\boldsymbol{J}_{k+1}$ and $\boldsymbol{L}_{\aleph_{k+1}}$.

(b) Calculate latest output weight $\boldsymbol{\beta}^{(k+1)}$ based on the recursive Equation (19).

4 Experiment Results

In this section, we will systematically evaluate the performance of our proposed SOS-ELM for regression and classification problems on some benchmark datasets by comparing with that of SS-ELM and OS-ELM. The benchmark datasets used in the paper is shown in Table 1. The number of labeled data, unlabeled data and testing data used in our experiments are also listed in the Table 1.

4.1 The Configuration of Parameters for Related ELM Algorithm

The tradeoff parameter λ and penalty coefficient C_0 in SOS-ELM and SS-ELM learning methods need to be specified in advance, and in our experiments the former is selected from the linear sequence $\{0.05, 0.1, ..., 1\}$, the latter is randomly selected from the exponential sequence $\{10^{-5}, 10^{-4}, ..., 10^{5}\}$. To our designed a 2-fold validation in labeled training data, the λ is selected based on the produced performance. In the initial phase, labeled chunk size N_0 in SOS-ELM and OS-ELM, and unlabeled chunk size M_0 in SOS-ELM, also need to be specified before learning, and referring to the experimental condition reported in [8], we set the $N_0 = 1/10 \times N_l$ for regression problems while N_0 and M_0 are set randomly within the ranges $[50, 200]$ for classification problems.

In our experiment, the training data are generated randomly for 10 times at the ratio of those shown in Table 1, and 10 trials are done on every combination of training data and testing data, the final results are the average of all trials (10×10 cross validation).

Table 1. Specification of benchmark datasets

Dataset	♯Attributes	♯Class	♯Labeled	♯Unlabeled	♯Testing
SinC	1	-	50	2500	2500
Abalone	8	-	50	2500	1627
Auto-MPG	7	-	30	300	62
g241c	241	2	500	500	500
g241n	241	2	500	500	500
Digit1	241	2	500	500	500
USPS(B)	241	2	500	500	500

4.2 Performance Evaluation for Regression Problems

For Regression, three UCI benchmark problems have been studied, namely: SinC, Auto-MPG, Abalone [15].

The regression performance of SOS-ELM, OS-ELM and SS-ELM in terms of testing RMSE and training time are shown in Fig.1.

We have a maximum number (shown in Table 1) of labeled and unlabeled samples in each dataset, and show the effect of increased numbers of unlabeled data. Fig.1 (a), (c) and (e) explore the training time at different numbers of

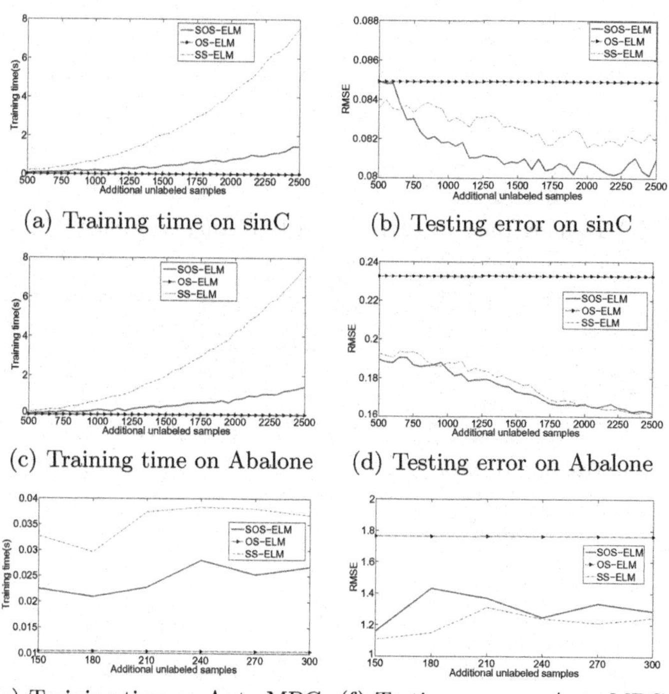

(a) Training time on sinC (b) Testing error on sinC

(c) Training time on Abalone (d) Testing error on Abalone

(e) Training time on Auto-MPG (f) Testing error on Auto-MPG

Fig. 1. Experiments for regression problems

unlabeled samples with labeled samples number being fixed and Fig.1 (b), (d) and (f) explore the corresponding testing error.

We can observe that our proposed SOS-ELM consistently achieves the best generalization performance with extremely fast learning speed.The accuracy of SOS-ELM is almost the same as SS-ELM even higher, while the speed of SOS-ELM is much faster than SS-ELM. As shown in Fig.1 (a), (c) and (e), as the number of unlabeled training samples increases, the training time of SS-ELM grows quickly, while the training time of SOS-ELM almost remain unchanged as OS-ELM.

4.3 Performance Evaluation for Classification Problems

For classification studies, four benchmark problems are considered, including g241c, g241n, Digit1, USPS(B) [18]. Specially, the USPS(B) classes is one of

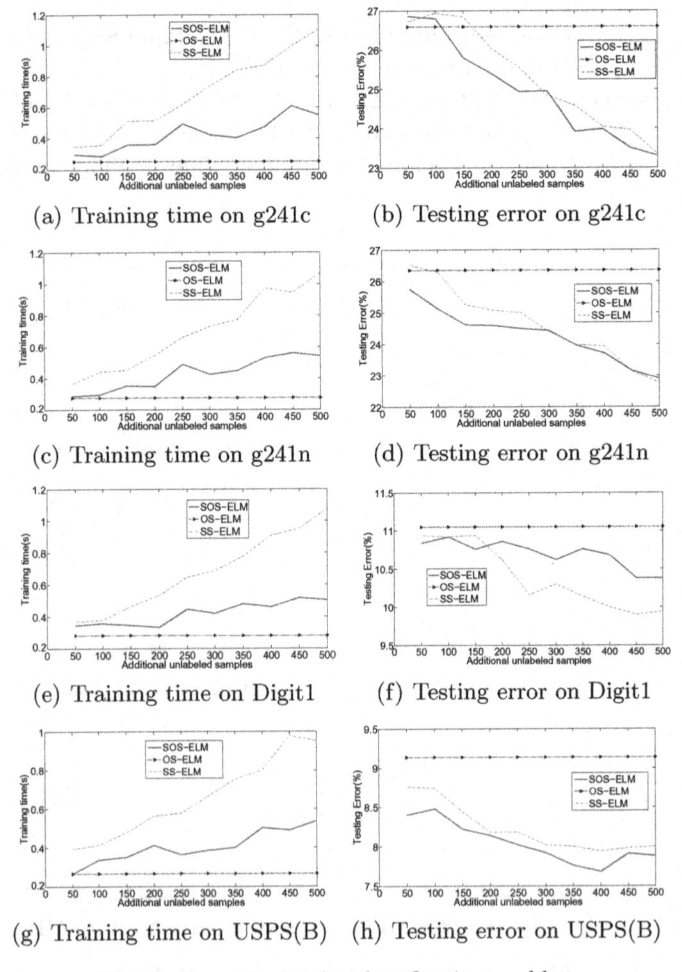

(a) Training time on g241c (b) Testing error on g241c

(c) Training time on g241n (d) Testing error on g241n

(e) Training time on Digit1 (f) Testing error on Digit1

(g) Training time on USPS(B) (h) Testing error on USPS(B)

Fig. 2. Experiments for classification problems

challenge one which has imbalanced instances between categories with relative sizes of 1:4.

The classification results using the proposed SOS-ELM, SS-ELM and OS-ELM respectively are shown in Fig.2. Fig.2 (a), (c), (e) and (g) give the corresponding training time along different number of unlabeled samples, from 50 to 500, and the labeled samples is fixed as 50 and Fig.2 (b), (d), (f) and (h) give the corresponding classification errors.

All the results on three balanced datasets and one imbalanced dataset (USPS(B)) demonstrate that our proposed SOS-ELM outperforms OS-ELM and SS-ELM with the leverage performance of classification accuracy and training speed. The accuracy of SOS-ELM is almost similar with that of SS-ELM or even higher, but the training speed of SOS-ELM is much faster than that of SS-ELM. And the superiority of our proposed SOS-ELM is much distinct when the training data size is large. Compared with OS-ELM, SOS-ELM obtained much higher accuracy especially when the unlabeled data was big enough, while the training speed was almost at the same level.

5 Conclusions

In this paper, a new algorithm in the ELM family called semi-supervised OS-ELM (SOS-ELM) is proposed. This algorithm can not only handle data arriving chunk-by-chunk like OS-ELM, but also reduce the requirement of labeled data and increases the performance with utilizing the unlabeled data as well. The performance of SOS-ELM is evaluated by comparing with that of OS-ELM and SS-ELM on real world benchmark datasets for regression and classification problems. The results demonstrate that proposed SOS-ELM shows stable good performance in different applications with less error with fast training speed.

Acknowledgments. This research is partly supported by the Beijing Natural Science Foundation under Grant (No.4122004), Specialized Research Fund for the Doctoral Program of Higher Education (20121103110031) and the Importation and Development of High-Caliber Talents Project of Beijing Municipal Institutions. The authors also thank A/Prof. Guangbin Huang for his invaluable suggestions.

References

1. Huang, G.-B., Zhu, Q.-Y., Siew, C.-K.: Extreme learning machine: a new learning scheme of feedforward neural networks. In: Proceedings of International Joint Conference on Neural Networks (IJCNN 2004), vol. 2, pp. 985–990 (2004)
2. Huang, G.-B., Zhu, Q.-Y., Siew, C.-K.: Extreme learning machine: theory and applications. Neurocomputing 70(1), 489–501 (2006)
3. Huang, G.-B., Chen, L., Siew, C.-K.: Universal approximation using incremental constructive feedforward networks with random hidden nodes. IEEE Trans Neural Netw. 17(4), 879–892 (2006)

4. Huang, G.-B., Chen, L.: Convex incremental extreme learning machine. Neurocomputing 70, 3056–3062 (2007)
5. Huang, G.-B., Zhou, H., Ding, X., et al.: Extreme learning machine for regression and multiclass classification. IEEE Transactions on Systems, Man, and Cybernetics, Part B: Cybernetics 42(2), 513–529 (2012)
6. Huang, G.-B., Ding, X., Zhou, H.: Optimization method based extreme learning machine for classification. Neurocomputing 74(1), 155–163 (2010)
7. Lan, Y., Soh, Y.C., Huang, G.-B.: Constructive hidden nodes selection of extreme learning machine for regression. Neurocomputing 73(16), 3191–3199 (2010)
8. Lan, Y., Soh, Y.C., Huang, G.-B.: Two-stage extreme learning machine for regression Neurocomputing 73(16), 3028–3038 (2010)
9. Huang, G.-B.: An insight into extreme learning machines: random neurons, random features and kernels. Cognitive Computation, pp. 1–15 (2014)
10. Liang, N.-Y., Huang, G.-B., Saratchandran, P., et al.: A fast and accurate online sequential learning algorithm for feedforward networks. IEEE Transactions on Neural Networks 17(6), 1411–1423 (2006)
11. Li, K., Zhang, J., Xu, H., et al.: A Semi-supervised extreme learning machine method based on co-training. J Comput. Inf. Syst. 9(1), 207–214 (2013)
12. Huang, G., Song, S., Gupta, J.N.D., et al.: Semi-supervised and unsupervised extreme learning machines. IEEE Transactions on Systems, Man, and Cybernetics, Part B: Cybernetics (2014)
13. Tang, X., Han, M.: A semi-supervised learning method based on extreme learning machine. Journal of Dalian University of Technology 50(5), 771–776 (2010)
14. Liu, J.-F., Chen, Y.-Q., Liu, M.-J., et al.: SELM: Semi-supervised ELM with application in sparse calibrated location estimation. Neurocomputing 74(16), 2566–2572 (2011)
15. Blake, C., Merz, C.: UCI repository of machine learning databases Dept. Inf. Comp. Sci. Univ.California, Irvine (1998)
16. Huang, G.-B., Zhu, Q.-Y., Siew, C.-K.: Extreme learning machine: theory and applications. Neurocomputing 70(1), 469–501 (2006)
17. Vapnik, V.: The nature of statistical learning theory. Springer (2000)
18. Chapelle, O., Sch, B., Zien, A.: Semi-Supervised Learning. MIT Press (2006), `http://www.kyb.tuebingen.mpg.de/ssl-book`
19. Belkin, M., Matveeva, I., Niyogi, P.: Regularization and semi-supervised learning on large graphs. In: Shawe-Taylor, J., Singer, Y. (eds.) COLT 2004. LNCS (LNAI), vol. 3120, pp. 624–638. Springer, Heidelberg (2004)

ELM Feature Mappings Learning: Single-Hidden-Layer Feedforward Network without Output Weight

Yimin Yang[1,2], Q.M. Jonathan Wu[1], Yaonan Wang[3], Dibyendu Mukherjee[1], and Yanjie Chen[3]

[1] Department of Electrical and Computer Engineering, University of Windsor, Windsor, N9B 3P4, Canada

[2] College of Electric Engineering, Guangxi University, Nanning, 530004, China

[3] College of Electrical and Information Engineering, Hunan University, Changsha, 410082, China

Abstract. According to conventional neural network theories, the feature of single-hidden-layer feedforward neural networks(SLFNs) resorts to parameters of the weighted connections and hidden nodes. SLFNs are universal approximators when at least the parameters of the networks including hidden-node parameter and output weight exist. Unlike above neural network theories, this paper indicates that in order to let SLFNs work as universal approximators, one may simply calculate only the hidden node parameter and the output weight is not required at all. In other words, this proposed neural network architecture can be considered as a standard SLFN without output weights. Furthermore, this paper presents experiments which show that the proposed learning method tends to extremely reduce network output error to a very small number with only several hidden nodes. Simulation results demonstrate that the proposed method can provide hundreds times faster learning speed compared to other learning algorithms including BP, SVM/SVR and other ELM methods.

Keywords: Bidirectional extreme learning machine, Feedforward neural network, Universal approximation, Learning effectiveness.

1 Introduction

The widespread popularity of neural networks in many fields is mainly due to their ability to approximate complex nonlinear mappings directly from the input samples. In the past two decades, due to their universal approximation capability, feedforward neural networks (FNNs) have been extensively used in classification and regression problem [1]. According to Jaeger's estimation, 95% of relevant literature is mainly on FNNs. As a specific type of FNN, the single-hidden-layer feedforward network (SLFNs) plays an important role in practical applications. For N arbitrary distinct samples $(\mathbf{x}_i, \mathbf{t}_i)$, where $\mathbf{x}_i = [x_{i1}, x_{i2}, \cdots, x_{in}]^T \in \mathbf{R}^n$ and $\mathbf{t}_i \in \mathbf{R}^m$, the input data is mapped to L-dimensional feature mapping space, and the network output is

$$f_L(\mathbf{x}) = \sum_{i=1}^{L} \beta_i g(\mathbf{a}_i \cdot \mathbf{x}_j + b_i) = \mathbf{H}(\mathbf{x})\beta, j = 1, \cdots, N \tag{1}$$

J. Cao et al. (eds.), *Proceedings of ELM-2014 Volume 1*,
Proceedings in Adaptation, Learning and Optimization 3, DOI: 10.1007/978-3-319-14063-6_27

where $h(\mathbf{a}_i, b_i, \mathbf{x})$ denotes the output of the ith hidden node with the hidden-node parameters $(\mathbf{a}_i, b_i) \in \mathbf{R}^n \times \mathbf{R}$ and $\beta_i \in \mathbf{R}$ is the output weight between the ith hidden node and the output node. $\mathbf{a}_i \cdot \mathbf{x}$ denotes the inner product of vector $\mathbf{a}_i$ and $\mathbf{x}$ in $\mathbf{R}^n$. $\mathbf{H}$ is called the feature mapping.

An active topic on the universal approximation capability of SLFNs is on how to determine the parameters $\mathbf{a}_i, b_i$, and $\beta_i (i = 1, \cdots, L)$ such that the network output $f_L(\mathbf{x})$ can approximate a given target $(\mathbf{T}, \mathbf{T} = [\mathbf{t}_1, \cdots, \mathbf{t}_N])$. The feature of SLFNs resorts to parameters of the output weight and hidden nodes parameters. According to conventional neural network theories, SLFNs are universal approximators when all the parameters of the networks including the hidden-node parameters $(\mathbf{a}, b)$ and output weight β are allowed to be adjustable [2].

Unlike above neural network theories claiming that all the parameters in networks are allowed to be adjustable, other researches proposed some semi-random network theories [3][4][5]. For example, Lowe [5] focus on a specific RBF network: the centers $\mathbf{a}$ in [5] can be randomly selected from the training data instead of tuning, but the impact factor b of RBF hidden node is not randomly selected and usually determined by users.

Unlike above semi-random network theories, in 2006, Huang *et. al.* [6][7] illustrated that iterative techniques are not required in adjusting all the parameters of SLFNs at all. Based on this idea, Huang *et. al.* proposed full-random learning method referred to as extreme learning machine (ELM). In [6][7], Huang *et al* have proved that SLFNs with randomly generated hidden node parameter can work as universal approximators by only calculating the output weights linking the hidden layer to the output nodes. Recent ELM development [8][9] shows that ELM unifies FNNs and SVM/LS-SVM. Compared to ELM, LS-SVM and PSVM achieve suboptimal solutions and have a higher computational cost. Also, applications of ELM have recently been presented in computer vision [10][11], feature selection [12][13], power system analysis [14], automation control [15], etc.

Above neural network theories indicate that SLFNs can work as universal approximation if at least hidden-node parameters[1] and output weight should be exist. However, in this paper we indicate that output weights do not need exist in SLFNs at all. During the recent 20 years, researchers who are trying to use NNs in their research are facing following questions [16]: 1) What neural network architecture should be used? 2) How many neurons should be used? Unfortunately, current neural network theories cannot easy answers these questions. For first question, current neural network theories indicate that NNs can work as universal approximation if at least hidden-node parameters and output weight should be exist. For second question, traditional theories demonstrate that with a correct number of hidden nodes with correct parameters, any network can approximate a given function to a certain degree of accuracy.

In [15] we proposed a learning algorithm, called bidirectional extreme learning machine (B-ELM) in which half of hidden-node parameters are not randomly selected and are calculated by pulling back the network residual error to input weight. Our recent experimental results indicate that in B-ELM[15], output weights play a very

[1] Hidden-node parameters can be generated randomly.

minion role in the network learning effectiveness. Furthermore, recently Huang and *et al* indicate that ELM feature mapping can also be used for feature selection and clustering. Inspired by these results, in this paper, we show that B-ELM feature mapping can be used as a universal approximators. We indicate that SLFNs without output weights can approximate any target continuous function and classify any disjoint region if one pulls back error to hidden-node parameters. Thus different from traditional network theories, in this paper, we answer these two questions as follow: 1) output weights do not need exist in SLFN at all; 2) this kind of network with m hidden nodes can still approximate any target continuous function and classify any disjoint regions. In particular, the following contributions have been made in this paper.

1) The learning speed of proposed learning method can be hundreds times faster learning speed compared to other learning algorithms including SVM, BP and other ELMs. Furthermore, it can provide good generalization performance and can be directly applied to regression and classification applications.

2) Contrary to conventional SLFNs which require the hidden node parameters and output weights, in this paper, we proved that SLFNs without output weight can still approximate any target continuous function and classify any disjoint region. Thus the architecture of this single parameter neural network is considerably simpler compared to traditional SLFNs.

3) One of the major difficulties facing researchers using NNs is the selection of the proper size of the networks. This paper shows that the proposed learning method with m hidden nodes can give significant improvements on accuracy instead of maintaining a large number of hidden nodes[2]. Experimental results show that the proposed method with only m hidden nodes can provide comparable or improved generalization performance in comparison to other networks with thousands of hidden nodes.

2 Preliminaries and Notation

2.1 Notations and Definitions

The sets of real, integer, positive real and positive integer numbers are denoted by $\mathbf{R}, \mathbf{Z}, \mathbf{R}^+$ and $\mathbf{Z}^+$, respectively. Similar to [6], let $F^2(X)$ be a space of functions f on a compact subset X in the n-dimensional Euclidean space $\mathbf{R}^n$ such that $|f|^2$ are integrable, that is, $\int_X |f(x)|^2 dx < \infty$. Let $F^2(\mathbf{R}^n)$ be denoted by F^2. For $u, v \in F^2(X)$, the inner product $< u, v >$ is defined by

$$< u, v >= \int_X u(\mathbf{x})\overline{v(\mathbf{x})}d\mathbf{x} \tag{2}$$

The norm in $F^2(X)$ space will be denoted as $\|\cdot\|$. L denotes the number of hidden nodes. For N training samples, $\mathbf{x}, \mathbf{x} \in \mathbf{R}^{N\times n}$ denotes the input matrix of network, $\mathbf{T} \in \mathbf{R}^{N\times m}$ denotes the desire output matrix of network. $\mathbf{H} \in \mathbf{R}^{N\times m}$ is called the hidden layer output matrix of the SLFNs; the ith column of $\mathbf{H}$ ($\mathbf{H}_i$) is the ith hidden node output with respect to inputs. The hidden layer output matrix $\mathbf{H}_i$ is said to

[2] m equal to the desired output dimensionality. Thus in real applications, m is a fixed value.

be randomly generated function sequence $\mathbf{H}_i^r$ if the corresponding hidden-node parameters $(\mathbf{a}_i, b_i)$ are randomly generated. $\mathbf{e}_n$ ($\mathbf{e}_n \in \mathbf{R}^{N\times m}$) denotes the residual error function for the current network f_n with n hidden nodes. $\mathbf{I}$ is unit matrix and $\mathbf{I} \in \mathbf{R}^{m\times m}$.

3 SLFNs without Output Weight

Basic idea: For fixed output weight $\boldsymbol{\beta}$ equal to unit matrix or vector ($\beta \in \mathbf{R}^{m\times m}$), to train an SLFN is simply equivalent to finding a least-square solution $\mathbf{a}^{-1}$ of the linear system $\mathbf{H}(\mathbf{a},\mathbf{x}) = \mathbf{T}$. If activation function can be invertible, to train an SLFN is simply equivalent to pulling back residual error to input weight. For example, for N arbitrary distinct samples $\{\mathbf{x},\mathbf{T}\}$, ($\mathbf{x} \in \mathbf{R}^{N\times n}, \mathbf{T} \in \mathbf{R}^{N\times m}, \mathbf{T} \in [0,1]$), if activation function is sine function, to train an SLFN is simply equivalent to finding a least-square solution $\hat{\mathbf{a}}$ of the linear system $\mathbf{a}\cdot\mathbf{x} = \arcsin(\mathbf{T})$:

$$\|\mathbf{H}(\hat{\mathbf{a}}_1,\cdots,\hat{\mathbf{a}}_n,\mathbf{x}) - \mathbf{T}\| = \min_{\mathbf{a}} \|\mathbf{H}(\mathbf{a}_1,\cdots,\mathbf{a}_n,\mathbf{x}) - \mathbf{T}\| \tag{3}$$

According to [16], the smallest norm least-squares solution of the above linear system is $\hat{\mathbf{a}_n} = \arcsin(\mathbf{e}_{n-1})\cdot\mathbf{x}^{-1}$. Based on this idea, we give the following theorem.

Lemma 1. *[6] Given a bounded nonconstant piecewise continuous function $H : \boldsymbol{R} \to \boldsymbol{R}$, we have*

$$\lim_{(\boldsymbol{a},b)\to(\boldsymbol{a}_0,b_0)} \|\boldsymbol{H}(\boldsymbol{a}\cdot\boldsymbol{x}+b) - \boldsymbol{H}(\boldsymbol{a}_0\cdot\boldsymbol{x}+b_0)\| = 0 \tag{4}$$

Theorem 1. *Given N arbitrary distinct samples $\{\boldsymbol{x},\boldsymbol{T}\}, \boldsymbol{x} \in \boldsymbol{R}^{N\times n}, \boldsymbol{T} \in \boldsymbol{R}^{N\times m}$, given the sigmoid or sine activation function h, for any continuous desired output $\boldsymbol{T}$, $\lim_{n\to\infty} \|\boldsymbol{T} - (\hat{\boldsymbol{H}}_1(\hat{\boldsymbol{a}}_1,\hat{b}_1,\boldsymbol{x})\boldsymbol{\beta}_1 + \cdots + \hat{\boldsymbol{H}}_n(\hat{\boldsymbol{a}}_n,\hat{b}_n,\boldsymbol{x})\boldsymbol{\beta}_n\| = 0$ holds with probability one if*

$$\begin{aligned} &\boldsymbol{H}_n^e = \boldsymbol{e}_{n-1} \\ &\hat{\boldsymbol{a}}_n = h^{-1}(u(\boldsymbol{H}_n^e))\cdot\boldsymbol{x}^{-1}, \hat{\boldsymbol{a}}_n \in \boldsymbol{R}^{n\times m} \\ &\hat{b}_n = \sqrt{mse(h^{-1}(u(\boldsymbol{H}_n^e)) - \hat{\boldsymbol{a}}_n\cdot\boldsymbol{x})}, \hat{b}_n \in \boldsymbol{R}^m \end{aligned} \tag{5}$$

$$\hat{\boldsymbol{H}}_n = u^{-1}(h(\hat{\boldsymbol{a}}_n\cdot\boldsymbol{x} + \hat{b}_n)) \tag{6}$$

where if activation function h is sin/cos, given a normalized function $u : \boldsymbol{R} \to [0,1]$; if activation function h is sigmoid, given a normalized function $u : \boldsymbol{R} \to (0,1]$. h^{-1} and u^{-1} represent their reverse function, respectively. If h is sine activation function, $h^{-1}(\cdot) = \arcsin(u(\cdot)$; if h is sigmoid activation function, $h^{-1}(\cdot) = -\log(\frac{1}{u(\cdot)} - 1)$, $\boldsymbol{x}^{-1} = \boldsymbol{x}^T(\boldsymbol{I} + \boldsymbol{x}\boldsymbol{x}^T)^{-1}$.

Proof. For an activation function $h(x) : \mathbf{R} \to \mathbf{R}$, $\mathbf{H}_n^e$ is given by

$$\mathbf{H}_n^e = h(\boldsymbol{\lambda}_n) \tag{7}$$

In order to let $\boldsymbol{\lambda}_{2n} \in \mathbf{R}^m$, here we give a normalized function $u(\cdot)$: $u(\mathbf{H}) \in [0,1]$ if activation function is sin/cos; $u(\mathbf{H}) \in (0,1)$ if activation function is sigmoid. Then for sine hidden node

$$\boldsymbol{\lambda}_{2n} = h^{-1}(u(\mathbf{H}_n^e)) = (arcsin(u(\mathbf{H}_n^e)) \tag{8}$$

For sigmoid hidden node

$$\boldsymbol{\lambda}_n = h^{-1}(u(\mathbf{H}_n^e)) = -\log(\frac{1}{u(\mathbf{H}_n^e)} - 1) \tag{9}$$

Let $\boldsymbol{\lambda}_n = \mathbf{a}_n \cdot \mathbf{x}$, for sine activation function, we have

$$\hat{\mathbf{a}}_n = h^{-1}(u(\mathbf{H}_n^e)) \cdot \mathbf{x}^{-1} = arcsin(u(\mathbf{H}_n^e)) \cdot \mathbf{x}^{-1} \tag{10}$$

For sigmoid activation function, we have

$$\hat{\mathbf{a}}_n = h^{-1}(u(\mathbf{H}_n^e)) \cdot \mathbf{x}^{-1} = -\log(\frac{1}{u(\mathbf{H}_n^e)} - 1) \cdot \mathbf{x}^{-1} \tag{11}$$

where $\mathbf{x}^{-1}$ is the Moore-Penrose generalized inverse of the given set of training examples [17]. Similar to [7], we have

1: $\hat{\mathbf{a}}_n = arcsin(u(\mathbf{H}_n^e)) \cdot \mathbf{x}^{-1}$ is one of the least-squares solutions of a general linear system $\mathbf{a}_n \cdot \mathbf{x} = \boldsymbol{\lambda}_n$, meaning that the smallest error can be reached by this solution:

$$\|\hat{\mathbf{a}}_n \cdot \mathbf{x} - \boldsymbol{\lambda}_n\| = \|\hat{\mathbf{a}}_n \mathbf{x}^{-1}\mathbf{x} - \boldsymbol{\lambda}_n\| = \min_{\mathbf{a}_n} \|\mathbf{a}_n \cdot \mathbf{x} - arcsin(u(\mathbf{H}_n^e))\| \tag{12}$$

2: the special solution $\hat{\mathbf{a}}_n = h^{-1}(u(\mathbf{H}_n^e)) \cdot \mathbf{x}^{-1}$ has the smallest norm among all the least-squares solutions of $\mathbf{a}_n \cdot \mathbf{x} = \boldsymbol{\lambda}_n$, which guarantees that that $\mathbf{a}_n \in [-1, 1]$. Although the smallest error can be reached by equation (10)-(11), we still can reduce its error by adding bias b_n. For sine activation function:

$$\begin{aligned}\hat{b}_n &= \sqrt{mse(h^{-1}(u(\mathbf{H}_n^e)) - \hat{\mathbf{a}}_n \cdot \mathbf{x})} \\ &= \sqrt{mse(arcsin(u(\mathbf{H}_n^e)) - \hat{\mathbf{a}}_n \cdot \mathbf{x})}\end{aligned} \tag{13}$$

For sigmoid activation function

$$\begin{aligned}\hat{b}_n &= \sqrt{mse(h^{-1}(u(\mathbf{H}_n^e)) - \hat{\mathbf{a}}_n \cdot \mathbf{x})} \\ &= \sqrt{mse((-\log(1/u(\mathbf{H}_n^e) - 1)) - \hat{\mathbf{a}}_n \cdot \mathbf{x})}\end{aligned} \tag{14}$$

According to equation (12) and *Lemma* 1, we have

$$\begin{aligned}&\min_{\mathbf{a}_n} \|u^{-1}(h(\mathbf{a}_n \cdot \mathbf{x})) - u^{-1}(h(\boldsymbol{\lambda}_n))\| \\ &= \|u^{-1}(h(\hat{\mathbf{a}}_n \cdot \mathbf{x})) - u^{-1}(h(\boldsymbol{\lambda}_n))\| \\ &> \|u^{-1}(h(\hat{\mathbf{a}}_n \cdot \mathbf{x} + \hat{b}_n)) - u^{-1}(h(\boldsymbol{\lambda}_n))\| = \|\boldsymbol{\sigma}\|\end{aligned} \tag{15}$$

We consider the residual error as

$$\begin{aligned}\Delta &= \|e_{n-1}\|^2 - \|e_{n-1} - \mathbf{H}_n^e\|^2 \\ &= 2\beta_n \langle e_{n-1}, \mathbf{H}_n^e \rangle - \|\mathbf{H}_n^e\|^2 \\ &= \|\mathbf{H}_n^e\|^2 (\frac{2\langle e_{n-1}, \mathbf{H}_n^e \rangle}{\|\mathbf{H}_n^e\|^2} - 1)\end{aligned} \tag{16}$$

Let

$$\begin{aligned}\hat{\mathbf{H}}_n^e &= u^{-1}(h(\hat{\mathbf{a}}_n \cdot \mathbf{x} + \hat{b}_n)) \\ &= \mathbf{e}_{n-1} \pm \boldsymbol{\sigma} \\ &= \hat{\mathbf{e}}_{n-1}\end{aligned} \tag{17}$$

Because $\|\hat{\mathbf{e}}_{n-1}\| \geq \|\boldsymbol{\sigma}\|$, we have $\Delta \geq 0$ still valid for

$$\begin{aligned}\Delta &= \|\hat{\mathbf{H}}_n^e\|^2 \left(\frac{2\|\|\langle \hat{\mathbf{e}}_{n-1} \pm \boldsymbol{\sigma}, \hat{\mathbf{e}}_{n-1}\rangle}{\|\hat{\mathbf{e}}_{n-1}\|^2} - 1\right) \\ &= \|\hat{\mathbf{H}}_n^e\|^2 \left(\frac{2(\|\hat{\mathbf{e}}_{n-1}\|^2 \pm \langle \boldsymbol{\sigma}, \hat{\mathbf{e}}_{n-1}\rangle)}{\|\hat{\mathbf{e}}_{n-1}\|^2} - 1\right) \\ &= \|\hat{\mathbf{H}}_n^e\|^2 \left(1 \pm \frac{\|\boldsymbol{\sigma} \cdot \hat{\mathbf{e}}_{n-1}^T\|}{\|\hat{\mathbf{e}}_{n-1}\|^2}\right) \\ &\geq \|\hat{\mathbf{H}}_n^e\|^2 \left(1 \pm \frac{\|\boldsymbol{\sigma}\|}{\|\hat{\mathbf{e}}_{n-1}\|}\right) \geq 0\end{aligned} \tag{18}$$

Now based on equation (18), we have $\|\mathbf{e}_{n-1}\| \geq \|\mathbf{e}_n\|$, so the sequence $\|\mathbf{e}_n\|$ is decreasing, bounded below by zero and converges.

Remark 1. According to Theorem 1, for N arbitrary distinct samples $(\mathbf{x}_i, \mathbf{t}_i)$ where $\mathbf{x}_i = [x_{i1}, x_{i2}, \cdots, x_{iN}]^T \in \mathbf{R}^n$ and $\mathbf{t}_i \in \mathbf{R}^m$, the proposed network with $d \times m$ hidden nodes and activation function $h(x)$ is mathematically modeled as

$$f_L(\mathbf{x}) = \sum_{c=1}^{d} \sum_{i=1}^{m} u^{-1}(h(\mathbf{a}_i \cdot \mathbf{x}_j + b_i)), j = 1, \cdots, N \tag{19}$$

where u is a normalized function, $\mathbf{a}_i \in \mathbf{R}^{n \times m}, b_i \in \mathbf{R}^m$.

Remark 2. Different from other neural network learning methods in which output weight parameter should be adjusted, in the proposed method, the output weight of SLFNs can be equal to unit matrix and thus the proposed neural network does not need output weight at all. Thus the architecture and computational cost of the proposed method are much smaller than other traditional SLFNs.

Remark 3. Experiment show that the proposed method at the early learning stage can give similar generalization performance as the proposed network with hundreds of hidden node (see Figure 2). Based on the experimental results, for N arbitrary distinct samples $(\mathbf{x}_i, \mathbf{t}_i)$ where $\mathbf{x}_i = [x_{i1}, x_{i2}, \cdots, x_{iN}]^T \in \mathbf{R}^n$ and $\mathbf{t}_i \in \mathbf{R}^m$, the proposed network is mathematically modeled as

$$f_L(\mathbf{x}) = \sum_{i=1}^{m} u^{-1}(h(\mathbf{a}_1 \cdot \mathbf{x}_j + b_1)), j = 1, \cdots, N \tag{20}$$

where u is a normalized function, $\mathbf{a}_1 \in \mathbf{R}^{n \times m}, b_1 \in \mathbf{R}^m$. Thus algorithm 1 can be modified as algorithm 2.

Algorithm 1. the proposed method

Initialization: Given a training set $\{(\mathbf{x}_i, \mathbf{t}_i)\}_{i=1}^N \subset \mathbf{R}^n \times \mathbf{R}^m$, the activation function $\mathbf{H}(\cdot)$, continues target function f. Set number of hidden nodes $L = m$.
Learning step:
Step 1) set $\mathbf{H}_L^e = \mathbf{T}$;
Step 2) calculate the input weights $\mathbf{a}_1, \cdots, \mathbf{a}_L$, bias $b_1, \cdots, b_L$ based on equation (5);

Table 1. Specification of regression problems

Datasets	Type	#Attri	#Train	# Test
Auto MPG	Regression	8	200	192
Machine CPU	Regression	6	100	109
Fried	Regression	11	20768	20000
Wine Quality	Regression	12	2898	2000
Puma	Regression	9	4500	3692
California Housing	Regression	8	16000	4000
House 8L	Regression	9	16000	6784
Parkinsons motor	Regression	26	4000	1875
Parkinsons total	Regression	26	4000	1875
Puma	Regression	9	6000	2192
Delta elevators	Regression	6	6000	3000
Abalone	Regression	9	3000	1477
A9a	Regression	123	32561	16281
colon-cancer	Classification	2000	40	22
USPS	Classification	256	7291	2007
Sonar	Classification	60	150	58
Hill Valley	Classification	101	606	606
Protein	Classification	357	17766	6621
Covtype.binary	Classification	54	300000	280000
Mushrooms	Classification	112	4000	4122
Gisette	Classification	5000	6000	1000
Leukemia	Classification	7129	38	34
Duke	Classification	7129	29	15
Connect-4	Classification	126	50000	17557
Mnist	Classification	780	40000	30000
DNA	Classification	180	1046	1186
w3a	Classification	300	4912	44837

4 Experimental Verification

In this section, aimed at examining the performance of our proposed learning method, we test the proposed method on 27 regression and classification problems. The experiments are conducted in Matlab 2013a with 32 GB of memory and an E3-1230 v2 (3.3G) processor. In the experiment, Neural networks are first tested on some SLFN methods including ELM, B-ELM, I-ELM, EM-ELM, PC-ELM, EI-ELM, SVR/SVM and BP.

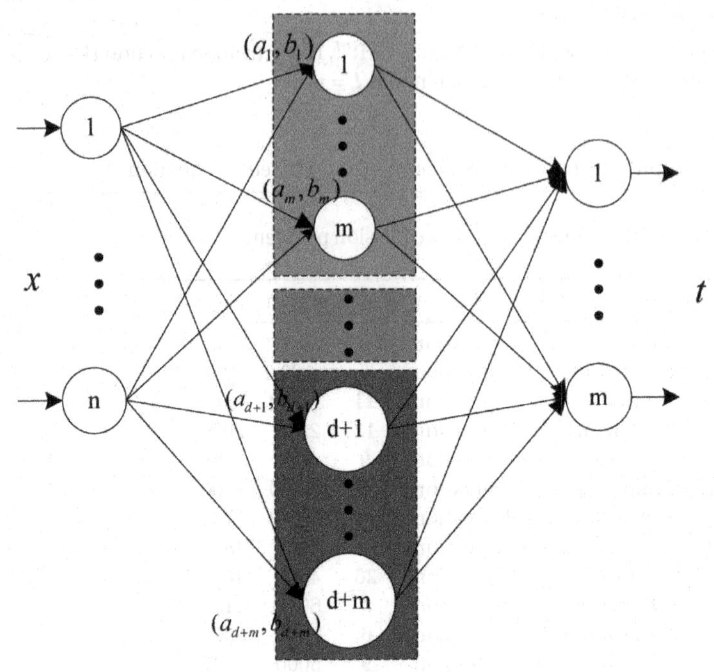

(a) the proposed network architecture

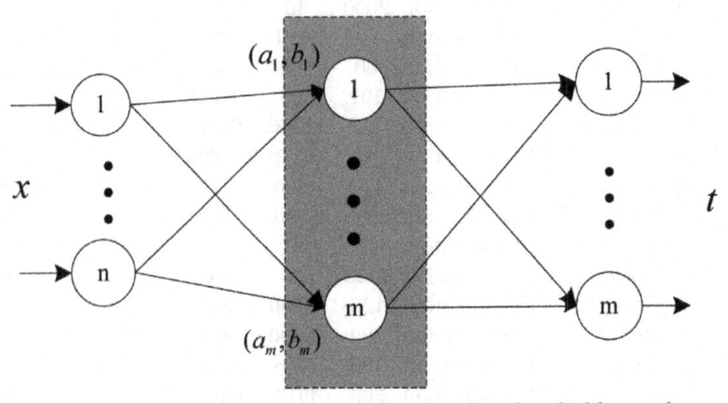

(b) the proposed network with m hidden nodes

Fig. 1. The proposed network architecture: The proposed method could be considered to simply standard feedforward neural networks without output weight β. Further more, the proposed network with m hidden nodes (m is the dimensionality of desired output $\mathbf{t}$) can provide compared performance than other SLFNs with hundreds of hidden nodes.

4.1 Experiment Environment Settings

In order to extensively verify the performance of different algorithms in our experiments, tested datasets are from wide type including *small size, medium dimensions, large size, and/or high dimensions.* These data sets include 13 regression problems, 14 classification problems and 4 image recognition problems (see Table I and VI). Most of the data sets are taken from UCI Machine Learning Repository[3] and LIBSVM DATA SETS[4].

In these data sets, the input data are normalized into $[-1, 1]$ while the output data for regression are normalized into the range $[0, 1]$. All data sets have been preprocessed in the same way (held-out method). Ten different random permutations of the whole data set are taken without replacement, and some(see in tables) are used to create the training set while the remaining is used for the test set. The average results are obtained over 10 trials for all problems. For SVM and SVR, in order to achieve good generalization performance, the cost parameter C and kernel parameter γ of SVM and SVR need to be chosen appropriately. We have tried a wide range of C and γ. For each data set, we have used 30 different value of C and γ, resulting in a total of 900 pairs of (C, γ). The 30 different value of C and γ are $\{2^{-15}, 2^{-14}, \cdots, 2^{14}, 2^{15}\}$. For ELM, EM-ELM and our proposed method, the parameter C is selected from $C \in \{2^{-15}, \cdots, 2^{15}\}$. The experimental results between the proposed method and these competing methods are given in Table II-Table VIII. In these tables, the similar results obtained by different algorithms are underlined and the apparent better results are shown in boldface.

Table 2. Performance comparison (mean-mean testing RMSE; time-training time)

Datasets	I-ELM (200 nodes)		B-ELM (200 nodes)		EI-ELM (200 nodes)		ELM (200 nodes)		EM-ELM (200 nodes)		proposed method (1 nodes)	
	Mean	time(s)	Mean	time(s)	Mean	time(s)	Mean	time(s)	Mean	time(s)	Mean	time(s)
House 8L	0.0946	1.1872	0.0818	3.8821	0.0850	10.7691	0.0718	0.8369	0.0763	7.0388	0.0819	**0.0020**
Auto MPG	0.1000	0.2025	**0.0920**	0.3732	0.0918	1.3004	0.0976	0.0156	0.0968	0.0075	0.0996	**<0.0001**
Machine CPU	0.0591	0.1909	0.0554	0.3469	0.0551	1.2633	0.0513	0.0069	0.0521	0.1385	**0.0439**	**<0.0001**
Fried	0.1135	0.8327	0.0857	5.5063	0.0856	7.4016	0.0619	1.3135	**0.0618**	18.0290	0.0834	**0.0051**
Delta ailerons	0.0538	0.4680	0.0431	1.3946	0.0417	3.5478	0.0431	0.0616	0.0421	0.1342	0.0453	**<0.0001**
PD motor	0.2318	0.4639	0.2241	4.7680	0.2251	3.9016	0.2190	0.2730	0.2196	0.7394	0.2210	**0.0037**
PD total	0.2178	0.4678	0.2137	4.9278	0.2124	3.7854	0.2076	0.2838	0.2094	0.5944	0.2136	**0.0023**
Puma	0.1860	0.5070	0.1832	2.1846	0.1830	4.2161	0.1602	0.3725	**0.1478**	4.8392	0.1808	**0.0012**
Abalone	0.0938	0.3398	0.0808	1.2549	0.0848	2.6676	0.0824	0.0761	0.0817	0.1638	**0.0790**	**0.0017**
Wine	0.1360	0.3516	0.1264	1.7098	0.1266	2.7126	0.1229	0.1950	0.1216	0.3806	0.1250	**0.0031**
California house	0.1801	1.1482	0.1450	7.2625	0.1505	12.0832	0.1354	0.9753	**0.1302**	3.5574	0.1420	**0.0078**

4.2 Real-World Regression Problems

In this subsection, all the incremental ELMs (I-ELM, B-ELM, EI-ELM) increase the hidden nodes one by one till nodes-numbers equal to 200, while for fixed ELMs (ELM, EM-ELM), 200-hidden-nodes are used. It can be seen that the proposed method can always achieve similar performance as other ELMs with much higher learning speed. In Table II and Table IV, for Machine CPU problem, the the proposed method runs 1900 times, 3400 times and 12000 times faster than the I-ELM, B-ELM and EI-ELM, respectively. For Abalone problem, the proposed method runs 200 times, 700 times and 1600 times faster than I-ELM, B-ELM and EI-ELM, respectively. In Table V, for Wine problem, the proposed method runs 120 times and 60 times faster than EM-ELM and ELM, respectively. Also the testing RMSE of EI-ELM

[3] `http://archive.ics.uci.edu/ml/datasets.html`
[4] `http://www.csie.ntu.edu.tw/~cjlin/libsvmtools/datasets/`

Table 3. Performance comparison (mean-mean testing RMSE; time-training time)

Datasets	ELM (1 nodes)		the proposed method (1 nodes)	
	Mean	time(s)	Mean	time(s)
House 8L	0.1083	**0.0009**	**0.0819**	0.0020
Auto MPG	0.2126	< **0.0001**	**0.0996**	<0.0001
Machine CPU	0.1331	<0.0001	**0.0439**	<0.0001
Fried	0.2207	**0.0031**	**0.0834**	0.0051
Delta ailerons	0.0864	<0.0001	**0.0453**	<0.0001
PD motor	0.2620	**0.0020**	**0.2210**	0.0037
PD total	0.2548	**0.0007**	**0.2136**	0.0023
Puma	0.2856	0.0012	**0.1808**	**0.0012**
Delta ele	0.1454	<0.0001	**0.1174**	<0.0001
Abalone	0.1363	**0.0007**	**0.0828**	0.0017
Wine	0.1750	**0.0006**	**0.1250**	0.0031
California house	0.2496	**0.0027**	**0.1420**	0.0078

Table 4. Performance comparison (mean-mean testing RMSE; time-training time)

Datasets	Eplison-SVR		BP		the proposed method (1 nodes)	
	Mean	time(s)	Mean	time(s)	Mean	time(s)
House 8L	0.0799	53.6531	0.0790	27.8462	0.0819	**0.0020**
Auto MPG	0.0985	0.0234	0.0953	1.6034	0.0996	**<0.0001**
Machine CPU	0.0727	0.0187	0.0843	0.7129	**0.0439**	**<0.0001**
Fried	0.0829	197.9534	**0.0591**	81.8774	0.0834	**0.0051**
Delta ailerons	**0.0402**	6.8718	0.0415	12.6735	0.0453	**<0.0001**
California house	0.1529	35.2250	0.1435	54.3081	0.1420	**0.0078**
PD total	0.2082	7.2540	0.2120	12.6438	0.2136	**0.0023**

is 2 times larger than that of B-ELM. The B-ELM runs 1.5 times faster than the I-ELM and the testing RMSE for the obained I-ELM is 5 times larger than that for B-ELM.

If only 1-hidden-node is used, ELM methods such as I-ELM, ELM, EM-ELM and B-ELM can be considered as the same as ELM. Thus in Table III, we carry out performance comparisons between the proposed method and ELM. As observed from Table III, the average testing RMSE obtained by the proposed method are much better than that of ELM. For California house problem, the testing RMSE obtained by ELM is 2 times larger than that of the proposed method. In real applications, SLFNs with only 1 hidden nodes are extremely small network structures, meaning that after training this small size network may respond to new external unknown stimuli much faster and much more accurately than other ELM algorithms in real deployment.

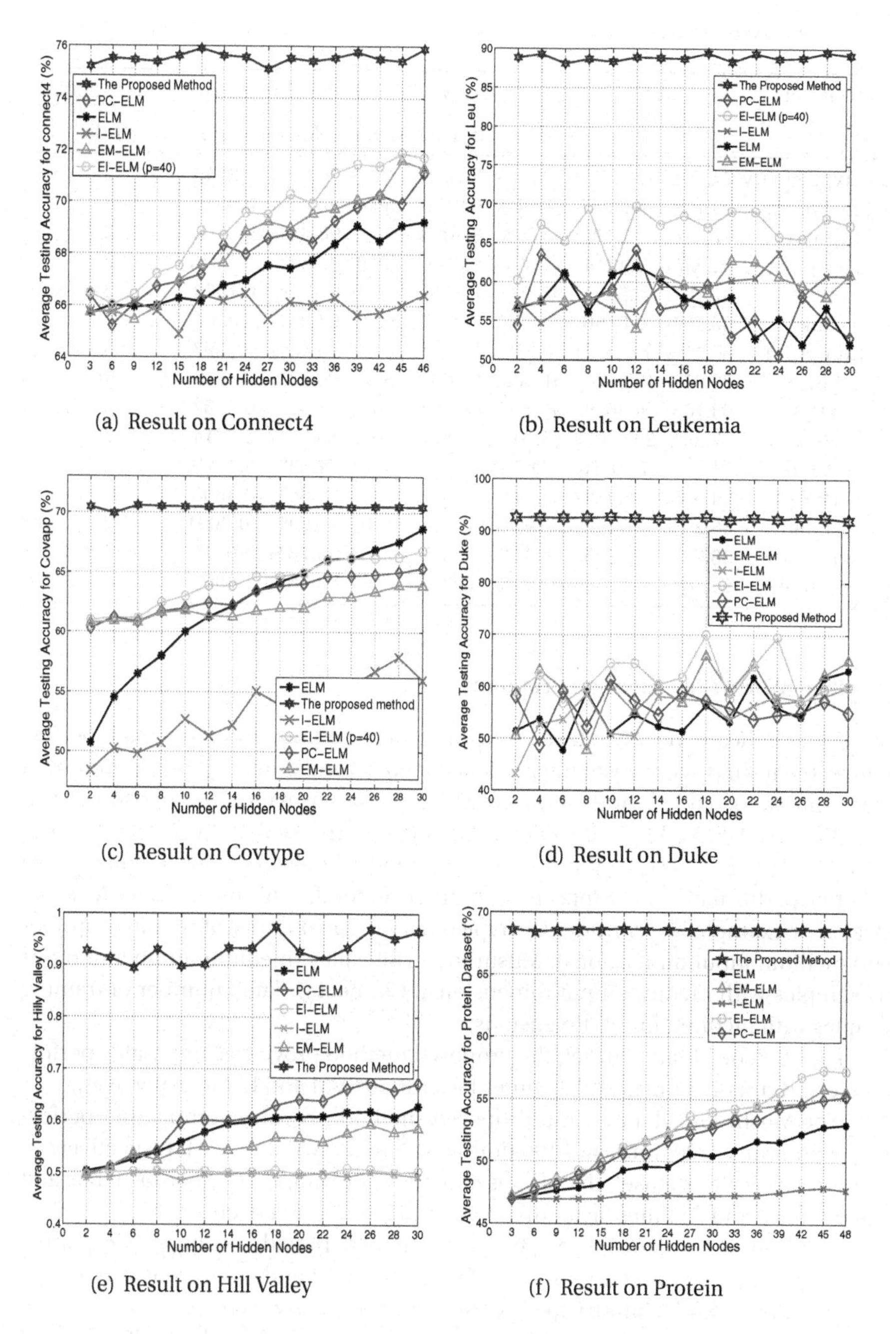

(a) Result on Connect4

(b) Result on Leukemia

(c) Result on Covtype

(d) Result on Duke

(e) Result on Hill Valley

(f) Result on Protein

Fig. 2. Performance of the proposed method with different learning methods

Table 5. Performance comparison (mean-mean testing RMSE; time-training time)

Datasets	SVM		ELM			the proposed method (m nodes)		
	Mean	time(s)	Mean	time(s)	#node	Mean	time(s)	node
Covtype.binary	74.84%	413.5275	77.27%	36.5947	500	76.55%	**1.2043**	2
Mushrooms	89.10%	38.6247	88.91%	0.9126	500	88.84%	**0.0047**	2
Gisette	77.68%	309.3968	88.69%	**6.4093**	500	**98.00%**	48.2027	2
Leukemia	82.58%	**2.3914**	77.06%	9.0340	5000	**88.24%**	20.9915	2
W3a	97.18%	4.5552	97.95%	0.9095	500	97.20%	**0.1872**	2
Duke	86.36%	**0.0156**	79.32%	**7.8437**	5000	**92.95%**	20.0352	2
Connect-4	66.01%	569.6221	76.89%	7.3757	500	75.40%	**0.7597**	3
Mnist	70.85%	478.4707	**91.60%**	8.1651	500	86.70%	8.8858	10
DNA	93.70%	0.4680	91.42%	0.2122	500	93.51%	**0.0187**	3
A9a	77.39%	295.0603	85.10%	4.5871	500	85.57%	**0.5714**	2
Colon	81.67%	10.0156	80.67%	11.6283	5000	**84.83%**	**0.9719**	2
USPS	94.65%	146.4942	93.54%	2.0639	500	93.86%	**0.4898**	10
Sonar	**86.29%**	0.0172	80.86%	0.0686	500	75.69%	**<0.0001**	2
Hill Valley	58.67%	0.1295	76.31%	0.1647	500	**98.78%**	**0.0047**	2
Protein	51.18%	253.5796	67.09%	5.0919	500	**68.76%**	**1.9953**	3

4.3 Real-World Classification Problems

In order to indicate the advantage of the proposed method on classification performance, the testing accuracy of the proposed method and other algorithms have also been compared. Table V and Figure 2 display the performance comparison of PC-ELM, EI-ELM, I-ELM, ELM, EM-ELM, SVM and the proposed method. As seen from these experimental results, the proposed method can always achieve comparable or better performance in comparison to other methods with much faster learning speed. For example, we consider Covtype.binary (large number of training samples with medium input dimensions), Mushroom, Hill Valley (medium number of training samples with medium input dimensions), Connect-4 (large number of training samples with medium input dimensions).

1) For Covtype.binary dataset, the proposed method provides comparable performance and runs 403 times and 35 times faster than SVM and ELM, respectively.

2) For Mushroom dataset, the proposed method provides comparable performance and runs 8200 times and 190 times faster than SVM and ELM, respectively.

3) For Hill Valley dataset, the proposed method provides better performance and runs 27 times and 35 times faster than SVM and ELM, respectively.

4) For Connect-4 dataset, the proposed method provides comparable performance and runs 758 times and 10 times faster than SVM and ELM, respectively.

On the other hand, contrary to other learning methods which are sensitive to the parameter (C) and number of hidden nodes, the proposed method with sigmoid nodes is not sensitive to the unique user-specified parameters (C and number of hidden nodes, see Fig.3) and is easy to use in the respective implementations.

5 Conclusion

Unlike other SLFN learning methods, in our new approach, one may simply calculate the hidden node parameter once and the output weight is not required at all. Also, it has been rigorously proved that the proposed method can greatly enhance the learning effectiveness, reduce the computation cost, and eventually further increase the learning speed. The simulation results on sigmoid type of hidden nodes show that compared to other learning methods including SVM/SVR, BP and ELMs, the new approach can significantly reduce the NN training time several to thousands of times and can applied to regression and classification problems. Thus this method can be used efficiently in many applications. Furthermore, experimental results show that this proposed learning method with m hidden node can achieve similar or better generalization performance than other learning method with hundreds of hidden nodes. This phenomenon of this proposed method bring about many advantages, and it maybe have far reaching consequences on the generalization ability of neural network.

References

1. Lin, H.T., Liang, T.J., Chen, S.M.: Estimation of battery state of health using probabilistic neural network. IEEE Transactions on Industrial Informatics 9(2), 679–685 (2013) [311]
2. Zhang, R., Lan, Y., Huang, G.B., Xu, Z.B.: Universal approximation of extreme learning machine with adaptive growth of hidden nodes. IEEE Transactions on Neural Networks and Learning Systems 23(2), 365–371 (2012) [312]
3. Igelnik, B., Pao, Y.H.: Stochastic choice of basis functions in adaptive function approximation and the functional-link net. IEEE Transactions on Neural Networks 6(6), 1320–1329 (1995) [312]
4. Pao, Y.H., Park, G.H., Sobajic, D.J.: Learning and generalization characteristics of the random vector functional-link net. Neurocomputing 6(2), 163–180 (1994) [312]
5. Broomhead, D.S., Lowe, D.: Multivariable functional interpolation and adaptive networks. Complex Systems 2, 321–355 (1988) [312]
6. Huang, G.B., Chen, L., Siew, C.K.: Universal approximation using incremental constructive feedforward networks with random hidden nodes. IEEE Transactions on Neural Networks 17(4), 879–892 (2006) [312, 313, 314]
7. Huang, G.B., Zhu, Q.Y., Siew, C.K.: Extreme learning machine: Theory and applications. Neurocomputing 70, 489–501 (2006) [312, 315]
8. Huang, G.B., Zhou, H.M., Ding, X.J., Zhang, R.: Extreme learning machine for regression and multiclass classification. IEEE Transactions on Systems Man and Cybernetics Part B-Cybernetics 42(2), 513–529 (2012) [312]
9. Huang, G.B.: An insight into extreme learning machines: Random neurons, random features and kernels. Cognitive Computation (2014), doi:10.1007/s12559-014-9255-2 [312]
10. Baradarani, A., Wu, Q.M.J., Ahmadi, M.: An efficient illumination invariant face recognition framework via illumination enhancement and dd-dtcwt filtering. Pattern Recognition 46(1), 57–72 (2013) [312]
11. Minhas, R., Mohammed, A.A., Wu, Q.M.J.: Incremental learning in human action recognition based on snippets. IEEE Transactions On Circuits and Systems For Video Technology 22(11), 1529–1541 (2012), doi:10.1109/TCSVT.2011.2177182 [312]

12. Huang, G.B., Kasun, L.L.C., Zhou, H., Vong, C.M.: Representational learning with extreme learning machine for big data. IEEE Intelligent Systems 28(6), 31–34 (2013) [312]
13. Mohammed, A.A., Minhas, R., Wu, Q.M.J., Sid-Ahmed, M.A.: Human face recognition based on multidimensional pca and extreme learning machine. Pattern Recognition 44(10-11), 2588–2597 (2011) [312]
14. Xu, Y., Dong, Z.Y., Xu, Z., Meng, K., Wong, K.P.: An intelligent dynamic security assessment framework for power systems with wind power. Ieee Transactions on Industrial Informatics 8(4), 995–1003 (2012) [312]
15. Yang, Y.M., Wang, Y.N., Yuan, X.F.: Parallel chaos search based incremental extreme learning machine. Neural Processing Letters 37(3), 277–301 (2013) [312]
16. Hunter, D., Yu, H., Pukish, M.S., Kolbusz, J., Wilamowski, B.M.: Selection of proper neural network sizes and architectures-a comparative study. IEEE Transactions on Industrial Informatics 8(2), 228–240 (2012) [312]
17. Hoerl, A.E., Kennard, R.W.: Ridge regression: Biased estimation for nonorthogonal problems. Technometrics 42(1), 80–86 (2000) [315]

ROS-ELM: A Robust Online Sequential Extreme Learning Machine for Big Data Analytics

Yang Liu[1], Bo He[1,*], Diya Dong[1], Yue Shen[1], Tianhong Yan[2,*], Rui Nian[1], and Amaury Lendasse[3]

[1] School of Information Science and Engineering, Ocean University of China, 238 Songling Road, Qingdao 266100, China
[2] School of Mechanical and Electrical Engineering, China Jiliang University, 258 Xueyuan Street, Xiasha High-Edu Park, Hangzhou 310018, China
[3] Department of Mechanical and Industrial Engineering and the Iowa Informatics Initiative, 3131 Seamans Center, The University of Iowa, Iowa City, IA 52242-1527, USA
bhe@ouc.edu.cn, thyan@163.com

Abstract. In this paper, a robust online sequential extreme learning machine (ROS-ELM) is proposed. It is based on the original OS-ELM with an adaptive selective ensemble framework. Two novel insights are proposed in this paper. First, a novel selective ensemble algorithm referred to as particle swarm optimization selective ensemble (PSOSEN) is proposed. Noting that PSOSEN is a general selective ensemble method which is applicable to any learning algorithms, including batch learning and online learning. Second, an adaptive selective ensemble framework for online learning is designed to balance the robustness and complexity of the algorithm. Experiments for both regression and classification problems with UCI data sets are carried out. Comparisons between OS-ELM, simple ensemble OS-ELM (EOS-ELM) and the proposed ROS-ELM empirically show that ROS-ELM significantly improves the robustness and stability.

Keywords: extreme learning machine, online learning, selective ensemble, PSOSEN, robustness.

1 Introduction

Due to the advancement of data acquisition, the amount of information in many fields of sciences increases very rapidly. The world is entering the age of big data. Large data set is helpful to analyze various phenomena because abundant information is available. However, it also raises many new problems. First, the computational time for big data analytics increasing rapidly. Second, various data sets require more robust learning algorithms.

Feedforward neural networks is one of the most prevailing neural networks, which is very popular for data processing in the past decades[1-2]. However, all the parameters in the networks need to be tuned iteratively. Moreover, the slow gradient descent

* Corresponding author.

J. Cao et al. (eds.), *Proceedings of ELM-2014 Volume 1*,
Proceedings in Adaptation, Learning and Optimization 3, DOI: 10.1007/978-3-319-14063-6_28

based learning methods are always used to train the networks[3]. Therefore, the learning speed of the feedforward neural networks is very slow, which limits its applications.

Recently, an original algorithm designed for single hidden layer feedforward neural networks (SLFNs) named extreme learning machine (ELM) was proposed by Huang et al.[4]. ELM is a tuning free algorithm for it randomly selects the input weights and biases of the hidden nodes instead of learning these parameters. And also, the output weights of the network are then analytically determined. ELM proves to be a few orders faster than traditional learning algorithms and obtains better generalization performance as well. It lets the fast and accurate big data analytics becomes possible and has been applied to many fields[5-7].

However, the algorithms mentioned above need all the training data available to build the model, which is referred to as batch learning. In many industrial applications, it is very common that the training data can only be obtained one by one or chunk by chunk. If batch learning algorithms are performed each time new training data is available, the learning process will be very time consuming. Hence online sequential learning is necessary for many real world applications.

An online sequential extreme learning machine is then proposed by Liang et al.[8]. OS-ELM can learn the sequential training observations online at arbitrary length (one by one or chunk by chunk). New arrived training observations are learned to modify the model of the SLFNs. As soon as the learning procedure for the arrived observations is completed, the data is discarded. Moreover, it has no prior knowledge about the amount of the observations which will be presented. Therefore, OS-ELM is an elegant sequential learning algorithm which can handle both the RBF and additive nodes in the same framework and can be used to both the classification and function regression problems. OS-ELM proves to be a very fast and accurate online sequential learning algorithm[9-11], which can provide better generalization performance in faster speed compared with other sequential learning algorithms such as GAP-RBF, GGAP-RBF, SGBP, RAN, RANEKF and MRAN etc.

However, various data sets require more robust learning algorithms. Due to the random generation of the parameters for the hidden nodes, the robustness and stability of OS-ELM sometimes cannot be guaranteed, similar to ELM. Some ensemble based methods and pruning based methods have been applied to ELM to improve its robustness[12-15]. Ensemble learning is a learning scheme where a collection of a finite number of learners is trained for the same task[16-17]. It has been demonstrated that the generalization ability of a learner can be significantly improved by ensembling a set of learners. In [18] a simple ensemble OS-ELM, i.e., EOS-ELM, has been investigated. However, Zhou et al.[19] proved that selective ensemble is better a choice. We apply this idea to OS-ELM. At first, a novel selective ensemble algorithm--PSOSEN, is proposed. PSOSEN adopts particle swarm optimization[20] to select the individual OS-ELMs to form the ensemble. It should be noted that PSOSEN is a general selective ensemble algorithm suitable for any learning algorithms.

Different from batch learning, online learning algorithms need to perform learning continually. Therefore the complexity of the learning algorithm should be taken into account. Obviously, performing selective ensemble learning each step is not a good choice for sequential learning. Thus we designed an adaptive selective ensemble framework for OS-ELM. A set of OS-ELMs are trained online, and the root mean square error

(RMSE) will always be calculated. The error will be compared with a pre-set threshold λ. If RMSE is bigger than the threshold, it means the model is not accurate. Then PSOSEN will be performed and a selective ensemble M is obtained. Otherwise, it means the model is relatively accurate and the ensemble will not be selected. Then the output of the system is calculated as the average output of the individuals in the ensemble set. And each individual OS-ELM will be updated recursively.

UCI data sets[21], which contain both regression and classification data, are used to verify the feasibility of the proposed ROS-ELM algorithm. Comparisons of three aspects including RMSE, standard deviation and running time between OS-ELM, EOS-ELM and ROS-ELM are presented. The results convincingly show that ROS-ELM significantly improves the robustness and stability compared with OS-ELM and EOS-ELM.

The rest of the paper is organized as follows: In section 2, previous work including ELM and OS-ELM are reviewed. A novel selective ensemble based on particle swarm optimization is presented in section 3. An adaptive selective ensemble framework is designed for OS-ELM referred to as ROS-ELM, is proposed in section 4. Experiments are carried out in section 4 and the comparison results are also presented. In section 5, we draw the conclusion of the paper.

2 Review of Related Works

In this section, both the basic ELM algorithm and the online version OS-ELM are reviewed in brief as the background knowledge for our work.

2.1 Extreme Learning Machine (ELM)

ELM algorithm is derived from single hidden layer feedforward neural networks (SLFNs). Unlike traditional SLFNs, ELM assigns the parameters of the hidden nodes randomly without any iterative tuning. Besides, all the parameters of the hidden nodes in ELM are independent with each other. Hence ELM can be seen as generalized SLFNs. The only problem for ELM is to calculate the output weights.

Given N training samples $(x_i, t_i) \in R^n \times R^m$, where x_i is an input vector of n dimensions and t_i is a target vector of m dimensions. Then SLFNs with $\tilde{N}$ hidden nodes each with output function $G(a_i, b_i, x)$ are mathematically modeled as

$$f_{\tilde{N}}(x_j) = \sum_{i=1}^{\tilde{N}} \beta_i G(a_i, b_i, x_j) = t_j, j = 1, \cdots, N. \quad (1)$$

Where (a_i, b_i) are parameters of hidden nodes, and β_i is the weight vector connecting the ith hidden node and the output node. To simplify, equation (1) can be written equivalently as:

$$H\beta = T \quad (2)$$

where

$$H\left(a_1,\cdots,a_N,b_1,\cdots,b_{\tilde{N}},x_1,\cdots,x_N\right)=\begin{bmatrix} G(a_1,b_1,x_1) & \cdots & G(a_{\tilde{N}},b_{\tilde{N}},x_1) \\ \vdots & \ddots & \vdots \\ G(a_1,b_1,x_N) & \cdots & G(a_{\tilde{N}},b_{\tilde{N}},x_N) \end{bmatrix}_{N\times\tilde{N}} \quad (3)$$

$$\beta=\begin{bmatrix} \beta_1^T \\ \vdots \\ \beta_{\tilde{N}}^T \end{bmatrix}_{\tilde{N}\times m} \qquad T=\begin{bmatrix} t_1^T \\ \vdots \\ t_N^T \end{bmatrix}_{N\times m} \quad (4)$$

H is called the hidden layer output matrix of the neural network, and the i th column of H is the output of the i th hidden node with respect to inputs $x_1,x_2,\cdots,x_N$.

In ELM, H can be easily obtained as long as the training set is available and the parameters (a_i,b_i) are randomly assigned. Then ELM evolves into a linear system and the output weights β are calculated as:

$$\hat{\beta}=H^{\dagger}T \quad (5)$$

where $H^{\dagger}$ is the Moore-Penrose generalized inverse of matrix H.

The ELM algorithm can be summarized in three steps as shown in Algorithm 1:

Algorithm 1

Input:

A training set $\mathbb{N}=\left\{(x_i,t_i)\middle|x_i\in R^n,t_i\in R^m,i=1,\cdots,N\right\}$, hidden node output function $G(a_i,b_i,x)$, and the number of hidden nodes $\tilde{N}$

Steps:

1. Assign parameters of hidden nodes (a_i,b_i) randomly, $i=1,\cdots,\tilde{N}$.
2. Calculate the hidden layer output matrix H.
3. Calculate the output weight β: $\hat{\beta}=H^{\dagger}T$, where $H^{\dagger}$ is the Moore-Penrose generalized inverse of hidden layer output matrix H.

2.2 OS-ELM

In many industrial applications, it is impossible to have all the training data available before the learning process. It is common that the training observations are sequentially inputted to the learning algorithm, i.e., the observations arrive one-by-one or chunk-by-chunk. In this case, the batch ELM algorithm is no longer applicable. Hence, a fast and accurate online sequential extreme learning machine was proposed to deal with online learning.

The output weight β obtained from equation (5) is actually a least-squares solution of equation (2). Given $rank(H) = \tilde{N}$, the number of hidden nodes, $H^{\dagger}$ can be presented as:

$$H^{\dagger} = \left(H^T H\right)^{-1} H^T \tag{6}$$

This can also be called the left pseudoinverse of H for it satisfies the equation $H^{\dagger}H = I_{\tilde{N}}$. If $H^T H$ tends to be singular, smaller network size $\tilde{N}$ and larger data number N_0 should be chosen in the initialization step of OS-ELM. Substituting equation (6) to equation (5), we can get

$$\hat{\beta} = \left(H^T H\right)^{-1} H^T T \tag{7}$$

which is the least-squares solution to equation (2). Then the OS-ELM algorithm can be deduced by recursive implementation of the least-squares solution of (7).

There are two main steps in OS-ELM, initialization step and update step. In the initialization step, the number of training data N_0 needed in this step should be equal to or larger than network size $\tilde{N}$. In the update step, the learning model is updated with the method of recursive least square (RLS). And only the newly arrived single or chunk training observations are learned, which will be discarded as soon as the learning step is completed.

The two steps for OS-ELM algorithm in general:

a. Initialization step: batch ELM is used to initialize the learning system with a small chunk of initial training data $\mathbb{N}_0 = \{(x_i, t_i)\}_{i=1}^{N_0}$ from given training set $\mathbb{N} = \{(x_i, t_i), x_i \in R^n, t_i \in R^m, i = 1, \cdots\}$ $N_0 \geq \tilde{N}$.

1. Assign random input weights a_i and bias b_i (for additive hidden nodes) or center a_i and impact factor b_i (for RBF hidden nodes), $i = 1, \cdots, \tilde{N}$.
2. Calculate the initial hidden layer output matrix:

$$H_0 = \begin{bmatrix} G(a_1, b_1, x_1) & \cdots & G(a_{\tilde{N}}, b_{\tilde{N}}, x_1) \\ \vdots & \ddots & \vdots \\ G(a_1, b_1, x_{N_0}) & \cdots & G(a_{\tilde{N}}, b_{\tilde{N}}, x_{N_0}) \end{bmatrix}_{N_0 \times \tilde{N}} \tag{8}$$

3. Calculate the initial output weight $\beta^{(0)} = P_0 H_0^T T_0$, where $P_0 = \left(H_0^T H_0\right)^{-1}$ and $T_0 = \left[t_1, \cdots, t_{NO}\right]^T$.
4. Set k=0. Initialization is finished.

b. Sequential learning step:

The $(k+1)$ th chunk of new observations can be expressed as:

$$\mathbb{N}_{k+1} = \{(x_i, t_i)\}_{i=\left(\sum_{j=0}^{k} N_j\right)+1}^{\sum_{j=0}^{k+1} N_j} \tag{9}$$

where $\mathbb{N}_{k+1}$ represents the number of observations in the (k+1)th chunk newly arrived.

1. Compute the partial hidden layer output matrix H_{k+1} for the $(k+1)$ th chunk.

$$H_{k+1} = \begin{bmatrix} G\left(a_1, b_1, x_{\left(\sum_{j=0}^{k} N_j\right)+1}\right) & \cdots & G\left(a_{\tilde{N}}, b_{\tilde{N}}, x_{\left(\sum_{j=0}^{k} N_j\right)+1}\right) \\ \vdots & \ddots & \vdots \\ G\left(a_1, b_1, x_{\sum_{j=0}^{k+1} N_j}\right) & \cdots & G\left(a_{\tilde{N}}, b_{\tilde{N}}, x_{\sum_{j=0}^{k+1} N_j}\right) \end{bmatrix}_{N_{k+1} \times \tilde{N}} \tag{10}$$

2. Set $T_{k+1} = \left[t_{\left(\sum_{j=0}^{k} N_j\right)+1}, \cdots, t_{\sum_{j=0}^{k+1} N_j}\right]^T$. And we have

$$K_{k+1} = K_k + H_{k+1}^T H_{k+1} \tag{11}$$

$$\beta^{(k+1)} = \beta^{(k)} + K_{k+1}^{-1} H_{k+1}^T \left(T_{k+1} - H_{k+1} \beta^{(k)}\right) \tag{12}$$

To avoid calculating inverse in the iterative procedure, K_{k+1}^{-1} is factored as the following according to Woodbury formula:

$$
\begin{aligned}
K_{k+1}^{-1} &= \left(K_k + H_{k+1}^T H_{k+1}\right)^{-1} \\
&= K_k^{-1} - K_k^{-1} H_{k+1}^T \left(I + H_{k+1} K_k^{-1} H_{k+1}^T\right)^{-1} H_{k+1} K_k^{-1}
\end{aligned}
\tag{13}
$$

Let $P_{k+1} = K_{k+1}^{-1}$.

3. Calculate the output weight $\beta^{(k+1)}$, according to the updating equations:

$$P_{k+1} = P_k - P_k H_{k+1}^T \left(I + H_{k+1} P_k H_{k+1}^T\right)^{-1} H_{k+1} P_k \tag{14}$$

$$\beta^{(k+1)} = \beta^{(k)} + P_{k+1} H_{k+1}^T \left(T_{k+1} - H_{k+1} \beta^{(k)}\right) \tag{15}$$

4. Set $k = k+1$. Go to step b.

3 Particle Swarm Optimization Selective Ensemble

In this section, a novel selective ensemble method referred to as particle swarm optimization selective ensemble (PSOSEN) is proposed. PSOSEN adopts particle swarm optimization to select the good learners and combine their predictions. Detailed procedures of the PSOSEN algorithm will be introduced in this section.

Zhou et al.[19] have demonstrated that ensembling many of the available learners may be better than ensembling all of those learners in both regression and classification. The detailed proof of this conclusion will not be presented in this paper. However, one important problem for selective ensemble is how to select the good learners in a set of available learners.

The novel approach--PSOSEN, is proposed to select good learners in the ensemble. PSOSEN is based on the idea of heuristics. It assumes each learner can be assigned a weight, which could characterize the fitness of including this learner in the ensemble. Then the learner with the weight bigger than a pre-set threshold λ could be selected to join the ensemble.

We will explain the principle of PSOSEN from the context of regression. We use ω_i to denote the weight of the i th component learner. The weight should satisfy the following equations:

$$0 \le \omega_i \le 1 \tag{16}$$

$$\sum_{i=1}^{N} \omega_i = 1 \tag{17}$$

Then the weight vector is:

$$\omega = (\omega_1, \omega_2, \ldots, \omega_N) \tag{18}$$

Suppose input variables $x \in R^m$ according to the distribution $p(x)$, the true output of x is $d(x)$, and the actual output of the i th learner is $f_i(x)$. Then the output of the simple weighted ensemble on x is:

$$\hat{f}(x) = \sum_{i=1}^{N} \omega_i f_i(x) \tag{19}$$

Then the generalization error $E_i(x)$ of the i th learner and the generalization error $\hat{E}(x)$ of the ensemble are calculated on x respectively:

$$E_i(x) = (f_i(x) - d(x))^2 \tag{20}$$

$$\hat{E}(x) = (\hat{f}(x) - d(x))^2 \tag{21}$$

The generalization error E_i of the i th learner and that of the ensemble $\hat{E}$ are calculated on $p(x)$ respectively:

$$E_i = \int dx p(x) E_i(x) \tag{22}$$

$$\hat{E} = \int dx p(x) \hat{E}(x) \tag{23}$$

We then define the correlation between the i th and the j th component learner as following:

$$C_{ij} = \int dx p(x)(f_i(x) - d(x))(f_j(x) - d(x)) \tag{24}$$

Obviously C_{ij} satisfies the following equations:

$$C_{ii} = E_i \tag{25}$$

$$C_{ij} = C_{ji} \tag{26}$$

Considering the equations defined above, we can get:

$$\hat{E}(x) = (\sum_{i=1}^{N} \omega_i f_i(x) - d(x))(\sum_{j=1}^{N} \omega_j f_j(x) - d(x)) \tag{27}$$

$$\hat{E} = \sum_{i=1}^{N} \sum_{j=1}^{N} \omega_i \omega_j C_{ij} \tag{28}$$

To minimize the generalization error of the ensemble, according to equation (28), the optimum weight vector can be obtained as:

$$\omega_{opt} = \arg\min_{\omega} (\sum_{i=1}^{N} \sum_{j=1}^{N} \omega_i \omega_j C_{ij}) \tag{29}$$

The k th variable of ω_{opt}, i.e., $\omega_{opt.k}$, can be solved by Lagrange multiplier:

$$\frac{\partial\ (\sum_{i=1}^{N} \sum_{j=1}^{N} \omega_i \omega_j C_{ij} \text{-} 2 * \lambda (\sum_{i=1}^{N} \omega_i - 1)\ )}{\partial \omega_{opt.k}} = 0 \tag{30}$$

The equation can be simplified to:

$$\sum_{\substack{j=1 \\ j \neq k}}^{N} \omega_{opt.k} C_{kj} = \lambda \tag{31}$$

Taking equation (2) into account, we can get:

$$\omega_{opt.k} = \frac{\sum_{j=1}^{N} C_{kj}^{-1}}{\sum_{i=1}^{N} \sum_{j=1}^{N} \omega_i \omega_j C_{ij}^{-1}} \tag{32}$$

Equation (32) gives the direct solution for ω_{opt}. But the solution seldom work well in real word applications. Due to the fact that some learners are quite similar in performance, when a number of learners are available, the correlation matrix C_{ij} may be irreversible or ill-conditioned.

Although we cannot obtain the optimum weights of the learner directly, we can approximate them in some way. Equation (29) can be viewed as an optimization problem. As particle swarm optimization has been proved to be a powerful optimization tool, PSOSEN is then proposed. The basic PSO algorithm is showed in Figure 1.

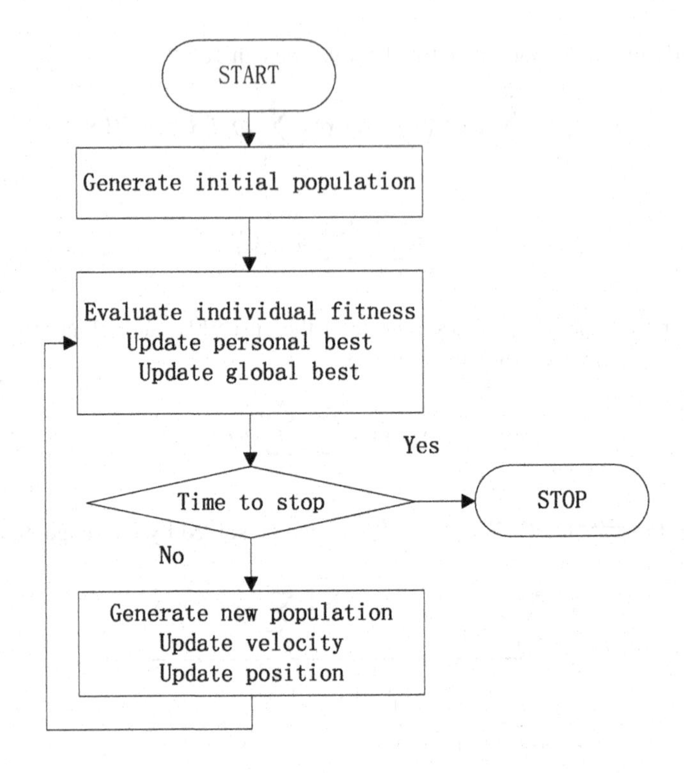

Fig. 1. Flowchart for particle swarm optimization algorithm

PSOSEN randomly assigns a weight to each of the available learners at first. Then it employs particle swarm optimization algorithm to evolve those weights so that the weights can characterize the fitness of the learners in joining the ensemble. Finally, learners whose weight is bigger than a pre-set threshold λ are selected to form the ensemble. Note that if all the evolved weights are bigger than the threshold λ, then all the learners will be selected to join the ensemble.

PSOSEN can be applied to both regression and classification problems for the purpose of the weights evolving process is only to select the component learners. In particular, the output of the ensemble for regression are combined via simple averaging instead of weighted averaging. The reason is that previous work [19] showed that using the weights both in selection of the component learners and combination of the outputs tends to suffer the overfitting problem.

In the process of generating population, the goodness of the individuals are evaluated via validation data bootstrap sampled from the training data set. We use $\hat{E}_{\omega}^{V}$ to denote the generalization error of the ensemble, which corresponds to individual ω on the validation data V . Obviously $\hat{E}_{\omega}^{V}$ can describe the goodness of ω .

The smaller $\hat{E}_{\omega}^{V}$ is, the better ω is. So, PSOSEN adopts $f(\omega)=1/\hat{E}_{\omega}^{V}$ as the fitness function.

The PSOSEN algorithm is summarized as follows. $S_1, S_2, ..., S_T$ are bootstrap samples generated from original training data set. A component learner N_t is trained from each S_T. And an selective ensemble N^* is built from $N_1, N_2, ..., N_T$. The output is the average output of the ensemble for regression, or the class label who receives the most number in voting process for classification.

PSOSEN

Input: training set S, learner L, trial T, threshold λ
Steps:

1. for t = 1 to T{
   S_T =bootstrap sample from S
   $N_T = L(S_T)$

}

2. generate a population of weight vectors
3. evolve the population by PSO, where the fitness of the weight vector ω is defined as $f(\omega)=1/\hat{E}_{\omega}^{V}$.
4. ω^* = the evolved best weight vector

Output: ensemble N^*:

$$N^*(x) = Ave \sum_{\omega_t^* > \lambda} N_t(x) \quad \text{for regression}$$

$$N^*(x) = \arg\max_{y \in Y} \sum_{\omega_t^* > \lambda, N_t(x)=y} 1 \quad \text{for classification}$$

---

4 Robust Online Sequential Extreme Learning Machine

In this section, the detailed procedure of the proposed robust online sequential learning algorithm is introduced. The novel selective ensemble algorithm--PSOSEN is applied to the original OS-ELM to improve the robustness. In order to reduce the complexity and employ PSOSEN flexibly, an adaptive framework is then designed. The new algorithm, which is based on OS-ELM and adaptive ensemble, is termed as robust online sequential extreme learning machine (ROS-ELM).

The flowchart of ROS-ELM is showed as follows:

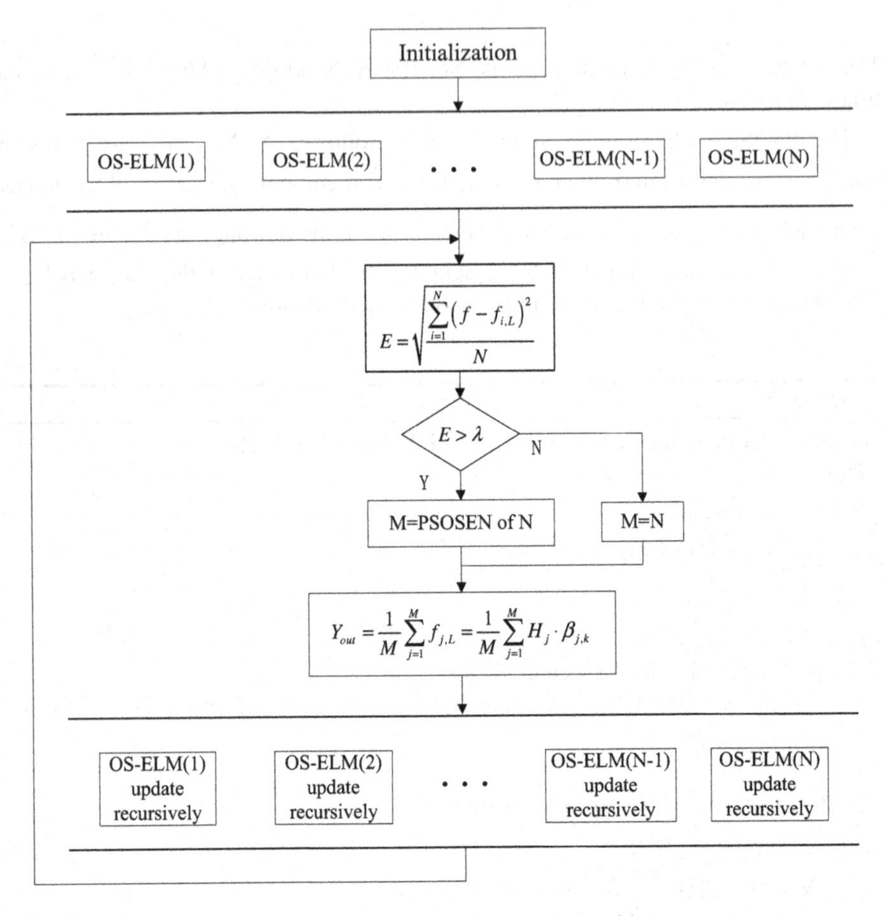

Fig. 2. Flowchart for the ROS-ELM algorithm

Online sequential learning is necessary in many industrial applications. In this situations, training data can only be obtained sequentially. Although OS-ELM is proposed as a fast and accurate online learning algorithm, it still suffers from the robustness problem, which results from the random generation of the input weights and biases, similar to ELM. Ensemble methods has been investigated in OS-ELM, i.e., the EOS-ELM algorithm[18]. However, it is only very simple ensemble method, which just calculates the average of all the N individual OS-ELMs. In this section, selective ensemble, which is superior to simple ensemble, is adopted to OS-ELM. The novel selective ensemble method--PSOSEN, proposed in section 3, is chose as the algorithm. Apparently, performing PSOSEN each step is a time consuming process. We design an adaptive framework to determine whether to perform PSOSEN or simple ensemble. Thus the robustness and the complexity can be balanced well. The ROS-ELM algorithm can be explained as follows:

First, N individual OS-ELMs are initialized. The number of nodes is same for each OS-ELM. While the input weights and biases for each OS-ELM are randomly generated.

Second, the RMSE error is calculated:

$$E = \sqrt{\frac{\sum_{i=1}^{N} \left(f - f_{i,L}\right)^2}{N}} \tag{33}$$

where f is the expected output, while $f_{i,L}$ is the actual output of the i th individual OS-ELM.

The RMSE will be compared with a pre-set threshold λ. If E is bigger than λ, which means simple ensemble is not accurate, PSOSEN is performed and a selective ensemble M is obtained. And if E is smaller than λ, which indicates that simple ensemble is relatively accurate, the ensemble will not be selected.

Third, the output of the system is calculated as the average output of the individual in the ensemble set:

$$Y_{out} = \frac{1}{M} \sum_{j=1}^{M} f_{j,L} = \frac{1}{M} \sum_{j=1}^{M} H_j \cdot \beta_{j,k} \tag{34}$$

where H_j is the output matrix of the j th OS-ELM, and $\beta_{j,k}$ is the output weight calculated by the j th OS-ELM at step k.

At last, each OS-ELM will update recursively according to the update equations presented in section 2.

5 Performance Evaluation of ROS-ELM

In this section, a series of experiments were conducted to evaluate the performance of the proposed ROS-ELM algorithm. OS-ELM and EOS-ELM are also compared with ROS-ELM in this section. All the experiments were carried out in the MatlabR2012b environment on a desktop of CPU 3.40GHz and 8GB RAM.

5.1 Model Selection

For OS-ELM, the number of hidden nodes is the only parameter needs to be determined. Cross-validation method are usually used to choose this parameter. Fifty trials of simulations are performed respectively for classification, regression and time-series problems. The number of hidden nodes is then determined by the validation error.

For EOS-ELM and ROS-ELM, there is another parameter that needs to be determined, i.e., the number of networks in the ensemble. The parameter is set from 5 to 30 with the interval 5. Finally, the optimal parameter is selected according to the RMSE for regression, testing accuracy for classification and standard deviation value. Under the same problem, the number of OS-ELMs is selected based on the lowest standard deviation and the comparable RMSE or accuracy compared with OS-ELM. Table 1 is an example of selecting the optimal number of networks for ROS-ELM with RBF hidden nodes on New-thyroid dataset. As illustrated by Table 1, the lowest standard deviation occurs when the number of OS-ELMs is 20. Meanwhile, the prediction accuracy of ROS-ELM is better than OS-ELM. Hence we set the number of networks to be 20 for the New-thyroid dataset. The numbers of OS-ELMs for other datasets are determined in the same way.

Both the Gaussian radial basis function (RBF) $G(a,b,x) = \exp\left(-\|x-a\|^2 / b\right)$ and the sigmoid additive $G(a,b,x) = 1/\left(1+\exp\left(-(a \cdot x + b)\right)\right)$ are adopted as activation function in OS-ELM, EOS-ELM and ROS-ELM.

Table 1. Network selection for New-thyroid dataset

Num of networks	1	5	10	15	20	25	30
Testing accuracy	90.73	91.25	90.65	90.18	92.24	91.79	91.8
Testing Dev	0.0745	0.0254	0.0316	0.0276	0.0138	0.024	0.0156

In the experiments, OS-ELM and EOS-ELM were compared with ROS-ELM. Some general information of the benchmark datasets used in our evaluations is listed in Table 2. Both regression and classification problems are included.

Table 2. Specification of benchmark datasets

	Dataset	Classes	Training data	Testing data	Attributes
Regression problems	Auto-MPG	-	320	72	7
	Abalone	-	3000	1177	8
	California housing	-	8000	12640	8
	Mackey-Glass	-	4000	500	4
Classification problems	Zoo	7	71	30	17
	Wine	3	120	58	13
	New-thyroid	3	140	75	5
	Monks-1	2	300	132	6
	Image segmentation	7	1500	810	19
	Satellite image	6	4435	2000	36

For OS-ELM, the input weights and biases with additive activation function or the centers with RBF activation function were all generated from the range [-1, 1]. For regression problems, all the inputs and outputs were normalized into the range [0, 1], while the inputs and outputs were normalized into the range [-1, 1] for classification problems.

The benchmark datasets studied in the experiments are from UCI Machine Learning Repository except California Housing dataset from the StatLib Repository. Besides, a time-series problem, Mackey-Glass, from UCI was also adopted to test our algorithms.

5.2 Algorithm Evaluation

To verify the superiority of the ROS-ELM, RMSE for regression problems and testing accuracy for classification problems are respectively computed. The evaluation results are presented in Table 3 and Table 4, which are respectively corresponding to the models with sigmoid hidden nodes and RBF hidden nodes. Each result is an average of 50 trials. And in every trial of one problem, the training and testing samples were randomly adopted from the dataset that was addressed currently.

From the comparison results of Table 2 and Table 3, we can easily find that ROS-ELM and EOS-ELM are more time consuming than OS-ELM, but they still keep relatively fast speed at most of the time. What's important, ROS-ELM and EOS-ELM attain lower testing deviation and more accurate regression or classification results than OS-ELM. In terms of the comparison between ROS-ELM and EOS-ELM, it can be observed that ROS-ELM takes a little more time than EOS-ELM, which results from the selective ensemble by PSOSEN in ROS-ELM instead of simply averaging the networks in EOS-ELM. It should be noted that the complexity of ROS-ELM is adjustable, which depends on the threshold λ. Nevertheless, ROS-ELM always outperforms EOS-ELM in terms of accuracy and testing deviation. Hence with the adaptive ensemble framework, ROS-ELM tends to generate more accurate and robust results.

To verify the reliability of the proposed ROS-ELM more convincingly, an artificial dataset is dissected for instance. The dataset was generated from the function $y = x^2 + 3x + 2$, comprising 4500 training data and 1000 testing data. Figure 3 and 4 explicitly depict the variability of training accuracy of ROS-ELM, EOS-ELM and OS-ELM with respect to the number of training data in the process of learning. It can be observed that with the increasing number of training samples, RMSE values of the three methods significantly decline. As the online learning progressed, the training models are continuously updated and corrected. We can then conclude that the more training data the system learns, the more precise the model is. Whether sigmoid or RBF the hidden nodes is, ROS-ELM always obtains smaller RMSE than EOS-ELM and OS-ELM, which indicates that the performance of ROS-ELM is considerably accurate and robust compared with the other methods. Moreover, the smaller testing dev of ROS-ELM in Table 3 and 4 also confirms the robust performance of ROS-ELM.

Table 3. Comparison of OS-ELM, EOS-ELM and ROS-ELM for sigmoid hidden nodes

Datasets	Algorithm	#Nodes	#Network	Training time (s)	RMSE or Accuracy		Testing Dev
					Training RMSE	Testing RMSE	
Auto-MPG	OS-ELM	25		0.0121	0.0695	0.0745	0.0087
	EOS-ELM	25	20	0.2385	0.0691	0.0751	0.0065
	ROS-ELM	25	20	1.9083	0.0683	0.0741	0.0053
Abalone	OS-ELM	25		0.1191	0.0758	0.0782	0.0049
	EOS-ELM	25	5	0.5942	0.0754	0.0775	0.0023
	ROS-ELM	25	5	4.1528	0.0742	0.0758	0.0015
Mackey-Glass	OS-ELM	120		0.9827	0.0177	0.0185	0.0018
	EOS-ELM	120	5	4.8062	0.0176	0.0183	0.0007
	ROS-ELM	120	5	25.1608	0.0173	0.0179	0.0006
California Housing	OS-ELM	50		0.6871	0.1276	0.1335	0.0035
	EOS-ELM	50	5	3.2356	0.1280	0.1337	0.0019
	ROS-ELM	50	5	15.6326	0.1238	0.1323	0.0014
Zoo	OS-ELM	35		0.0042	100%	93.09%	0.0498
	EOS-ELM	35	25	0.0986	100%	93.68%	0.0375
	ROS-ELM	35	25	0.8768	100%	94.51%	0.0315
Wine	OS-ELM	30		0.0053	99.83%	97.24%	0.0251
	EOS-ELM	30	5	0.0247	99.79%	97.49%	0.0117
	ROS-ELM	30	5	0.1628	99.88%	98.01%	0.0094
New-thyroid	OS-ELM	20		0.0043	93.18%	89.66%	0.1138
	EOS-ELM	20	15	0.0627	94.32%	90.92%	0.02765
	ROS-ELM	20	15	0.5012	95.23%	91.78%	0.01986
Monks-1	OS-ELM	80		0.0378	89.34%	78.77%	0.0325
	EOS-ELM	80	15	0.5432	89.18%	78.79%	0.0187
	ROS-ELM	80	15	4.2804	90.24%	79.85%	0.0138
Image segmentation	OS-ELM	180		1.8432	97.07%	94.83%	0.0078
	EOS-ELM	180	20	36.2458	97.08%	94.79%	0.0055
	ROS-ELM	180	20	254.0721	97.56%	95.21%	0.0043
Satellite image	OS-ELM	400		42.2503	92.82%	88.92%	0.0058
	EOS-ELM	400	20	853.2675	92.80%	89.05%	0.0026
	ROS-ELM	400	20	6928.0968	93.96%	90.16%	0.0018

Table 4. Comparison of OS-ELM, EOS-ELM and ROS-ELM for RBF hidden nodes

Datasets	Algorithm	#Nodes	#Net-work	Training time (s)	RMSE or Accuracy		Testing Dev
					Training RMSE	Testing RMSE	
Auto-MPG	OS-ELM	25		0.0302	0.0685	0.0763	0.0081
	EOS-ELM	25	20	0.5986	0.0681	0.0754	0.0072
	ROS-ELM	25	20	4.1862	0.0672	0.0741	0.0063
Abalone	OS-ELM	25		0.3445	0.0753	0.0775	0.0027
	EOS-ELM	25	25	8.5762	0.0752	0.0773	0.0023
	ROS-ELM	25	25	49.3562	0.0741	0.0761	0.0017
Mackey-Glass	OS-ELM	120		1.6854	0.0181	0.0185	0.0092
	EOS-ELM	120	5	8.4304	0.0171	0,0171	0.0028
	ROS-ELM	120	5	55.1469	0.0159	0.0156	0.0016
California Housing	OS-ELM	50		1.8329	0.1298	0.1317	0.0017
	EOS-ELM	50	5	9.0726	0.1296	0.1316	0.0011
	ROS-ELM	50	5	64.9625	0.1202	0.1243	0.0009
Zoo	OS-ELM	35		0.0074	99.91%	91.15%	0.0508
	EOS-ELM	35	15	0.1028	99.87%	90.47%	0.0429
	ROS-ELM	35	15	0.8543	99.93%	91.26%	0.0315
Wine	OS-ELM	30		0.0132	99.73%	97.09%	0.0225
	EOS-ELM	30	5	0.6015	99.76%	97.18%	0.0138
	ROS-ELM	30	5	4.9028	99.84%	98.14%	0.0087
New-thyroid	OS-ELM	20		0.0118	93.45%	89.92%	0.0702
	EOS-ELM	20	15	0.1682	93.87%	89.86%	0.0428
	ROS-ELM	20	15	1.2315	94.68%	91.02%	0.0315
Monks-1	OS-ELM	80		0.1024	94,58%	87.28%	0.0882
	EOS-ELM	80	20	2.1567	93.69%	86.34%	0.0324
	ROS-ELM	80	20	15.2896	95.71%	88.47%	0.0195
Image segmenta-tion	OS-ELM	180		2.6702	94.98%	91.92%	0.0324
	EOS-ELM	180	5	13.2174	94.39%	91.35%	0.0148
	ROS-ELM	180	5	90.2856	96.02%	953.24%	0.0079
Satellite image	OS-ELM	400		45.2702	93.62%	89.54%	0.0056
	EOS-ELM	400	10	448.1347	93.86%	89.37%	0.0034
	ROS-ELM	400	10	3145.8528	94.75%	90.48%	0.0019

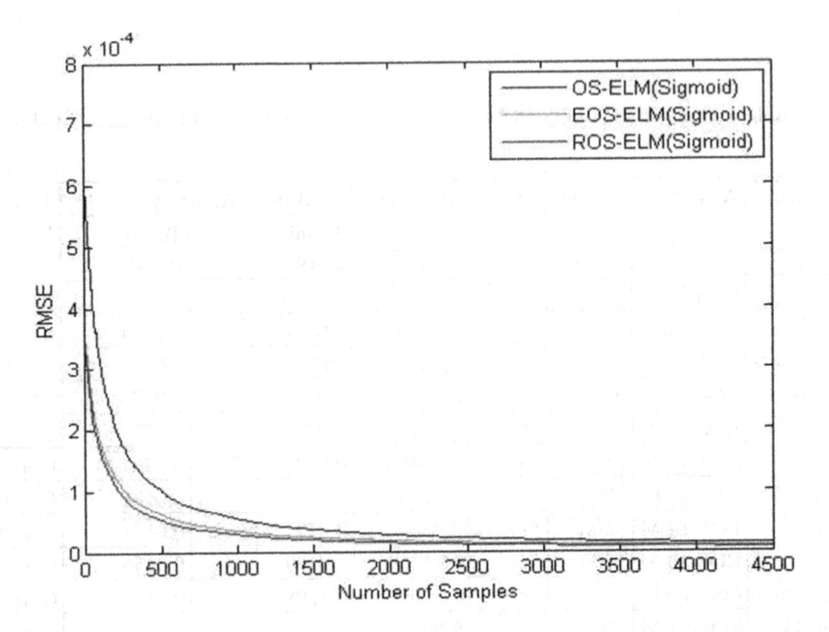

Fig. 3. RMSE with respect to the number of training samples for sigmoid hidden nodes

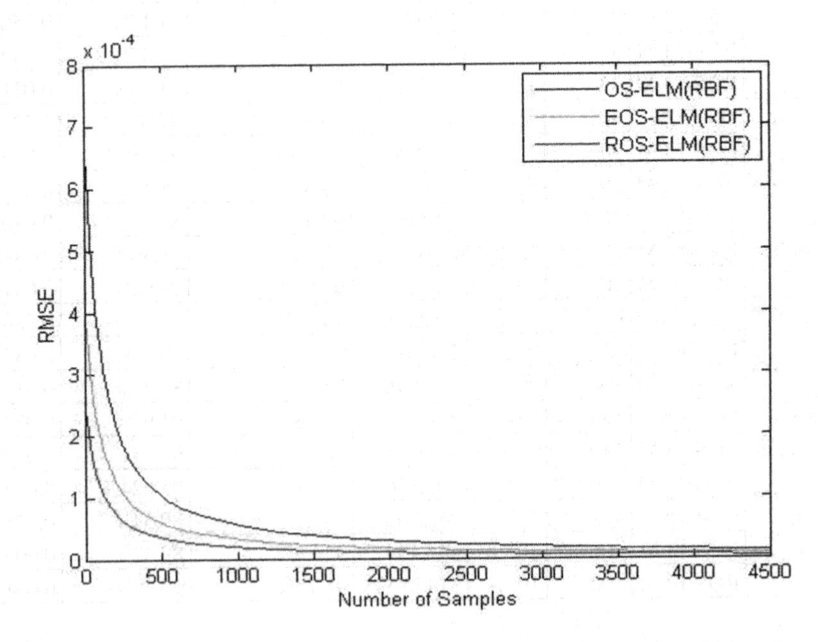

Fig. 4. RMSE with respect to the number of training samples for RBF hidden nodes

Hence, by analyzing the results in Figure 3, Figure 4, Table 3 and Table 4 comprehensively, we can draw the conclusion that ROS-ELM improves the accuracy and robustness of the online sequential learning algorithm significantly for both regression and classification applications, with a still relative fast speed.

6 Conclusion

In this paper, a robust online sequential extreme learning machine algorithm is proposed. To improve the robustness and stability of the sequential learning algorithm, we apply the selective ensemble method to OS-ELM. And in purpose of balancing the complexity and accuracy, an adaptive selective ensemble framework for OS-ELM is designed, which is referred to as ROS-ELM. In addition, before building the ROS-ELM system, a novel selective ensemble algorithm is proposed which is suitable for any learning methods, both batch learning and sequential learning. The proposed selective ensemble algorithm-PSOSEN, adopts particle swarm optimization method to select individual learner to form the ensemble. Experiments were carried out on UCI data set. The results convincingly show that ROS-ELM improves the robustness and stability of OS-ELM, while also keeps balance on complexity.

Acknowledgments. This work is partially supported by Natural Science Foundation of China (41176076, 31202036, 51075377).

References

1. Haykin, S., Network, N.: A comprehensive foundation. Neural Networks 2 (2004)
2. Hornik, K., Stinchcombe, M., White, H.: Multilayer feedforward networks are universal approximators. Neural Networks 2(5), 359–366 (1989)
3. Rumelhart, D.E., Hinton, G.E., Williams, R.J.: Learning internal representations by error propagation. California Univ San Diego La Jolla Inst For Cognitive Science (1985)
4. Huang, G.B., Zhu, Q.Y., Siew, C.K.: Extreme learning machine: theory and applications. Neurocomputing 70(1), 489–501 (2006)
5. Zhang, R., Huang, G.B., Sundararajan, N., et al.: Multicategory classification using an extreme learning machine for microarray gene expression cancer diagnosis. IEEE/ACM Transactions on Computational Biology and Bioinformatics (TCBB) 4(3), 485–495 (2007)
6. Xu, Y., Dong, Z.Y., Zhao, J.H., et al.: A reliable intelligent system for real-time dynamic security assessment of power systems. IEEE Transactions on Power Systems 27(3), 1253–1263 (2012)
7. Choi, K., Toh, K.A., Byun, H.: Incremental face recognition for large-scale social network services. Pattern Recognition 45(8), 2868–2883 (2012)
8. Liang, N.Y., Huang, G.B., Saratchandran, P., et al.: A fast and accurate online sequential learning algorithm for feedforward networks. IEEE Transactions on Neural Networks 17(6), 1411–1423 (2006)
9. Uçar, A., Demir, Y., Güzeliş, C.: A new facial expression recognition based on curvelet transform and online sequential extreme learning machine initialized with spherical clustering. Neural Computing and Applications, 1–12 (2014)

10. Yin, J.C., Zou, Z.J., Xu, F., et al.: Online ship roll motion prediction based on grey sequential extreme learning machine. Neurocomputing 129, 168–174 (2014)
11. Wang, X., Han, M.: Online sequential extreme learning machine with kernels for nonstationary time series prediction. Neurocomputing (2014)
12. Liu, N., Wang, H.: Ensemble based extreme learning machine. IEEE Signal Processing Letters 17(8), 754–757 (2010)
13. Xue, X., Yao, M., Wu, Z., et al.: Genetic ensemble of extreme learning machine. Neurocomputing 129, 175–184 (2014)
14. Rong, H.J., Ong, Y.S., Tan, A.H., et al.: A fast pruned-extreme learning machine for classification problem. Neurocomputing 72(1), 359–366 (2008)
15. Miche, Y., Sorjamaa, A., Bas, P., et al.: OP-ELM: optimally pruned extreme learning machine. IEEE Transactions on Neural Networks 21(1), 158–162 (2010)
16. Hansen, L.K., Salamon, P.: Neural network ensembles. IEEE Transactions on Pattern Analysis and Machine Intelligence 12(10), 993–1001 (1990)
17. Krogh, P.S.A.: Learning with ensembles: How over-fitting can be useful. In: Proceedings of the 1995 Conference, vol. 8, p. 190 (1996)
18. Lan, Y., Soh, Y.C., Huang, G.B.: Ensemble of online sequential extreme learning machine. Neurocomputing 72(13), 3391–3395 (2009)
19. Zhou, Z.H., Wu, J., Tang, W.: Ensembling neural networks: many could be better than all. Artificial Intelligence 137(1), 239–263 (2002)
20. Kennedy, J.: Particle swarm optimization. Encyclopedia of Machine Learning, pp. 760–766. Springer, US (2010)
21. Blake, C., Merz, C.J.: Repository of machine learning databases (1998)

Deep Extreme Learning Machines for Classification

Migel D. Tissera and Mark D. McDonnell

Computational and Theoretical Neuroscience Laboratory,
Institute for Telecommunications Research,
University of South Australia, Mawson Lakes, SA 5095, Australia
migel.tissera@mymail.unisa.edu.au, mark.mcdonnell@unisa.edu.au

Abstract. We present a method for synthesising deep neural networks using Extreme Learning Machines (ELMs) as a stack of supervised autoencoders. We show that the network achieves comparable performance to an ELM with a single hidden layer with a size equal to the total number of hidden-layer neurons in the deep network. The main advantage of our method is in its significantly reduced network training time and memory usage. These favourable properties suggest that our method can be applied to a resource-constrained hardware implementation to increase the network performance.

Keywords: Extreme Learning Machine, Supervised learning, Autoencoder, MNIST, Classifier.

1 Introduction

In recent years the computational neuroscience field has seen multiple, independent and parallel emergence of hardware platforms optimised for neural network implementation. These implementations range from massively-parallel custom-built System-on-Chip (SoC) silicon microprocessor arrays (e.g. SpiNNaker [1]), to analog VLSI processors directly emulating the ion channels of the neurons as leakage currents in CMOS subthreshold region (e.g. Neurogrid [2]). The emergence of these platforms has been accompanied by a parallel effort to develop algorithms which mimic the computational capability of the human brain, particularly in developing synthesised (trained) neural networks.

These algorithms are now widely utilised for both investigating brain function in computational neuroscience (for example, models of controlling eye position and working memory [3]), and for implementing computing systems in machine learning. In machine learning, an emerging algorithm is the Extreme Learning Machine (ELM) [4], which is fast to train and performs with similar accuracy to Support Vector Machines (SVMs) [5]. ELM, first introduced in 2006 [6], consists of a standard three layer feedforward structure. Its first layer acts as the input layer, and the second (or the hidden) layer projects the input layer to a higher dimensionality using a very large number of non-linear sigmoid neurons. Its third

J. Cao et al. (eds.), *Proceedings of ELM-2014 Volume 1*,
Proceedings in Adaptation, Learning and Optimization 3, DOI: 10.1007/978-3-319-14063-6_29

and final layer acts as the output and consists of neurons with linear input-output characteristics.

In ELM, the connection weights between the input and the hidden layer neurons are randomly set and fixed for the entire duration of the network. That is, they are not altered during the training phase. In most cases, a distribution between -0.5 and +0.5 is used for the connection weights. This is analogous with biology, in the sense that a negative connection weight inhibits the network, and a positive weight excites the network. After projecting the input to a higher dimensionality at the hidden layer, a non-linear sigmoid function is used to generate the outputs of these hidden layer neurons. Then using training data, the connection weights between this hidden and the output layers is trained in a single pass by mathematical optimisation. Only this connection weight matrix is altered during training. It is calculated by a least squares regression method such as the Moore-Penrose pseudoinverse [7].

The methodology used in the above approach can be summarised as follows:

1. Using random and fixed weights, connect an input layer to a higher dimensional and large hidden layer.
2. Using training data, numerically solve the output weights between the large hidden layer and the output layer by calculating the pseudoinverse of the product of the hidden layer values and the desired output.

This class of methods are now referred to as Linear Solutions of Higher Dimensional Interlayer (LSHDI) networks [5]. It denotes a significant deviation from classical artificial neural network training methods. In classical artificial neural networks, the input weights are iteratively trained, rather than computing the output with the above approach. This interesting property can significantly enhance the efficiency of training since the full and final solution is obtained by mathematical optimisation in one single step. This LSHDI method also solves a significant problem in bio-inspired neural network simulations. Although widely accepted and very capable models exist at the single neuron level to mimic neurobiology, it had been extremely difficult to implement neural networks to model specific relationships. Until the emergence of LSHDI, there had been no widely applicable method to synthesise (train) a network to solve a given task. This class of methods are now emerging as the core of a generic neural compiler for creating silicon neural systems [8].

However, classical ELMs use a very large number of neurons in its single hidden layer, hence training the network can be computationally heavy given a large dataset. It also makes use of batch training, meaning that the network is trained using the entire dataset at once, which usually requires large memory and processing power. In [5] the authors have proposed an on-line training method (as opposed to batch training) to overcome this, but due to the large number of neurons typically used in the single hidden layer, training still largely depends on the size of the network.

Here we propose a network structure which takes inspiration from biology and the recent advances in deep learning architectures. We show that by constructing a deep ELM network as a stack of supervised autoencoders and training layer-by-layer, the network training time and memory usage can be significantly improved. The system therefore is well suited for a resource-constrained hardware implementation, since it offers potential for particular hardware components to be time-multiplexed.

2 Methodology

We start by creating an input layer that takes as input a test or training vector, $\mathbf{I} \in \mathbb{R}^{N\times 1}$. This input layer is connected to the first hidden layer by an all-to-all weight matrix $\mathbf{W}_{\text{in}} \in \mathbb{R}^{M\times N}$ We write the input vector to this layer as $\mathbf{H}_1 \in \mathbb{R}^{M\times 1}$. The output of the hidden layer is given by

$$\text{out}[\mathbf{H}_1] = \frac{1}{1+\exp(-0.5\mathbf{W}_{\text{in}}\mathbf{I})} \tag{1}$$

Using training data, we then solve for a weight matrix $\mathbf{W}_{h1}$ to reconstruct the input layer, as follows. For k training data, we can form a matrix $\mathbf{A} \in \mathbb{R}^{M\times k}$ in which each column contains the output of the hidden layer $\mathbf{H}_1$ at one training point. Then using the training data itself as another matrix $\mathbf{Y} \in \mathbb{R}^{N\times k}$ in which each column contains training vectors, we solve for $\mathbf{W}_{h1} \in \mathbb{R}^{N\times M}$ that minimises the mean square error between

$$\mathbf{Y} = \mathbf{W}_{h1}\mathbf{A} \tag{2}$$

and the original training image. We solve this problem numerically by taking the Moore-Penrose pseudoinverse of $\mathbf{A}$, which is denoted as $\mathbf{A}^+ \in \mathbb{R}^{k\times M}$. Therefore we have

$$\mathbf{W}_{h1} = \mathbf{Y}\mathbf{A}^+. \tag{3}$$

We now use this trained weight matrix $\mathbf{W}_{h1}$ to create a new vector $\hat{\mathbf{I}}$, which is the autoencoded version of the original data. Then using a new projection weight matrix $\mathbf{W}_p \in \mathbb{R}^{L\times N}$, we connect $\hat{\mathbf{I}}$ to the second hidden layer $\mathbf{H}_2 \in \mathbb{R}^{L\times 1}$.

The above explanation provides an intuitive description of the training procedure. After the network has been trained however, it is natural to combine the two weight matrices $\mathbf{W}_{h1}$ and $\mathbf{W}_p$ to form one single weight matrix $\mathbf{W}_{h12} \in \mathbb{R}^{L\times M}$, which fully connects the two hidden layers, i.e.

$$\mathbf{W}_{h12} = \mathbf{W}_p\mathbf{W}_{h1}. \tag{4}$$

We then repeat the procedure described by Equations (1) to (4) to obtain $\mathbf{W}_{h23} \in \mathbb{R}^{Q\times L}$ and $\mathbf{W}_{h3}$, where Q is the number of hidden neurons in the third hidden layer, and W_{h3} represents the connection weight matrix from the third hidden layer to the output layer vector $\mathbf{T} \in \mathbb{R}^{N\times 1}$. The system is depicted in Figure 1.

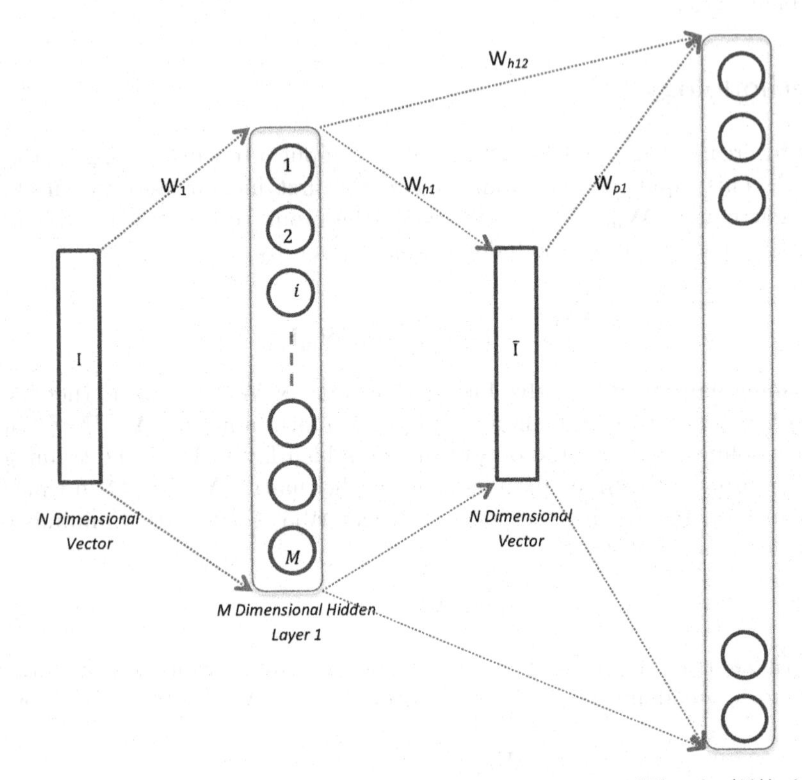

Fig. 1. The first two hidden-layers of our supervised deep extreme learning machine network are shown. From left-to-right in feedforward: an N dimensional input vector is projected to the M dimensional first hidden-layer using a random weight matrix $\mathbf{W}_{\text{in}}$. Then using Equation (3), the weight matrix $\mathbf{W}_{h1}$ is obtained using training data. This results in a vector $\hat{\mathbf{I}}$ of the same dimensionality as the input. Then the vector $\hat{\mathbf{I}}$ is projected to an L dimensional second hidden-layer using a random weight matrix $\mathbf{W}_{p1}$. Next we calculate the hidden-layer-1 to hidden-layer-2 weight matrix $\mathbf{W}_{h12}$ using Equation (4).

3 Simulation Details

For the purpose of this paper, we have constructed a network with 3 hidden layers and an output layer. The parameters of the network were chosen to be of the following form, where d is a real, positive, non-zero integer:

- First Layer: $2d$ neurons;
- Second Layer: d neurons;
- Third Layer: $5d$ neurons.

These layer sizes were chosen to illustrate the ability of the network to have fan-in or fan-out capability. The comparison single layer ELM has a single hidden layer with $8d$ neurons.

We train and test our network on the MNIST hand-written digit classification dataset, and therefore the input layer consist of 784 units. Since we perform autoencoding, the size of the output layer is the same as the input layer, which comprises of 784 units.

Our Deep ELM system differs to a deep ELM described previously [9] in two significant ways:

1. We use supervised autoencoding. We have found that supervised autoencoding performs better than un-supervised autoencoding.
2. We use a hidden-vector ($\hat{\mathbf{I}}$) to compute the weight matrix between the hidden layers. The dimension of this hidden-vector is equal to the dimension of training vectors, and represents an auto encoding of the input.

Optionally, we have carried out simulations where we add a pre-training step for the projection weights using Gibbs sampling similar to a Restricted Boltzmann Machine (RBM). This has improved the network performance, although with an added memory and processing power penalty. However, the focus of this paper is to introduce our Deep ELM method, therefore we have not included the simulations we have carried out using pre-training.

Before presenting some results, we first describe the supervised aspect of the autoencoding aspect in more detail.

3.1 Supervised Autoencoding

We use supervised autoencoding to train our weight matrices. To state explicitly, the training labels are embedded in the training images. We do this the following way for the MNIST database. We take the first 10 pixels of the training images, and reset them to zero (if they are not already at zero). Then we use the label of the respective image, to set the corresponding pixel to 1. For example, for label '0', pixel 1 is set to one; for label 9, pixel 10 is set to one. This is illustrated in Figure 2.

We use the resulting combined "images + labels" training set as the target used to train each autoencoding stage in our network. Note that the first layer receives the images only, without the supervised bits. In the classification step (i.e. at the output layer), we use the highest valued pixel in the first 10 pixels as the classified output.

Despite the fact that we use a 3 hidden layer network for the purpose of this paper, note that the method stated in Section 2 can be repeated for any number of hidden layers.

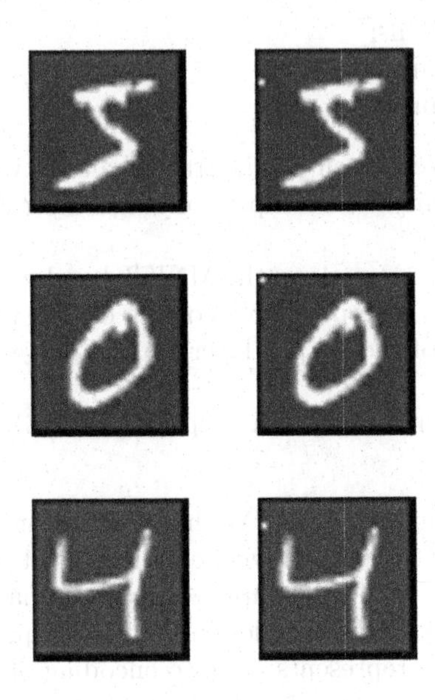

Fig. 2. Illustration of the introduction of supervised label bits to training data images. Left hand side: Examples of the original MNIST Training images. Right hand side: Combined "image + label" training data. Top Right: Note the 6th pixel has been set to 1 to indicate the label 5, and similar for the other exempts.

4 Results

We now show the results of applying our Deep ELM network to classifying the MNIST hand-written digit classification dataset. We vary the parameter d stated in Section 3 between 750 and 1250 with increments of 50. Therefore the total number of hidden neurons in the network varied between 6,000 and 10,000 with increments of 400 neurons.

We trained the network on the full 60,000 training images, and tested the network on the 10,000 image test-set. Figure 3 shows the network performance on the test-set, compared to a large single hidden-layer network comprised from $8d$ neurons.

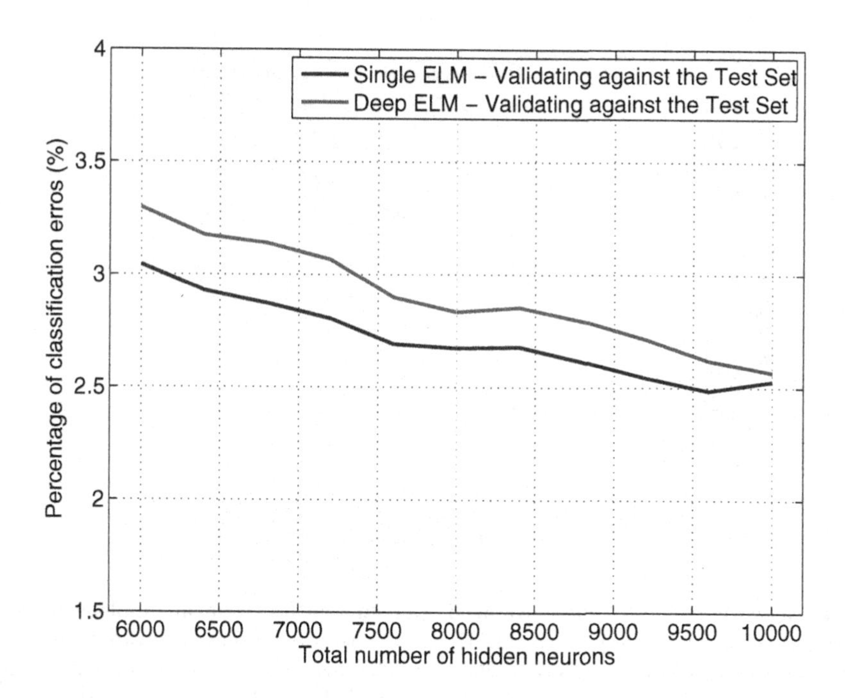

Fig. 3. Illustration of the classification error rate as the number of hidden neurons increases, averaged over 10 repeats. Our supervised Deep ELM method achieves comparable performance to the single-layer ELM.

We compare the network training times in Figure 4. Even though the Deep ELM network performs slightly below the classification rate of the Single ELM network (Figure 3), its strongest property is in the reduced network training time and memory usage. This is potentially highly favourable in time critical applications where the training time is important, such as for on-line training.

Also, our approach can be applied to a network comprising of many fixed-size hidden layers where time-multiplexing of hardware can be utilised. In that case, even though the hardware resources may be limited, the performance of the network can be greatly enhanced. We illustrate this property of increasing network performance with added hidden layers of a fixed-size in Figure 5. It can be clearly seen that the network performance improves with the number of hidden layers. However, we note that this improvement plateaus after a certain number of hidden-layers.

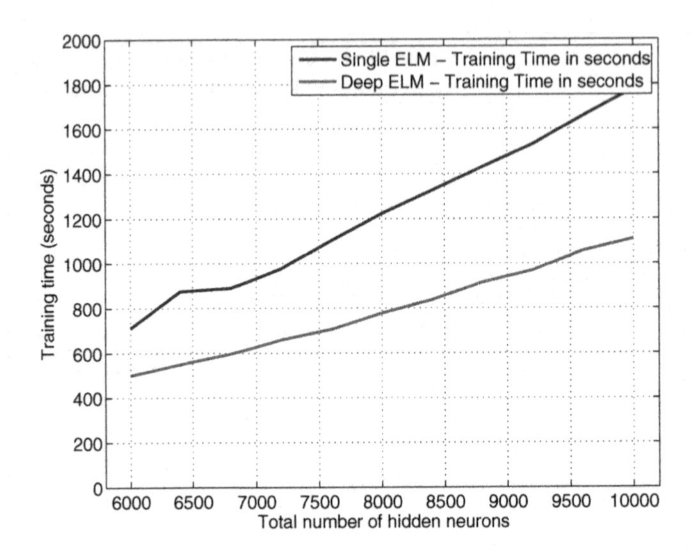

Fig. 4. Network training times, averaged over 10 repeats, from an implementation performed on a single-core processor. Hence the training times illustrated here for Single and Deep ELM is much slower than a modern multicore desktop computer, and here the point to highlight is the difference between the single and deep ELMs.

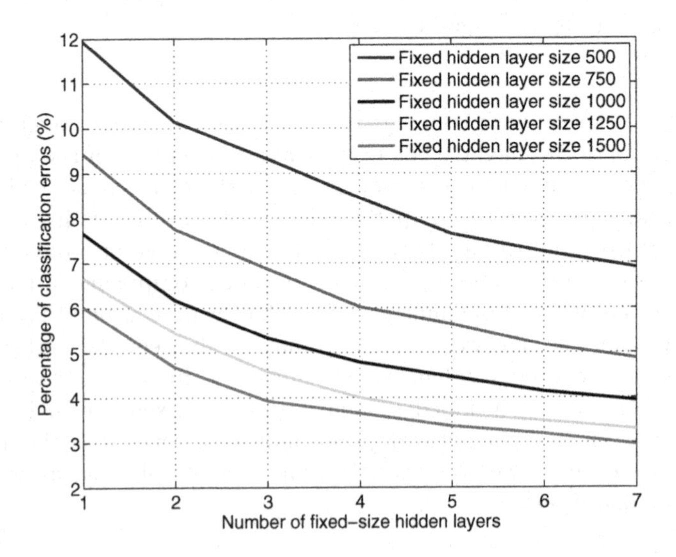

Fig. 5. Illustration of the change in classification error rate as the number of fixed-size hidden layers increases. It can be seen that the network performance is enhanced with added hidden-layers.

5 Discussion

Utilising Extreme Learning Machines (ELMs) as layers in a deep network has attracted recent attention [9]. Here we introduce a novel method for creating such a network, with comparable performance to a large single hidden-layer network with equal number of hidden neurons. We list the advantages of our method below.

1. We use supervised autoencoding, as opposed to unsupervised autoencoding. We have found that this has increased the performance of our network, relative to unsupervised autoencoding (data not shown here). A possible reason for this is that since the network is trained with the label in every layer, it creates an internal filtered representation of the input at each layer. This internal filter is enhanced by the subsequent layers. Therefore at the classification step, (i.e. when the network receives an input it has never seen before), it automatically creates an internal representation which is similar to its training data. This can be seen by plotting the outputs of layers, as shown below in Figure 6.

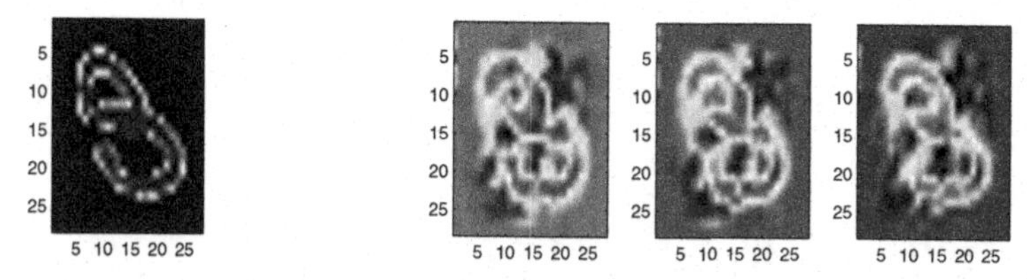

Fig. 6. The leftmost image shows an example the input to the network, taken from the MNIST test dataset. The three images to the right shows the output of the 1st hidden-layer, 4th hidden layer and the 7th hidden layer of a 7 layer Deep ELM.

2. Using our method, we have reduced the network training time compared with a single hidden-layer network of equivalent hidden neurons. This becomes apparent as the size of the hidden-layer neurons increases, as shown in Figure 4.
3. This method of constructing deep neural networks is potentially favourable for hardware implementations. In resource constrained environments, our method can be applied to increase the network performance using time-multiplexed hardware.

References

1. Furber, S.B., Galluppi, F., Temple, S., Plana, L.A.: The SpiNNaker Project. Proceedings of the IEEE 102, 652–665 (2014)
2. Benjamin, B.V., Gao, P., McQuinn, E., Choudhary, S., Chandrasekaran, A.R., Bussat, J.-M., Alvarez-Icaza, R., Arthur, J.V., Merolla, P.A., Boahen, K.: Neurogrid: A mixed analog-digital multi chip system for large-scale neural simulations. Proceedings of the IEEE 102, 699–716 (2014)

3. Eliasmith, C., Anderson, C.H.: Neural Engineering: Computation, Representation, and Dynamics in Neurobiological Systems. The MIT Press (2004)
4. Huang, G.-B., Wang, D.H., Lan, Y.: Extreme Learning Machines: A Survey. International Journal of Machine Learning and Cybernetics 2, 107–122 (2011)
5. Tapson, J., van Schaik, A.: Learning the pseudoinverse solution to network weights. Neural Networks 45, 94–100 (2013)
6. Huang, G.-B., Zhu, Q.-Y., Siew, C.-K.: Extreme Learning Machine: Theory and applications. Neurocomputing 70, 489–501 (2006)
7. Penrose, R.: A generalized inverse for matrices. Mathematical Proceedings of the Cambidge Philosophical Society 51, 406–413 (1955)
8. Galluppi, F., Davies, S., Eliasmith, C., Stewart, T., Furber, S.: Real Time On-Chip Implementation of Dynamical Systems with Spiking Neurons. In: The International Joint Conference on Neural Networks IJCNN, pp. 1–8 (2012)
9. Kasun, L.L.C., Zhou, H., Huang, G.-B.: Representational Learning with ELMs for Big Data. IEEE Intelligent Systems 28, 31–34 (2013)

C-ELM: A Curious Extreme Learning Machine for Classification Problems

Qiong Wu and Chunyan Miao

Nanyang Technological University, Singapore
{wuqi0005,ASCYMiao}@ntu.edu.sg

Abstract. In psychology, curiosity is generally known as the critical intrinsic motivation for learning. It drives human beings to explore for novel and interesting information that can elicit the feeling of pleasure. This paper proposes such a curiosity driven algorithm for Extreme Learning Machine, which is referred to as Curious Extreme Learning Machine (C-ELM). C-ELM follows the psychological theory of curiosity proposed by Berlyne and performs curiosity appraisal towards each input data based on four collative variables: *novelty*, *uncertainty*, *conflict*, and *surprise*. The collative variables reflect the level of curiosity stimulation in the input data. Based on the level of curiosity stimulation, the network decides on the strategies for learning, including *neuron addition*, *neuron deletion*, and *parameter update*. During *neuron addition*, a new neuron is added based on the input data, thereby reducing the randomization effect of ELM. The *parameter update* is conducted using recursive least squares method and *neuron deletion* aims at deleting the most conflicting knowledge. The empirical performance study of the proposed method on benchmark classification problems clearly highlights the learning and generalization ability of C-ELM.

1 Introduction

An extremely fast learning neural algorithm referred to as Extreme Learning Machine (ELM) has been developed for single-hidden layer feedforward networks (SLFNs) by Huang et al. [1, 2]. The essence of ELM is that the hidden layer of SLFNs need not be tuned [3]. ELM randomly assigns hidden neuron parameters and find the output weights analytically. It has been shown to generate good generalization performance at extremely high learning speed [1, 2, 4, 5] and has been successfully applied to many real world applications [4–7].

Although ELM has shown advanced generalization performance with extremely high learning speed, several major issues still remain in ELM:

1. Manually set the number of hidden neurons: The number of hidden neurons needs to be set a priori to training [8]. The number of hidden neurons is usually chosen by trial-and-error.
2. Fixed structure: The network structure is fixed once the number of hidden neurons is set [9]. It can not evolve, i.e., add or delete hidden neurons, based on the training data.

J. Cao et al. (eds.), *Proceedings of ELM-2014 Volume 1*,
Proceedings in Adaptation, Learning and Optimization 3, DOI: 10.1007/978-3-319-14063-6_30

3. Randomization effect: The random assignment of hidden neuron parameters induces high randomization effect in the generated results.

To address issue 1), several algorithms have been proposed, such as incremental ELM (I-ELM) [10], enhanced incremental ELM (EI-ELM) [11], pruning ELM (P-ELM) [9], optimally-pruned ELM(OP-ELM) [12], and error minimized ELM (EM-ELM) [8]. However, all these algorithms can either add neurons (I-ELM, EI-ELM, EM-ELM) or delete neurons (P-ELM, OP-ELM) without being able to adjust network structure based on the incoming data, in other words, the evolving capability. Recently, a meta-cognitive ELM (McELM) has been proposed [13], which addresses issue 1) and partially issue 2). McELM can decide network structure based on the training data, but it can only add neurons without pruning capability. To our knowledge, few works have been done towards issue 3).

To address all the three issues mentioned above, we propose a curious extreme learning machine (C-ELM) algorithm for classification problems. It is a psychologically inspired algorithm based on the theory of curiosity [14]. In psychology, curiosity is commonly known as the important intrinsic motivation that drives human exploration and learning [15]. The psychological concept of curiosity has been applied in many computational systems to enhance their learning capability (e.g., intrinsically motivated reinforcement learning) [16, 17] and believability (e.g., curious companions) [18–20]. To our knowledge, this is the first attempt to introduce curiosity in an ELM framework.

C-ELM is inspired by the psychological theory of curiosity by Berlyne [21]. Berlyne interpreted curiosity as a process of stimulus selection, i.e., when several conspicuous stimuli are introduced at once, to which stimulus will human respond. He identified several key collative variables, e.g., *novelty*, *uncertainty*, *conflict*, and *surprise*, that govern the stimulus selection process. Based on this theory, C-ELM classifier treats each training data as a stimulus and decides its learning strategy based on the appraisal of collative variables. There are three learning strategies for C-ELM: *neuron addition*, *neuron deletion*, and *parameter update*.

When a new neuron is added, conventional incremental ELM algorithms will randomly assign the center and impact factor of the RBF kernel (other kernels such as the linear kernel can also apply) in the new neuron. However, random center selection may require more number of hidden neurons to approximate the decision function accurately [22]. Hence, C-ELM use data-driven center selection which use the current training data that triggers the *neuron addition* strategy as the center of the new neuron. It removes partially the random effect of the traditional ELM algorithms. Data-driven center selection also allows the class label of the new neuron to be apparent, which enables further analysis of the hidden neurons. During *neuron deletion*, the most conflicting neuron for the current training data is removed from the network. In literature, various neuron deletion schemes for ELM have been proposed such as pruning based on relevance [9] or based on leave-one-out cross-validation [12]. These techniques, although effective, might render the system slow. Hence, we propose the *neuron deletion* strategy based on conflict resolution, which helps the system attain fast and efficient convergence. The *parameter update* is conducted using recursive least squares method.

A simulation study of the C-ELM on several commonly used classification benchmark problems shows that the proposed approach leads to compact network classifiers that generate fast response and better generalization performance on unseen data when compared with the popular classifiers such as SVM, the traditional ELM and its recently enhanced variant McELM [13].

2 C-ELM

In this section, we provide a detailed description of the C-ELM architecture and the learning algorithm for classification problems. The classification problem is defined as follows:

Given: a stream of training data $\{(\mathbf{x}^1, c^1), \cdots, (\mathbf{x}^t, c^t), \cdots\}$, where $\mathbf{x}^t = [x_1^t, \cdots, x_M^t]^T \in \Re^M$ is a M-dimensional input vector of the tth input data, $c^t \in [1, 2, \cdots, N]$ is its class label, and N represents the total number of distinct classes. The coded class label $\mathbf{y}^t = [y_1^t, \cdots, y_j^t, \cdots, y_N^t] \in \Re^N$ is obtained by converting the class label (c^t) as follows:

$$y_j^t = \begin{cases} 1 & \mathit{If}\ j = c^t \\ -1 & \mathit{otherwise} \end{cases} \quad j = 1, 2, \cdots, N. \tag{1}$$

Find: a decision function $\mathbb{F}$ that maps the input features ($\mathbf{x}^t$) to the coded class labels ($\mathbf{y}^t$), i.e., $\mathbb{F} : \Re^M \to \Re^N$, as close as possible.

To solve this problem, we propose a fast evolving data-driven neural algorithm referred to as curious extreme learning machine (C-ELM). C-ELM has two components: a unified single layer feed-forward neural network (SLFN) and a curiosity driven extreme learning algorithm. The curiosity driven extreme learning algorithm consists of the definitions for four collative variables, i.e., *novelty*, *uncertainty*, *conflict*, and *surprise*, and three learning strategies, i.e., *neuron addition*, *neuron deletion*, and *parameter update*. Next, we will describe these components in detail.

2.1 Unified SLFN

The architecture of the C-ELM is an SLFN with M input neurons, K hidden neurons and N output neurons. For RBF hidden neuron with activation function $g(x) : \Re \to \Re$ (e.g., Gaussian), the output of the kth hidden neuron with respect to the input $\mathbf{x}^t$ is given by:

$$G(\mathbf{x}^t, \mathbf{a}_k, b_k) = g(b_k||\mathbf{x}^t - \mathbf{a}_k||), b_k \in \Re^+, \tag{2}$$

where $\mathbf{a}_k$ and b_k are the center and impact factor of the kth RBF neuron. $\Re^+$ denotes the set of all positive real values.

The predicted output for the input $\mathbf{x}^t$ is denoted by:

$$\hat{\mathbf{y}}^t = \left[\hat{y}_1^t, \cdots, \hat{y}_i^t, \cdots, \hat{y}_N^t\right]. \tag{3}$$

Here, the output of the ith neuron in the output layer is given by:

$$\hat{y}_i^t = \sum_{k=1}^{K} G(\mathbf{x}^t, \mathbf{a}_k, b_k) w_{ki}, \tag{4}$$

where w_{ki} is the output weight connecting the kth hidden neuron to the ith output neuron. The output for a chunk of t input data can be written by:

$$\hat{\mathbf{Y}} = \mathbf{HW}, \tag{5}$$

where $\mathbf{H}$ is the hidden layer output matrix and $\mathbf{W}$ is the weight matrix connecting the hidden neurons to the output neurons as shown below:

$$\mathbf{H} = \begin{bmatrix} G(\mathbf{x}^1, \mathbf{a}_1, b_1) & \cdots & G(\mathbf{x}^1, \mathbf{a}_K, b_K) \\ \vdots & \ddots & \vdots \\ G(\mathbf{x}^t, \mathbf{a}_1, b_1) & \cdots & G(\mathbf{x}^t, \mathbf{a}_K, b_K) \end{bmatrix}_{t \times K}, \tag{6}$$

and

$$\mathbf{W} = \begin{bmatrix} w_{1,1} & w_{1,2} & \cdots & w_{1,N} \\ w_{2,1} & w_{2,2} & \cdots & w_{2,N} \\ \vdots & \vdots & \ddots & \vdots \\ w_{K,1} & w_{K,2} & \cdots & w_{K,N} \end{bmatrix}_{K \times N}. \tag{7}$$

With the architecture of C-ELM described above, next, we will introduce the curiosity driven extreme learning algorithm.

2.2 Definitions of Collative Variables

C-ELM employs an intrinsically motivated learning paradigm transposed from the psychological theory of curiosity proposed by Berlyne [21]. Learning is regulated based on the curiosity appraisal of input data. For each input data, the curiosity appraisal is governed by four collative variables, i.e., *novelty*, *uncertainty*, *surprise*, and *conflict*. In this section, we will introduce the definitions of the four collative variables in C-ELM.

Novelty: Novelty reflects how much the input data differs from the network's current knowledge. In kernel methods, spherical potential is often used to determine the novelty of data [23]. The spherical potential of an input data $\mathbf{x}^t$ is defined by (a detailed derivation can be found in [23]):

$$\psi(\mathbf{x}^t) = \frac{1}{K} \sum_{k=1}^{K} G\left(\mathbf{x}^t, \mathbf{a}_k, b_k\right). \tag{8}$$

A higher potential indicates the input data is similar to the existing knowledge, while a smaller potential indicates that the input data is novel. Hence, the novelty $\mathcal{N}$ of an input data $\mathbf{x}^t$ is determined by:

$$\mathcal{N}(\mathbf{x}^t) = 1 - \frac{1}{K}\sum_{k=1}^{K} G\left(\mathbf{x}^t, \mathbf{a}_k, b_k\right). \tag{9}$$

Uncertainty: Uncertainty reflects how not confident the network is in its predictions. The confidence of a network is often measured by the posterior probability of the prediction. It has been proven theoretically that hinge-loss function can accurately estimate the posterior probability for a classification problem [24]. Hence, we use the truncated hinge-loss error ($\mathbf{e}^t = \left[e_1^t, \cdots, e_j^t, \cdots, e_N^t\right]^T \in \Re^N$) to measure the prediction error, where each element is defined by:

$$e_j^t = \begin{cases} 0 & \textit{if } \hat{y}_j^t y_j^t > 1 \\ min(max(\hat{y}_j^t - y_j^t, -1), 1) & \textit{otherwise.} \end{cases} \tag{10}$$

With the truncated hinge-loss error, the estimated posterior probability of input $\mathbf{x}^t$ belonging to class c is given by [22]:

$$p(c|\mathbf{x}^t) = \frac{e_j^t + 1}{2}, c = 1, 2, \cdots, N. \tag{11}$$

Since uncertainty measures how not confident a network is in its predictions, we define uncertainty $\mathcal{U}$ of the prediction to an input data $\mathbf{x}^t$ by:

$$\mathcal{U}(\mathbf{x}^t) = 1 - p(\hat{c}|\mathbf{x}^t), \tag{12}$$

where $\hat{c}$ is the predicted class for $\mathbf{x}^t$.

Conflict: In psychology, conflict occurs when a stimulus arouses two or more incompatible responses in an organism [14]. The degree of conflict depends on the competing strengths of those incompatible responses. For a classifier, conflict can be reflected by the competing strengths of the most fired two output neurons.

Given an input $\mathbf{x}^t$, let $\hat{y}_{j1}^t$ and $\hat{y}_{j2}^t$ be the outputs of the top two highly activated output neurons. The more closer $\hat{y}_{j1}^t$ is to $\hat{y}_{j2}^t$, the higher competing strength is between the two output neurons, which indicates a higher conflict between the network's decisions. Hence, the conflict $\mathcal{C}$ induced by an input $\mathbf{x}^t$ is defined by:

$$\mathcal{C}(\mathbf{x}^t) = \begin{cases} 1 - \frac{|\hat{y}_{j1}^t - \hat{y}_{j2}^t|}{|\hat{y}_{j1}^t + \hat{y}_{j2}^t|} & \textit{if } \hat{y}_{j1}^t \hat{y}_{j2}^t > 0 \\ 0 & \textit{otherwise.} \end{cases} \tag{13}$$

Surprise: In psychology, surprise indicates a violation of expectation [14]. For a classifier, surprise occurs when the predicted output differs from the true class label. The degree of surprise is determined by prediction errors for both the true class and the predicted class. As we adopt hinge-loss error in this work to measure prediction error, the surprise $\mathcal{S}$ induced by an input $\mathbf{x}^t$ is defined by:

$$\mathcal{S}(\mathbf{x}^t) = \begin{cases} |e_c^t \cdot e_{\hat{c}}^t| & \textit{if } \hat{c} \neq c \\ 0 & \textit{otherwise}, \end{cases} \tag{14}$$

where $\hat{c}$ and c represent the predicted class and the true class, respectively; and e is the hinge-loss error calculated according to Eq (10).

It can be analyzed that all the four collative variables are within the range of $[0, 1]$. The collative variables determine the level of curiosity arousal and the learning strategy selection. With the collative variables defined as above, we will introduce the learning strategies in the following section.

2.3 Learning Strategies

C-ELM has three learning strategies: *neuron addition*, *neuron deletion*, and *parameter update*. C-ELM begins with zero hidden neurons, add or delete hidden neurons and update parameters of the existing neurons to achieve an optimal network structure with optimal parameters. The decision on whether to update network structure or update parameters is made based on the appraisal of collative variables. Intuitively, high values of collative variables induce a high level of curiosity towards the input data, which require more efforts in learning, i.e., updating network structure, to incorporate the new knowledge; otherwise, simply update parameters of the existing neurons to reinforce the 'familiar' knowledge. Next, we will introduce the three learning strategies of C-ELM in detail.

Neuron Addition Strategy: Intuitively, for an input data, if novelty is high and uncertainty is high and surprise is high, it indicates that a misclassification (i.e., surprise high) with high uncertainty in its prediction (i.e., uncertainty high) is caused by the newness of the knowledge (i.e., novelty high). In this case, the network should add new neurons to capture this new knowledge. Hence, given an input $\mathbf{x}^t$, the neuron addition condition is:

$$\mathcal{N}(\mathbf{x}^t) > \theta_{\mathcal{N}^{add}} \textbf{ AND } \mathcal{U}(\mathbf{x}^t) > \theta_{\mathcal{U}} \textbf{ AND } \mathcal{S}(\mathbf{x}^t) > \theta_{\mathcal{S}}, \tag{15}$$

where $\theta_{\mathcal{N}^{add}}$, $\theta_{\mathcal{U}}$, and $\theta_{\mathcal{S}}$ are neuron addition thresholds for novelty, uncertainty, and surprise, respectively. If these parameters are chosen close to 1, then very few input data can trigger the neuron addition strategy and the network cannot approximate the decision function accurately. If these parameters are chosen close to 0, then many input data can trigger the neuron addition strategy, leading to poor generalization ability. In general, $\theta_{\mathcal{N}^{add}}$ is chosen in the range of [0.1,0.5], $\theta_{\mathcal{U}}$ is chosen in the range of [0.1,0.3], and $\theta_{\mathcal{S}}$ is chosen in the range of [0.2,0.9].

A typical ELM randomly chooses hidden neuron parameters and find the output weights analytically. However, random exploration of the feature space may need more hidden neurons to accurately approximate the decision function. In C-ELM, we propose data-driven center selection for the hidden neurons without compromising the extreme learning capability. When a new neuron is added, instead of randomly assign the center $\mathbf{a}_k$, assign the input features of the current input data as the center, i.e., $\mathbf{a}_k = \mathbf{x}_t$. Since the center selection is data-driven, we can label the class of the new neuron using the target value of the input data, i.e., $l_k = c^t$. Data-driven center selection allows fast hidden neuron clustering using their class labels and provides class specific information when deleting neurons. The values of the impact factors in the hidden neurons are randomly assigned. Hence, with the new hidden neuron, the dimension of the hidden layer output matrix $\mathbf{H}$ increases from $(t-1) \times (k-1)$ to $t \times k$. The target values of the t input data is represented by:

$$\mathbf{Y} = \begin{bmatrix} \mathbf{y}^1 \\ \vdots \\ \mathbf{y}^t \end{bmatrix}_{t \times N.} \tag{16}$$

The output weight for $\mathbf{W}$ can be analytically find by:

$$\mathbf{W} = \mathbf{H}^{\dagger}\mathbf{Y}, \tag{17}$$

where $\mathbf{H}^{\dagger}$ is the Moore-Penrose generalized inverse of the hidden layer output matrix $\mathbf{H}$.

Neuron Deletion Strategy: Intuitively, for an input data, if surprise is high and conflict is high and novelty is low, it indicates that a misclassification (i.e., surprise high) occurs for a familiar stimulus (i.e., novelty low) due to high competing strengths between two decisions (i.e., conflict). In this case, the network should adjust its decision making by strengthening the correct decision and weakening the wrong decision, i.e., deleting the most contributing neuron in the wrong class. Hence, given an input $\mathbf{x}^t$, the neuron deletion condition is:

$$\mathcal{S}(\mathbf{x}^t) > \theta_{\mathcal{S}} \ \ \textbf{AND} \ \ \mathcal{C}(\mathbf{x}^t) > \theta_{\mathcal{C}} \ \ \textbf{AND} \ \ \mathcal{N}(\mathbf{x}^t) < \theta_{\mathcal{N}^{del}}, \tag{18}$$

where $\theta_{\mathcal{S}}$, $\theta_{\mathcal{C}}$, and $\theta_{\mathcal{N}^{del}}$ are neuron deletion thresholds for surprise, conflict, and novelty, respectively. When $\theta_{\mathcal{S}}$ and $\theta_{\mathcal{C}}$ are chosen close to 1 and $\theta_{\mathcal{N}^{del}}$ is chosen close to 0, then very few input data can trigger neuron deletion, leading to poor generalization ability. When $\theta_{\mathcal{S}}$ and $\theta_{\mathcal{C}}$ are chosen close to 0 and $\theta_{\mathcal{N}^{del}}$ is chosen close to 1, then many input data can trigger neuron deletion and the network cannot approximate the decision function accurately. In general, $\theta_{\mathcal{S}}$ is chosen in the range of [0.2,0.9], $\theta_{\mathcal{C}}$ is chosen in the range of [0.1,0.3], and $\theta_{\mathcal{N}^{del}}$ is chosen in the range of [0.1,0.8].

When neuron deletion is triggered, C-ELM will remove the most fired hidden neuron k belonging to the predicted class:

$$\{k | \max(G(\mathbf{x}^t, a_k, b_k)) \ \ \text{AND} \ \ l_k = \hat{c}\}. \tag{19}$$

After the kth neuron is removed, the network will re-calculate the output weight $\mathbf{W}$ with the t input data.

Parameter Update Strategy: When both the neuron addition strategy and the neuron deletion strategy are not triggered, it indicates that the new input data is a 'familiar' one. Hence, the network will update the output weight using recursive least squares to reinforce the familiar knowledge.

For the new input data $\mathbf{x}^t$, let the partial hidden layer output be represented by $\mathbf{h}^t = [G(\mathbf{x}^t, \mathbf{a}_1, b_1), \cdots, G(\mathbf{x}^t, \mathbf{a}_K, b_K)]$. The output weights are updated according to [25] by:

$$\mathbf{W}^t = \mathbf{W}^{t-1} + \mathbf{P}^t \mathbf{h}^t ((\mathbf{y}^t)^T - (\mathbf{h}^t)^T \mathbf{W}^{t-1}), \tag{20}$$

where

$$\mathbf{P}^t = \mathbf{P}^{t-1} - \frac{\mathbf{P}^{t-1}\mathbf{h}^t (\mathbf{h}^t)^T \mathbf{P}^{t-1}}{1 + (\mathbf{h}^t)^T \mathbf{P}^{t-1} \mathbf{h}^t}. \tag{21}$$

A pseudo-code for the C-ELM learning algorithm is given in Algorithm 1.

Step 1. Present a sample $(\mathbf{x}^t, c^t)$.
Step 2. Perform curiosity appraisal towards the sample:
compute novelty $\mathcal{N}(\mathbf{x}^t)$ using Eq (9), compute uncertainty $\mathcal{U}(\mathbf{x}^t)$ using Eq (12), compute conflict $\mathcal{C}(\mathbf{x}^t)$ using Eq (13), and compute surprise $\mathcal{S}(\mathbf{x}^t)$ using Eq (14).
Step 3. Select learning strategy based on the collative variables:
if $\mathcal{N}(\mathbf{x}^t) > \theta_{\mathcal{N}^{add}}$ *AND* $\mathcal{U}(\mathbf{x}^t) > \theta_{\mathcal{U}}$ *AND* $\mathcal{S}(\mathbf{x}^t) > \theta_{\mathcal{S}}$ **then**
  add hidden neuron with $\mathbf{x}^t$ as the center and randomize the impact factor of the hidden neuron. Update the hidden neuron output matrix $\mathbf{H}$ and compute the output weight $\mathbf{W}$ according to Eq (17).
if $\mathcal{S}(\mathbf{x}^t) > \theta_{\mathcal{S}}$ *AND* $\mathcal{C}(\mathbf{x}^t) > \theta_{\mathcal{C}}$ *AND* $\mathcal{N}(\mathbf{x}^t) < \theta_{\mathcal{N}^{del}}$ **then**
  delete the most fired hidden neuron that belonging to the predicted class following Eq (19).
else
  update the network parameter using Eq (20).
end
Step 4. Increment $t = t + 1$, go to Step 1.

Algorithm 1. Pseudo-code for C-ELM classifier

3 Performance Evaluation

The performance of C-ELM is evaluated on the benchmark problems described in Table 1 from the UCI machine learning repository, which contains three multi-category classification problems (vehicle classification, iris, and wine) and three binary classifcation problems (liver disorder, PIMA, and breast cancer). The performance of C-ELM is evaluated in comparison with other popular classifiers such as SVM, ELM and McELM. The results of SVM, ELM and McELM are reproduced from [13]. For simulating the results of C-ELM, MATLAB 2014b with 3.2 ghz and 16gb ram was used. The parameters were optimized using grid search.

Table 1. Specification of benchmark data sets

Data set	# Features	# Classes	# Training Data	# Testing Data
Vehicle classification	18	4	424	422
Iris	4	4	45	105
Wine	13	3	60	118
Liver disorder	6	2	200	145
PIMA	8	2	400	368
Breast cancer	9	2	300	383

3.1 Performance Measures

The performance of C-ELM is measured against other popular classifiers using two types of performance measures:

Average Classification Efficiency: The average classification efficiency η_a is defined by:

$$\eta_a = \frac{1}{C}\sum_{i}^{C}\frac{q_i}{N_i} \times 100, \tag{22}$$

where q_i is the number of data in class i that have been correctly classified, and N_i is the total number of data in class i. It reflects the average ratio of correctly classified data in each class.

Overall Classification Efficiency: The overall classification efficiency η_o is defined by:

$$\eta_o = \frac{\sum_i^C q_i}{N_T} \times 100 \tag{23}$$

where N_T is the total number of data in the testing data set. It reflects the overall ratio of correctly classified data in the whole testing data set.

3.2 Performance Study on Multi-category Classification Problems

The performance of the C-ELM on multi-category benchmark classification problems is shown in Table 2. It can be observed from Table 2 that the generalization performance of C-ELM is better than other classifiers used for comparison on all the multi-category classification problems. Also, the number of hidden neurons added during the evolving process is comparable with other algorithms. For example, C-ELM added 140 hidden neurons for Vehicle classification problem, 6 hidden neurons for Iris problem, and 8 hidden neurons for Wine problem. Another advantage of C-ELM in comparison with other self-regulated learning algorithms such as McELM is that it takes a substantially small amount of time for training. For example, McELM takes 40 seconds to train the Vehicle classification problem whereas C-ELM only takes 15.2 seconds. Hence, it shows that C-ELM achieves better performance than other classifiers on multi-category classification problems due to the intrinsically motivated learning mechanism of curiosity.

Table 2. Performance study on the multi-category classification problems

Data set	classifier	# hidden neurons	Training time	Testing η_o	Testing η_a
Vehicle classification	SVM	340[a]	550	70.62	68.51
	ELM	150	0.4	77.01	77.59
	McELM	120	40	81.04	81.3
	C-ELM	140	15.2	81.99	82.42
Iris	SVM	13[a]	0.02	96.19	96.19
	ELM	10	0.01	96.19	96.19
	McELM	6	0.03	98.1	98.1
	C-ELM	6	0.0001	99.05	99.05
Wine	SVM	13[a]	0.1	97.46	98.04
	ELM	10	0.25	97.46	98.04
	McELM	9	0.02	98.31	98.69
	C-ELM	8	0.015	99.15	99.35

[a] # support vectors

3.3 Performance Study on Binary Classification Problems

The performance of C-ELM on three binary classification problems is shown in Table 3. Table 3 shows that C-ELM achieves better generalization performance than other classifiers used for comparison on all the binary classification problems. Also, the total number of hidden neurons added during the evolving process

Table 3. Performance study on the binary classification problems

Data set	classifier	# hidden neurons	Training time	Testing η_o	Testing η_a
Liver disorder	SVM	141[a]	0.1	71.03	70.21
	ELM	100	0.17	72.41	71.41
	McELM	50	0.95	74.48	73.83
	C-ELM	31	0.73	76.55	76.5
PIMA	SVM	221[a]	0.21	77.45	76.33
	ELM	400	0.29	76.63	75.25
	McELM	25	0.47	80.43	78.49
	C-ELM	33	1.42	81.25	80.31
Breast cancer	SVM	24[a]	0.11	96.6	97.06
	ELM	66	0.14	96.36	96.5
	McELM	10	0.05	97.39	97.84
	C-ELM	9	0.09	97.65	98.04

[a] # support vectors

is comparable with other algorithms. For example, C-ELM requires 31 hidden neurons for Liver disorder problem, 33 hidden neurons for PIMA problem, and 9 hidden neurons for Breast cancer problem. For binary classification problems, the training time of C-ELM is comparable with other self-regulated learning algorithms such as McELM. For example, it requires 0.73 seconds for C-ELM and 0.95 seconds for McELM to train the Liver disorder problem. Hence, it shows that C-ELM achieves better generalization ability than other classifiers on binary classification problems without compromising the extreme leaning ability of ELM, due to the intrinsically motivated learning mechanism of curiosity.

4 Conclusion

In this paper, a curious extreme learning machine (C-ELM) algorithm is proposed based on the psychological theory of curiosity by Berlyne. C-ELM treats each input data as a curious stimulus and performs curiosity appraisal towards each input data based on four collative variables: *novelty*, *uncertainty*, *conflict*, and *surprise*. Three learning strategies can be chosen from based on the curiosity appraisal results, including *neuron addition*, *neuron deletion*, and *parameter update*. C-ELM enhances traditional ELM algorithms with the evolving capability, which determines optimal network structure dynamically based on training data. Also, C-ELM reduces partially the random effect of traditional ELM algorithms by selecting RBF centers based on data instead of random assignment. Moreover, C-ELM employs a novel neuron deletion strategy which is based on conflict resolution. Empirical study of C-ELM shows that the proposed approach leads to compact network structures and generates better generalization performance with fast response, comparing with traditional ELM and other popular classifiers.

References

1. Huang, G.-B., Zhu, Q.-Y., Mao, K.Z., Siew, C.-K., Saratchandran, P., Sundararajan, N.: Can threshold networks be trained directly? IEEE Transactions on Circuits and Systems II: Express Briefs 53(3), 187–191 (2006)
2. Huang, G.-B., Zhu, Q.-Y., Siew, C.-K.: Extreme learning machine: theory and applications. Neurocomputing 70(1), 489–501 (2006)
3. Huang, G.-B., Wang, D.H., Lan, Y.: Extreme learning machines: a survey. International Journal of Machine Learning and Cybernetics 2(2), 107–122 (2011)
4. Huang, G.-B., Zhu, Q.-Y., Siew, C.-K.: Real-time learning capability of neural networks. IEEE Transactions on Neural Networks 17(4), 863–878 (2006)
5. Liang, N.-Y., Saratchandran, P., Huang, G.-B., Sundararajan, N.: Classification of mental tasks from EEG signals using extreme learning machine. International Journal of Neural Systems 16(01), 29–38 (2006)
6. Xu, J.-X., Wang, W., Goh, J.C.H., Lee, G.: Internal model approach for gait modeling and classification. In: Annual International Conference of the IEEE Engineering in Medicine and Biology Society, pp. 7688–7691 (2006)

7. Yeu, C.-W.T., Lim, M.-H., Huang, G.-B., Agarwal, A., Ong, Y.-S.: A new machine learning paradigm for terrain reconstruction. IEEE Geoscience and Remote Sensing Letters 3(3), 382–386 (2006)
8. Feng, G., Huang, G.-B., Lin, Q., Gay, R.: Error minimized extreme learning machine with growth of hidden nodes and incremental learning. IEEE Transactions on Neural Networks 20(8), 1352–1357 (2009)
9. Rong, H.-J., Ong, Y.-S., Tan, A.-H., Zhu, Z.: A fast pruned-extreme learning machine for classification problem. Neurocomputing 72(1), 359–366 (2008)
10. Huang, G.-B., Chen, L.: Convex incremental extreme learning machine. Neurocomputing 70(16), 3056–3062 (2007)
11. Huang, G.-B., Chen, L.: Enhanced random search based incremental extreme learning machine. Neurocomputing 71(16), 3460–3468 (2008)
12. Miche, Y., Sorjamaa, A., Lendasse, A.: OP-ELM: Theory, experiments and a toolbox. In: Kůrková, V., Neruda, R., Koutník, J. (eds.) ICANN 2008, Part I. LNCS, vol. 5163, pp. 145–154. Springer, Heidelberg (2008)
13. Savitha, R., Suresh, S., Kim, H.J.: A meta-cognitive learning algorithm for an extreme learning machine classifier. Cognitive Computation 6(2), 253–263 (2014)
14. Wu, Q., Miao, C.: Curiosity: From psychology to computation. ACM Computing Surveys (CSUR) 46(2), 18 (2013)
15. Loewenstein, G.: The psychology of curiosity: A review and reinterpretation. Psychological Bulletin 116(1), 75–98 (1994)
16. Barto, A.G., Singh, S., Chentanez, N.: Intrinsically motivated learning of hierarchical collections of skills. In: Proc. of International Conference on Development Learn, 112–119 (2004)
17. Schmidhuber, J.: Formal theory of creativity, fun and intrinsic motivation. IEEE Transaction on Autonomous Mental Development 2(3), 230–247 (2009)
18. Wu, Q., Miao, C., Shen, Z.: A curious learning companion in virtual learning environment. In: Proc. of IEEE International Conference on Fuzzy Systems, pp. 1–8 (2012)
19. Wu, Q., Miao, C.: Modeling curiosity-related emotions for virtual peer learners. Computational Intelligence Magazine 8(2), 50–62 (2013)
20. Wu, Q., Miao, C., An, B.: Modeling curiosity for virtual learning companions. In: Proc. of the International Joint Conference on Autonomous Agents and Multiagent Systems (2014)
21. Berlyne, D.E.: Conflict, arousal, and curiosity. McGraw-Hill (1960)
22. Suresh, S., Sundararajan, N., Saratchandran, P.: Risk-sensitive loss functions for sparse multi-category classification problems. Information Sciences 178(12), 2621–2638 (2008)
23. Subramanian, K., Suresh, S., Sundararajan, N.: A metacognitive neuro-fuzzy inference system (McFIS) for sequential classification problems. IEEE Transaction on Fuzzy Systems 21(6), 1080–1095 (2013)
24. Zhang, T.: Statistical behavior and consistency of classification methods based on convex risk minimization. Annals of Statistics 32, 56–85 (2004)
25. Liang, N.-Y., Huang, G.-B., Saratchandran, P., Sundararajan, N.: A fast and accurate online sequential learning algorithm for feedforward networks. IEEE Transactions on Neural Networks 17(6), 1411–1423 (2006)

Review of Advances in Neural Networks: Neural Design Technology Stack

Stanisław Woźniak[1,2], Adela-Diana Almási[1,3], Valentin Cristea[3], Yusuf Leblebici[2], and Ton Engbersen[1]

[1] IBM Research - Zurich, Rüschlikon, Switzerland
{stw,dlm,apj}@zurich.ibm.com
[2] Microelectronic Systems Laboratory, EPFL, Lausanne, Switzerland
yusuf.leblebici@epfl.ch
[3] Computer Science and Engineering Department, UPB, Bucharest, Romania
valentin.cristea@cs.pub.ro

Abstract. This review provides a high-level synthesis of significant recent advances in artificial neural network research. We assume that a global view of the field can benefit researchers by providing alternative viewpoints. Therefore, we present different network and neuron models, we discuss model parameters and the means to obtain them, and we draw a quick outline of information encoding, before proceeding to an overview of the relevant learning mechanisms, ranging from established approaches to novel ideas. We specifically focus on comparing the classical artificial model with the biologically-feasible spiking neuron.

Keywords: neural networks, machine learning, spiking neurons, artificial neurons, neuromorphic systems.

1 Introduction and Background

Recently, we have been witnessing increasing interest in understanding how the brain functions and how it could be modeled. This momentum is fueled by major research initiatives, including the Human Brain Project, the BRAIN Initiative, as well as significant commercial efforts, such as IBM Watson.

In this paper we provide an overview of the state-of-the-art in neural networks, the main building block of the brain. Computer scientists typically focus on artificial neural networks; however, numerous alternative models and learning approaches are studied within various brain-related disciplines. We discuss not only the most popular and successful approaches, but also niche ideas, which offer an interesting perspective on the field. We aim to provide the readers with a global view of the field, in order for them to consider a wider array of possibilities in neural network design.

2 Neural Network Definitions

The brain consists of a densely interconnected network of neurons and in this section we present artificial models attempting to recreate this structure.

J. Cao et al. (eds.), *Proceedings of ELM-2014 Volume 1*,
Proceedings in Adaptation, Learning and Optimization 3, DOI: 10.1007/978-3-319-14063-6_31

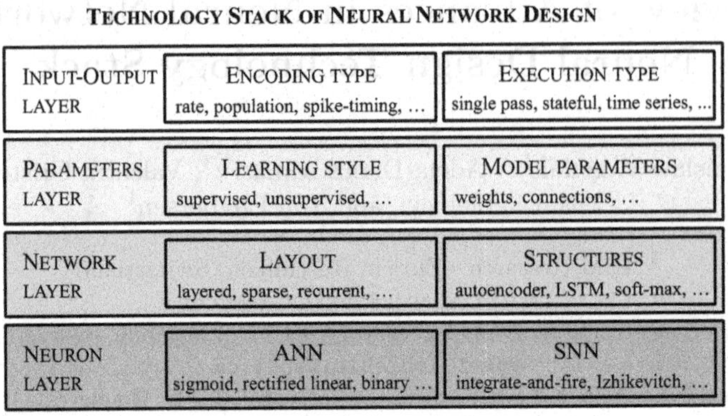

Technology Stack of Neural Network Design

Input-Output Layer	Encoding Type rate, population, spike-timing, ...	Execution Type single pass, stateful, time series, ...
Parameters Layer	Learning Style supervised, unsupervised, ...	Model Parameters weights, connections, ...
Network Layer	Layout layered, sparse, recurrent, ...	Structures autoencoder, LSTM, soft-max, ...
Neuron Layer	ANN sigmoid, rectified linear, binary ...	SNN integrate-and-fire, Izhikevitch, ...

Fig. 1. Neural network designers can choose from a wider variety of neuron models, network architectures and learning approaches, than they are often aware of

2.1 Neuron Model

The discovery of the principles driving biological neurons has provided the theoretical basis for modeling neurons. The complex functionality of the biological neuron was abstracted to simpler theoretical models. Depending on the level of simplification, these models have led to spiking neural models - which preserve more biological behavior, by communicating via spikes. A further abstraction level are the artificial neurons, such as the perceptron. This development is schematically represented in figure 2.

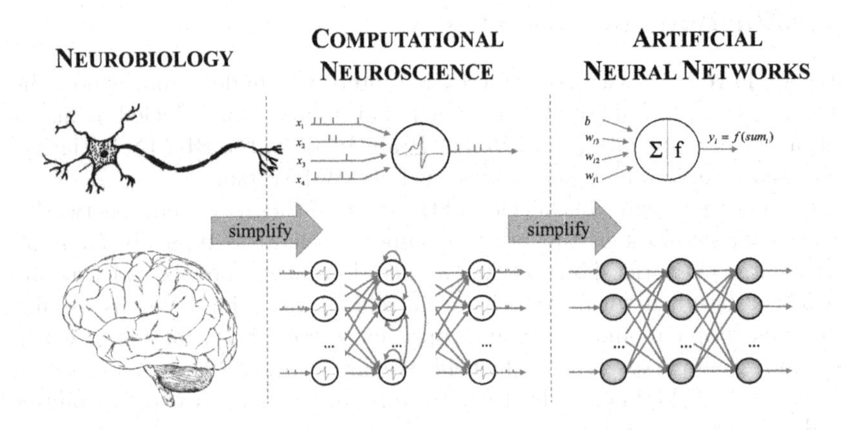

Fig. 2. The transition from neurological to spiking to artificial neurons is achieved through consecutive model simplifications: a) biological, b) spiking, c) artificial neuron

Artificial Neuron Model. The basic computational unit for an artificial neural network (ANN) is the artificial neuron, described by equation (1). It consists of three main components - firstly, an array of input synapses, through which the neuron receives input from other neurons. Secondly, the cell body - which contains the neuron model, the function that the neuron uses to process the inputs and obtain an output. Thirdly, the output value of the neuron. Each of these concepts has its corresponding biological counterpart.

$$f(x) = f_{act}(w \cdot x + b) \tag{1}$$

Spiking Neuron Model. A bottom-up approach to functional neuron modeling begins with the work of Hodgkin and Huxley [16] and the equation (2) for the current flowing in and out of the cell through a membrane with a potential V_m, capacitance C_m and ion channels with conductance g_i and potential V_i.

$$I = C_m \frac{\mathrm{d}V_m}{\mathrm{d}t} + \sum_{i=1}^{n} g_i(V_m - V_i) \tag{2}$$

Activities of other neurons cause charge accumulation through the ion channels. Then, if a threshold voltage is reached, the cell will emit a strong uniform pulse, called spike. Simultaneously the charge of the cell is reset and cell remains disabled for a time called refractory period.

Besides the description of the flow of the synaptic current, biological neurons have numerous other properties. Based on the level of detail, dozens of different models were proposed [14] , which is in contrast to the rather standardized ANN approach.

Out of this wide variety of models, the most popular implementation is the simple leaky integrate-and-fire (LIF) neuron. It is an approximation of biological behavior and its parameters correspond to the properties of real neurons. When integrated charge surpasses a given threshold a dirac signal, spike, is emitted [8].

Relationship between Artificial and Spiking Neurons. The equation of the artificial neuron can be derived from the equation of the spiking neuron [7] under certain assumptions, which include deterministic ion channels operation, constant stable input and continuous stable network behavior. Then, the floating point values of the artificial neurons correspond to the averaged spiking rate of the spiking neurons after their convergence to a stable state.

The response rate characteristics of spiking neurons and artificial sigmoid neurons are shown in figure 3. The initial shift of the spiking neuron's plot is caused by leakage and non-zero current required to reach the threshold and to actually start emitting spikes. The refractory period of the neuron determines the maximum spiking frequency. Equation 3 explains the plots of the neuronal response - spiking rate $\nu_{spiking}$ [11], where Δ^{abs} - absolute refractory period, τ_m - membrane time constant, R - membrane resistance, I_0 - current input, ϑ - threshold voltage.

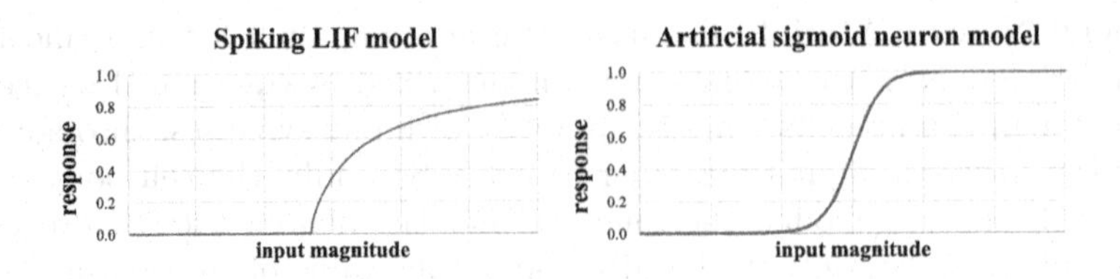

Fig. 3. 0-1 normalized response rate based on the internal neuron's stimulation rate of a spiking neuron and artificial neuron. Horizontal axes deliberately omitted, as zero is determined by a particular neuron's bias and units are not directly comparable.

$$\nu_{spiking} = \left(\Delta^{abs} + \tau_m \, ln \frac{R I_0}{R I_0 - \vartheta} \right)^{-1}, \quad \nu_{sigmoid} = \frac{1}{1+e^{-x}} \tag{3}$$

In artificial neurons, the response is defined by an activation function chosen by the network designer. The function should be non-linear in order to ensure the expressiveness of the model [22]. Very often, the learning method imposes additional constraints, such as the monotonous and differentiable function required by the back-propagation algorithm, which makes sigmoid a popular choice.

2.2 Network Model

An artificial neural network (ANN) is a mathematical and computational model, which consists of a series of computational units interconnected by uni-directional communication channels. Each synapse has an associated numerical weight, which is used to determine the relative influence of each input connection. In order to obtain the desired outputs for given inputs, weights are modified through an optimization process that we refer to as learning.

From the point of view of connectivity, a very popular approach is to have a layered feed-forward neural network, with full connectivity between adjacent layers, an acyclic directed graph. If a network exhibits loops, the model is that of a recurrent neural network (RNN), which at present are an active field of research [5] [12].

2.3 Information Encoding

One of the challenges in neurobiological studies is tracking and understanding the way information is encoded in the brain, analogously to the way it is stored in an artificial neural network.

The distributed nature of information storage, as well as the difficulty of obtaining experimental data, have hindered the construction of a unified and exact model of how information is encoded.

Aspects of information encoding include what the input to a single neuron is, how information is distributed accross the network, and how the inputs from the senses are processed by the network.

Neural Encoding. In the brain, information is encoded via trains of action potentials. Spikes can be interpreted as Boolean, all-or-nothing values, despite the variations in voltage of the spike traveling throughout the neuron body.
Depending on the brain region and the neuron type,
this information can be encoded using various types of codes, such as:

- *Rate code* - In this case, the translation of an analog value into a train of spikes is done via the frequency of consecutive spikes, the firing rate. This is the mechanism in the case of motor neurons [1].
- *Population code* - is a method of representing information by the joint activities of a larger number of neurons. [8]
- *Spike-time dependent code* – Another possible encoding uses a spike-time dependent code in order to translate relative spike positions to numeric values, as more of a computational solution than a neuroscientific hypothesis [3].

3 Establishing Model Parameters

A neural network is a meta-model for computation. Applying it to solve a concrete problem requires tuning of parameters presented in this section.

3.1 Types of Parameters

The term *parameters* is often used interchangeably with *weights*, because they are typically the central focus of ANN researchers. However, looking broader into the frontier of what constitutes the neural model and what is a human design, the following aspects may be treated as parameters:

- *Weight* w_{ij} - value controlling the transmission level of synaptic signal between (i, j) pair of neurons.
- *Connection* c_{ij} - synapses between neurons can be created or removed. One might argue that this is equivalent to setting a weight to zero, but thinking in terms of connection removal is more natural for problems such as the construction of sparse connections in non-layered models, where connections can form an arbitrary graph.
- *Neuron model* n_i - neuron types are changed more often than one might expect. For instance, the neurons in the brain are either inhibitory or excitatory (Dale's principle), so a very common change of ANN weight polarity is, from a biological standpoint, a change of neuron type. In fact, such change is not feasible in the brain, but this is considered allowable as it is possible to transform a neural network model with mixed synapses into a network with inhibitory or excitatory neurons only [25].
- *Network model* s - the number of neurons and their layout are typically fixed parameters, defined during the design stage. However, it can be also optimized as a part of the problem, e.g. by using evolutionary genetic algorithms to construct the network [9].

3.2 Obtaining Parameter Values

The aforementioned parameters can be obtained using the following methods:

- *Discretionary* - in this case, the network and the neuron model are chosen based on individual experience, rules of thumb or trial-and-error. This approach is typically used for most of the parameters, except the weights.
- *Analytic* - is based on mathematical theories that can potentially provide explicit solutions quickly and without a need for multiple iterations and fine-tuning. Early works on neural networks considered calculating all of the weights from the training examples by solving network equations, but this is feasible only in very simple cases. However, it is possible to solve the weight equations for a single layer. ELMs (Extreme Learning Machines) [17] build on top of a random network and analytically calculate only the weights in the last layer. Then, a divide-and-conquer approach can be used to stack them into multiple layers (ML-ELM). For more information see section 4.2.
- *Data-driven* - learned from the training data, i.e. derived from a certain set of problem examples using a parameter optimization algorithm, called the learning algorithm. This approach is the most widely used for weight tuning and is described in the section 4.

4 Learning

Learning is a data-driven mechanism for obtaining model parameters. Algorithms, in the classic approach, are developed by programmers. The size of the code and its complexity grow non-linearly with the size of the problem it models, which constitutes the unpleasant specificity of the software industry.

In contrast, machine learning techniques use relatively compact formulations, no matter the problem size. The complexity of the problem is converted into the values of neural network parameters, which are not directly programmed, but rather determined as a function of a training set used as problem formulation.

4.1 Learning Approaches - Classification

Supervised. This class of learning algorithms comprises of methods that use labeled training data in order to generalize the relationship between data and labels to unknown examples. In this case, one specifies a cost function related to the desired output, and then the learning consists of using a gradient descent based optimization method in order to minimize this cost function.

The most important method is the backpropagation algorithm [26]. It is a generalized delta rule , which imposes a set of limitations on the network, as the activation function of the neuron needs to be monotonous and differentiable.

It is feasible to also use other general supervised approaches, not specifically designed to be applied in an ANN context, such as simulated annealing [4].

Unsupervised Learning. Most of the data generally available is not labeled, as labeling is usually a manual and time-consuming process. Unsupervised approaches include clustering and hidden Markov models (HMMs), and it has proven to be useful when it comes to neural networks, especially when used in conjunction with supervised learning, in the case of deep learning.

Unsupervised clustering occurs in the mammalian brain in the primary visual cortex. It is therefore possible to develop localized receptive fields, similar to those found in the primary visual cortex, by using an unsupervised learning algorithm [24].

Biological Models. There are also alternative bio-inspired algorithms, valuable from the point of view of simulation insight, with sometimes mixed computational performance. The application of biologically inspired mechanisms can be done at different levels of the system:

- *whole organism level* - In this case, the algorithm is at the level of the whole system, such as genetic algorithms (GA). It is possible to apply GA to ANNs, by considering each potential ANN weight/structure configuration as an individual. Another possibility is to evolve the entire structure of the ANN, as is the case for evolutionary ANNs (EANNs) [29]. In this case, evolutionary algorithms are used to evolve the network parameters to obtain an optimal network architecture for a given problem. By eliminating redundancies, this often results in a sparser network, which therefore requires less resources during execution.
  This can be used, for instance, for anomaly detection [13], or in a deep network context, by using a GA-assisted method for a deep autoencoder [6].
- *nervous system level or neuron level* - These include Hebbian learning, synaptic plasticity - more information in section 4.2.

Reinforcement. Reinforcement learning is based on the idea of having an agent which interacts with its environment. The agent chooses a task from a probability distribution, and this task will yield a response from the environment. The goal of the learning is to maximize the total reward function. Applying these principles to artificial neural networks, an interesting result has been obtained in [23], where a deep network learned to play several Atari2600 games using as input only what is visible on video screen.

4.2 Learning Trends

Deep Learning. One of the remarkable advances in recent neural network research has been the advent of deep learning. In 2006, Hinton et al. [15] proposed a fast learning algorithm for deep belief networks (DBN), which uses layer-wise unsupervised learning for each layer of the DBN. It has been proven that this mechanism of combining supervised learning with unsupervised layer-wise pre-training yields very good experimental results, and this field has developed into a very active field of research [27].

By helping the network to automatically develop features, this approach addresses some of the the limitations of backpropagation - the results are not so dependent anymore on the initial random initialization of the weights, which makes it less likely for the algorithm to get stuck in a local minimum. Using only backpropagation for a multi-layered network is not computationally effective, and it does not yield good results for more than 1-2 hidden layers, mainly due to the higher likelihood of local minima. A comparison of the results obtained with different flavors of deep learning can be found at [19].

Extreme Learning Machines. ELM (Extreme Learning Machine) is an analytic approach to weight calculation. The equation (4) is the network model used by an ELM in the neural network case [17]. It is a single-hidden-layer feed-forward architecture, where β_i are the trained weights of the output layer, g represents the activation function of the nodes in the hidden layer and x is the input vector.

$$f_n(\mathbf{x}) = \sum_{i=1}^{n} \beta_i \, g(\mathbf{a}_i \cdot \mathbf{x} + b_i) \tag{4}$$

For a set of target labels T, the equation can be rewritten as $T = \beta H$. Then, the core step of the ELM is to calculate the output weights from $\beta = H^{\dagger}T$, where $H^{\dagger}$ is the Moore-Penrose generalized inverse of matrix H [18].

As an analytic method, it is orders of magnitude faster than an iterative approach, e.g. for MNIST dataset after 7.5 minutes of training it outperforms Deep Belief Network trained for 5.7 hours [18].

Other advantages of ELM include lack of additional algorithm parameters and less constraints on the activation functions than in the case of gradient following approaches. It is enough to have a bounded nonconstant piecewise continuous function, which makes it possible to use for instance the threshold function [17]. Alltogether, this makes ELM a very interesting alternative research direction for neural networks.

STDP. As more insights had been collected from biological neurons, it was discovered that the relative timing of the spikes in spiking neurons alters the weights, in a process named Spike-Timing Dependent Plasticity [10] [21].

Synaptic plasticity explains biological learning through the ongoing changes in synapse strength. [11] If the post-synaptic neuron will spike after the pre-synaptic neuron, then the synapse between them is strengthened; conversely, if the post-synaptic neuron will spike before the pre-synaptic one, the connection is weakened. STDP is therefore a local weight change mechanism.

Is then SDTP a more universal biologically-feasible approach, which could replace back-propagation? Despite huge hope and effort to refine STDP as the standard learning mechanism for spiking networks, there are many questions about the extend of its validity. On one hand, it is intuitively appealing and confirmed in various experiments, but on the other hand, there are numerous neuroscientific experiments in which STDP fails. In effect, it is proposed to treat it as a one dimensional approximation of a high-dimensional learning rule [28].

5 Summary

In this paper, we have provided a high-level overview of current trends in neural network research. By starting from the definitions of neural network models, we compare the artificial neuron model with the biologically-plausible spiking neuron. Furthermore, we explain the basic concepts behind the encoding and decoding of information in a neural network model, be it an artificial mathematical abstraction, or a biological neuroscience model. We continue by presenting an outline of the main learning approaches in neural network design, starting from classical methods and ranging all the way to novel promising paradigms.

Recently, neural networks have regained mainstream attention due to breakthroughs in neuroscience research, advances in artificial models, as well as novel hardware. Firstly, hardware advances have made it possible to verify concepts at a much larger scale, e.g. by simulating a system with 10^9 synapses [2], which is comparable in scale to a cat's brain. Secondly, the unsupervised feature learning paradigm offered a logic that could utilize large unpreprocessed datasets providing results such as a self-emerging cat image detector [20]. Those ideas led to improvements over many individual computational tasks.

However, there is still much to be done in order to develop a system or an algorithm that can exhibit human-like cognition and intelligence. The interesting question now is how to build more versatility into these systems. Disciplines such as neuroscience and cognitive sciences define foundations and concepts from which AI researchers could greatly benefit. The key to further progress in neural networks does not require only incremental changes, but rather a qualitative change, which is likely to come from a surprising combination of existing work and insights from one or more of the connected fields.

References

1. Adrian, E.D., Zotterman, Y.: The impulses produced by sensory nerve-endings part ii. the response of a single end-organ. J. Physiol. 61(2), 151–171 (1926)
2. Ananthanarayanan, R., Esser, S.K., Simon, H.D., Modha, D.S.: The cat is out of the bag. In: Proceedings of the Conference on High Performance Computing Networking, Storage and Analysis, pp. 1–12. IEEE (2009)
3. Bohte, S.M., Kok, J.N., La Poutre, H.: Error-backpropagation in temporally encoded networks of spiking neurons. Neurocomputing 48(1), 17–37 (2002)
4. Chen, L., Aihara, K.: Chaotic simulated annealing by a neural network model with transient chaos. Neural Networks 8(6), 915–930 (1995)
5. Cireşan, D., Meier, U., Masci, J., Schmidhuber, J.: Multi-column deep neural network for traffic sign classification. Neural Networks 32, 333–338 (2012)
6. David, O.E., Greental, I.: Genetic algorithms for evolving deep neural networks. In: Proceedings of the 2014 Conference Companion on Genetic and Evolutionary Computation Companion, pp. 1451–1452. ACM (2014)
7. Dayan, P., Abbott, L.F.: Theoretical Neuroscience: Computational and Mathematical Modeling of Neural Systems. MIT Press (2005)
8. Eliasmith, C., Anderson, C.H.: Neural engineering: Computation, representation, and dynamics in neurobiological systems. MIT Press (2004)

9. Ferreira, C.: Designing neural networks using gene expression programming. Applied Soft Computing Technologies: The Challenge of Complexity, 517–535 (2006)
10. Gerstner, W., Kempter, R., van Hemmen, J.L., Wagner, H.: A neuronal learning rule for sub-millisecond temporal coding. Nature 383(6595), 76–78 (1996)
11. Gerstner, W., Kistler, W.M.: Spiking neuron models: Single neurons, populations, plasticity. Cambridge University Press (2002)
12. Graves, A., Mohamed, A.R., Hinton, G.: Speech recognition with deep recurrent neural networks. In: 2013 IEEE International Conference on Acoustics, Speech and Signal Processing (ICASSP), pp. 6645–6649. IEEE (2013)
13. Han, S.-J., Cho, S.-B.: Evolutionary neural networks for anomaly detection based on the behavior of a program. IEEE Transactions on Systems, Man, and Cybernetics, Part B: Cybernetics 36(3), 559–570 (2005)
14. Hines, M.L., Morse, T., Migliore, M., Carnevale, N.T., Hines, M.L.: ModelDB: A database to support computational neuroscience. Journal of Computational Neuroscience 17(1), 7–11 (2004)
15. Hinton, G., Osindero, S., Teh, Y.-W.: A fast learning algorithm for deep belief nets. Neural Computation 18(7), 1527–1554 (2006)
16. Hodgkin, A.L., Huxley, A.F.: A quantitative description of membrane current and its application to conduction and excitation in nerve. The Journal of Physiology 117(4), 500 (1952)
17. Huang, G.-B., Chen, L., Siew, C.-K.: Universal approximation using incremental constructive feedforward networks with random hidden nodes. IEEE Transactions on Neural Networks 17(4), 879–892 (2006)
18. Kasun, L.L.C., Zhou, H., Huang, G.-B., Vong, C.-M.: Representational learning with ELMs for big data. IEEE Intelligent Systems 28(6), 31–34 (2013)
19. Larochelle, H., Bengio, Y., Louradour, J., Lamblin, P.: Exploring strategies for training deep neural networks. JMLR 10, 10:1–10:40 (2009)
20. Le, Q.V., Ranzato, M., Monga, R., Devin, M., Chen, K., Corrado, G.S., Dean, J., Ng, A.Y.: Building high-level features using large scale unsupervised learning. arXiv preprint arXiv:1112.6209 (2011)
21. Markram, H., Lbke, J., Frotscher, M., Sakmann, B.: Regulation of synaptic efficacy by coincidence of postsynaptic APs and EPSPs. Science 275(5297), 213–215 (1997)
22. Minsky, M., Papert, S.: Perceptrons. MIT Press (1969)
23. Mnih, V., Kavukcuoglu, K., Silver, D., Graves, A., Antonoglou, I., Wierstra, D., Riedmiller, M.: Playing atari with deep reinforcement learning. arXiv preprint arXiv:1312.5602 (2013)
24. Olshausen, B.A., Field, D.J.: Natural image statistics and efficient coding. Network: Computation in Neural Systems 7(2), 333–339 (1996)
25. Parisien, C., Anderson, C.H., Eliasmith, C.: Solving the problem of negative synaptic weights in cortical models. Neural Comput. 20(6), 1473–1494 (2008)
26. Rumelhart, D.E., Hinton, G.E., Williams, R.J.: Learning representations by back-propagating errors. Cognitive Modeling (1988)
27. Schmidhuber, J.: Deep learning in neural networks: An overview. arXiv preprint arXiv:1404.7828 (2014)
28. Shouval, H.: Spike timing dependent plasticity: a consequence of more fundamental learning rules. Frontiers in Computational Neuroscience (2010)
29. Yao, X.: Evolving artificial neural networks. P. IEEE 87(9), 1423–1447 (1999)

Applying Regularization Least Squares Canonical Correlation Analysis in Extreme Learning Machine for Multi-label Classification Problems

Yanika Kongsorot, Punyaphol Horata, and Khamron Sunat

Department of Computer Science, Khon Kaen University,
Friendship Road, Khon Kaen, Thailand 40002
yanika_k@kkumail.com, punhor1@kku.ac.th, khumron_sunat@yahoo.com

Abstract. Multi-label classification is a type of classification where each instance is associated with a set of labels. Many methods such as BP-MLL, rank-SVM, and MLRBF have been proposed for multi-label classification but their learning abilities are too slow. Extreme Learning Machine (ELM) is a well known algorithm for SLFNs that can learn faster than the traditional gradient-base neural networks and it also provides better generalization performance. However, the classification performance of ELM involving multi-label classification may not be good enough despite its advantage in fast training. Therefore, this paper proposes two multi-label classification approaches in ELM. The first approach uses the 1-norm regularized Least-square for Canonical Correlation Analysis (1-norm LSCCA) to obtain the projection vectors, which in turn uses the vectors to provide the new information. Then, ELM is then used to learn this new information in the new space. The second approach applies the ensemble method to the first approach to reduce the random effects of ELM. The experimental results show that the two proposed methods can improve the performance of ELM in multi-label classification and are also faster than the previous multi-label classification methods.

Keywords: Extreme Learning Machine, Canonical Correction Analysis, A Least Squares Formulation for Canonical Correction Analysis, Multi-Label Classification.

1 Introduction

Traditional single-label classification is learning from examples that are associated with one label l_i from a set of labels $L = \{l_1, l_2, ..., l_m\}, m > 1$. However, many problems can be associated with more than one category and these are called multi-label problem [1], [2]. The task for solving the multi-label problem is called multi-label classification. This task involves learning to map an example to a set of labels $y \subset L$ [3]. There are three main approaches for solving

J. Cao et al. (eds.), *Proceedings of ELM-2014 Volume 1*,
Proceedings in Adaptation, Learning and Optimization 3, DOI: 10.1007/978-3-319-14063-6_32

multi-label problems [1],[2]: a) Problem Transformation approaches, b) Algorithm Adaptation approaches, and c) Ensemble approaches. Problem transformation approaches are related to the transformation of the multi-label problem into a single-label then the binary or multi class learning methods are used for classifying. Algorithm adaptation approaches are associated with the application of the algorithms to handle the multi-label problem directly. The ensemble approaches are hybrid of the problem transformation approaches and algorithm adaptation approaches.

The correlation between the examples set X, and labels set L is a matrix to indicate a characteristic of each data. The Canonical Correlation Analysis (CCA) is a well known technique for analyzing the correlation between two sets of variables. The technique is computed to find the pairs of projection vectors such that the correlation is maximized. CCA is applied to many learning methods and it is also applied to multi-label classification [4]. CCA can be formulated to the least-square formulation for very large datasets, which is called LSCCA. Moreover, CCA and LSCCA can be extended using the regularization technique, which is a 1-norm and 2-norm of the extension. Research by [4] showed that the 1-norm LSCCA is better than CCA.

Extreme Learning Machine (ELM) was proposed by Huang et al.[5]. The learning speed of ELM is faster than traditional gradient-based approaches because its parameters are randomly generated. ELM is used in both binary and multiclass classification and it has been applied to several applications. In multiclass classification, ELM uses multi-output nodes instead of a single-output node [6]. Only the element with the highest output value from the output nodes is considered as the label of the input. However, ELM has a shortcoming for multi-label classification. For example, in the Scene multi-label problem, Fig.1 shows that the training times of ELM is still fast but the F-measure of ELM on every number of hidden nodes is nearly with zero. The research [7] presented an approach of ELM for working on multi-label classification using CCA but its performance depended on the random effects of ELM. Also, the random of generating the parameters is the one cause of the fluctuating performance of ELM. There are many ways to protect this random effect, as showed by [8].

Therefore, this paper proposes two approaches for multi-label classification in ELM named as LSCCA$_1$-ELM and LSCCA$_1$-ELM Ensemble. These approaches are based on the extension of L_1-LSCCA [4], [9] and ELM multi-output nodes. In LSCCA$_1$-ELM, the L_1-LSCCA is used to compute the projection vectors and ELM is adapted to learn the multi-label problem in new space. For LSCCA$_1$-ELM Ensemble is the ensemble of LSCCA$_1$-ELM for mitigating the random parameters effect in ELM and also obtaining good results from ELM learning.

The remainder of this paper is organized as follows: section 2 reviews background and related works. Section 3 presents the proposed method, while section 4 describes the experimental results. And the final section is the conclusion of this work.

2 Related Works

2.1 Multi-label Classification

Multi-label classification is widely used in several real-world problems such as text categorization, medical diagnosis, biology categorization, music categorization, and scene categorization. For example in scene categorization, one scene could be assigned to many kinds of picture, such as sunset, sea, forest, or mountain. Multi-label learning is concerned with learning from a set of examples, where each example is related to more than one labels. The task of multi-label learning is summarized in the following section.

Given the d-dimensional example space $X \in \mathbb{R}^d$ and label space $Y \in \{0,1\}^L$, where L is number of labels and given a set of N examples $\{(\mathbf{x}_i, \mathbf{y}_i) \mid i = 1, ..., N\}$ $\subset (X \times Y)$, the task of multi-label learning is to learn a function $f : \mathbf{X} \rightarrow \mathbf{Y}$ to find a set of labels $\widehat{\mathbf{y}} = f(\mathbf{x})$.

Tsoumakas and Katakis [1] grouped the multi-label approaches into two categories: a) problem transformation approaches and b) algorithm adaptation approaches. Problem transformation approaches are related to the transformation of multi-label problems to one or more single-label classification. Example of it included, the binary relevance method [1], the classifier chain method (CC) [10], the label power-set method (LP) [1],[11],[12], HOMER [13], and the pairwise method [14],[15]. The algorithm adaptation approaches extended the machine learning algorithms to handle the multi-label problem directly. Examples of algorithm adaptation methods include, Adaboost for multi-label data [16], k-Nearest Neighbor (ML-KNN) [17], ML-C4.5 [18], BP-MLL [19], ML-RBF [20], and rank-SVM [21].

Madjarove et al. [2] presented the third category, ensemble approaches. The ensemble approaches are based on the problem transformation methods and algorithm adaptation methods. RAKEL system [12] is one of the most popular methods, which learns a single-label classifier to predict each label. Another popular method is the ensemble of classification chain (ECC) [10], which uses classifier chains (CC) as base classifiers and applies threshold to select the final set of labels.

2.2 The 1-Norm Regularized Least-Squares Formulation for Canonical Correlation Analysis (LSCCA$_1$)

This section, we describe the Canonical Correlation Analysis (CCA), the Least-Squares Formulation for CCA, and LSCCA$_1$, respectively.

Canonical Correlation Analysis (CCA)

Canonical Correlation Analysis (CCA) [22] is a technique for finding the correlations between two sets of multidimensional variables and it is used for reducing multidimensional of data to smaller. The computing of CCA is to find the project vectors, which provide the maximum correlation between the inputs and labels.

Given the n samples: $(\mathbf{X}, \mathbf{Y}) = \{(x_i, y_i) \mid i = 1, ..., n\}$, where $\mathbf{X}$ is a $n \times d$ data matrix and $\mathbf{Y}$ is a $n \times k$ label matrix. CCA computes two projection vectors, $w_x \in \mathbb{R}^d$ and $w_y \in \mathbb{R}^k$, such that the correlation is maximized. The correlation coefficient r is written as follows:

$$r = \frac{w_x^T \mathbf{X}\mathbf{Y}^T w_y}{\sqrt{(w_x^T \mathbf{X}\mathbf{X}^T w_x)(w_y^T \mathbf{Y}\mathbf{Y}^T w_y)}}. \tag{1}$$

To get the maximum correlation r in Eq.(1), CCA can be formulated equivalently as

$$\begin{array}{rl} \max_{w_x, w_y} & w_x^T \mathbf{X}\mathbf{Y}^T w_y, \\ \text{subject to} & w_x^T \mathbf{X}\mathbf{X}^T w_x = 1, \\ & w_y^T \mathbf{Y}\mathbf{Y}^T w_y = 1. \end{array} \tag{2}$$

Focusing on the projection vector w_x, it can be assumed that $\mathbf{Y}\mathbf{Y}^\mathbf{T}$ is a nonsingular matrix. w_x can be obtained by solving the following equation:

$$\begin{array}{rl} \max_{w_x, w_y} & w_x^T \mathbf{X}\mathbf{Y}^T (\mathbf{Y}\mathbf{Y}^T)^{-1} \mathbf{Y}\mathbf{X}^T w_x, \\ \text{subject to} & w_x^T \mathbf{X}\mathbf{X}^T w_x = 1. \end{array} \tag{3}$$

The above formula is solved by using the Largrangian formulation and tends to be a generalized eigenvalues problem :

$$\mathbf{X}\mathbf{Y}^T (\mathbf{Y}\mathbf{Y}^T)^{-1} \mathbf{Y}\mathbf{X}^T w_x = \eta \mathbf{X}\mathbf{X}^T w_x, \tag{4}$$

where η is the eigenvalue corresponding to the eigenvector w_x. The projection vectors w_x can be obtained by computing the eigendecomposition of matrix $(\mathbf{X}\mathbf{X}^T)^{-1}\mathbf{X}\mathbf{Y}^T(\mathbf{Y}\mathbf{Y}^T)^{-1}\mathbf{Y}\mathbf{X}^T$ and selecting the first l eigenvectors of the projection vectors as follows:

$$\mathbf{W}_{CCA} = (\mathbf{X}\mathbf{X}^T)^{-1}\mathbf{X}\mathbf{Y}^T(\mathbf{Y}\mathbf{Y}^T)^{-1}. \tag{5}$$

The Least-Squares Formulation for CCA

The research [4] has shown that CCA can be formulated as a least-squares problem. The projection vector w_x of CCA can be obtained by the least-squares solution using the following equation:

$$\min_W f(W) = \left\|\mathbf{W}^T\mathbf{X} - \mathbf{T}\right\|_F^2. \tag{6}$$

The matrix $\mathbf{W}$ in Eq.(6) is obtained by

$$\mathbf{W} = (\mathbf{X}\mathbf{X}^T)^{\dagger}\mathbf{X}\mathbf{T}^T, \tag{7}$$

where $(\mathbf{X}\mathbf{X}^T)^{\dagger}$ is the Moore-Penrose generalized inverse of $\mathbf{X}\mathbf{X}^T$.

The class indicator $\mathbf{T}$ is defined as follows:

$$\mathbf{T} = (\mathbf{Y}\mathbf{Y}^T)^{-1}\mathbf{Y}^T. \tag{8}$$

The solution to the least-squares problem for $\mathbf{T}$ from Eq.(7) is

$$\mathbf{W}_{LS} = (\mathbf{X}\mathbf{X}^T)^{\dagger}\mathbf{X}\mathbf{Y}^T(\mathbf{Y}\mathbf{Y}^T)^{-1}. \tag{9}$$

Sun and Ji [9],[4] showed an equivalent least-squares formulation for CCA under a mind condition, which tends to hold for high-dimensional data. They established that, CCA can work on very large dataset using the least-square solutions. The equivalent least-squares CCA formulation is called *LS-CCA*.

The 1-Norm Regularized Least-Squares Formulation for Canonical Correlation Analysis

The least-square CCA formulation can be extended using the regularization technique, which is used to control the complexity and prevent the over fitting of model. Research by [23] showed that L_1 norm regularization can produce sparse model that can then be presented into the least-square formulation; the resulting model is called *lasso*. Moreover, L_2 norm regularization is the most common variant, which is often applied in liner models with the resulting model being called *ridge regression*.

Both regularized L_1 and L_2 can be applied in least-square CCA formulation as named LS-CCA$_1$ and LS-CCA$_2$, respectively. The experimental result in [9],[4] showed that the performance of LS-CCA$_1$ is better than LS-CCA$_2$ so we used the LS-CCA$_1$ in this work to find the projection vectors. The minimized objective function of 1-norm regularized least-square CCA formulation by a target using matrix $\mathbf{T}^T$ as follows:

$$L_1(\mathbf{W}, \lambda) = \sum_{j=1}^{k}(\sum_{i=1}^{n}(x_i^T w_j - \mathbf{T}_{ij}^T)^2 + \lambda \left\| w_j \right\|_1). \tag{10}$$

2.3 Extreme Learning Machine

ELM [5] works on the single hidden layer feedforward neural network (SLFNs), where the hidden parameters (input weights and biases) are randomly chosen and the output weights are calculated analytically using the Moore-Penrose generalized inverse [24], and tend to the smallest norm by the least square solution.

For N arbitrary distinct samples $\left\{(\mathbf{x}_i, \mathbf{t}_i) \in \mathbb{R}^d \times \mathbb{R}^m\right\}$, the SLFNs with $\widetilde{N}$ hidden nodes and an activation function $g(x)$, are mathematically modeled as

$$\sum_{i=1}^{\widetilde{N}} \beta_i g\left(\mathbf{w_i} \cdot \mathbf{x_j} + b_i\right) = t_j, j = 1, ..., N, \tag{11}$$

where $\mathbf{w}_i = \left[w_{i1}, w_{i2}, ..., w_{id}\right]^T$ is the weight vector for connecting the ith hidden node and input node, $\mathbf{w}_i \cdot \mathbf{x}_j$ denotes the inner product between $\mathbf{w}_i$ and $\mathbf{x}_j$, b_i is the bias of the ith hidden node, and $\beta_i = \left[\beta_{i1}, \beta_{i2}, ..., \beta_{im}\right]^T$ is the output weight vector for connecting the ith hidden node and the output node. Eq.(11) can be rewritten in a compact form as follows:

$$\mathbf{H}\beta = \mathbf{T}, \tag{12}$$

where

$$\mathbf{H} = \begin{bmatrix} g(\mathbf{w}_1 \cdot \mathbf{x}_1 + b_1) & \cdots & g(\mathbf{w}_{\widetilde{N}} \cdot \mathbf{x}_1 + b_{\widetilde{N}}) \\ \vdots & \ddots & \vdots \\ g(\mathbf{w}_1 \cdot \mathbf{x}_N + b_1) & \cdots & g(\mathbf{w}_{\widetilde{N}} \cdot \mathbf{x}_N + b_{\widetilde{N}}) \end{bmatrix}_{N \times \widetilde{N}}, \tag{13}$$

$$\beta = \begin{bmatrix} \beta_1^T \\ \vdots \\ \beta_{\widetilde{N}}^T \end{bmatrix}_{\widetilde{N} \times m} \quad \mathbf{T} = \begin{bmatrix} \mathbf{t}_1^T \\ \vdots \\ \mathbf{t}_N^T \end{bmatrix}_{N \times m}. \tag{14}$$

$\mathbf{H}$ is the hidden layer output matrix of the neural network. $\mathbf{T}$ is the target matrix of the output layer. Eq.(12) becomes a linear system and the output weight $\widehat{\beta}$ can be obtained by a least-square solution of Eq. (12), as follows:

$$\left\| \mathbf{H}\widehat{\beta} - \mathbf{T} \right\| = \min_{\beta} \left\| \mathbf{H}\beta - \mathbf{T} \right\| \tag{15}$$

$$\widehat{\beta} = \mathbf{H}^{\dagger}\mathbf{T}, \tag{16}$$

where $\mathbf{H}^{\dagger}$ is the Moore-Penrose generalized inverse of $\mathbf{H}$.

ELM has been applied in various applications and used in both regression and classification tasks. Looking at the classification task, ELM can be applied in both binary and multiclass classification. ELM for multi-categories classification applications has been reported by Rong et al [25] who presented the study of multiclass classification problem using ELM. They compared the performance of a single ELM classifier and an ELM binary classifier (ELM-OAO and ELM-OAA). Their study results showed that the classification performances of single ELM, ELM-OAO, and ELM-OAA are similar.

ELM with regression and multiclass classification was proposed by Huang et al. [6]. The output layer was composed of multi-output nodes. m-class of classifiers have m output nodes. This means that only the largest output value is used to represent the predicted class label of the input data. There is the over view of several applications of ELM in [26].

Fig.1 shows the performance of ELM in the scene multi-label problem with every number of hidden nodes. It can be seen that ELM has a shortcoming for multi-label classification but its learning speed is still fast. Our previous work with CCAELM [7], used a common CCA and ELM for working on multi-label classification but it provided a lesser measure value and had many steps for implementation. And also, its performance depended on its input weight and bias, which were generated randomly.

The random of generating the parameters is the one cause of the fluctuating performance of ELM. Thus, the ensemble approaches of ELM [27],[28] were proposed to control and protect the random parameters effect of ELM. The common structure of ELM consists of P individual ELMs where the input weights and biases of hidden nodes $(\mathbf{w}^k, b^k), k \in [1, P]$ are randomly generated and their output weights β^k are analytically determined through inverse operation of their hidden layer outputs. Although, many methods exist for obtaining the final output, the common one is the average of each individual ELM'result [27] as follows:

$$f(\mathbf{x}_i) = \frac{1}{P}\sum_{k=1}^{P} f^{(k)}(\mathbf{x}_i), \tag{17}$$

where $f^{(k)}(\mathbf{x}_i), k = 1, ..., P$ is the output of each ELM and $f(\mathbf{x}_i)$ is the output of the whole system with the input $\mathbf{x}_i$. The implementation of Eq. (17) is fast and easy. Other methods of ensemble ELM are presented in [29],[30],[8], and [28].

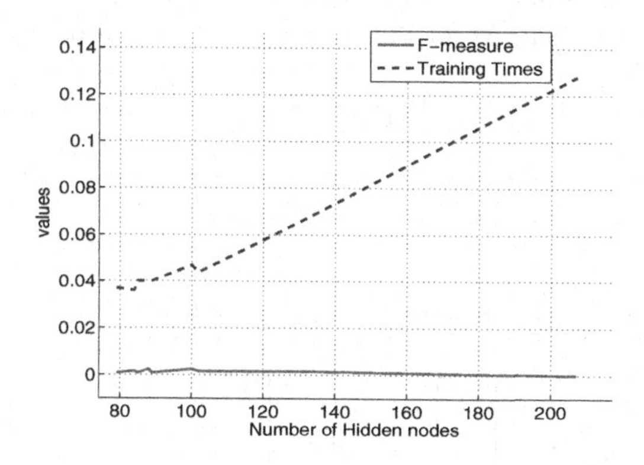

Fig. 1. The F-measure of ELM in multi-label Scene problems

3 Proposes Methods

An applying 1-Norm Regularization Least-Squares Canonical Correlation in Extreme Learning Machine

This section presents the extensions of ELM and LS-CCA$_1$ to handle the multi-label classification. In the multi-label problem, the correlation between dataset and label set is a key idea in obtaining good information. In this work, LS-CCA$_1$ is used to find the projection vectors, which produce the maximum correlation. These projection vectors are then used to reduce the dimension of data. Next, ELM multi-output is used to learn the information from LS-CCA$_1$ and classify the multi-label problem to predict the label set of input data.

Let $(\mathbf{X}, \mathbf{Y}) = \{(\mathbf{x}_i, \mathbf{y}_i)|i = 1, ..., N\}$ be the set of N training samples, where $\mathbf{X}$ is a $N \times d$ data matrix, $\mathbf{Y}$ is a $N \times L$ label matrix and $\mathbf{y}_i = \{-1, 1\}^L$. Each $\mathbf{x}_i$ can belong to one or more label of $\mathbf{y}_i$. In this work we used L output nodes for

L labels for constructing the structure of ELM. The proposed methods named as LSCCA$_1$-ELM and LSCCA$_1$-ELM Ensemble can be summarized as follows:

First, LS-CCA$_1$ is used to compute the projection matrix $\mathbf{W}_x$ for $\mathbf{X}$ as in following formula:

$$\mathbf{W}_x = (\mathbf{X}\mathbf{X}^T + \lambda\mathbf{I})^\dagger \mathbf{X}\mathbf{Y}^T. \tag{18}$$

Then, $\mathbf{X}$ is mapped from the input space to low dimension by the projection $\mathbf{W}_x$ as follows:

$$\mathbf{Z} = \mathbf{X}\mathbf{W}_x, \tag{19}$$

where $\mathbf{Z}$ is a new matrix in low dimension of $\mathbf{X}$ and its size is $N \times L$.

Next, ELM with multi-outputs is used for learning in low dimension space of $\mathbf{X}$. Thus, $\mathbf{Z}$ is the input information matrix for feeding into ELM. The approximation equation of ELM for output target $\mathbf{T}$ can be rewritten as follows:

$$\mathbf{H}_z\beta = \mathbf{T}, \tag{20}$$

where

$$\mathbf{H}_z = \begin{bmatrix} g(\mathbf{w}_1 \cdot \mathbf{z}_1 + b_1) & \cdots & g(\mathbf{w}_{\tilde{N}} \cdot \mathbf{z}_1 + b_{\tilde{N}}) \\ \vdots & \ddots & \vdots \\ g(\mathbf{w}_1 \cdot \mathbf{z}_N + b_1) & \cdots & g(\mathbf{w}_{\tilde{N}} \cdot \mathbf{z}_N + b_{\tilde{N}}) \end{bmatrix}_{N \times \tilde{N}}, \tag{21}$$

$$\beta = \begin{bmatrix} \beta_1^T \\ \vdots \\ \beta_{\tilde{N}}^T \end{bmatrix}_{\tilde{N} \times L}, \mathbf{T} = \begin{bmatrix} \mathbf{T}_1^T \\ \vdots \\ \mathbf{T}_N^T \end{bmatrix}_{N \times L}. \tag{22}$$

When ELM is finished learning,the outputs of ELM output layer are classified by the decision function and finish the LSCCA$_1$-ELM.

In the case of multi-label classification, we define the decision function of ELM classifier is

$$\widehat{y}_i = \text{sign}(\mathbf{t}_i), \tag{23}$$

where $\mathbf{t}_i \in \mathbf{T}_i$. The approach of the second proposal, LSCCA$_1$-ELM Ensemble, is the ensemble of LSCCA$_1$-ELM. The LSCCA$_1$-ELM Ensemble uses the ensemble approach in [27] and constructs with P individual ELMs. The final output is obtained by the average of each individual ELM as follows:

$$f(\mathbf{z}_i) = \frac{1}{P}\sum_{k=1}^{P} f^{(k)}(\mathbf{z}_i), \tag{24}$$

where $f^{(k)}(\mathbf{z}_i), k = 1, ..., P$ is the output of each ELM and $f(\mathbf{z}_i)$ is the output of the whole system with the input $\mathbf{z}_i$.

Finally, the decision function Eq.(23) is used to classify $\mathbf{t}_i$ from Eq.(24).

4 Experimental Results

4.1 Dataset and Configuration of Experiments

Six multi-label data problems were used in the experiments: *Scene*, *Emotion*, *Cal500*, *Yeast*, *Flags*, and *Bird* dataset. The number of data, feature dimensions, and labels are shown in Table 1. The training and testing sets of all dataset were obtained using cross validation technique. The experiments perform 5 runs of 10-fold cross validation and the averaged performances were reported. All experiments were derived by MATLAB 2011a and run on a Pentium Core2 Duo PC, 2G of RAM.

Table 1. Summary of Statistics of the dataset. n is the number of data points, d is the number of dimensionality, and l is the number of labels.

Dataset	n	d	l
scene	2407	294	6
yeast	2417	103	14
emotion	593	72	6
CAL500	502	68	49
birds	654	260	19
flag	194	19	7

4.2 Evaluations Measure

Precision, Recall, f-Measure, and AUC

In multi-label classification, the most commonly used techniques for performance measuring are precision, recall, and the average F-measure. The confusion matrix of them is shown in Table 2.

Table 2. Confusion matrix

		Actual Value	
Predicted Value		1	-1
	1	TP True Positive	FP False Prositive
	-1	FN False Nagative	TN True Negative

Let L be the number of labels in label space, the precision and recall are defined as follows:

$$Precision = \frac{\sum_{i=1}^{L} TP_i}{\sum_{i=1}^{L}(TP_i + FP_i)}, \tag{25}$$

$$Recall = \frac{\sum_{i=1}^{L} TP_i}{\sum_{i=1}^{L}(TP_i + FN_i)}, \tag{26}$$

and

$$F - meause = \frac{2 \times Precision \times Recall}{Precision + Recall}, \tag{27}$$

where $i = 1, ..., L$, and Precision is the number of the correct returned labels. Recall is the number of the positive labels that are returned and F-measure is a measure of accuracy, which considers both the precision and recall of the results.

In addition to this, the Receiver Operator Characteristic (ROC) is widely used to measure the performance of model. ROC [31] [32] [33] is a graphical plot of the fraction of true positive rate and false positive rate. It is widely used to estimate the performance of the model. The results of ROC analysis can be represented as the number of Areas Under the ROC Curve (AUC), which its value is between 0 and 1. If the AUC score of the specific classifier is close to 1 then its performance is good.

4.3 Experimental Results

The performance of our proposed methods are compared with eight other methods, namely: ELM [5], Binary ELM [25], ELM Kernel [6], rank SVM [21], BP-MLL [19], MLRBF [20], CCA-OC [34], and CCA-ELM [7]. The cost parameters of each method are obtained by searching the value of the parameters for the best performance score, which shows in Table 3, and we used 5 individual ELMs for $LSCCA_1$-ELM Ensemble. The performances of all methods are reported as the average of precision, recall, F-measure and the times scales. Then, the ROC is employed to evaluate the classification performance. The mean AUC score is reported as the average Area Under the ROC Curve (AUC) over all labels and all splitting for each data set.

Table 3. The value of the cost parameter of ELM, ELM-Binary, ELM kernel, CCAELM, and two proposes methods

Datasets	Algorithms	Number of hidden nodes	Regularization value	Kernel parameter value
Scene	ELM	56	-	-
	ELM-Binary	62	-	-
	ELM kernel	-	2^4	2^{-8}
	CCAELM	102	-	-
	$LSCCA_1$-ELM	120	2^{-3}	-
	$LSCCA_1$-ELM Ensemble	120	2^{-3}	-
Yeast	ELM	409	-	-
	ELM-Binary	421	-	-
	ELM kernel	-	2^{-2}	2^{-8}
	CCAELM	78	-	-
	$LSCCA_1$-ELM	103	2^{-4}	-
	$LSCCA_1$-ELM Ensemble	103	2^{-4}	-
Emotions	ELM	213	-	-
	ELM-Binary	217	-	-
	ELM kernel	-	2^{-2}	2^{-6}
	CCAELM	36	-	-
	$LSCCA_1$-ELM	40	2^{-3}	-
	$LSCCA_1$-ELM Ensemble	40	2^{-3}	-
CAL500	ELM	257	-	-
	ELM-Binary	262	-	-
	ELM kernel	-	2^4	2^1
	CCAELM	123	-	-
	$LSCCA_1$-ELM	99	2^{-5}	-
	$LSCCA_1$-ELM Ensemble	99	2^{-5}	-
Birds	ELM	166	-	-
	ELM-Binary	171	-	-
	ELM kernel	-	2^1	2^{-4}
	CCAELM	33	-	-
	$LSCCA_1$-ELM	45	2^{-2}	-
	$LSCCA_1$-ELM Ensemble	45	2^{-2}	-
Flags	ELM	23	-	-
	ELM-Binary	25	-	-
	ELM kernel	-	2^1	2^{-2}
	CCAELM	29	-	-
	$LSCCA_1$-ELM	21	2^{-3}	-
	$LSCCA_1$-ELM Ensemble	21	2^{-3}	-

Comparison Performance on the Precision, Recall, and F-Measure

Table 4 shows the comparative performances of the proposed methods with ELM, ELM-Binary, ELM kernel, CCAELM on the precision, recall, and F-measure. The precision of LSCCA$_1$-ELM Ensemble is the highest in the Scene, Emotions, CAL500, Birds, and Flags problem and comparable to ELM-kernel in one problem as 0.71818 and 0.71822 in Yeast problem. The recall of LSCCA$_1$-ELM Ensemble is the highest the Scene, Yeast, and Flags problem. Also, the recall of LSCCA$_1$-ELM and LSCCA$_1$-ELM Ensemble is comparable to ELM-kernel, ELM-binary, and CCAELM in the Yeast, Emotions, and Flags problem, respectively. It is however, less than ELM in the CAL500 problem. The F-measure of LSCCA$_1$-ELM Ensemble is the highest in the Scene, Yeast, Emotions, CAL500, and Flags problem and the F-measure LSCCA$_1$-ELM is comparable to ELM-kernel in the Yeast problem.

Table 5 shows the comparative performances of the proposed methods with CCA-OC, RankSVM, MLRBF on the precision, recall, and F-measure. The precision of LSCCA$_1$-ELM Ensemble provide the highest precision in the Yeast, Emotions, CAL500, and Flags problem and less than MLRBF in the Scene, and Birds problem. The recall LSCCA$_1$-ELM Ensemble is the highest in the Scene problem and the Emotions problem, while the recall of LSCCA$_1$-ELM is the highest in Birds problem. Both the LSCCA$_1$-ELM Ensemble and the LSCCA$_1$-ELM are less than BPML and MLRBF in the Yeast, Flags, and CAL500 problem, respectively. The F-measure of LSCCA$_1$-ELM Ensemble is the highest in the Scene, Emotions, CAL500, and Flags problem, while LSCCA$_1$-ELM is the highest in the Birds problem. Both the F-measure of the LSCCA$_1$-ELM Ensemble and the LSCCA$_1$-ELM are less than MLRBF in the Yeast problem.

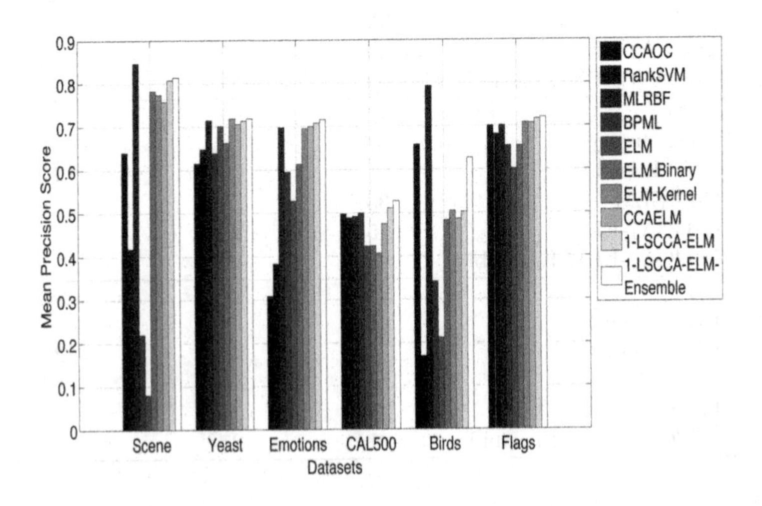

Fig. 2. Comparison of the average Precision of all algorithms on all dataset

Fig.2 shows the precision performances of all methods in all problems. In this figure, $LSCCA_1$-ELM Ensemble provides the highest precision in the Emotions, CAL500, and Birds probles and is also comparable to $LSCCA_1$-ELM in the Flags problem. Both the $LSCCA_1$-ELM Ensemble and the $LSCCA_1$-ELM are less than MLRBF in the Scene and Bird problem.

Fig.3 shows the recall of all methods in all problems. In this figure, the recall of $LSCCA_1$-ELM Ensemble is higher than $LSCCA_1$-ELM in the flags problem and is also comparable to $LSCCA_1$-ELM in the Scene, Yeast, and Emotion problem.

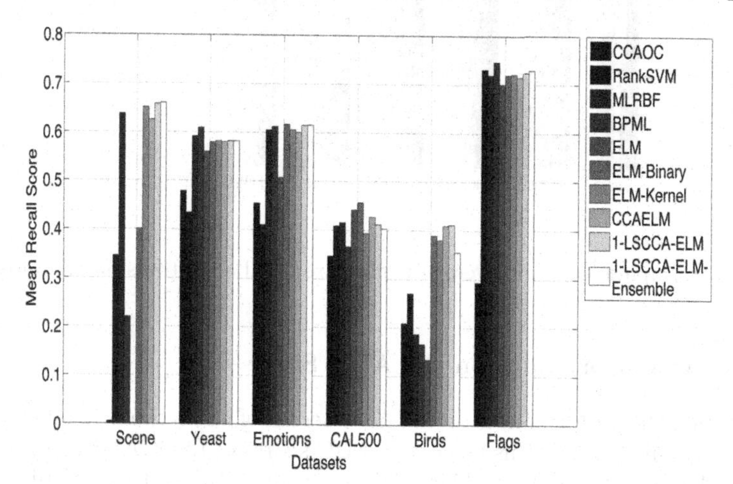

Fig. 3. Comparison of the average Recall of all algorithms on all dataset

$LSCCA_1$-ELM Ensemble is less than BPML and MLRBF in the Yeast, Flags and CAL500 problem and is also less than ELM kernel in CAL500 and Birds problem.

Finally, Fig.4 shows the F-measure performances of all methods in all problems. The F-measure of $LSCCA_1$-ELM Ensemble is higher than that of $LSCCA_1$-ELM in the Scene, Yeast, Emotions, CAL500, and Birds problem and comparable to $LSCCA_1$-ELM in the Flags problem. The F-measure of these methods are highest in most problems and comparable to MLRBF in some problems. The F-measure of both $LSCCA_1$-ELM Ensemble and $LSCCA_1$-ELM is the highest in the Emotions, CAL500, and Flags problem and less than that of MLRBF in Scene and Birds problem. It can be seen that, the $LSCCA_1$-ELM Ensemble can provide the good performance on the precision and F-measure and provide the low scores of recall. The $LSCCA_1$-ELM provided a comparable performance to $LSCCA_1$-ELM Ensemble on the recall and less than $LSCCA_1$-ELM Ensemble on precision and F-measure.

When comparing $LSCCA_1$-ELM Ensemble with the other eight methods in the experiment, we can see that $LSCCA_1$-ELM Ensemble is comparable to $LSCCA_1$-ELM and MLRBF on the precision, recall, and F-measure. Moreover, the experimental results show that both $LSCCA_1$-ELM Ensemble and $LSCCA_1$-ELM can improve ELM performance in multi-label classification.

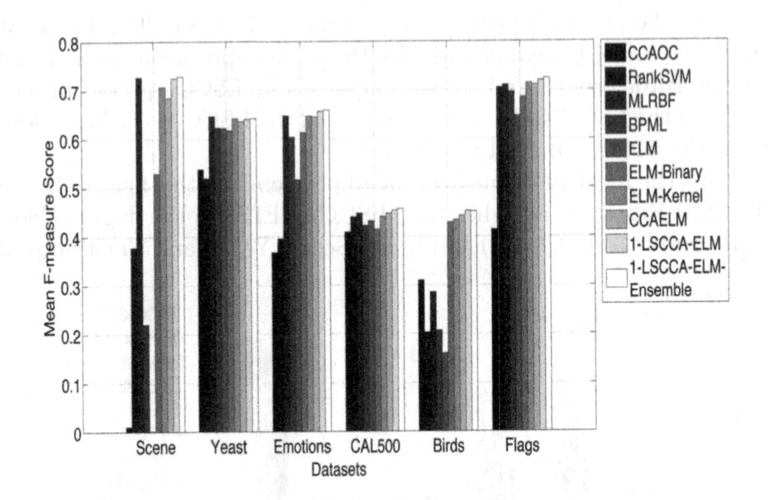

Fig. 4. Comparison of the average F-measure of all algorithms on all dataset

Comparison Performance on the AUC Score

Table 6 shows the performances on the AUC scores of all methods in all problems, the highest is the best. In this table, the AUC scores of LSCCA_1-ELM Ensemble of all problem are the highest. The AUC scores of LSCCA_1-ELM of all problem are higher than all of the comparative methods, except for the LSCCA_1 -ELM Ensemble. In Fig.5, the AUC scores of LSCCA_1-ELM are much lower than that of LSCCA_1-ELM Ensemble.

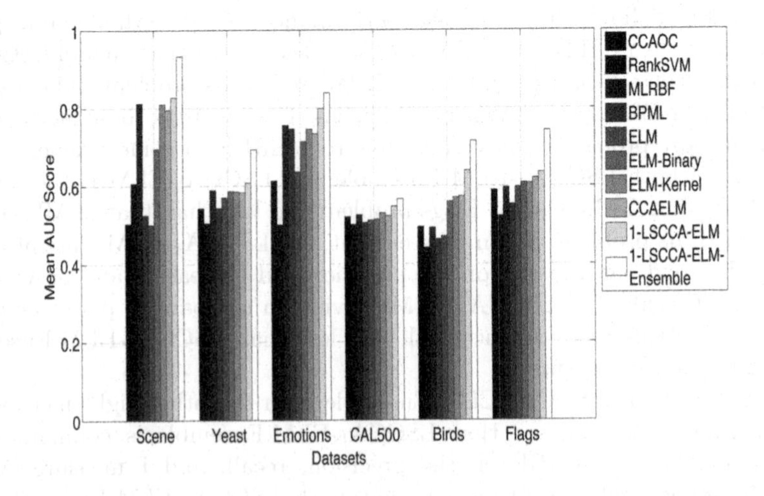

Fig. 5. Comparison of the average AUC score of all algorithms on all dataset

It can be seen that, the ensemble approach can reduce the random effect of ELM in $LSCCA_1$-ELM and provide the best performance in the experiment. Moreover, the AUC score in both Table 6 and Fig.5 shows that, the proposed methods outperform the comparative methods (ELM [5], Binary ELM [25], ELM Kernel [6], rank SVM [21], BP-MLL [19], MLRBF [20], CCA-OC [34], and CCA-ELM [7]).

Table 4. Comparison of Precision, Recall, and F-measure of ELM, ELM-Binary, ELM kernel, CCAELM, and two proposes methods in real world problems. The two-trailed T-test at 0.05 significance level is performed between $LSCCA_1$-ELM Ensemble and others methods. The plus symbol(+) is indicated that the performance of $LSCCA_1$-ELM Ensemble is better than the comparative algorithms, while the minus symbol(-) is indicated that the performance of $LSCCA_1$-ELM Ensemble is lower than the comparative algorithms and the approx symbol($\approx$) is means that the performance of $LSCCA_1$-ELM Ensemble is comparable to the comparative algorithms.

Datasets	Algorithms	The performance estimators		
		Precision (sd.)	Recall (sd.)	F-measure (sd.)
Scene	ELM	0.08000(0.09273) +	0.00054(0.00050) +	0.00106(0.00098) +
	ELM-Binary	0.78209(0.01029) +	0.40057(0.00875) +	0.52899(0.00894) +
	ELM kernel	0.77391(0.01151) +	0.65089(0.00868) +	0.70694(0.01004) +
	CCAELM	0.75718(0.00892) +	0.62510(0.00910) +	0.68457(0.00779) +
	$LSCCA_1$-ELM	0.80743(0.00696) $\approx$	0.65769(0.00310) +	0.72465(0.00420) $\approx$
	$LSCCA_1$-ELM Ensemble	**0.81325(0.00803)**	**0.66003(0.00643)**	**0.72844(0.00605)**
Yeast	ELM	0.70114(0.00948) +	0.56016(0.00511) +	0.62258(0.00517) +
	ELM-Binary	0.66223(0.00703) +	0.57959(0.00527) +	0.61802(0.00416) +
	ELM kernel	**0.71822(0.00907)** $\approx$	0.58165(0.00467) $\approx$	0.64259(0.00543) $\approx$
	CCAELM	0.70529(0.00926) +	0.57981(0.00344) +	0.63623(0.00350) +
	$LSCCA_1$-ELM	0.71295(0.01027) $\approx$	0.58194(0.00456) +	0.64063(0.00488) $\approx$
	$LSCCA_1$-ELM Ensemble	0.71818(0.00910)	**0.58202(0.00417)**	**0.64261(0.00430)**
Emotions	ELM	0.52788(0.00719) +	0.50695(0.01262) +	0.51600(0.00603) +
	ELM-Binary	0.61270(0.00681) +	**0.61675(0.00928)** $\approx$	0.61355(0.00542) +
	ELM kernel	0.69513(0.00586) +	0.60538(0.01105) +	0.64716(0.00677) +
	CCAELM	0.69982(0.00420) +	0.60045(0.01338) +	0.64560(0.00630) +
	$LSCCA_1$-ELM	0.70744(0.00496) +	0.61483(0.01448) $\approx$	0.65704(0.00889) $\approx$
	$LSCCA_1$-ELM Ensemble	**0.71551(0.00561)**	0.61528(0.01522)	**0.65954(0.00916)**
CAL500	ELM	0.42228(0.00687) +	**0.44140(0.00433)** -	0.43133(0.00355) +
	ELM-Binary	0.42251(0.00524) +	0.40620(0.00320) +	0.41419(0.00520) +
	ELM kernel	0.40620(0.00168) +	0.39323(0.00359) +	0.44058(0.00168) +
	CCAELM	0.47365(0.00278) +	0.42671(0.00501) +	0.44853(0.00116) +
	$LSCCA_1$-ELM	0.51074(0.00580) +	0.41148(0.00427) -	0.45296(0.00149) $\approx$
	$LSCCA_1$-ELM Ensemble	**0.52739(0.00365)**	0.40293(0.00738)	**0.45639(0.00575)**
Birds	ELM	0.21160(0.01867) +	0.13424(0.00811) +	0.16002(0.00708) +
	ELM-Binary	0.48250(0.01118) +	0.38920(0.01274) -	0.42831(0.01005) +
	ELM kernel	0.50409(0.00795) +	0.38009(0.01102) -	0.43339(0.01063) +
	CCAELM	0.48604(0.01449) +	0.40948(0.01314) -	0.44186(0.00875) +
	$LSCCA_1$-ELM	0.50174(0.00580) -	**0.41148(0.00427)** -	**0.45176(0.00149)** $\approx$
	$LSCCA_1$-ELM Ensemble	**0.62658(0.01909)**	0.35540(0.01088)	0.45065(0.01331)
Flags	ELM	0.60136(0.01218) +	0.70086(0.01580) +	0.64731(0.01289) +
	ELM-Binary	0.65432(0.01895) +	0.71968(0.01436) +	0.68545(0.01541) +
	ELM kernel	0.70739(0.00884) $\approx$	0.72164(0.00886) $\approx$	0.71400(0.00609) +
	CCAELM	0.70648(0.00674) $\approx$	0.71547(0.00736) $\approx$	0.71054(0.00561) +
	$LSCCA_1$-ELM	0.71647(0.00830) $\approx$	0.72409(0.00974) $\approx$	0.71986(0.00772) +
	$LSCCA_1$-ELM Ensemble	**0.71992(0.01206)**	**0.72920(0.01103)**	**0.72404(0.00998)**

Table 5. Comparison of Precision, Recall, and F-measure of CCAOC, RankSVM, ML-RBF, and two proposes methods in real world problems. The two-trailed T-test at 0.05 significance level is performed between $LSCCA_1$-ELM Ensemble and other methods. The plus symbol(+) is indicated that the performance of $LSCCA_1$-ELM Ensemble is better than the comparative algorithms, while the minus symbol(-) is indicated that the performance of $LSCCA_1$-ELM Ensemble is lower than the comparative algorithms and the approx symbol(≈) is means that the performance of $LSCCA_1$-ELM Ensemble is comparable to the comparative algorithms.

Problems	Algorithms	The performance estimators		
		Precision (sd.)	Recall (sd.)	F-measure (sd.)
Scene	CCAOC	0.64000(0.02152) +	0.00496(0.00079) +	0.00984(0.00154) +
	RankSVM	0.41770(0.00837) +	0.34556(0.00806) +	0.37767(0.00833) +
	MLRBF	**0.84685(0.00816)** -	0.63756(0.00902) +	0.72712(0.00705) ≈
	BPML	0.21978(0.00822) +	0.21961(0.00941) +	0.21962(0.00883) +
	$LSCCA_1$-ELM	0.80743(0.00696) ≈	0.65769(0.00310) +	0.72465(0.00420) ≈
	$LSCCA_1$-ELM Ensemble	0.81325(0.00803)	**0.66003(0.00643)**	**0.72844(0.00605)**
Yeast	CCAOC	0.61487(0.00761) +	0.47804(0.00397) +	0.53789(0.00483) +
	RankSVM	0.64757(0.01021) +	0.43405(0.00295) +	0.51934(0.00427) +
	MLRBF	0.71457(0.00614) ≈	0.59153(0.00468) -	**0.64711(0.00544)** ≈
	BPML	0.63874(0.00546) +	**0.60915(0.00434)** -	0.62344(0.00329) +
	$LSCCA_1$-ELM	0.71295(0.01027) ≈	0.58194(0.00456) +	0.64063(0.00488) ≈
	$LSCCA_1$-ELM Ensemble	**0.71818(0.00910)**	0.58202(0.00417)	0.64261(0.00430)
Emotions	CCAOC	0.30840(0.01178) +	0.45407(0.01043) +	0.36732(0.01253) +
	RankSVM	0.38248(0.01017) +	0.41023(0.01162) +	0.39563(0.01107) +
	MLRBF	0.69744(0.01396) +	0.60498(0.00576) +	0.64793(0.00854) +
	BPML	0.59482(0.00840) +	0.61204(0.00727) ≈	0.60349(0.00360) +
	$LSCCA_1$-ELM	0.70744(0.00496) +	0.61483(0.01448) ≈	0.65704(0.00889) ≈
	$LSCCA_1$-ELM Ensemble	**0.71551(0.00561)**	**0.61528(0.01522)**	**0.65954(0.00916)**
CAL500	CCAOC	0.49755(0.00638) +	0.34688(0.00181) +	0.40877(0.00245) +
	RankSVM	0.48770(0.06934) +	0.40881(0.03492) ≈	0.43988(0.05140) +
	MLRBF	0.49099(0.00405) +	**0.41538(0.00445)** -	0.44969(0.00366) +
	BPML	0.49899(0.00540) +	0.36599(0.00588) +	0.42226(0.00439) +
	$LSCCA_1$-ELM	0.51074(0.00580) +	0.41148(0.00427) -	0.45296(0.00149) ≈
	$LSCCA_1$-ELM Ensemble	**0.52739(0.00365)**	0.40293(0.00738)	**0.45639(0.00575)**
Birds	CCAOC	0.65724(0.02379) -	0.20891(0.01101) +	0.30954(0.01294) +
	RankSVM	0.16913(0.01866) +	0.26999(0.01474) +	0.20260(0.01016) +
	MLRBF	**0.79229(0.03437)** -	0.18664(0.00868) +	0.28497(0.01263) +
	BPML	0.34110(0.01902) +	0.16602(0.01144) +	0.20669(0.01453) +
	$LSCCA_1$-ELM	0.50174(0.00580) -	**0.41148(0.00427)** -	**0.45176(0.00149)** ≈
	$LSCCA_1$-ELM Ensemble	0.62658(0.01909) +	0.35540(0.01088)	0.45065(0.01331)
Flags	CCAOC	0.70020(0.04185) +	0.29350(0.02517) +	0.41362(0.03580) +
	RankSVM	0.68093(0.01244) +	0.73096(0.00585) -	0.70506(0.00714) +
	MLRBF	0.70112(0.01279) +	0.71949(0.00764) +	0.71019(0.00946) +
	BPML	0.65424(0.01044) +	**0.74638(0.00900)** -	0.69632(0.01044) +
	$LSCCA_1$-ELM	0.71647(0.00830) ≈	0.72409(0.00974) ≈	0.71986(0.00772) +
	$LSCCA_1$-ELM Ensemble	**0.71992(0.01206)**	0.72920(0.01103)	**0.72404(0.00998)**

Comparison the Computation Times

Table 7 shows the performance on the computation time of all methods, the lowest is the best. The training times of $LSCCA_1$-ELM is the lowest because its number of hidden node is lower than ELM and ELM-Binary which shows in Table 3. The computation times of $LSCCA_1$-ELM Ensemble are slower than ELM, Binary ELM, ELM Kernel, and CCA-ELM but much faster than rank SVM, BP-MLL, MLRBF and CCA-OC.

Table 6. Comparison of all methods on all dataset in terms of average AUC score. (*) represents the highest of AUC scores.

Datasets	Algorithms	ROC-AUC Score
Scene	CCAOC	0.50298
	RankSVM	0.60654
	MLRBF	0.81199
	BPML	0.52506
	ELM	0.50017
	ELM-Binary	0.69573
	ELM kernel	0.80969
	CCAELM	0.79519
	$LSCCA_1$-ELM	0.82678
	$LSCCA_1$-ELM Ensemble	**0.93238** *
Yeast	CCAOC	0.54194
	RankSVM	0.50418
	MLRBF	0.58841
	BPML	0.54275
	ELM	0.56995
	ELM-Binary	0.58581
	ELM kernel	0.58429
	CCAELM	0.5824
	$LSCCA_1$-ELM	0.60718
	$LSCCA_1$-ELM Ensemble	**0.69323** *
Emotions	CCAOC	0.61251
	RankSVM	0.50042
	MLRBF	0.75529
	BPML	0.7461
	ELM	0.63648
	ELM-Binary	0.71423
	ELM kernel	0.74528
	CCAELM	0.73344
	$LSCCA_1$-ELM	0.7982
	$LSCCA_1$-ELM Ensemble	**0.83834** *
CAL500	CCAOC	0.52037
	RankSVM	0.50067
	MLRBF	0.52514
	BPML	0.50546
	ELM	0.51137
	ELM-Binary	0.51439
	ELM kernel	0.52962
	CCAELM	0.52257
	$LSCCA_1$-ELM	0.54705
	$LSCCA_1$-ELM Ensemble-Ensemble	**0.56551** *
Birds	CCAOC	0.49465
	RankSVM	0.44147
	MLRBF	0.49228
	BPML	0.46321
	ELM	0.47025
	ELM-Binary	0.55896
	ELM kernel	0.56917
	CCAELM	0.57238
	$LSCCA_1$-ELM	0.63906
	$LSCCA_1$-ELM Ensemble	**0.7149** *
Flags	CCAOC	0.58838
	RankSVM	0.52201
	MLRBF	0.59586
	BPML	0.55217
	ELM	0.59708
	ELM-Binary	0.60822
	ELM kernel	0.60704
	CCAELM	0.62045
	$LSCCA_1$-ELM	0.6349
	$LSCCA_1$-ELM Ensemble	**0.74222** *

Table 7. Comparison of all methods on all dataset in terms of average Training Times. (*) represents the lowest of Training Times.

Datasets	Algorithms	Training Times
Scene	CCAOC	441.69905
	RankSVM	470.04817
	MLRBF	1.61773
	BPML	2481.0215
	ELM	**0.03588** *
	ELM-Binary	0.06744
	ELM kernel	1.96218
	CCAELM	0.55505
	$LSCCA_1$-ELM	0.1301
	$LSCCA_1$-ELM Ensemble	1.56063
Yeast	CCAOC	171.07662
	RankSVM	3380.37444
	MLRBF	9.10578
	BPML	2550.62914
	ELM	0.6421
	ELM-Binary	1.67824
	ELM kernel	1.3023
	CCAELM	0.14758
	$LSCCA_1$-ELM	**0.10951** *
	$LSCCA_1$-ELM Ensemble	1.55408
Emotions	CCAOC	18.17599
	RankSVM	114.2414
	MLRBF	0.29016
	BPML	633.06267
	ELM	0.08299
	ELM-Binary	0.24742
	ELM kernel	0.12199
	CCAELM	0.02777
	$LSCCA_1$-ELM	**0.0078** *
	$LSCCA_1$-ELM Ensemble	0.11482
CAL500	CCAOC	116.7942
	RankSVM	4558.45468
	MLRBF	0.93445
	BPML	542.92778
	ELM	0.1039
	ELM-Binary	0.29887
	ELM kernel	0.09641
	CCAELM	0.06427
	$LSCCA_1$-ELM	**0.05023** *
	$LSCCA_1$-ELM Ensemble-Ensemble	0.05398
Birds	CCAOC	384.75899
	RankSVM	196.68887
	MLRBF	0.37253
	BPML	380.73822
	ELM	0.06427
	ELM-Binary	0.18068
	ELM kernel	0.44055
	CCAELM	0.15975
	$LSCCA_1$-ELM	**0.05023** *
	$LSCCA_1$-ELM Ensemble	0.26271
Flags	CCAOC	1.64425
	RankSVM	49.37588
	MLRBF	0.10951
	BPML	109.7336
	ELM	0.00562
	ELM-Binary	0.00753
	ELM kernel	0.02465
	CCAELM	0.01342
	$LSCCA_1$-ELM	**0.00281** *
	$LSCCA_1$-ELM Ensemble	0.04586

5 Conclusion

This paper proposed an extension to ELM using LS-CCA_1 for handling multi-label classification called $LSCCA_1$-ELM. The LS-CCA_1 is used to compute the projection matrix of input space, which is used to transform the input to new information in low dimension space. Next, ELM learns this new information and obtains the predicted label. The ensemble method is then used to reduce the random effects of ELM, which is proposed as $LSCCA_1$-ELM Ensemble. The performance on the precision, and F-measure reveal that the $LSCCA_1$-ELM Ensemble outperform ELM, Binary ELM, ELM Kernel, rank SVM, BP-MLL, CCA-OC, and CCA-ELM and it is comparable to $LSCCA_1$-ELM and MLRBF. The performance on recall $LSCCA_1$-ELM Ensemble is comparable to $LSCCA_1$-ELM and MLRBF. The computation time of $LSCCA_1$-ELM is the fastest in all of problems. The AUC scores show that the $LSCCA_1$-ELM Ensemble provided the highest performance of its model. It can be seen that, both $LSCCA_1$-ELM Ensemble and $LSCCA_1$-ELM can improve ELM performance in multi-label classification and also provide good performance for multi-label classification.

Acknowledgment. This work was supported by the Graduate Education of Computer and Information Science Interdisciplinary Research Grant from Department of Computer Science, Faculty of Science, Khon Kaen University, 2012.

References

1. Tsoumakas, G., Katakis, I.: Multi Label Classification: An Overview. Int. J. Data Warehousing and Mining. 3, 1–13 (2007)
2. Madjarov, G., Kocev, D., Gjorgjevikj, D., Deroski, S.: An extensive experimental comparison of methods for multi-label learning. Pattern Recogn. 45, 3084–3104 (2012)
3. Zhang, Y., Schneider, J.G.: Multi-label output codes using canonical correlation analysis. In: International Conference on Artificial Intelligence and Statistics, USA, pp. 873–882 (2011)
4. Sun, L., Ji, S., Ye, J.: A Least Squares Formulation for Canonical Correlation Analysis. In: Proceedings of the 25th International Conference on Machine Learning, pp. 1024–1031. ACM, New York (2008)
5. Huang, G.B., Zhu, Q.Y., Siew, C.K.: Extreme learning machine: Theory and applications. Neurocomputing 70, 489–501 (2006)
6. Huang, G.B., Zhou, H., Ding, X., Zhang, R.: Extreme learning machine for regression and multiclass classification. IEEE Trans Syst Man Cybern B Cybern 42, 513–529 (2012)
7. Kongsorot, Y., Horata, P.: Multi-label classification with extreme learning machine. In: 6th International Conference on Knowledge and Smart Technology (KST), Thailand, pp. 81–86 (2014)
8. Dietterich, T.G.: Ensemble methods in machine learning. In: Multiple classifier systems, pp. 1–15. Springer (2000)
9. Sun, L., Ji, S., Ye, J.: Canonical Correlation Analysis for Multilabel Classification: A Least-Squares Formulation, Extensions, and Analysis. IEEE Trans. Pattern Anal. Mach. Intell. 33, 194–200 (2011)
10. Read, J., Pfahringer, B., Holmes, G., Frank, E.: Classifier Chains for Multi-label Classification. In: Proceedings of the European Conference on Machine Learning and Knowledge Discovery in Databases: Part II, pp. 254–269. Berlin (2009)
11. Read, J., Pfahringer, B., Holmes, G.: Multi-label Classification Using Ensembles of Pruned Sets. In: 8th IEEE International Conference on Data Mining (ICDM 2008), Pisa, pp. 995–1000 (2008)
12. Tsoumakas, G., Vlahavas, I.P.: Random *k*-Labelsets: An Ensemble Method for Multilabel Classification. In: Kok, J.N., Koronacki, J., Lopez de Mantaras, R., Matwin, S., Mladenič, D., Skowron, A. (eds.) ECML 2007. LNCS (LNAI), vol. 4701, pp. 406–417. Springer, Heidelberg (2007)

13. Tsoumakas, G., Katakis, I., Vlahavas, I.: Effective and Efficient Multilabel Classification in Domains with Large Number of Labels. In: Proc. ECML/PKDD 2008 Workshop on Mining Multidimensional Data (MMD 2008), Belgium (2008)
14. Fürnkranz, J.: Round Robin Classification. J. Mach. Learn. Res. 2, 721–747 (2002)
15. Wu, T.F., Lin, C.J., Weng, R.C.: Probability Estimates for Multi-class Classification by Pairwise Coupling. J. Mach. Learn. Res. 5, 975–1005 (2004)
16. Freund, Y., Schapire, R.E.: A Decision-Theoretic Generalization of On-Line Learning and an Application to Boosting. J. Comput. System Sci. 55, 119–139 (1997)
17. Zhang, M.L., Zhou, Z.H.: Ml-knn: A lazy learning approach to multi-label learning. Pattern Recogn. 40, 2038–2048 (2007)
18. Clare, A.J., King, R.D.: Knowledge Discovery in Multi-label Phenotype Data. In: Siebes, A., De Raedt, L. (eds.) PKDD 2001. LNCS (LNAI), vol. 2168, pp. 42–53. Springer, Heidelberg (2001)
19. Zhang, M.L., Zhou, Z.H.: Multilabel Neural Networks with Applications to Functional Genomics and Text Categorization. IEEE Trans. Knowl. Data Eng. 18, 1338–1351 (2006)
20. Zhang, M.L.: Ml-rbf: RBF Neural Networks for Multi-Label Learning. Neural Process. Lett. 29, 61–74 (2009)
21. Elisseeff, A., Weston, J.: A Kernel Method for Multi-Labelled Classification. In: Advances in Neural Information Processing Systems, vol. 14, pp. 681–687. MIT Press (2001)
22. Hotelling, H.: Relations between two sets of variates. Biometrika 28, 321–377 (1936)
23. Tibshirani, R.: Regression Shrinkage and Selection Via the Lasso. J. Roy. Statist. Soc. Ser. B. 58, 267–288 (1994)
24. Penrose, R.: A generalized inverse for matrices. In: Mathematical Proceedings of the Cambridge Philosophical Society, pp. 406–413. Cambridge Univ Press (1955)
25. Rong, H.J., Huang, G.B., Ong, Y.S.: Extreme learning machine for multi-categories classification applications. In: IEEE International Joint Conference on Neural Networks (IEEE World Congress on Computational Intelligence), pp. 1709–1713. IEEE (2008)
26. Huang, G.B., Wang, D.H., Lan, Y.: Extreme learning machines: a survey. Int. J Mach. Learn. and Cyber. 2, 107–122 (2011)
27. Lan, Y., Soh, Y.C., Huang, G.B.: Ensemble of online sequential extreme learning machine. Neurocomputing 72, 3391–3395 (2009)
28. Liu, N., Wang, H.: Ensemble Based Extreme Learning Machine. IEEE Signal Process. Lett. 17, 754–757 (2010)
29. Zhu, Q.Y., Qin, A.K., Suganthan, P.N., Huang, G.B.: Evolutionary extreme learning machine. Pattern Recogn. 38, 1759–1763 (2005)
30. Freund, Y.: Boosting a weak learning algorithm by majority. Inform. Comput. 121, 256–285 (1995)
31. Fawcett, T.: ROC graphs: Notes and practical considerations for researchers. Mach. Learn. 31, 1–38 (2004)
32. Brown, C.D., Davis, H.T.: Receiver operating characteristics curves and related decision measures: A tutorial. Chemometr. Intell. Lab. 80, 24–38 (2006)
33. Zou, K.H., OMalley, A.J., Mauri, L.: Receiver-operating characteristic analysis for evaluating diagnostic tests and predictive models. Circulation 115, 654–657 (2007)
34. Zhang, Y., Schneider, J.G.: Multi-label output codes using canonical correlation analysis. In: International Conference on Artificial Intelligence and Statistics, USA, pp. 873–882 (2011)

Least Squares Policy Iteration Based on Random Vector Basis

Lei Zuo and Xin Xu

The College of Mechatronics and Automation,
National University of Defense Technology, Changsha 410073, China
zuo1986lei@163.com, xinxu@nudt.edu.cn

Abstract. Basis functions construction is a fundamental problem in value function approximation for reinforcement learning. Inspired by extreme learning machine (ELM), we construct the basis functions by using single-hidden layer feedforward neural networks (SLFNs) with random input weights and hidden layer biases. We call such basis function as random vector basis (RVB). The algorithm of least squares policy iteration based on random vector basis (RVB-LSPI) is proposed. One advantage of RVB is that it can be constructed automatically. The performance of the proposed method is compared with the traditional least squares policy iteration (LSPI) with radial basis functions (RBFs). The results indicate that RVB-LSPI has better performance.

Keywords: Reinforcement learning, least squares policy iteration, extreme learning machine, random vector basis.

1 Introduction

Reinforcement learning (RL) has been widely studied in recent years [1]. In reinforcement learning, the agent interacts with the environment and modifies its action policies to maximize its cumulative payoffs. Reinforcement learning is not only receiving attention from the neural network community, but also from the fields of decision theory, operations research, and control engineering [2]. Reinforcement learning has been applied in backgammon [3], job-shop scheduling [4], elevator scheduling [5], helicopter flight control [6] and so on. However, it is still difficult for RL to solve MDPs with large or continuous spaces. In such cases, RL has difficulty with the curse of dimensionality and many RL algorithms cannot converge to an optimal or near-optimal policy and require numerous training samples.

Value function approximation (VFA) [7-11] has been widely studied to conquer the curse of dimensionality and improve the generalization ability of Reinforcement learning algorithms. Approximate policy iteration (API) is one class of VFA methods. A model-free API algorithm called least squares policy iteration (LSPI) was presented in [9]. Xu et al proposed a kernel version of LSPI called KLSPI[10]. Later, a framework for VFA called proto-reinforcement learning (PRL) was proposed in [11]. This approach yields a control learning algorithm called representation policy iteration

J. Cao et al. (eds.), *Proceedings of ELM-2014 Volume 1*,
Proceedings in Adaptation, Learning and Optimization 3, DOI: 10.1007/978-3-319-14063-6_33

(RPI) where both the underlying representations (basis functions) and policies are simultaneously learned. However, there are still many parameters in these methods need to be selected, which is not convenient to construct basis functions.

In the past decades, feedforward neural networks have been widely used in many different fields. They can approximate complex nonlinear mapping directly from the input samples. To benefit from the approximation ability of single-layer hidden feedforward neural networks (SLFNs), Huang et al proposed the ELM algorithm [12], which is efficient in supervised classification and regression. ELM is slightly different from traditional SLFNs. In traditional SLFNs, the input weights and hidden layer biases need to be learned, whereas those in ELM are randomly assigned. ELM has been widely studied and used in pattern recognition in recent years[13, 14]. But little work has been done to use this method in reinforcement learning.

Motivated by ELM, we propose least squares policy iteration based on random vector basis (RVB-LSPI), in which the basis functions are constructed by using single-hidden layer feedforward neural networks (SLFNs) with random input weights and hidden layer biases. One advantage of RVB is that it can be constructed fast and automatically. The rest of this paper is organized as follows. Section 2 briefly provides some background information about Markov decision process and value function approximation in reinforcement learning. Then, the RVB-LSPI is presented in Section 3. In Section 4, simulations on the mountain-car problem are carried out and the results illustrate the effectiveness of the proposed method. Section 5 draws conclusions.

2 Background

2.1 Markov Decision Process

The Markov decision process (MDP) is the underlying formalism for reinforcement learning algorithms. An MDP is defined as a tuple $\{S, A, R, P\}$, where S is the state space, A is the action space, P is the state transition probability, and R is the reward function. The policy of the MDP is defined as a function π: $S \rightarrow Pr(A)$, where $Pr(A)$ is a probability distribution in the action space. The quality of a policy π is quantified by the value function $V^{\pi}(s)$, defined as the expected discounted cumulative reward starting in a state s and then following the policy π

$$V^{\pi}(s) = E_{\pi}\left[\sum_{i=0}^{\infty} \gamma^i r_i \mid s_0 = s\right] \tag{1}$$

where γ is the discount factor.

Closely related to the value function $V^{\pi}(s)$ is the so-called state-action value function, which is usually used to improve the policy. The state-action value function $Q^{\pi}(s,a)$ is defined as the expected cumulative reward of performing action a in state s and then following policy π thereafter:

$$Q^{\pi}(s,a) = E_{\pi}\left[\sum_{i=0}^{\infty} \gamma^{i} r_{i} \mid s_{0} = s, a_{0} = a\right] \tag{2}$$

The state-action value function satisfies the following Bellman equation:

$$Q^{\pi}(s_{i},a_{i}) = E_{\pi}\left[r(s_{i},a_{i}) + \gamma Q^{\pi}(s_{i+1},a_{i+1})\right] \tag{3}$$

The optimal state-action value function is

$$Q^{*}(s,a) = \max_{\pi} Q^{\pi}(s,a) \tag{4}$$

Then, the optimal policy can be obtained by a greedy strategy

$$\pi^{*}(s) = \arg\max_{a} Q^{*}(s,a) \tag{5}$$

2.2 Approximate Policy Iteration and Basis Functions Construction

For many practical control problems, the transition model and the reward function of the underlying MDPs are usually unknown. However, relying on information that comes from interaction with the process, it is still possible to evaluate, or, even better, find good policies for such problems. Such information includes observations of states, actions, and rewards, which are commonly known as samples.

Approximate policy iteration (API) is a class of algorithms that learns decision policies from samples. The samples can be obtained from actual episodes of interaction with the process or from queries to a generative model of the process. One popular approximate policy iteration algorithm based on linear value function approximation is the least-squares policy iteration (LSPI), which has been widely studied in recent years [10, 11]. In LSPI, the state-action value function $Q^{\pi}(s,a)$ can be approximated using a linearly weighted combination of k basis functions (features):

$$\hat{Q}^{\pi}(s,a) = \phi^{T}(s,a) w^{\pi} = \sum_{i=1}^{k} \phi_{i}(s,a) w_{i} \tag{6}$$

where $w^{\pi} = (w_1, w_2, \ldots, w_k)^T$ is the weight vector and $\phi(s,a)$ is the basis function vector computed at (s, a), denoted by $\phi(s,a) = (\phi_1(s,a), \phi_2(s,a), \ldots, \phi_k(s,a))^T$.

The choice of basis functions is a fundamental problem in LSPI. In [9], the polynomial and radial basis functions were used. The KLSPI algorithm proposed by Xu et.al [10] describes basis functions by kernel-based features.

ELM has been proven efficient in supervised learning [12-14]. This algorithm is very fast because the input weights and hidden layer biases in ELM are randomly assigned rather than learned. Inspired by this, we consider using this method to construct the basis functions in value function approximation for reinforcement learning. If the input weights and hidden layer biases can be randomly assigned, then the process of basis functions construction will be fast and there will be fewer parameters need to be adjusted by hand.

3 Least Squares Policy Iteration Based on Random Vector Basis

Feedforward neural networks have been extensively used in machine learning due to their powerful approximation ability from the input samples. The value function approximator based on random vector basis (RVB) in this paper can be considered as a single-layer hidden feedforward neural network (SLFN). However, the proposed method is different from traditional SLFNs. Motivated by ELM [12], the input weights and hidden layer biases in the proposed method are randomly assigned. The architecture of the value function approximator based on RVB is shown in Fig. 1.

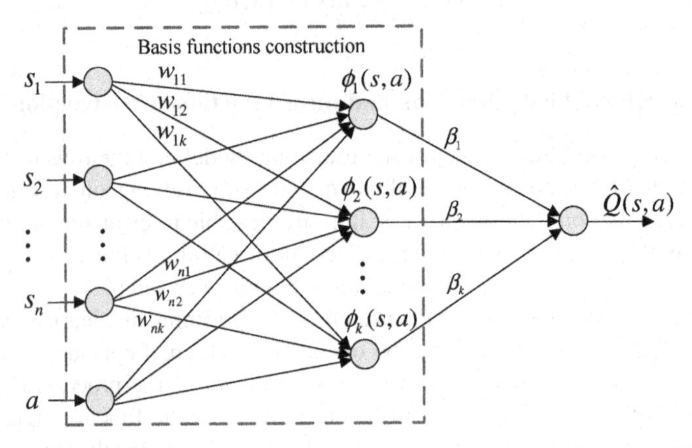

Fig. 1. The architecture of the value function approximator based on RVB

In such a neural network as shown in Fig. 1, it takes states and actions as inputs, and the output is the approximate state-action value function. The process of basis functions construction is performed in the input layer and the hidden layer, where the state-action pairs are mapped to the state-action features. The state basis function can be written as

$$\phi_i(s) = g(w_i \cdot s + b_i) \tag{7}$$

where w_i and b_i are random vectors, $w_i \cdot s$ is an inner product of the vector w_i and s, $g(\cdot)$ is the activation function. Note that the activation functions in the hidden layer are assumed infinitely differentiable.

The state-action basis function φ(s, a) can be generated by duplicating the state bases φ(s) |A| times, and setting all the elements of this vector to 0 except for the ones corresponding to the chosen action.

Thus, the state-action value function $Q^{\pi}(s, a)$ can be approximated as

$$\hat{Q}^{\pi}(s,a) = \sum_{i=1}^{k} \beta_i \cdot \phi_i(s,a) \tag{8}$$

Because the input weights and hidden layer biases are assigned at random and not to be learned, the learning in this method is a linear one. If the true values of $Q^{\pi}(s, a)$ are known in advance, then it is easy to compute a set of parameters that makes the approximate value function close enough to the true one. Unfortunately, the true values are not known in most reinforcement learning problems, thus, the learning algorithms in supervised learning, like ELM, can not be used directly in reinforcement learning. However, it is still possible to solve this problem. In this paper, we use the least-squares fixed-point approximation method.

Algorithm 1: RVB-LSPI
Input : D, π_0, ε **Output:** β, w, b

1 **Initialization:** $\tilde{A} \leftarrow 0$, $\tilde{c} \leftarrow 0$, $\beta' \leftarrow \beta_0$, $\pi' \leftarrow \pi_0$;
   $w \leftarrow$random( );
   $b \leftarrow$random( );
2 **while** $\| \beta - \beta' \| < \varepsilon$ **do**
3   $\beta \leftarrow \beta'$; $\pi \leftarrow \pi'$;
4   **for each** $(s_i, a_i, s_{i+1}, r_i) \in D$ **do**
5     $\phi(s_i, a_i) = e_I(a_i) \otimes \sum_{j=1}^{k} g(w_j \bullet s_i + b_j)$;
6     $\tilde{A} \leftarrow \tilde{A} + \phi(s_i, a_i)(\phi(s_i, a_i) - \gamma\phi(s_{i+1}, \pi(s_{i+1})))^T$;
7     $\tilde{c} \leftarrow \tilde{c} + \phi(s_i, a_i) r_i$;
8   **end for**
9   $\beta' \leftarrow \tilde{A}^{-1}\tilde{c}$;
10 **return** β, w, b;

One method to learn the parameters is the least-squares fixed-point projection. The state-action value function is the fixed point of the Bellman operator:

$$T_{\pi} Q^{\pi} = Q^{\pi}.$$

Thus, force the approximate value function to be a fixed point under the Bellman operator:

$$T_{\pi} \hat{Q}^{\pi} \approx \hat{Q}^{\pi}$$

Then, the solution is to minimize $\| T_{\pi}\hat{Q}^{\pi} - \hat{Q}^{\pi} \|$:

$$\min_{\beta} \| T_{\pi}\hat{Q}^{\pi} - \hat{Q}^{\pi} \| = \min_{\beta} \| R^{\pi} + \gamma P \Pi_{\pi} \Phi \beta^{\pi} - \Phi \beta^{\pi} \| \tag{9}$$

Then, the least-squares fixed-point solution and the corresponding improved policy can be obtained as follows [9]:

$$\begin{aligned} \beta^{\pi} &= (\Phi^T(\Phi - \gamma\Phi'))^{-1}\Phi^T R = A^{-1}c \\ \pi(s) &= \arg\max_{a\in A} \phi^T(s,a)\beta^{\pi} \end{aligned} \tag{10}$$

where

$$A = \Phi^T(\Phi - \gamma P \Pi_{\pi} \Phi), \; c = \Phi^T R .$$

Unbiased estimates of the matrix A and vector c can be obtained from samples by the following updates:

$$\begin{aligned} \tilde{A} &= \tilde{A} + \phi(s,a)(\phi(s,a) - \gamma\phi(s',\pi(s')))^T \\ \tilde{c} &= \tilde{c} + \phi(s,a)r \end{aligned} \tag{11}$$

Algorithm 1 specifies the pseudocodes of the RVB-LSPI algorithm.

4 Experimental Results

We explored the effectiveness of RVB-LSPI in the mountain car problem, which has been viewed as one of the benchmarks in the field. The goal of the mountain car task is to get a simulated car to the top of a hill as quickly as possible [1]. The car does not have enough power to get there immediately, and so must oscillate on the hill to build up the necessary momentum (as shown in Fig. 2). The reward is -1 per step until the car reaches the goal.

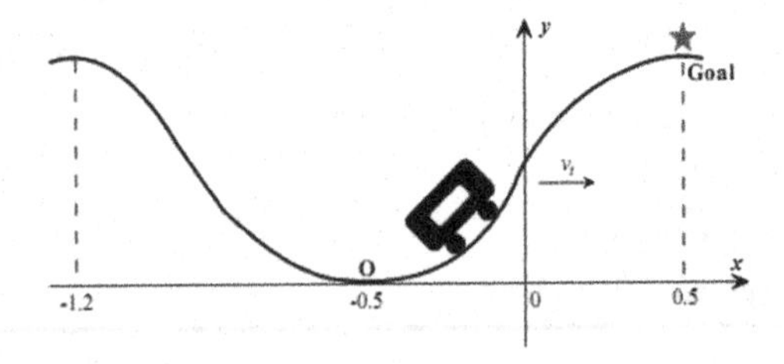

Fig. 2. The mountain-car problem

The state space includes the position and velocity of the car. There are three actions: full throttle forward (+1), full throttle reverse (-1), and zero throttle (0). Its position, p_t and velocity v_t, are updated by

$$\begin{aligned} v_{t+1} &= bound[v_t + 0.001a_t - 0.0025\cos(3p_t)] \\ p_{t+1} &= bound[p_t + v_{t+1}] \end{aligned} \tag{12}$$

where the bound operation enforces $-1.2 \le p_{t+1} \le 0.6$ and $-0.07 \le v_{t+1} \le 0.07$. The episode ends when the car successfully reaches the top of the mountain, defined as position $p_t >= 0.5$. In our experiments we allow a maximum of 300 steps, after which the task is terminated without success. The discount factor was set to 0.95.

Both LSPI and RVB-LSPI were evaluated and the performances of the two algorithms were compared. The same sample sets with different episodes were used for both algorithms, in which the samples were collected using random walk sampling policy. We collected six different sample sets to test the performance of the algorithms. Some information about these sample sets is listed in Table 1.

Table 1. Information of sample sets

Sample set order	Sample episodes	Max steps per episode	Number of samples
#1	50	70	2348
#2	50	70	2569
#3	100	70	3952
#4	100	70	4955
#5	150	70	6080
#6	150	70	7521

In all simulations, the states as inputs have been normalized into the range [0, 1] and the discount factor was set to 0.95. Every simulation was run for 10 times and results were median averages over 10 runs.

In the first simulation, we selected sigmoid as the activation function in RVB-LSPI. Different numbers of hidden nodes were selected, 10 and 15, respectively. The results are listed in Table 2 and Table 3.

Table 2. Results for 10 hidden nodes and sigmoid activation function

Sample set	Average iteration times	Average convergent time (s)	Average steps	Dev	Minimum steps
#1	6.5	4.2609	160.3	5.74	153
#2	7	4.9578	169.9	34.8	149
#3	6	6.1844	165.8	27.3	140
#4	6.9	8.6672	152.6	22.1	114
#5	5.4	8.2828	172.2	2.7	170
#6	5.7	10.7469	158.6	23.9	116

Table 3. Results for 15 hidden nodes and sigmoid activation function

Sample set	Average iteration times	Average convergent time (s)	Average steps	Dev	Minimum steps
#1	6	4.1797	165.3	12.2	152
#2	9	7.0938	216	0	216
#3	18.5	18.8641	234	69.2	160
#4	14.5	18.3359	226	61.9	159
#5	15.1	22.8328	165.5	13.2	149
#6	18.6	34.5453	207.3	49.3	172

As can be seen in the two tables above, the RVB-LSPI algorithm can converge to a policy based on every sample set. Compare the results in Table 2 with that in Table 3, we can find that the RVB-LSPI algorithm with 10 hidden nodes has better performance than that with 15 hidden nodes.

Then, we tested the performance of traditional LSPI with radial basis functions (RBFs). In LSPI, 39 basis functions were used, a set of 13 state basis functions for each of the 3 actions. These 13 basis functions included a constant term and 12 radial basis functions. The centers were 12 points of the grid {0, 1/3, 2/3, 1} × {0, 0.5, 1} and the width σ^2=0.5. Table 4 shows the results. Then, we changed the width to σ^2=0.1 and the results are shown in Table 5.

Table 4. Results for 13 RBFs with the width σ^2=0.5

Sample set	Average iteration times	Average convergent time (s)	Average steps
#1	13.6	8.1281	failed
#2	failed		failed
#3	6.1	5.5969	failed
#4	5.9	6.5500	failed
#5	6	8.1063	failed
#6	7.7	12.6422	259

Table 5. Results for 13 RBFs with the width σ^2=0.1

Sample set	Average iteration times	Average convergent time (s)	Average steps
#1	failed		failed
#2	6.5	4.0922	failed
#3	5.7	5.1703	failed
#4	5.9	6. 4859	202
#5	5.5	7.2484	163
#6	5.4	8.8500	failed

As can be seen in the results listed in Table 4 and Table 5, most policies learned by the traditional LSPI algorithm with RBFs have failed to move the car to the goal in 300 steps. Compared with traditional LSPI with RBFs, the RVB-LSPI algorithm has better convergent performance and also the final policies learned by RVB-LSPI have better control performance.

5 Conculsions

In this paper, we have propose a RVB based LSPI algorithm for reinforcement learning. In this method, the basis functions are constructed by using single-hidden layer feedforward neural networks (SLFNs) with random input weights and hidden layer biases, which are called random vector basis (RVB). One advantage of RVB is that it can be constructed automatically. The performance of the proposed method is compared with the traditional least squares policy iteration (LSPI) with radial basis functions (RBFs). The results indicate that RVB-LSPI has better performance.

Acknowledgment. This paper is supported by National Natural Science Foundation of China under Grant 60175072, & 90820302, the Program for New Century Excellent Talents in University under Grant NCET-10-0901.

References

1. Sutton, R., Barto, A.G.: Reinforcement Learning: An Introduction. MIT Press, Cambridge (1998)
2. Wang, F.Y., Zhang, H., Liu, D.: Adaptive dynamic programming: An introduction. IEEE Comput. Intell. Mag. 4(2), 39–47 (2009)
3. Tesauro, G.: TD-gammon a self-teaching backgammon program achieves master-level play. Neural Computing 6(2), 215–219 (1994)
4. Zhang, W., Dietterich, T.: A reinforcement learning approach to job-shop scheduling. In: Proceedings of the 14th International Joint Conference on the Artificial Intelligence, pp. 1114–1120. San Francisco (1995)
5. Crites, R.H., Barto, A.G.: Elevator group control using multiple reinforcement learning agents. Mach. Learn. 33(2–3), 235–262 (1998)
6. Ng, A.Y., Kim, H.J., Jordan, M., Sastry, S.: Autonomous helicopter flight via reinforcement learning. In: Advances in Neural Information Processing Systems, vol. 16, MIT Press, Cambridge (2004)
7. Sutton, R.: Generalization in reinforcement learning: Successful examples using sparse coarse coding. In: Advances in Neural Information Processing Systems, vol. 8, pp. 1038–1044. MIT Press, Cambridge (1996)
8. Xu, X., He, H.G.: Residual-gradient-based neural reinforcement learning for the opti-mal control of an acrobat. In: Proc. IEEE Int. Symp. Intell, pp. 758–763. Vancouver, Canada (2002)
9. Lagoudakis, M.G., Parr, R.: Least-squares policy iteration. J. Mach. Learn. Res. 4, 1107–1149 (2003)
10. Xu, X., Hu, D.W., Lu, X.C.: Kernel-based least squares policy iteration for reinforcement learning. IEEE Trans on Nerual Networks 18(4), 973–992 (2007)
11. Mahadevan, S., Maggioni, M.: Proto-Value Functions: A Laplacian Framework for Learning Representation and Control in Markov Decision Processes. Journal of Machine Learning Research 8, 2169–2231 (2007)
12. Huang, G.B., Zhu, Q.Y., Siew, C.K.: Extreme learning machine: Theory and applications. Neurocomputing 70, 489–501 (2006)
13. Huang, G.B., Chen, L., Siew, C.-K.: Universal approximation using incremental constructive feedforward networks with random hidden nodes. IEEE Transactions on Neural Networks 17(4), 879–892 (2006)
14. Huang, G.B., Zhou, H., Ding, X., et al.: Extreme learning machine for regression and multiclass classification. IEEE Transactions on System, Man, and Cybernetics-Part B: Cybernetics 42(2), 513–529 (2012)

Identifying Indistinguishable Classes in Multi-class Classification Data Sets Using ELM

Felis Dwiyasa and Meng-Hiot Lim

School of Electrical and Electronic Engineering,
Nanyang Technological University,
50 Nanyang Avenue, Singapore 639798
http://www.ntu.edu.sg

Abstract. For many years, various neural network models have been used to solve regression, binary classification, and multi-class classification problems. Their performance has been extensively compared against each other in terms of testing accuracy and training time. For multi-class classification problem, testing accuracy does not always give comprehensive information about the performance. It only shows the number of false detections without any clues on the false detection distribution. In this work we propose a cross-validation based on Extreme Learning Machine to identify classes that are found to have high number of false detections. These classes are treated as indistinguishable classes that need further processing or information. Our simulation shows that our proposed method is able to detect indistinguishable classes in three data sets. We also found that when indistinguishable classes exist, the training accuracy can be higher if each pair of those classes are marked as one merged class.

Keywords: ELM, classification, indistinguishable classes, cross-validation.

1 Introduction

Artificial neural network approach has been widely used to solve various regression and classification problems. As black box non-parametric model that is able to adjust its own weights, artificial neural network is easy to use and requires minimum amount of parameters adjustment.

There are many different architectures of artificial neural networks, including Radial Basis Function (RBF) [2], Multi Layer Perceptron (MLP) [9], Support Vector Machine (SVM) [3], *etc.* With architectures of differing complexity and performance, there has been extensive works in comparing performances of various architectures which are generally measured in learning time and testing accuracy.

Extreme Learning Machine (ELM) introduced by Huang et. al. [5] has been found to have learning speed up to thousand times faster compared to other artificial neural network architectures. In contrast to back propagation learning

J. Cao et al. (eds.), *Proceedings of ELM-2014 Volume 1*,
Proceedings in Adaptation, Learning and Optimization 3, DOI: 10.1007/978-3-319-14063-6_34

method, which finds the best weights for its hidden layers and output layer in an iterative manner, the ELM randomly assigns the weights of hidden layers and simply solves the output layer weights. Despite its quasi-optimization, ELM has shown good accuracy in solving various regression and classification problems [4].

Although ELM has empirically achieved 90 percent accuracy and more in several data sets, there are still some data sets that produce training accuracy less than 90 percent or even much lower. Lower accuracy means higher misclassification rates which lessens the reliability of the classification.

We observe that performance comparison tables from previous works [4] [6] show that the problems in those data sets can be further classified into two types. Attached to the first type are data sets which have good accuracy for other neural network algorithms but not for ELM. For the second type are data sets which have low accuracy for any neural network algorithms including ELM.

Optimally-Pruned Extreme Learning Machine (OP-ELM) [6] addresses low accuracy which is only experienced by ELM. Being tested on Iris, Wisconsin's Breast Cancer, Pima Indian Diabetes, and Wine [1] classification data sets, the OP-ELM approach shows that removing irrelevant input variables can increase the accuracy of two data sets so that all four data sets becomes comparable to MLP, SVM, and Gaussian Process (GP) accuracy. OP-ELM achieves 95.0 and 90.7 percent accuracy for Iris and Wine data sets respectively, which are significantly higher than 72.2 and 81.8 percent accuracy achieved by the regular ELM.

As for the second type of problem, we have not found any previous works which try to handle data sets in which any neural network algorithms experience low accuracy. In this paper we propose a strategy to handle them by using cross validation to identify indistinguishable classes in the data.

This approach is not aimed at achieving better classification capability. However, it allows the system to be more realistic and understand its own limitation, which are important in practical implementations.

By being alerted that some data are beyond the system capability to classify, users may try to modify the input data or to feed data that belong to those classes into other classification systems. It allows users to take further actions required to handle those data instead of blindly accepting results with high misclassification rate.

The organization of this paper is as follows. Section 2 explains the concept of the proposed method. Section 3 describes how the experiments were conducted, the data sets we were using and the test configurations. Section 4 shows the results and our brief analysis on the results. Finally we present our conclusions in Section 5.

2 Methodology

As shown in Fig. 1, we applied ELM training and testing as usually used in the original algorithm. As indicated by the dashed lines in Fig. 1, our proposed blocks

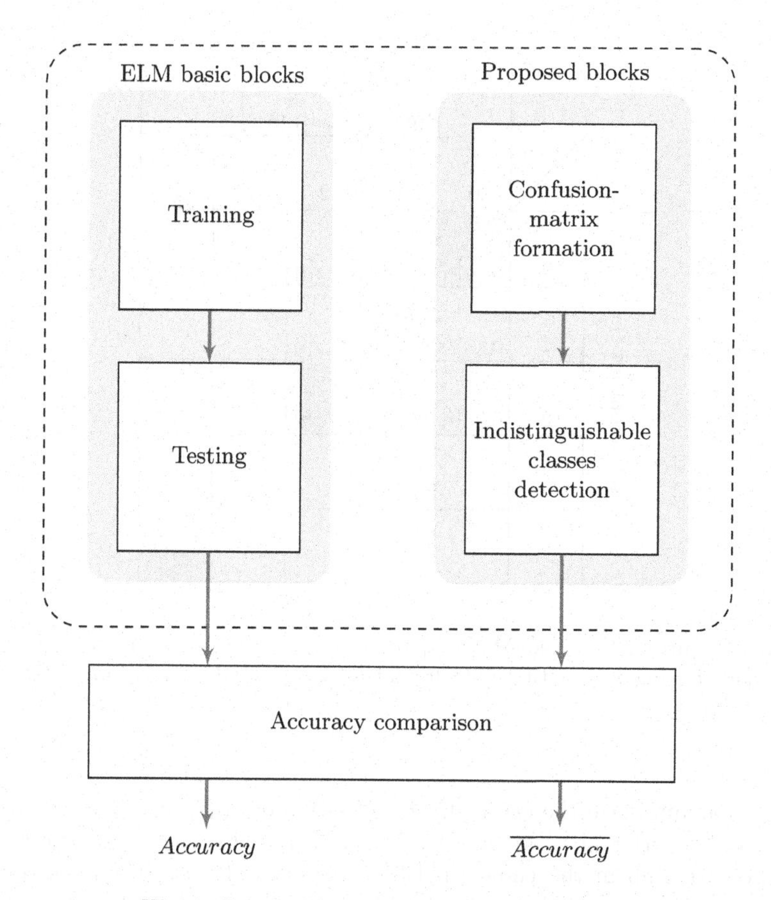

Fig. 1. System Flow and Accuracy Comparison

complement ELM basic blocks by informing if there is any indistinguishable classes.

The proposed blocks consist of two major parts: 1) confusion-matrix formation through cross validation, 2) indistinguishable classes detection. There are also minor changes in how to measure the testing accuracy in case our proposed blocks exist. The details of the proposed blocks and the accuracy comparison method are explained in the following subsections.

2.1 Confusion-Matrix Formation through Cross Validation

Our proposed method applies cross validation into the ELM by separating training data into two parts: training data and validation data, which is common in other neural network architectures such as MLP and RBF [7]. However, in this work we do not use cross validation in the same way as how it is usually used in MLP and RBF. In those architectures, validation data are used to confirm whether the learning process has found the best weights, which are defined as

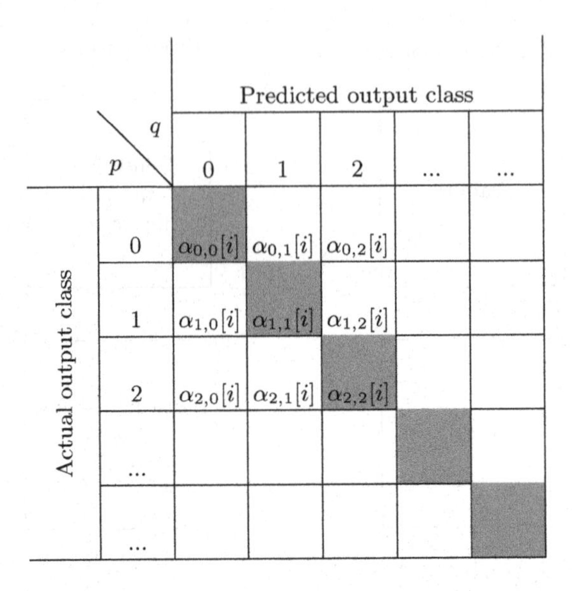

Fig. 2. Structure of ConfusionMatrix$[i]$ where $\alpha_{p,q}[i]$ represents the number of validation data having p-th actual output and q-th predicted output, and i represents repetition index

weights that minimize the error of the validation data [7]. As opposed to MLP and RBF, the original ELM architecture does not use cross validation. Cross validation to confirm the choice of ELM weights is found in a work done by Sovilj et.al. [10], but this is not applicable in our proposed method.

In this work the weights obtained from ELM training are not affected by the validation data. Instead, we use the validation data to update the confusion matrix, which is a tool introduced by Provost et. al. [8] for analyzing classifier systems.

Confusion matrix is a $n_o \times n_o$ matrix where n_o is the number of output classes. Each time we compare the actual output of the validation data against the predicted output, which is the ELM output when an input vector from the validation data is being fed into the ELM, we update the value of the confusion matrix accordingly.

Each element of the confusion matrix $\alpha_{p,q}[i]$ shown in Fig. 2 represents the number of validation data having p-th actual output and q-th predicted output in the ConfusionMatrix$[i]$. Diagonal elements of the matrix show the number of correct predictions for each output class, whereas the non-diagonal elements show the number of incorrect predictions.

After all of the input vectors from the validation data have been fed into the ELM to update the confusion matrix, the procedure above is repeated N_r times. Therefore we obtain a set of ConfusionMatrix$[i]$, where repetition index $i = 0, 1, ..., N_r - 1$. The complete flowchart of this process is shown in Fig. 3.

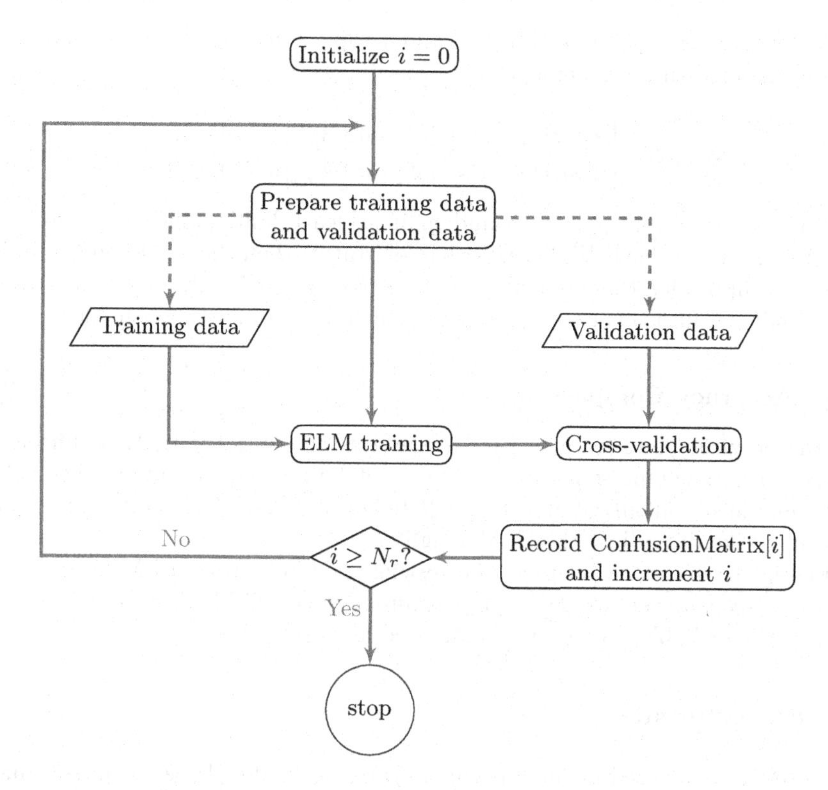

Fig. 3. Flowchart of confusion-matrix formation through cross validation

2.2 Indistinguishable Classes Detection

We analyze the ConfusionMatrix[i] for $i = 0, 1, ..., N_r - 1$ to find whether there is a tendency that incorrectness often occurs at particular pairs of output classes.

To see incorrectness rate, we normalize each matrix element $\alpha_{p,q}[i]$ by dividing it with the average of the diagonal elements. For each $p = 0, 1, ..., n_o - 1$ and $q = 0, 1, ..., n_o - 1$, the normalization equation is given by

$$A_{p,q}[i] = \frac{\alpha_{p,q}[i]}{\frac{1}{n_o}\sum_{j=0}^{n_o-1} \alpha_{j,j}} \quad (1)$$

where $A_{p,q}[i]$ is the normalized confusion matrix element representing p-th actual output and the q-th predicted output.

The upper triangular and the lower triangular elements of the **A** matrix can be further simplified by summing $A_{p,q}[i]$ and $A_{q,p}[i]$ because they refer to the same pair of classes: p and q. For example, both $A_{1,4}[i]$ and $A_{4,1}[i]$ refer to how distinguishable the output class 1 and output class 4. Any values that is greater than or equal to a certain threshold ϵ indicates that the particular actual output class and the particular predicted output class are not easily distinguishable.

Therefore for every $p = 0, 1, ..., n_o - 1$ and $q = p+1, p+2, ..., n_o - 1$ we formulate detection status matrix elements $\Psi_{p,q}[i]$ as

$$\Psi_{p,q}[i] = \begin{cases} Distinguishable & \text{if } (A_{p,q}[i] + A_{q,p}[i]) < \epsilon \\ Indistinguishable & \text{if } (A_{p,q}[i] + A_{q,p}[i]) \geq \epsilon \end{cases} \quad (2)$$

For every possible pair of p and q, if at least half of the $\Psi_{p,q}[i]$ for each $i = 0, 1, ..., N_r - 1$ have "Indistinguishable" status, then there is a tendency that the data inputs for that pair have high degree of misclassification. Any classes pairs that fall into this category can be concluded as indistinguishable.

2.3 Accuracy Comparison

Testing accuracy and training time are used to measure the ELM performance. Testing accuracy represents the percentage of data which have unmatched values between the actual output and the predicted output, whereas training time shows the time taken to adjust the ELM weights.

In Fig. 1, *Accuracy* refers to the testing accuracy of the ELM basic blocks, whereas $\overline{Accuracy}$ shows the testing accuracy of the ELM when false detections at indistinguishable classes are not treated as misclassified data.

3 Experiments

Data sets that we used in our experiments are specified in Table 1. Glass, vehicle, and satellite data sets are taken from University of California, Irvine (UCI) Machine Learning Repository [1]. Those data sets are among the data sets that achieve training accuracy less than 90 percent for either SVM, LS-SVM, and ELM [4].

Table 1. Specification of Data Sets

Data set	# train	# validate	# test	# features	# classes
Glass	60	62	72	9	6
Vehicle	364	200	282	18	4
Satellite	3435	1000	2000	36	6
Room	1000	500	16737	2	6

The room data set was gathered from our experimental work in a university environment by using RFID tags and readers[1] which are shown in Fig. 4. We arranged 2 readers and 16 tags at 6 different rooms, with the room layout as shown in Fig. 5.

We divided the 16 tags into two groups and moved the groups of tags from one room to another every 10 minutes. Tags were always facing up and placed

[1] Tags and readers are from 1Rwave LLP.

Fig. 4. RFID Reader (left) and Tag (right)

	TR-89			TR-92	
TR-95		TR-97	TR-98		TR-100

Fig. 5. Room Layout

in the middle of the room. One reader was placed in TR-95 and another one was placed in TR-100. The readers measured and recorded signal strength of tags that transmit ping signal every 3 seconds. After applying moving average to the recorded data, we fed the data into the ELM and our proposed blocks.

Our simulations used regularization factor and hidden layer size L=1000 as recommended in [4]. We applied the sigmoid function as we do not see significant performance difference between the activation functions presented in the previous work. All input features are normalized to [0,1] range. We set the number of repetitions in Fig. 1 to be $N_r = 4$ and detection status threshold in Eq. 2 to be $\epsilon = 1$.

4 Results and Discussion

As shown in Table 2, our proposed blocks successfully detect indistinguishable classes pairs in room, glass and vehicle data sets. There are significant accuracy improvement on those data sets. The room and vehicle data sets achieve accuracy more than 90 percent when false predictions on indistinguishable classes are ignored. As for the glass data set, we achieved 80.50 percent accuracy, which is much better than the original testing accuracy which is only 57.88 percent.

However, our method could not detect any indistinguishable class in satellite data set and we have not found any solid reason on why this data set behaves differently. Our best guess is that the error of satellite data set is uniformly distributed, or in other words no pair of classes has significantly high error as compared to the other pairs.

From Fig. 3 we can infer that detection time is affected by the number of repetitions we chose. This is in line with our result which shows that the detection time is approximately 4 times that of ELM training time, noting that we chose $N_r = 4$ in this experiment. Although the detection time is longer than ELM training time, we do not see this as a downside. Instead, the fast training time of ELM allows us to have fast detection time. If we had used slower neural network algorithms instead of ELM as the base algorithm of the detection, the detection time would surely be much longer.

Table 2. Performance of ELM and the proposed indistinguishable class detection blocks

Data sets	ELM			ELM and Proposed Blocks			
	Accuracy		Training	Indistinguishable	*Accuracy*		Detection
	Rate (%)	Std (%)	Time (s)	Classes Pairs	Rate (%)	Std (%)	Time (s)
Room	75.01	0.64	0.6084	TR97/TR89	91.08	1.38	2.3868
Glass	57.88	7.18	0.0936	Type 2/Type 1	80.50	3.37	0.4836
Satellite	86.00	0.18	1.7472	-	86.55	0.22	7.1292
Vehicle	74.38	1.77	0.3276	Saab/Opel	95.06	1.51	1.2792

5 Conclusion

This work shows that ELM can be used to identify indistinguishable classes in data sets. Out of four data sets we have tested, three data sets have a pair of classes that is detected indistinguishable by our proposed method. We achieved accuracy improvement on those three data sets when we ignored false predictions in indistinguishable classes. As a trade-off to the accuracy, we obtained a reduced number of output classes because each pair of indistinguishable classes is merged into one class.

Identifying indistinguishable classes informs us the limitation of ELM classification capability as well as improving the system reliability. For example, if we are given a task to identify 6 types of glass in glass data set for crime investigation purpose, a system that are able to classify 5 types with 80.50 percent accuracy is more reliable than a system that are able to classify 6 types with 57.88 percent accuracy. Furthermore, when we are informed that ELM is not able to distinguish the Type 2 and Type 1 glass, we can take further actions required to differentiate those classes, such as applying other methods or collecting more input features.

We also had reasonable detection time for all tested data sets, which is always below 8 seconds. This short detection time is made possible by the fast training time of ELM.

There are two tunable parameters in our proposed algorithm which have not been explored in this work: 1) the number of repetitions, N_r, and 2) detection status threshold, ϵ. Future works might need to be done to investigate whether these parameters are data-dependent and how to find their optimum values.

References

1. Asuncion, A., Newman, D.: UCI machine learning repository (2007), `https://archive.ics.uci.edu/ml/datasets.html`
2. Broomhead, D.S., Lowe, D.: Radial basis functions, multi-variable functional interpolation and adaptive networks. Tech. rep., Defense Technical Information Center (1988)
3. Cortes, C., Vapnik, V.: Support-vector networks. Machine learning 20(3), 273–297 (1995)
4. Huang, G.-B., Zhou, H., Ding, X., Zhang, R.: Extreme learning machine for regression and multiclass classification. IEEE Transactions on Systems, Man, and Cybernetics, Part B: Cybernetics 42(2), 513–529 (2012)
5. Huang, G.-B., Zhu, Q.-Y., Siew, C.-K.: Extreme learning machine: a new learning scheme of feedforward neural networks. In: IEEE International Joint Conference on Neural Networks, vol. 2, pp. 985–990 (2004)
6. Miche, Y., Sorjamaa, A., Bas, P., Simula, O., Jutten, C., Lendasse, A.: OP-ELM: Optimally pruned extreme learning machine. IEEE Transactions on Neural Networks 21(1), 158–162 (2010)
7. Poole, D.L., Mackworth, A.K.: Artificial Intelligence: foundations of computational agents. Cambridge University Press (2010), `http://artint.info/html/ArtInt.html`
8. Provost, F.J., Fawcett, T., Kohavi, R.: The case against accuracy estimation for comparing induction algorithms. In: Proceedings of the Fifteenth International Conference on Machine Learning, pp. 445–453. Morgan Kaufmann Publishers Inc. (1998)
9. Rojas, R.: Neural networks: a systematic introduction. Springer (1996), `http://page.mi.fu-berlin.de/rojas/neural/`
10. Sovilj, D., Lendasse, A., Simula, O.: Extending extreme learning machine with combination layer. In: Rojas, I., Joya, G., Gabestany, J. (eds.) IWANN 2013, Part I. LNCS, vol. 7902, pp. 417–426. Springer, Heidelberg (2013)

Effects of Training Datasets on Both the Extreme Learning Machine and Support Vector Machine for Target Audience Identification on Twitter

Siaw Ling Lo, David Cornforth, and Raymond Chiong

School of Design, Communication and Information Technology, The University of Newcastle, Callaghan, NSW 2308, Australia
siawling.lo@uon.edu.au,
{david.cornforth,raymond.chiong}@newcastle.edu.au

Abstract. The ability to identify or predict a target audience from the increasingly crowded social space will provide a company some competitive advantage over other companies. In this paper, we analyze various training datasets, which include Twitter contents of an account owner and its list of followers, using features generated in different ways for two machine learning approaches - the Extreme Learning Machine (ELM) and Support Vector Machine (SVM). Various configurations of the ELM and SVM have been evaluated. The results indicate that training datasets using features generated from the owner tweets achieve the best performance, relative to other feature sets. This finding is important and may aid researchers in developing a classifier that is capable of identifying a specific group of target audience members. This will assist the account owner to spend resources more effectively, by sending offers to the right audience, and hence maximize marketing efficiency and improve the return on investment.

Keywords: Extreme learning machine, Support vector machine, Machine learning, Target audience, Twitter, Social media.

1 Introduction

With the prevalence of social media and openness of information sharing, the ability to analyze the contents of public social media posts and to uncover the underlying insights is valuable to any organization. Doing business on social media is becoming common, when one considers that 77% of the Fortune 500 companies have active Twitter accounts and 70% have active Facebook accounts [1]. It is understandable that most companies are making efforts to engage their customers in more than one social platform, as it can be rewarding to reach out to potential customers from the huge user base of over 1.28 billion from both Twitter and Facebook [2].

In view of the increasingly crowded social space, it is no longer feasible for a company to depend on gimmicks (such as incentive referrals) that may only provide short-term gain. With the growing "sophistication" of social media users, approaches like mass marketing may not justify the effort and amount of money spent. Furthermore, there is a thin line between broadcasting a general message and spamming, so instead

J. Cao et al. (eds.), *Proceedings of ELM-2014 Volume 1*,
Proceedings in Adaptation, Learning and Optimization 3, DOI: 10.1007/978-3-319-14063-6_35

of attracting a greater audience, there is a high risk of alienating and therefore losing current customers. Hence, it makes sense to identify a target audience in order to maximize marketing efficiency and improve the return of investment (ROI).

While there are many guidelines or tips on the web that suggest how to find a target audience on social media, most of these concentrate on searching specific keywords related to products or brands. This list of keywords is usually prepared by domain experts, and there is a need to ensure that the keywords are "up-to-date" or stay relevant, due to the dynamism of the business world. Furthermore, deciding which keywords to use may not be obvious to a non-expert and this may lead to inaccurate information extraction and hence a misunderstood market analysis. On top of this, there is a need to manually consolidate the list of members of a social audience found and to ensure that the contents shared by the audience match the keywords.

Prior work [3][4] has proposed various approaches such as translating both social networks and semantic information into Resource Description Framework (RDF) formats and using RDF methods for correlation, or making use of semantic tagging to correlate the current social tagging approach to make sense of the social media data. These approaches, however, require additional efforts of translating and tagging of current social media data, which can be a daunting task considering the huge amount of data and the possible manual procedure involved.

While the ability to predict a target audience from a list of Twitter followers will be beneficial to any company or organization, limited studies have investigated the effects of different training datasets used in supervised machine learning approaches for this purpose. The most relevant work by Yang et el. [5] used temporal effects of Twitter contents and a list of category-specific keywords to classify users' interests in the sports and politics domains. Both Term Frequency-Inverse Document Frequency (TF-IDF) [6] and Latent Dirichlet Allocation (LDA) [7] have been used by Yang and co-workers to generate features in two classification approaches – Naïve Bayes (NB) [8] and the Support Vector Machine (SVM) [9]. In comparison, our work focuses on predicting the target audience from a list of followers of a Twitter account owner, instead of analyzing a domain specific Twitter user interest. In addition, we extract a set of seed words using the contents of the owner's tweets, which reflects the key topics or terms at a specific point of time. This can aid in generating various training datasets, while eliminating the need for manual tagging.

It is well known that one of the biggest challenges of using a supervised machine learning approach is the constructing of its training dataset. Due to the vast amount and diverse nature of the followers' tweets, it is not feasible to manually annotate the tweets for training the machine learning model. As such, we reasoned that tweets from an account owner can be used to build a positive training dataset as the group of followers who are tweeting similar contents (within a similar period of time) are more likely to comprise the target audience compared to others who are not sharing similar contents. This saves us from the need to manually annotate the vast amount of tweets from the followers and is more practical if the approach is to be adopted in a real-world application.

Machine learning approaches, especially the SVM, have been used in various text categorization tasks [9] and are found to have obtained better performances compared to other methods [10][11]. As we are analyzing the textual content of tweets, it is of interest to study how the SVM would perform in predicting the target audience from a list of followers. Recently, biologically inspired Natural Language Processing (NLP) [12] has started to gain popularity and a new approach known as the Extreme Learning Machine (ELM) [13] has achieved good results in unstructured text analysis. This research will therefore use both the SVM and ELM for target audience classification. As the performance of a machine learning approach depends heavily on its training dataset, it is important to construct features that can represent the dataset in an appropriate manner. It is logical to consider the suitability of the features according to specific domain knowledge and human ingenuity. Hence, we derive our training datasets from the contents of an owner's tweets. The hypothesis here is based on the idea that the owner's tweets will contain the necessary information or features that the followers are interested and so they choose and take action to follow the account owner.

In this paper, our main objective is to investigate the effects of using different training datasets on both the ELM and SVM in order to make use of available resources to predict and identify the target audience without utilizing a considerable amount of human annotation effort. There are three types of training datasets: 1) tweets from the owner, 2) tweets from followers clustered using statistical topic modeling - Twitter LDA [14], and 3) tweets from followers generated through fuzzy matching using a list of seed words extracted from the owner's tweets.

The major contributions of this work are as follows:

- To the best of our knowledge, our work is the first attempt to predict and identify a target audience from a list of followers (of an account owner) on Twitter using the ELM with minimum manual annotation required.
- From our observation of the results, features generated using the content of the owner's tweets in the training dataset are more useful than training datasets that use the followers' tweets.
- We find that it is essential to remove all the duplicates after the data cleaning process, in order to improve the classification result.

2 Related Work

The aims of any business are to increase profit, build a long lasting brand name, and to grow the customer base or engage current customers. It is therefore essential for a company to understand the needs and behavior of its customers. This understanding can be achieved through different means and at different levels of detail. Most companies segment their customers according to their traits and behavior so that marketing activities are targeted and measured according to the segmentation.

However, this kind of segmentation is typically restricted to customer relationship management (CRM) or transaction data obtained either through customer surveys or tracking of product purchases to understand the customer demand. Demographic variables, RFM (recency, frequency, monetary) and LTV (lifetime value) are the most common input variables used in the literature for customer segmentation and clustering [15, 16]. While CRM or organizational transaction data can be coupled with geographical data to obtain additional information, the segmentation remains limited to a company's internal system and does not leverage on the sharing and activities on social media where customers tend to reveal more about themselves such as personal preferences and perception of brands.

There have been efforts in deriving or estimating demographics information [17, 18] from available social media data, but this set of information may not be suitable to be used directly in targeted marketing, as temporal effects and types of products are usually not considered. Besides that, demographic attributes such as age, gender and residence areas may not be updated and hence may result in a misled conclusion. Recently, eBay has expressed that, due to the viral campaigns and major social media activities, marketing and advertising strategies are evolving. Targeting specific demographics through segmentation, although this still has value, is being superseded by content-based approaches: eBay is focusing on "connecting people with the things they need and love, whoever they are" [19]. In other words, contents shared by individuals are more important than demographics for predicting the target audience on social media. Other research has also shown that using Facebook categories, such as likes and n-grams, for predicting purchase behavior from social media is better than using demographic features shared on Facebook [20]. Due to the privacy policy of Facebook profiles, our work focuses on Twitter, where most of the contents and activities shared online are open and available.

There are many approaches in understanding the preferences of Twitter users, which can provide important opportunities for businesses to improve their services (such as through targeted marketing or personalized services). The majority of these approaches focus on classifying Twitter users using the textual features (e.g., contents of the tweets) [21] or network features (e.g., follower/followee network) [22]. However, there is also work in which the researchers have adopted various sociolinguistic features such as emoticons and character repetition [22], and they used a SVM to classify latent attributes such as gender, age, regional origin and political orientation. Ikeda et al. [23] developed some demographic estimation algorithms for profiling Japanese Twitter users based on their tweets and community relationships, where characteristic biases in the demographic segments of users are detected by clustering their followers and followees.

As most of the Twitter users' basic demographic information (e.g., gender, age) is unknown or incomplete (as compared to Facebook), Yang et al. [5] examined the temporal effect of Twitter contents or tweets in classifying users' interests. Instead of using tweets directly, temporal information is derived from the word usage within the streams to boost the accuracy of the classification. Both binary- and multi-class classifications have been tested and found to outperform other methods within the sports and politics domains. Another approach by Hong et al. [24] modeled a user's interest and behavior by focusing on retweet actions in Twitter, which can be used to model user decisions and user-generated contents simultaneously. Even though tweets can be a rich source of information, the huge volume and real-time nature of tweets can sometimes result in noisy posting about daily lives. Hence, it is essential to extract relevant information from tweets for user profiling tasks. Michelson and Macskassy [25] presented work in discovering topics of interest by examining the entities in tweets. A "topic profile" is then developed to characterize the users. Besides that, statistical techniques that extract term- and concept- based user profiles are used to analyze customers' conversational data to provide insights on a user's interest so that commercial services can use these profiles for targeted marketing [26].

3 Methods

The focus of this work is to understand the effect of using different training datasets on predicting the target audience from a list of followers via two machine learning methods, namely the ELM and SVM.

3.1 Data Collection

We use the Twitter Search API [27] for our data collection. As the API is constantly evolving with different rate limiting settings, our data gathering is done through a scheduled program that requests a set of data for a given query. The subject or brand selected for this research is *Samsung Singapore* or "samsungsg" (its Twitter username). At the time of data collection, there were 3,727 samsungsg followers. In order to analyze the content of the account owner's tweets, the last 200 tweets by samsungsg have been extracted. The time of tweets ranges from 2 Nov 2012 to 3 Apr 2013. For each of the followers, the API is used to extract their tweets, giving a total of 187,746 records, and 2,449 unique users having at least 5 tweets in their past 100 tweets of the same period. We reasoned that those with fewer than 5 tweets were inactive in Twitter, as it implied that these user were tweeting an average of less than one tweet in a month (since the period was of 6 months).

3.2 Data Cleaning and Preparation

Tweets are known to be noisy and often mixed with linguistic variations. It is hence very important to clean up the tweet content prior to any content extraction:

- Non-English tweets are removed using the Language Detection Library for Java [28];
- URLs, any Twitter's username found in the content (which is in the format of @username) and hashtags (with the # symbol) are removed;
- Each tweet is pre-processed to lower case.

As tweets are usually informal and short (up to 140 characters), abbreviation and misspelling are often part of the content and hence the readily available Named Entity Recognition (NER) package may not be able to extract relevant entities properly. Due to this, we derive an approach called Entities Identification, which uses Part-of-Speech (POS) [29] tags to differentiate the type of words. All the single nouns are identified as possible entities. If the tag of the first fragment detected is 'N' or 'J' and the consecutive word(s) is of the 'N' type, these words will be extracted as phrases. This approach is then complemented by another process using the comprehensive stop words list used by search engines (http://www.webconfs.com/stop-words.php) in addition to a list of English's common words (preposition, conjunction, determiners) as well as Twitter's common words (such as "rt", "retweet", etc.) to identify any possible entity. In short, the original tweet is sliced into various fragments by using POS tags, stop words, common words and punctuations as separators or delimiters. For example, if the content is "Samsung is holding a galaxy contest!", two fragments will be generated for the content as follows: (samsung) | (galaxy contest).

3.3 Extreme Learning Machine (ELM)

The ELM [30] is a single-hidden layer feedforward neural networks (SLFN) where all the node parameters in the hidden layer are randomly generated without tuning. Through the replacement of a computationally costly procedure of training the hidden layer by using random initialization, the method is proven to have both universal approximation and classification capabilities [31][32].

Consider a set of N distinct samples (x_i, y_i) with $x_i \in \Re^D$ and $y_i \in \Re^d$. An ELM with K hidden neurons is modeled as

$$\sum_{k=1}^{K} \beta_k \phi(w_k x_i + b_k), \quad i \in [1, N] \tag{1}$$

where ϕ is the matrix activation function, w the input weights, b the biases and β the output weights.

A MATLAB implementation of ELM from *http://www.ntu.edu.sg/home/egbhuang/elm_random_hidden_nodes.html* is adopted in this study. A sigmoidal function is used as the activation function instead of the alternatives, as it has performed better on the various training datasets mentioned in Section 3.6. A range of numbers from 50-400 are used as the hidden neuron parameters, with an interval of 50.

As the number of tweets shared by each follower is different, an e score is calculated by aggregating the classification results from each individual tweet of each follower's tweet set. The final assignment of the e score is based on the following representation:

$$e = n_p/n_t \tag{2}$$

where n_p is the total number of tweets that are classified as positive by the ELM and n_t is the total number of tweets shared by the follower. If 5 tweets out of a total of 50 tweets of a particular follower are classified as positive, then the e score assigned is 5/50 = 0.01. The total number of tweets is used to normalize the score instead of an average value of all the tweets. This is due to the fact that the resulted score is more capable of representing the true interest of the follower. For example, if follower1 tweeted 2 related tweets out of a total of 10 tweets, the e score assigned will be 0.2, while the e score for follower2 is 0.02 if only 2 related tweets are classified as positive out of a total of 100 tweets. This is in contrast to using an average value, as both follower1 and follower2 will be assigned the same e score that may not fully represent the interests of the followers.

3.4 Support Vector Machine (SVM)

The SVM is a supervised learning approach for two- or multi-class classification and it has been used successfully in many applications, including text categorization [9]. It separates a given known set of {+1, -1} labeled training data via a hyperplane that is maximally distant from the positive and negative samples respectively. This optimally separating hyperplane in the feature space corresponds to a nonlinear decision boundary in the input space. More details of the SVM can be found in [33].

Consider a set of N distinct samples (x_i, y_i) with $x_i \in \Re^D$ and $y_i \in \Re^d$. An SVM is modeled as

$$\sum_i \alpha_i K(x, x_i) + b, \quad i \in [1, N] \tag{3}$$

where $K(x, x_i)$ is the kernel function, and α and b are the parameter and threshold of the SVM respectively.

The LibSVM implementation of RapidMiner [34] is used in this study, and the sigmoid kernel type is selected as it produces higher precision prediction than other kernels, such as RBF (Radial Basis Function) and polynomial.

Similar to the e score specified in Section 3.3, a v score is assigned for each follower according to individual tweet classification based on the SVM. The v score is generated using the following formula:

$$v = n_s/n_t \tag{4}$$

where n_s is the number of tweets that are classified as positive by the SVM and n_a is the total number of tweets shared by the follower.

3.5 Performance Metrics

The typical accuracy metric in statistical analysis of binary classification, which takes into consideration the true positive (TP) and true negative (TN), does not reflect the performance of a classifier well [35]. Therefore, we have used the precision, recall and F1 score as performance metrics when comparing the ELM and SVM.

The formulas of precision, recall and F1 score are as follows:

$$precision = \mathrm{TP}/(\mathrm{TP} + \mathrm{FP}) \tag{5}$$

$$recall = \mathrm{TP}/(\mathrm{TP} + \mathrm{FN}) \tag{6}$$

$$F1\ score = 2 * \frac{precision * recall}{precision + recall} \tag{7}$$

where TP, TN, FP and FN represent true positive, true negative, false positive and false negative respectively.

3.6 Generation of Training Datasets

As our main intention is to find an approach to predict the target audience from a list of followers without the need to manually annotate the vast amount of tweet contents for training purposes, we have designed the following three procedures:

i. Using tweets from the account owner (which logically should be tweeting contents that will attract followers of similar interest) as the positive dataset and tweets of account owners from other domains as the negative dataset;
ii. Using an unsupervised topic modeling approach to cluster relevant tweets from all followers as the positive dataset and other clusters as the negative dataset;
iii. Using the Fuzzy match approach with seed words extracted from tweets of the account owner to identify relevant tweets from all the followers. Those tweets that are matched with a certain threshold are assigned as the positive dataset and those below the threshold are assigned as the negative dataset.

Details of the various types of training datasets and their corresponding number of features can be found in Table 1.

Table 1. Types of training datasets and the number of features

Training datasets	Size of the datasets	Number of features	Notation
Tweets of owners	200 positive and 200 negative	245	owner
Tweets of followers generated using Twitter LDA	13,989 positive and 99,831 negative (7 sets of training datasets are created)	397	follower TLDA
Tweets of followers generated using Fuzzy match with seed words extracted from the owner	13,989 positive and 99,831 negative (7 sets of training datasets are created)	38	follower FV

3.6.1 Features Generated from the Owner's Tweets

The positive dataset is generated using processed tweets from the account owner (i.e., samsungsg). The negative dataset is randomly generated from account owners of 10 different domains, which include ilovedealssg (online shopping deals), hungrygowhere (food), joannepeh (celebrity), kiasuparents (parents), MOEsg (education), mtvasia (music), tiongbahruplaza (shopping), tocsg (TheOnlineCitizen/politics), SGnews (Singapore news) and sgdrivers (news on traffic) respectively. These domains are chosen as they represent the main topics discovered based on the analysis of Twitter LDA using the list of tweets from all the followers. The respective account owners are selected as they are the popular Twitter accounts in Singapore according to online Twitter analytic tools such as wefollow.com.

As all the tweets have been cleaned and preprocessed (see Section 3.2), only word stemming using Porter [36] is done on the tweets before forming a term frequency word vector. A total of 245 features are identified and used in creating both the training and testing feature vectors.

3.6.2 Seed Words Generation

In order to minimize the need to annotate the huge amount of followers' tweets for classification, seed words are extracted from the owner's tweets to assist in identifying relevant topic clusters in the unsupervised topic modeling approach (i.e., Twitter LDA, see Section 3.6.3) as well as the Fuzzy match approach (see Section 3.6.4).

All the tweets extracted from samsungsg are subjected to the data cleaning and preparation process described in Section 3.2. Each tweet is now represented by the identified fragments or words and phrases. This set of data is further processed using term frequency analysis to obtain a list of seed words (which include "samsung", "galaxy s iii", "galaxy camera", etc.). The words in a phrase are joined by '_' so that they can be identified as a single term but the '_' is filtered in all the matching processes. A total of 38 words and phrases are identified.

3.6.3 Features Generated Using Twitter Latent Dirichlet Allocation (LDA)

LDA, a renowned generative probabilistic model for topic discovery, has recently been used in various social media studies [14][37]. LDA uses an iterative process to build and refine a probabilistic model of documents, each containing a mixture of topics. However, standard LDA may not work well with Twitter as tweets are typically very short. If one aggregates all the tweets of a follower to increase the size of the documents, this may diminish the fact that each tweet is usually about a single topic. As such, we have adopted the implementation of Twitter LDA [14] for unsupervised topic discovery among all the followers.

As the volume of the tweet set from all the followers is within 200,000 tweets, only a small number of topics from Twitter LDA have been used. Specifically, we have used five topic models from 10 to 50 (with an interval of 10) in this study. We ran these five different topic models for 100 iterations of Gibbs sampling while keeping the other model parameters or Dirichlet priors constant: $\alpha = 0.5$, $\beta_{word} = 0.01$, $\beta_{background} = 0.01$ and $\gamma = 20$. Suitable topics are chosen automatically via comparison with the list of seed words.

The list of topic words under the selected topics are checked for duplication and a total of 397 words are identified for creation of training and testing datasets.

3.6.4 Features Generated Using Fuzzy Match with Seed Words Extracted from the Owner's Tweets

3.6.4.1 Fuzzy Match

It is not uncommon for Twitter users to use abbreviations or interjections or different forms of expression to represent similar terms. For example, "galaxy s iii" can be represented by "galaxy s 3", which is understandable by a human but cannot be captured by direct keyword match. As such, fuzzy matching based on the seed words derived is implemented.

The comparison here is based on a Dice coefficient string similarity score [38] using the following expression

$$s = 2*n_c/(n_x+n_y) \quad (8)$$

where n_c is the number of characters found in both strings, n_x is the number of characters in string x and n_y is the number of characters in string y. For example, to calculate the similarity between "process" and "proceed":

x = process bigrams for x = {pr ro oc ce es ss}
y = proceed bigrams for y = {pr ro oc ce ee ed}

Both x and y have 6 bigrams each, of which 4 of them are the same. Hence, the Dice coefficient string similarity score is 2*4/(6+6) = 0.667.

3.6.4.2 Features Generated from Fuzzy Match

As the Fuzzy match method is dependent on the list of seed words extracted from the owner's tweets, the total number of features for it is the same as the number of seed words, which is 38. Each of the tweets from each follower is compared with every seed word using Fuzzy match. The highest similarity score of each seed word match for the tweet is used to create the value of the feature for that seed word of the tweet.

3.7 Generation of Testing Datasets

In order to assess the performance of both the ELM and SVM, the contents of a total of 300 followers (which were randomly sampled) were annotated manually as either a potential target audience or not a target audience based on the contents shared by the account owner.

Even though the original tweet contents from the annotated followers were mostly different, the contents of the testing dataset after the data cleaning and preparation process (see Section 3.2) resulted in a fair amount of duplication (as shown in Table 2). Hence, it will be of interest to analyze if the duplication in the testing dataset has any effect on the performance of the classifiers.

Table 2. Types of testing datasets

Testing datasets	Size of the datasets	Notation
All tweets from annotated followers	21,297	nil
Unique tweets from annotated followers	13,550	noDup

4 Experiments and Results

4.1 Training Accuracy and *e* Scores of Various ELM Configurations

As all the hidden node parameters are randomly generated in the ELM, 10 runs using different ranges of hidden nodes or neuron numbers have been carried out. It is observed that neuron numbers within the range of 150 and 250 produced better results. The average training accuracy and time with experiments on 150, 200 and 250 neuron numbers are listed in Table 3.

Table 3. Training accuracy of various ELM configurations

Training datasets	Neuron numbers	Training accuracy*	Training time (s)*
ELM_owner	150	0.96	0.07
	200	0.99	0.13
	250	0.99	0.19
ELM_followerTLDA	150	0.67	2.38
	200	0.68	4.05
	250	0.69	5.42
ELM_followerFV	150	0.59	2.02
	200	0.60	3.49
	250	0.60	5.36

*The results are based on the average of 10 runs

The training model using 250 neuron numbers has consistently performed well compared to other configurations and hence we used it for testing the two different types of testing datasets generated using the tweets of the 300 randomly annotated followers. As indicated in Table 4, there are two testing datasets for each training model - the complete set and the no-duplicate set. An *e* score is generated for each follower and the top 10 and top 30 *e* scores are listed in the table. These two sets of scores have been selected to assess how well the ELM performs in identifying and predicting a target audience. It is beneficial to know how many of the top 10 followers as predicted by the ELM are true target audience members in addition to looking at scores from performance metrics such as precision, recall and F1. Detailed results of these can be found in Section 4.3 and 4.4.

Table 4. *e* scores of various ELM configurations

Training – Testing	Top 10 *e* scores	Top 30 *e* scores
ELM_owner	0.50	0.33
ELM_owner_noDup	0.35	0.21
ELM_followerTLDA	0.68	0.43
ELM_followerTLDA_noDup	0.60	0.41
ELM_followerFV	0.72	0.56
ELM_followerFV_noDup	0.70	0.58

4.2 Training Accuracy and *v* Scores of Various SVM Configurations

The 10 fold cross-validation result for the SVM has yielded an accuracy of 0.88 when owner contents are used as the training dataset. As shown in Table 5, the other training datasets are not doing as well as the SVM_owner training dataset.

Table 5. Training accuracy of various SVM configurations

Training datasets	Training accuracy*
SVM_owner	0.88
SVM_followerTLDA	0.70
SVM_followerFV	0.57

*The results are from 10 fold cross-validation

Similar to the ELM, the top 10 and top 30 v scores were generated for the assessment of various SVM configurations. It is interesting to observe that there is an increasing trend for the v score from using the owner tweets as the training dataset to followerTLDA and finally followerFV (see Table 6). This observation is consistent with the results obtained for various ELM configurations (as shown in Table 4) and it implies that the identification of target audience becomes less specific in other training datasets as compared to the owner training dataset. Besides that, the scores (both the e and v scores) are lower when the no-duplicate testing dataset is used.

Table 6. v scores of various SVM configurations

Training – Testing	Top 10 v scores	Top 30 v scores
SVM_owner	0.42	0.24
SVM_owner_noDup	0.33	0.18
SVM_followerTLDA	0.69	0.42
SVM_followerTLDA_noDup	0.71	0.39
SVM_followerFV	0.80	0.63
SVM_followerFV_noDup	0.75	0.60

4.3 Results of Using Top 10 Scores as Cut Off

Table 7 and Table 8 show the results of using top 10 scores as cut off for the ELM and SVM respectively. In general, the numbers of true positive (TP) identified decrease from using the owner training dataset to the followerFV dataset. However, it is worth highlighting that using the no-duplicate testing dataset has yielded better results compared to the complete set of test data here.

Table 7. Results of the ELM (based on the average of 10 runs, using 250 neurons) – top 10 score cut off

Training – Testing	Precision	Recall	F1 score	TP identified	Accuracy*
ELM_owner	0.40	0.06	0.11	4/10	0.40
ELM_owner_noDup	0.80	0.13	0.22	8/10	0.80
ELM_followerTLDA	0.40	0.06	0.11	4/10	0.40
ELM_followerTLDA_noDup	0.40	0.06	0.11	4/10	0.40
ELM_followerFV	0.30	0.05	0.08	3/10	0.30
ELM_followerFV_noDup	0.36	0.06	0.11	4/10	0.40

*The accuracy is based on the true positive (TP) identified

Table 8. Results of the SVM – top 10 score cut off

Training – Testing	Precision	Recall	F1 score	TP identified	Accuracy*
SVM_owner	0.54	0.10	0.16	6/10	0.60
SVM_owner_noDup	0.70	0.10	0.19	7/10	0.70
SVM_followerTLDA	0.40	0.06	0.11	4/10	0.40
SVM_followerTLDA_noDup	0.40	0.06	0.11	4/10	0.40
SVM_followerFV	0.20	0.03	0.05	2/10	0.20
SVM_followerFV_noDup	0.25	0.05	0.08	3/10	0.30

*The accuracy is based on the true positive (TP) identified

4.4 Results of Using Top 30 Scores as Cut Off

Tables 9 and 10 show the results of using top 30 scores as cut off for the ELM and SVM respectively. From the tables, a similar trend but with higher accuracy can be observed when using the owner as the training dataset and when no-duplicate test data is used.

Table 9. Results of the ELM (based on the average of 10 runs, using 250 neurons) – top 30 score cut off

Training – Testing	Precision	Recall	F1 score	TP identified	Accuracy*
ELM_owner	0.50	0.24	0.32	15/30	0.50
ELM_owner_noDup	0.53	0.29	0.37	18/30	0.60
ELM_followerTLDA	0.39	0.21	0.27	13/30	0.43
ELM_followerTLDA_noDup	0.32	0.16	0.21	10/30	0.33
ELM_followerFV	0.23	0.11	0.15	7/30	0.23
ELM_followerFV_noDup	0.26	0.14	0.19	9/30	0.30

*The accuracy is based on the true positive (TP) identified

Table 10. Results of the SVM – top 30 score cut off

Training – Testing	Precision	Recall	F1 score	TP identified	Accuracy*
SVM_owner	0.47	0.22	0.30	14/30	0.47
SVM_owner_noDup	0.53	0.25	0.34	16/30	0.53
SVM_followerTLDA	0.43	0.21	0.28	13/30	0.30
SVM_followerTLDA_noDup	0.33	0.16	0.22	10/30	0.33
SVM_followerFV	0.25	0.13	0.17	8/30	0.27
SVM_followerFV_noDup	0.27	0.13	0.17	8/30	0.27

*The accuracy is based on the true positive (TP) identified

4.5 Comparing the ELM and SVM

The following two figures (Fig. 1 and Fig. 2) clearly show that the trend of F1 scores obtained for both the ELM and SVM is similar. As can be seen in the figures, the highest scores are found using the same configuration – owner as the training dataset and no-duplicate as the testing dataset. In general, configurations based on owner as training datasets have yielded better results in predicting the target audience for both the ELM and SVM.

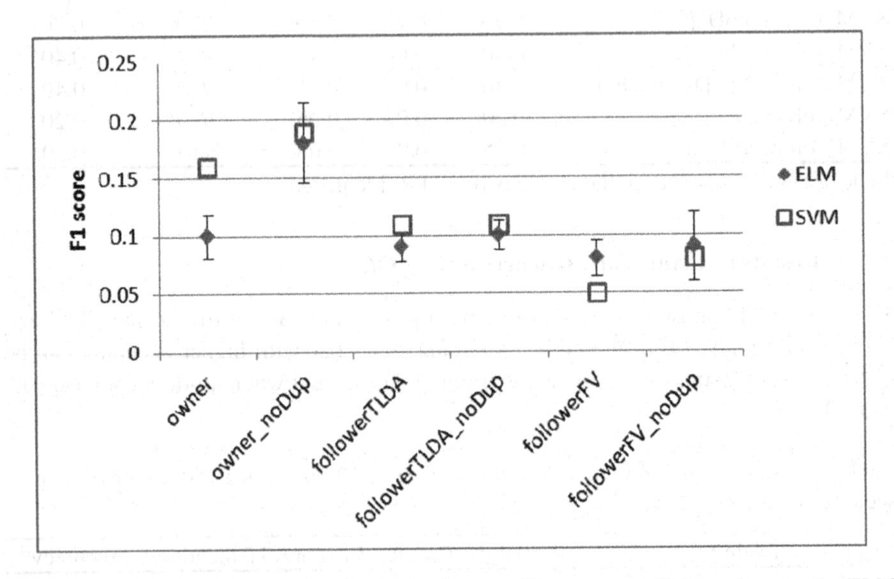

Fig. 1. F1 scores for the ELM and SVM based on top 10 score cut off. Error bars on the ELM indicate the 95% confidence intervals based on the Student T distribution of 10 runs.

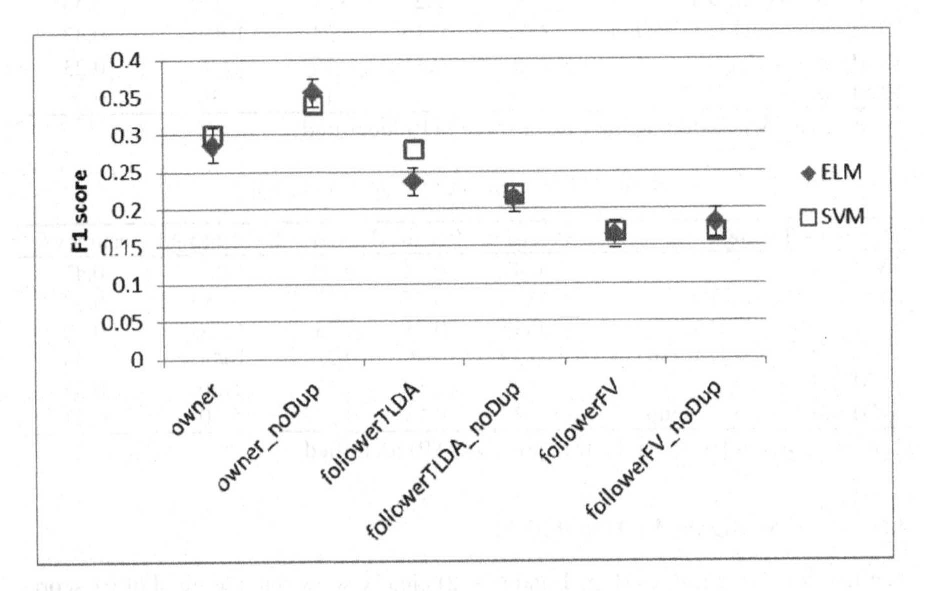

Fig. 2. F1 scores for the ELM and SVM based on top 30 score cut off. Error bars on the ELM indicate the 95% confidence intervals based on the Student T distribution of 10 runs.

5 Discussion

It is interesting to observe that, while traditionally a training dataset of the same source is often used for testing purposes, the F1 score results from top 10 and top 30 cut offs showed that using tweets from the account owner can yield better performances compared to tweets extracted from the list of followers when predicting the target audience. This finding is important as it eliminates the need to manually annotate the vast amount of tweets from the followers. Using tweets from the owner (which is well categorized within its domain) is more practical if it is to be adopted in a real-world application for target audience prediction.

The main objective of this study is to ascertain the effects of using training datasets built from either the owner or the list of followers for prediction. The results indicate that the types of training datasets have a clear impact on the outcome of the prediction process. Also, the preprocessing of the test dataset (i.e., having no duplication) is equally important in yielding better performances.

We have explored two approaches, namely Fuzzy match and Twitter LDA, for extracting representative features from the followers' tweets in this paper. The main reason of using these two approaches is because of their ability in enabling the annotation of followers' tweets through minimum manual efforts. Moreover, Fuzzy match and Twitter LDA (when used in conjunction with seed words extracted from the owner's tweets) have been shown to perform better than other methods in identifying the potential high-value social audience in our preliminary investigation.

It can be argued that the short list of seed words generated from the owner's tweets may not be able to form a feature vector that is representative of the tweets from the followers under the followerFV training dataset (using Fuzzy match). However, as the account chosen - "samsungsg" - is a technology and mobile company, its tweet contents tend to share specific terms such as products or events, which are essentially keywords that can be found in the tweets of the followers. In addition, we have analyzed both the training and testing datasets during the same period of time. It is therefore likely that the target audience who are interested in the content tweeted by the account owner will be tweeting similar terms or text. Having said that, this may not be the case for more generic accounts such as parent groups or current affairs, as the contents shared can be rather diverse and conceptual. As such, a more sophisticated feature generation approach based on domain-specific and common-sense knowledge may be required to enrich the bag of words [39] with new, more informative features.

While the results from both the ELM and SVM show similar trends, it is worthwhile to note that the computational time required by the ELM for training and testing is within the range of seconds. As this work focuses on the effect of using different training datasets for the prediction, however, we did not do a comprehensive comparison on the time spent in generating the model and the result between the two approaches.

We have compared the performances of the ELM and SVM mainly based on precision, recall and the F1 score. Considering the fact that companies and organizations are generally more interested in knowing which followers would more likely be interested in their products, we also used *e* scores and *v* scores generated from both the

ELM and SVM to identify the top 10 and 30 followers (see Tables 7, 8, 9 and 10). The ELM has performed slightly better than the SVM in this regard, as it succeeded in identifying more top followers by using tweets from the owner as the training dataset and no-duplicate data for testing.

6 Conclusion and Future Work

In this paper, we have used various training and testing datasets on both the ELM and SVM to predict and identify the target audience from a list of followers of a Twitter account owner "samsungsg". Our main purpose was to study the effect of using different training datasets to ascertain an approach for classifying the target audience with minimum annotation efforts.

From the results, we have observed that using the owner's tweets as the training dataset can better predict or classify the target audience than using the followers' tweets. In addition, it is essential to remove all the duplicates from the testing dataset, as this has shown to be able to improve the classification results. Equipping a company or organizer with the ability to predict the target audience enables the Twitter account owner to devise their marketing or engagement plan accordingly, in order to maximize the use of allocated budget and successfully reach out to potential customers in the crowded social media space.

Our future work will concentrate on enriching the features that can be identified from tweets, as the poor performance of Fuzzy match using the list of seed words extracted from the owner's tweets may be due to the limitation of bag of words. While Fuzzy match is able to identify terms without the need to have an exact match, it is not able to identify terms that it has not seen before. For example, Fuzzy match can identify "galaxy s iii" with "galaxy s3" but it is not capable of associating that "note 2" is also a related product. As such, it is essential to enrich the bag of words by incorporating online domain knowledge [40] (such as Wikipedia), integrating community-curated online databases [41] (such as Freebase) or combining entities from e-commerce sites (such as eBay) in order to form a more comprehensive view and thus improve on the prediction outcome.

References

1. 2013 Fortune 500 - UMass Dartmouth, http://www.umassd.edu/cmr/socialmediaresearch/2013fortune500/
2. How Many People Use Facebook, Pinterest, Twitter and 500 of the Top Social Media?, http://expandedramblings.com/index.php/resource-how-many-people-use-the-top-social-media/#.U6kUVxAy1vA
3. Breslin, J.G., Passant, A., Vrandečić, D.: Social semantic web. In: Handbook of Semantic Web Technologies, pp. 467–506. Springer (2011)
4. Torres, D., Diaz, A., Skaf-Molli, H., Molli, P.: Semdrops: A social semantic tagging approach for emerging semantic data. In: Proceedings of the IEEE/WIC/ACM International Conference on Web Intelligence and Intelligent Agent Technology (WI-IAT), pp. 340–347. IEEE (2011)

5. Yang, T., Lee, D., Yan, S.: Steeler nation, 12th man, and boo birds: classifying Twitter user interests using time series. Presented at the Proceedings of the 2013 IEEE/ACM International Conference on Advances in Social Networks Analysis and Mining (2013)
6. Ramos, J.: Using tf-idf to determine word relevance in document queries. Presented at the Proceedings of the First Instructional Conference on Machine Learning (2003)
7. Blei, D.M., Ng, A.Y., Jordan, M.I.: Latent dirichlet allocation. J. Mach. Learn. Res. 3, 993–1022 (2003)
8. Manning, C.D., Schütze, H.: Foundations of statistical natural language processing. MIT Press (1999)
9. Joachims, T.: Text categorization with support vector machines: learning with many relevant features. In: Nédellec, C., Rouveirol, C. (eds.) Machine Learning: ECML 1998. LNCS, vol. 1398, pp. 137–142. Springer, Heidelberg (1998)
10. Dumais, S., Platt, J., Heckerman, D., Sahami, M.: Inductive learning algorithms and representations for text categorization. Presented at the Proceedings of the Seventh International Conference on Information and Knowledge Management (1998)
11. Yang, Y., Liu, X.: A re-examination of text categorization methods. Presented at the Proceedings of the 22nd Annual International ACM SIGIR Conference on Research and Development in Information Retrieval (1999)
12. Cambria, E., Mazzocco, T., Hussain, A.: Application of multi-dimensional scal-ing and artificial neural networks for biologically inspired opinion mining. Biol. Inspired Cogn. Archit. 4, 41–53 (2013)
13. Cambria, E., Huang, G.-B., Kasun, L.L.C., Zhou, H., Vong, C.-M., Lin, J., Yin, J., Cai, Z., Liu, Q., Li, K.: Extreme learning machines. IEEE Intell. Syst. 28, 30–59 (2013)
14. Zhao, W.X., Jiang, J., Weng, J., He, J., Lim, E.-P., Yan, H., Li, X.: Comparing twitter and traditional media using topic models. In: Clough, P., Foley, C., Gurrin, C., Jones, G.J.F., Kraaij, W., Lee, H., Mudoch, V. (eds.) ECIR 2011. LNCS, vol. 6611, pp. 338–349. Springer, Heidelberg (2011)
15. Mo, J., Kiang, M.Y., Zou, P., Li, Y.: A two-stage clustering approach for multi-region segmentation. Expert Syst. Appl. 37, 7120–7131 (2010)
16. Namvar, M., Khakabimamaghani, S., Gholamian, M.R.: An approach to opti-mised customer segmentation and profiling using RFM, LTV, and demographic features. Int. J. Electron. Cust. Relatsh. Manag. 5, 220–235 (2011)
17. Mislove, A., Viswanath, B., Gummadi, K.P., Druschel, P.: You are who you know: inferring user profiles in online social networks. In: Proceedings of the Third ACM International Conference on Web Search and Data Mining, pp. 251–260. ACM (2010)
18. Kosinski, M., Stillwell, D., Graepel, T.: Private traits and attributes are predictable from digital records of human behavior. Proc. Natl. Acad. Sci. 110, 5802–5805 (2013)
19. How Ebay Uses Twitter, Smartphones and Tablets to Snap Up Shoppers, http://www.ibtimes.co.uk/how-ebay-uses-twitter-smartphones-tablets-snap-shoppers-1443441
20. Zhang, Y., Pennacchiotti, M.: Predicting purchase behaviors from social media. In: Proceedings of the 22nd International Conference on World Wide Web, pp. 1521–1532. International World Wide Web Conferences Steering Committee (2013)
21. Pennacchiotti, M., Popescu, A.-M.: Democrats, republicans and starbucks afficionados: user classification in twitter. In: Proceedings of the 17th ACM SIGKDD International Conference on Knowledge Discovery and Data Mining, pp. 430–438. ACM (2011)
22. Rao, D., Yarowsky, D., Shreevats, A., Gupta, M.: Classifying latent user attrib-utes in twitter. In: Proceedings of the 2nd International Workshop on Search and Mining User-generated Contents, pp. 37–44. ACM (2010)

23. Ikeda, K., Hattori, G., Ono, C., Asoh, H., Higashino, T.: Twitter user profiling based on text and community mining for market analysis. Knowl. Based Syst. 51, 35–47 (2013)
24. Hong, L., Doumith, A.S., Davison, B.D.: Co-factorization machines: modeling user interests and predicting individual decisions in twitter. In: Proceedings of the Sixth ACM International Conference on Web Search and Data Mining, pp. 557–566. ACM (2013)
25. Michelson, M., Macskassy, S.A.: Discovering users' topics of interest on twitter: a first look. In: Proceedings of the Fourth Workshop on Analytics for Noisy Un-structured Text Data, pp. 73–80. ACM (2010)
26. Konopnicki, D., Shmueli-Scheuer, M., Cohen, D., Sznajder, B., Herzig, J., Raviv, A., Zwerling, N., Roitman, H., Mass, Y.: A statistical approach to mining customers' conversational data from social media. IBM J. Res. Dev. 57, 14:1–14:13 (2013)
27. Using the Twitter Search API | Twitter Developers, `https://dev.twitter.com/docs/using-search`
28. Nakatani, S.: language-detection - Language Detection Library for Java - Google Project Hosting, `http://code.google.com/p/language-detection/`
29. Toutanova, K., Manning, C.D.: Enriching the knowledge sources used in a max-imum entropy part-of-speech tagger. Presented at the Proceedings of the 2000 Joint SIGDAT Conference on Empirical Methods in Natural Language Processing and Very Large Corpora: Held in Conjunction with the 38th Annual Meeting of the Association for Computational Linguistics, vol. 13 (2000)
30. Huang, G.-B., Zhu, Q.-Y., Siew, C.-K.: Extreme learning machine: a new learning scheme of feedforward neural networks. In: Proceedings of the 2004 IEEE International Joint Conference on Neural Networks, pp. 985–990. IEEE (2004)
31. Huang, G.-B., Chen, L., Siew, C.-K.: Universal approximation using incremental constructive feedforward networks with random hidden nodes. IEEE Trans. on Neural Netw. 17, 879–892 (2006)
32. Huang, G.-B., Zhou, H., Ding, X., Zhang, R.: Extreme learning machine for re-gression and multiclass classification. IEEE Trans. on Syst. Man Cybern. Part B Cybern. 42, 513–529 (2012)
33. Burges, C.J.: A tutorial on support vector machines for pattern recognition. Data Min. Knowl. Discov. 2, 121–167 (1998)
34. Predictive Analytics, Data Mining, Self-service, Open source - RapidMiner, `http://rapidminer.com/`
35. Sokolova, M., Japkowicz, N., Szpakowicz, S.: Beyond accuracy, F-score and ROC: a family of discriminant measures for performance evaluation. In: Sattar, A., Kang, B.-H. (eds.) AI 2006. LNCS (LNAI), vol. 4304, pp. 1015–1021. Springer, Heidelberg (2006)
36. Willett, P.: The Porter stemming algorithm: then and now. Program Electron. Libr. Inf. Syst. 40, 219–223 (2006)
37. Yang, M.-C., Rim, H.-C.: Identifying interesting Twitter contents using topical analysis. Expert Syst. Appl. 41, 4330–4336 (2014)
38. Kondrak, G., Marcu, D., Knight, K.: Cognates can improve statistical translation models. Presented at the Proceedings of the 2003 Conference of the North American Chapter of the Association for Computational Linguistics on Human Language Technology: companion volume of the Proceedings of HLT-NAACL 2003–short papers-Volume 2 (2003).
39. Harris, Z.S.: Distributional structure. Word (1954)
40. Gabrilovich, E., Markovitch, S.: Feature generation for text categorization using world knowledge. Presented at the IJCAI (2005)
41. Lo, S.L., Mei, S.D., Liew, V.: Use of Semantic Co-relation in Target Audience Profiling. Presented at the Fourth Global Congress on Intelligent Systems, GCIS (2013)

Extreme Learning Machine for Clustering

Chamara Kasun Liyanaarachchi Lekamalage[1], Tianchi Liu[1,*], Yan Yang[2], Zhiping Lin[1], and Guang-Bin Huang[1]

[1] School of Electrical and Electronic Engineering, Nanyang Technological University, Nanyang Avenue, Singapore 639798
{chamarak001,tcliu,ezplin,egbhuang}@ntu.edu.sg
[2] Energy Research Institute @ NTU (ERI@N), Nanyang Technological University, Nanyang Avenue, Singapore 639798
y.yang@ntu.edu.sg

Abstract. Extreme Learning Machine (ELM) is originally introduced for regression and classification. This paper extends ELM for clustering using Extreme Learning Machine Auto Encoder (ELM-AE) which learn key features of the input data. The embedding created by multiplying the input data with the output weights of ELM-AE is shown to produce better clustering results than clustering the original data space. Furthermore, ELM-AE is used to find the starting cluster points for k-means clustering, which produces better results than randomly assigning the cluster start points. The experimental results show that the proposed clustering algorithm Extreme Learning Machine Auto Encoder Clustering (ELM-AEC) is better than k-means clustering and is competitive with Unsupervised Extreme Learning Machine (USELM).

Keywords: Extreme Learning Machine, Auto Encoders, Clustering, K-means.

1 Introduction

Supervised learning algorithms such as Extreme Learning Machine (ELM) [1–6], Support Vector Machines (SVM) and neural networks learn the mapping between input data and output data by using a training dataset. In contrast, clustering algorithms learn the mapping between input data and output data by grouping similar input data points to represent output classes and do not require a training dataset. Hence, clustering algorithms do not require label information of the input data which is expensive as it requires a human to assign labels.

ELM [1–6] is initially introduced to train "generalized" single-hidden layer feedforward neural networks (SLFNs) with fast learning speed and good generalization capability. ELM can approximate any target function [1–3] and can be used for regression [6], classification [6], clustering [7] and feature learning [8, 9]. The hidden parameters of ELM are chosen randomly independent from the input

* Tianchi Lius work was supported by a grant from Singapore Academic Research Fund (AcRF) Tier 1 under Project RG 80/12 (M4011092).

J. Cao et al. (eds.), *Proceedings of ELM-2014 Volume 1*,
Proceedings in Adaptation, Learning and Optimization 3, DOI: 10.1007/978-3-319-14063-6_36

data. In contrast to the commonly used Back Propergation (BP) [10] training algorithm for neural networks which only minimizes the training error, ELM minimizes the both training error and norm of the output weights. According to Bartlett's theory [11], the minimum norm output weights produce better generalization performance.

ELM based clustering algorithm Un supervised Extreme Learning machine (USELM) [7] uses graph Laplacian for clustering. However, calculating graph Laplacian for large datasets is computationally expensive. The main objective of this paper is to show the following: 1) clustering the embedding created by projecting the input data with the output weights of ELM-AE produces better results than clustering the input data space; 2) ELM-AE output weights can be used to find starting points of the clusters.

2 Preliminaries

2.1 Extreme Learning Machine

Extreme Learning Machine (ELM) proposed by Huang, et.al [1,2,5,6] for SLFN shows that the hidden nodes can be randomly generated independent from input data. The input data is mapped to L dimensional ELM random feature space and the network output is given by Equation (1):

$$f_L(\mathbf{x}) = \sum_{i=1}^{L} \boldsymbol{\beta}_i h_i(\mathbf{x}) = \mathbf{h}(\mathbf{x})\boldsymbol{\beta} \tag{1}$$

where $\boldsymbol{\beta} = [\beta_1, \cdots, \beta_L]^T$ is the output weight matrix between the hidden nodes and the output nodes, while $\mathbf{h}(\mathbf{x}) = [h_1(\mathbf{x}), \cdots, h_L(\mathbf{x})]$ is the hidden node output (random hidden feature) for the input $\mathbf{x}$, and $h_i(\mathbf{x})$ is the output of the i-th hidden node. In ELM, the input data will be mapped to L dimensional ELM random feature space $\mathbf{h}(\mathbf{x})$. Given N training samples $\{(\mathbf{x}_i,\mathbf{t}_i)\}_{i=1}^{\mathrm{N}}$, ELM is to resolve the following learning problems:

$$\text{Minimize} : ||\boldsymbol{\beta}||_p^{\sigma_1} + C||\mathbf{H}\boldsymbol{\beta} - \mathbf{T}||_q^{\sigma_2} \tag{2}$$

where $\sigma_1 > 0$, $\sigma_2 > 0$, $p, q = 0, \frac{1}{2}, 1, 2, \cdots, +\infty$, $\mathbf{T} = [\mathbf{t}_1, \cdots, \mathbf{t}_N]^T$ consists of the target labels and $\mathbf{H} = [\mathbf{h}^{\mathrm{T}}(\mathbf{x}_1), \cdots, \mathbf{h}^{\mathrm{T}}(\mathbf{x}_N)]^{\mathbf{T}}$. According to ELM learning theory, many type of feature mappings such as sigmoid, hard-limit, gaussian and multi-quadric can be implemented in ELM. Further in theory, ELM can approximate any continuous target functions [1–3] and the output weights $\boldsymbol{\beta}$ can be calculated by Equation (3):

$$\boldsymbol{\beta} = \mathbf{H}^{\dagger}\mathbf{T} \tag{3}$$

where $\mathbf{H}^{\dagger}$ is the Moore-Penrose generalized inverse [12,13] of matrix $\mathbf{H}$, which tends to achieve the smallest norm of $\boldsymbol{\beta}$ while keeping the training error to the minimum. In theory, the least square solution of minimum norm ($\boldsymbol{\beta}$) is supposed to be unique and according to Bartlett's theory [11], ELM produces a good generalization performance.

Equality Optimization Constrains based ELM was introduced [6], to achieve a better generalization performance and a more robust solution. The output weights $\boldsymbol{\beta}$ from the ELM hidden layer to the output layer can be calculated as

$$\boldsymbol{\beta} = \left(\frac{\mathbf{I}}{C} + \mathbf{H}^T\mathbf{H}\right)^{-1} \mathbf{H}^T\mathbf{T} \tag{4}$$

or as:

$$\boldsymbol{\beta} = \mathbf{H}^T \left(\frac{\mathbf{I}}{C} + \mathbf{H}\mathbf{H}^T\right)^{-1} \mathbf{T} \tag{5}$$

2.2 Extreme Learning Machine Auto Encoder

Extreme Learning Machine Auto-Encoder (ELM-AE) [9] adopts an unsupervised learning as follows: input data are used as the output data $\mathbf{t}$=$\mathbf{x}$, random input weights and random biases of the hidden nodes are chosen to be orthogonal. ELM-AE can represent features of the input data in three different architectures: 1) compressed architecture; 2) sparse architecture; 3) equal dimension architecture.

Compressed architecture
: The number of input neurons is larger than the number of neurons in the hidden layer. In this case, ELM-AE learns to capture features from the ELM feature space, which has a lower dimensionality than the input data space.

Sparse architecture
: The number of input neurons is smaller than the number of neurons in the hidden layer. In this case, ELM-AE learns to capture features from the ELM feature space, which has a higher dimensionality than the input data space.

Equal dimension architecture
: The number of input neurons is equal to the number of neurons in the hidden layer. In this case, ELM-AE learns to capture features from the ELM feature space, which has the same dimensionality as to the input data.

The network structure of ELM-AE is shown in Figure 1. For compressed ELM-AE architecture, the hidden orthogonal random parameters project the input data to a lower dimension space. The euclidean distances between each input data points and the euclidean distances between each lower dimension space data points is equal; as shown by Johnson-Lindenstrauss Lemma [14] and calculated by Equation (6):

$$\begin{aligned} \mathbf{h} &= g(\mathbf{a} \cdot \mathbf{x} + \mathbf{b}) \\ \mathbf{a}^T\mathbf{a} &= \mathbf{I}, \mathbf{b}^T\mathbf{b} = 1 \end{aligned} \tag{6}$$

The orthogonal hidden random parameters of sparse ELM-AE architecture are calculated by Equation (7):

$$\begin{aligned} \mathbf{h} &= g(\mathbf{a} \cdot \mathbf{x} + \mathbf{b}) \\ \mathbf{a}\mathbf{a}^T &= \mathbf{I}, \mathbf{b}^T\mathbf{b} = 1 \end{aligned} \tag{7}$$

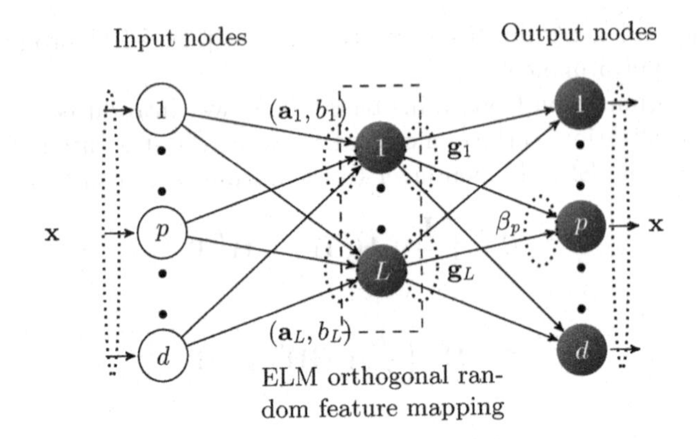

Fig. 1. ELM-AE has the same solution as the original ELM except for: 1) The target output of ELM-AE is the same as input $\mathbf{x}$; 2) The hidden node parameters $(\mathbf{a}_i, b_i)$ are made orthogonal after randomly generated. $g_i(\mathbf{x}) = g(\mathbf{a}_i, b_i, \mathbf{x})$ in the ith hidden node for input $\mathbf{x}$.

where $\mathbf{a} = [\mathbf{a}_1, \cdots, \mathbf{a}_L]$ is the orthogonal random weight and $\mathbf{b} = [b_1, \cdots, b_L]$ is the orthogonal random bias between the input nodes and the hidden nodes.

The output weights $\boldsymbol{\beta}$ of ELM-AE is responsible for the transformation from the feature space to input data. ELM-AE output weights $\boldsymbol{\beta}$ are calculated by Equation (8):

$$\boldsymbol{\beta} = \left(\frac{\mathbf{I}}{C} + \mathbf{H}^T\mathbf{H}\right)^{-1} \mathbf{H}^T\mathbf{X} \tag{8}$$

or by Equation (9):

$$\boldsymbol{\beta} = \mathbf{H}^T \left(\frac{\mathbf{I}}{C} + \mathbf{H}\mathbf{H}^T\right)^{-1} \mathbf{X} \tag{9}$$

3 Extreme Learning Machine Auto Encoder for Clustering

This paper shows that performing k-means clustering [15–17] on the embedding $\mathbf{X}\boldsymbol{\beta}^T$ produces better results than performing k-means clustering in the original data space $\mathbf{X}$. Furthermore, the embedding $\mathbf{X}\boldsymbol{\beta}^T$ is used to determine the initial cluster centroids of k-means which produces better results than using random cluster centroids. K-means clustering essentially groups the data points to k clusters and the objective function is given by Equation (10):

$$J_k = \sum_{i=1}^{k} \sum_{\mathbf{x}_j \in C_k} (\mathbf{x}_j - \mathbf{m}_k)^2 \tag{10}$$

where $\mathbf{m}_k$ is the cluster centroid of the k-th cluster C_K. K-mean clustering finds a centroid matrix $\mathbf{m}_k$ for k clusters and assigns each data point $\mathbf{x}_j$ to an centroid vector $\mathbf{m}_i$ based on the distance. We will consider the Singular Value Decomposition (SVD) [18] of input data as $\mathbf{X} = \mathbf{U}\boldsymbol{\Sigma}\mathbf{V}$, where $\mathbf{U} = [\mathbf{u}_1, \cdots, \mathbf{u}_N]$, $\boldsymbol{\Sigma} = [\boldsymbol{\sigma}_1, \cdots, \boldsymbol{\sigma}_d]$ and $\mathbf{V} = [\mathbf{v}_1, \cdots, \mathbf{v}_d]$. $\mathbf{U}$ are the eigenvectors of the gram matrix $\mathbf{X}\mathbf{X}^T$, $\mathbf{V}$ is the eigenvectors of the covariance matrix $\mathbf{X}^T\mathbf{X}$ and $\boldsymbol{\Sigma}$ is the singular value of $\mathbf{X}$.

3.1 Clustering Embedding $\mathbf{X}\boldsymbol{\beta}^T$

Lemma 1 shows that, projecting the input data $\mathbf{X}$ along the eigenvectors of the covariance matrix $\mathbf{V}^T$ clusters the data points similar to k-means and performing k-means on the embedding $\mathbf{X}\mathbf{V}^T$ produces better results than performing k-means in the original data space $\mathbf{X}$.

Lemma 1. *[19] The embedding $\mathbf{X}\mathbf{V}^T$ reduces the distances between the data points in the same cluster, while the distances between data points in different clusters are not changed.*

It has been shown that ELM-AE learn the variance information [9]. Hence, Proposition 1 shows that performing k-means clustering on the embedding $\mathbf{X}\boldsymbol{\beta}^T$ produces better results than performing k-means in the original data space $\mathbf{X}$.

Proposition 1. *The output weights $\boldsymbol{\beta}$ of ELM-AE learn the variance information of input data $\mathbf{X}$ [9]. Then by Lemma 1, the embedding $\mathbf{X}\boldsymbol{\beta}^T$ reduces the distances between the data points belonging to the same cluster, while distances in different clusters are not changed.*

Figure 2(a) shows the output of embedding $\mathbf{X_{IRIS}}\boldsymbol{\beta}^\mathbf{T}$ created by multiplying the IRIS input data $\mathbf{X_{IRIS}}$ by ELM-AE output weights $\boldsymbol{\beta}$. A ELM-AE with two hidden neurons was used to create the output weights $\boldsymbol{\beta}$. Figure 2(b) shows the original IRIS data points (output of $\mathbf{X_{IRIS}}$). Figure 2 illustrates that the embedding $\mathbf{X_{IRIS}}\boldsymbol{\beta}^\mathbf{T}$ reduces the intra-cluster euclidean distances but not the inter-cluster euclidean distances as shown in Proposition 1. Figure 2(a) shows that performing k-mean clustering in the embedding $\mathbf{X_{IRIS}}\boldsymbol{\beta}^\mathbf{T}$ is easier than performing clustering in the original data $\mathbf{X_{IRIS}}$ as the embedding $\mathbf{X_{IRIS}}\boldsymbol{\beta}^\mathbf{T}$ contains three distinct clusters.

3.2 Finding the Starting Cluster Centroids

K-means algorithm assigns the cluster centroids randomly and converges to a local minimum [20–22]. By assigning the cluster centroids to represent the actual clusters of the input data it is possible to achieve better performance. Lemma 2 shows that the eigenvectors of the gram matrix $\mathbf{u}_i$ clusters the data points. Each eigenvector of the gram matrix $\mathbf{u}_i$ represents one cluster.

Lemma 2. *[19] $C_{i1} = \{j|\mathbf{u}_i(j) \leq 0\}, C_{i2} = \{j|\mathbf{u}_i(j) > 0\}$, where C_{i1} is the first sub-cluster and C_{i2} is the second sub-cluster for the i-th cluster.*

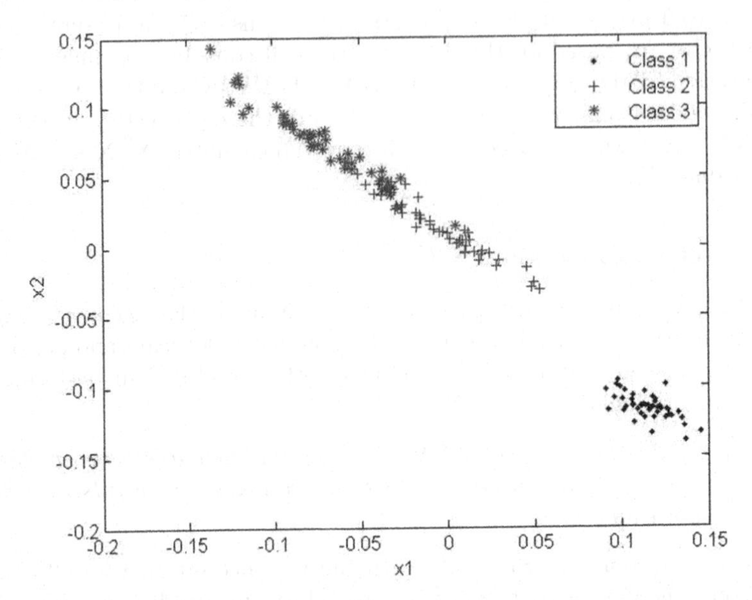

(a) Output of embedding $\mathbf{X}_{\mathrm{IRIS}}\boldsymbol{\beta}^{\mathbf{T}}$ created by multiplying the input data $\mathbf{X}_{\mathrm{IRIS}}$ by ELM-AE output weights $\boldsymbol{\beta}$.

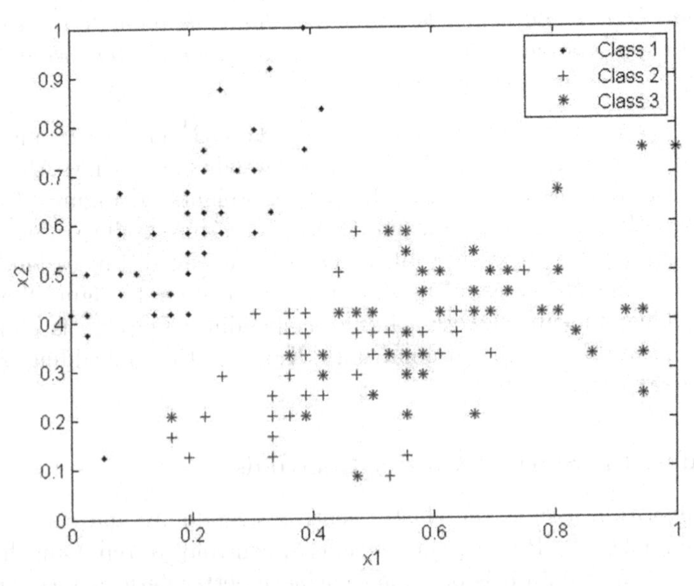

(b) Output of IRIS data points $\mathbf{X}_{\mathrm{IRIS}}$.

Fig. 2. The embedding $\mathbf{X}_{\mathrm{IRIS}}\boldsymbol{\beta}^{\mathbf{T}}$ reduces the intra-cluster euclidean distances but not the inter-cluster euclidean distances

Lemma 2 shows that the positive values of $\mathbf{u}_i$ represents the data points in sub-cluster C_{i2} while the negative values of $\mathbf{u}_i$ represents the data points belonging to sub-cluster C_{i1} for the i-th cluster. One sub-cluster (C_{i1} or C_{i2}) represents the data points belonging to the i-th cluster while the other sub-cluster represents the data points not belonging to the i-th cluster. Hence, we must decide which sub-cluster (C_{i1} or C_{i2}) represents the data points belonging to i-th cluster. If we consider that data points for each class i is equally distributed then each class will contain N/k data points. Hence for the i-th cluster, the sub-cluster with the smallest number of data points in C_{i1} or C_{i2} will represent data points belonging to cluster i. The mean value of the sub-cluster with the smallest number of data points represents the initial cluster centroid for the i-th cluster. For imbalance datasets the ratio of data points belonging to each cluster can be used to determine the sub-cluster (C_{i1} or C_{i2}).

When $d < N$, calculating the eigenvectors $\mathbf{V}$ of the covariance matrix $\mathbf{X}^T\mathbf{X}$ is computationally less expensive than the eigenvectors $\mathbf{U}$ of the gram matrix $\mathbf{X}\mathbf{X}^T$ as the covariance matrix is a $L \times L$ matrix and the gram matrix $\mathbf{X}^T\mathbf{X}$ is a $N \times N$ matrix. Proposition 2 shows that the eigenvectors of the covariance matrix $\mathbf{V}$ instead could be used to cluster the data points.

Proposition 2. *The SVD decomposition of* $\mathbf{X} = \mathbf{U}\boldsymbol{\Sigma}\mathbf{V}$ *and* $\mathbf{U} = \boldsymbol{\Sigma}^{-1}\mathbf{X}\mathbf{V}^T$. *By Lemma 2 embedding* $\mathbf{X}\mathbf{V}^T$ *clusters the data points.*

Proposition 2 shows that, as the embedding $\mathbf{X}\mathbf{V}^T$ clusters the data and can be used determine the initial cluster centroids for k-means.

Proposition 3. *As the output weights of ELM-AE* $\boldsymbol{\beta}$ *capture the variance of input data* $\mathbf{X}$ *and by Proposition 2 the projected space* $\mathbf{X}\boldsymbol{\beta}^T$ *represents eigenvectors of the gram matrix* $\mathbf{X}$.*By Lemma 2 the projected space* $\mathbf{X}\boldsymbol{\beta}^T$ *clusters the data points.* $C_{i1} = \{j|\mathbf{X}\boldsymbol{\beta}_i^T(j) \leq 0\}, C_{i2} = \{j|\mathbf{X}\boldsymbol{\beta}_i^T(j) > 0\}$, *where* C_{i1} *is the first sub-cluster and* C_{i2} *is the second sub-cluster of the i-th cluster.*

Proposition 3 shows that, the projected space $\mathbf{X}\boldsymbol{\beta}^T$ clusters the data it can be used determine the initial cluster centroids for k-means. Figure 2(a) shows that the positive values of embedding $\mathbf{X}_{\mathrm{IRIS}}\boldsymbol{\beta}^{\mathbf{T}}$ represents one cluster while the negative values represent the other cluster.

Algorithm 1 describes Extreme Learning Machine Auto Encoder Clustering (ELM-AEC) algorithm.

4 Experiments

Experiments were carried out in a laptop with 740QM 1.7 Ghz processor and 12 GB RAM. The values of the datasets were normalized in between 0 and 1. Three UCI datasets (IRIS, WINE, SEGMENT) [23] and face dataset (YALE) [24] is used to test the performance of the proposed ELM-AEC. The number of dimensions and number of samples of the datasets are shown in Table 1.

In order to verify whether initializing the cluster centroids of k-means using the embedding $\mathbf{X}\boldsymbol{\beta}^T$ aids in increasing the performance of clustering, two

Algorithm 1. Extreme Learning Machine Auto Encoder Clustering (ELM-AEC)

Input: Input Data $\mathbf{X}$; k is the number of clusters
Output: The cluster labels $\mathbf{T}$
1 Calculate ELM-AE output weights $\boldsymbol{\beta}$ for input data $\mathbf{X}$;
2 Normalize the ELM-AE output weights $\boldsymbol{\beta}$;
3 Create the embedding $\mathbf{X}\boldsymbol{\beta}^T$;
4 Normalize the embedding $\mathbf{X}\boldsymbol{\beta}^T$;
5 Calculate ELM-AE output weights $\boldsymbol{\beta}_{class}$ for input data $\mathbf{X}$ with k hidden neurons, where k are the clusters to be found;
6 Normalize the ELM-AE output weights $\boldsymbol{\beta}_{class}$;
7 Create the class embedding $\mathbf{X}\boldsymbol{\beta}^T_{class}$;
8 Normalize the class embedding $\mathbf{X}\boldsymbol{\beta}^T_{class}$;
9 For all clusters i of $\mathbf{X}\boldsymbol{\beta}^T_{class}(:, i)$ find the best sub-cluster. The best sub-cluster among sub-clusters C_{i1} and C_{i2} of the i-th cluster is the sub-cluster which have the smallest number of data points in one sub-cluster;
10 Calculate the mean of the data points found in the best sub-cluster;
11 Cluster the embedding $\mathbf{X}\boldsymbol{\beta}^T$ using K-means with the calculated cluster centroids;

Table 1. Specification of the Datasets

Dataset	# Dimensions	# Samples	# Classes
iris	4	150	3
wine	13	178	3
segment	19	2310	7
yale	1024	165	15

versions of ELM-AEC is tested named as ELM-AEC(Initial Cluster) and ELM-AEC (without Initial Cluster). ELM-AEC(Initial Cluster) algorithm initializing the k-means initial cluster centroids using the embedding $\mathbf{X}\boldsymbol{\beta}^T$ and the ELM-AEC(without Initial Cluster) algorithm uses random initial cluster centroids for k-means. ELM-AEC(Initial Cluster), ELM-AEC(without Initial Cluster) algorithm USELM and K-means was executed 100 times and the average clustering performance is shown in Table 2. Table 2 shows that the performance of ELM-AEC(Initial Cluster) is better than both k-means and USELM for IRIS, WINE and SEGMENT datasets. However, for YALE face recognition dataset ELM-AEC(Initial cluster) performs lower than USELM but better than k-means. ELM-AEC(Initial Cluster) algorithm performs similar to ELM-AEC(without Initial Cluster) for IRIS and WINE datasets and performs better for SEGMENT and YALE datasets. Hence, initializing the k-mean cluster centroids using embedding $\mathbf{X}\boldsymbol{\beta}^T$ increases the clustering performance. Furthermore, ELM-AEC(without initial cluster) performs better than k-means for all datasets.

Table 2. Clustering Performance of ELM-AEC(Initial Cluster), ELM-AEC(without initial cluster), US-ELM and K-means

Dataset	ELM-AEC (Initial Cluster) (std)	ELM-AEC (without Initial Cluster) (std)	USELM (std)	K-means (std)
iris	95.94(± 1.16)	95.43 (± 1.33)	89.63(± 12.13)	82.18(± 12.45)
wine	97.26(± 0.52)	97.48 (± 0.28)	96.07(± 0)	94.76 (± 0.48))
segment	71.09(± 3.42)	66.99 (± 5.49)	68.29 (± 4.74)	61.14 (± 5.91)
yale	45.63(± 3.58)	43.53(± 3.82)	47.25 (± 3.75)	38.76(± 3.57)

5 Conclusion

This paper shows that k-means clustering in the embedding $\mathbf{X}\boldsymbol{\beta}^T$ produces better results than clustering in the original space $\mathbf{X}$. Performance of k-means can be increased by initializing the initial centroids of k-means to represent the actual clusters, instead of using random cluster centroids. To this end, this paper shows that the embedding $\mathbf{X}\boldsymbol{\beta}^T$ finds the actual clusters of the input data and can be used to find the initial cluster centroids for k-means. The results show that the proposed ELM-AEC is better than USELM.

References

1. Huang, G.-B., Chen, L., Siew, C.-K.: Universal Approximation Using Incremental Constructive Feedforward Networks with Random Hidden Nodes. IEEE Transactions on Neural Networks 17(4), 879–892 (2006)
2. Huang, G.-B., Chen, L.: Convex Incremental Extreme Learning Machine. Neurocomputing 70, 3056–3062 (2007)
3. Zhang, R., Lan, Y., Huang, G.-B., Xu, Z.-B.: Universal Approximation of Extreme Learning Machine With Adaptive Growth of Hidden Nodes. IEEE Transactions on Neural Networks and Learning Systems 23(2), 365–371 (2012)
4. Huang, G.-B., Chen, L.: Enhanced random search based incremental extreme learning machine. Neurocomputing 71(16-18), 3460–3468 (2008)
5. Huang, G.-B., Zhu, Q.-Y., Siew, C.-K.: Extreme Learning Machine: Theory and Applications. Neurocomputing 70, 489–501 (2006)
6. Huang, G.-B., Zhou, H., Ding, X., Zhang, R.: Extreme Learning Machine for Regression and Multiclass Classification. IEEE Transactions on Systems, Man, and Cybernetics, Part B 42(2), 513–529 (2012)
7. Huang, G., Song, S., Gupta, J.N.D., Wu, C.: Semi-Supervised and Unsupervised Extreme Learning Machines. IEEE Transactions on Cybernetics 99, 1 (2014)
8. Huang, G.-B.: An Insight into Extreme Learning Machines: Random Neurons, Random Features and Kernels. Cognitive Computation, 1–15 (2014)
9. Kasun, L.L.C., Zhou, H., Huang, G.-B., Vong, C.M.: Representational Learning with Extreme Learning Machines for Big Data. IEEE Intelligent Systems 28(6), 31–34 (2013)
10. Rumelhart, D.E., Hinton, G.E., Williams, R.J.: Learning representations by back-propagating errors. Nature 323(6088), 533–536 (1986)

11. Bartlett, P.L.: The Sample Complexity of Pattern Classification with Neural Networks: The Size of the Weights is More Important than the Size of the Network. IEEE Transactions on Information Theory 44(2), 525–536 (1998)
12. Serre, D.: Matrices: Theory and Applications. Springer-Verlag New York, Inc. (2002)
13. Rao, C.R., Mitra, S.K.: Generalized Inverse of Matrices and its Applications. John Wiley, New York (1971)
14. Johnson, W., Lindenstrauss, J.: Extensions of Lipschitz mappings into a Hilbert space. In: Conference in Modern Analysis and Probability, vol. 26, pp. 189–206. American Mathematical Society (1984)
15. Hartigan, J.A., Wong, M.A.: Algorithm AS 136: A K-means clustering algorithm. Applied Statistics, 100–108 (1979)
16. Lloyd, S.: Least Squares Quantization in PCM. IEEE Trans. Inf. Theor. 28(2), 129–137 (2006)
17. MacQueen, J.B.: Some Methods for Classification and Analysis of MultiVariate Observations. In: Cam, L.M.L., Neyman, J. (eds.) Proc. of the fifth Berkeley Symposium on Mathematical Statistics and Probability, vol. 1, pp. 281–297. University of California Press (1967)
18. Eckart, C., Young, G.: The approximation of one matrix by another of lower rank. Psychometrika 1(3), 211–218 (1936)
19. Ding, C., He, X.: K-means Clustering via Principal Component Analysis. In: Proceedings of the Twenty-First International Conference on Machine Learning, ICML 2004, p. 29. ACM (2004)
20. Bradley, P.S., Fayyad, U.M.: Refining initial points for k-means clustering. In: Proceedings of the Fifteenth International Conference on Machine Learning, ICML 1998, pp. 91–99. Morgan Kaufmann Publishers Inc., San Francisco (1998)
21. Grim, J., Novovicova, J., Pudil, P., Somol, P., Ferri, F.J.: Initializing normal mixtures of densities. In: Proceedings of the Fourteenth International Conference on Pattern Recognition, vol. 1, pp. 886–890 (1998)
22. Moore, A.W.: Very Fast EM-based Mixture Model Clustering Using Multiresolution Kd-trees. In: Proceedings of the 1998 Conference on Advances in Neural Information Processing Systems II, pp. 543–549. MIT Press, Cambridge (1999)
23. Bache, K., Lichman, M.: UCI machine learning repository (2013)
24. Samaria, F.S., Harter, A.C.: Parameterisation of a stochastic model for human face identification. In: Proceedings of the Second IEEE Workshop on Applications of Computer Vision, pp. 138–142 (December 1994)

Author Index

Almási, Adela-Diana 367

Ban, Xiaojuan 293
Bisio, Federica 61

Cao, Ke-yan 193
Cao, Le-le 141
Chang, Xiaohui 293
Chen, Yanjie 311
Chetty, Girija 163
Chiong, Raymond 417
Cornforth, David 417
Cristea, Valentin 367

de Chazal, Philip 41, 183
Decherchi, Sergio 61
Deng, Zhaohong 203
Denk, Cornelia 245
Ding, Linlin 91
Dong, Diya 325
Dwiyasa, Felis 407

Engbersen, Ton 367

Feng, Lin 51

Gao, Hang 81
Gao, Qun 203
Gastaldo, Paolo 61
George, Koshy 215

Han, Bo 273
Han, Donghong 193
He, Bo 273, 325
He, Qing 151

Horata, Punyaphol 377
Huang, Guang-Bin 245, 435
Huang, Shan 31
Huang, Wen-bing 141

Jia, Xibin 301
Jin, Qibing 121

Klanner, Felix 245
Kongsorot, Yanika 377

Leblebici, Yusuf 367
Lekamalage, Chamara Kasun Liyanaarachchi 435
Lendasse, Amaury 273, 325
Li, Dazi 121
Li, Jiajia 71
Li, Ping 131
Lim, Meng-Hiot 407
Lin, Zhiping 435
Liu, Junfa 301
Liu, Shenglan 51
Liu, Tianchi 435
Liu, Xinwang 81
Liu, Yang 325
Liu, Yu 91
Lo, Siaw Ling 417
Luo, Xiong 293

Ma, Mengmeng 273
Mao, Wentao 263
McDonnell, Mark D. 345
Miao, Chunyan 355
Ming, Chen 225
Mukherjee, Dibyendu 311

Mutalik, Prabhanjan 215
Muzhou, Hou 225

Nian, Rui 325

Peng, Yuxing 81
Powers, David M.W. 301
Prabhu, Sachin 215

Qiu, Junhao 31

Shang, Tianfeng 151
Shen, Yue 325
Shi, Zhongzhi 151
Singh, Lavneet 163
Song, Baoyan 91
Song, Shiji 1
Sourina, Olga 245
Sun, Fu-chun 141
Sun, Haoqi 245
Sun, Tingting 273
Sunat, Khamron 377

Tapson, Jonathan 41, 183
Tian, Mei 263
Tissera, Migel D. 345

van Schaik, André 41, 183

Wang, Botao 31, 71
Wang, Guoren 31, 71, 193
Wang, Haocheng 151
Wang, Jing 51
Wang, Jinwan 263
Wang, Jun 203
Wang, Liyun 263
Wang, Runyuan 301
Wang, Shitong 203
Wang, Yaonan 311
Woźniak, Stanisław 367
Wu, Qiong 355
Wu, Q.M. Jonathan 311

Xiao, Yao 51
Xie, Qianwen 121
Xin, Junchang 91
Xu, Xin 397
Xu, Zhixin 15

Yan, Tianhong 273, 325
Yang, Yan 245, 435
Yang, Yimin 311
Yang, Zhixin 237
Yangchun, Zhang 225
Yao, Jitao 31
Yao, Min 15
Yu, Ge 31

Zhang, David 103
Zhang, Jiannan 1
Zhang, Lei 103
Zhang, Lin 131
Zhang, Pengbo 237
Zhang, Xunan 1
Zhang, Yu 131
Zhuang, Fuzhen 151
Zunino, Rodolfo 61
Zuo, Lei 397

Printed in the United States
By Bookmasters